Microbial Approaches for Insect Pest Management

Omkar
Editor

Microbial Approaches for Insect Pest Management

Springer

Editor
Omkar
Department of Zoology
University of Lucknow
Lucknow, Uttar Pradesh, India

ISBN 978-981-16-3594-6 ISBN 978-981-16-3595-3 (eBook)
https://doi.org/10.1007/978-981-16-3595-3

This Springer imprint is published by the registered company Springer Nature Singapore Pte Ltd.
The registered company address is: 152 Beach Road, #21-01/04 Gateway East, Singapore 189721, Singapore

Preface

Insects comprise almost 80% of the entire world fauna (almost 1 million species) and are present in all parts of the biosphere except the oceans. Many more genera and species of insects are still being reported, and these discoveries are not only bringing out new facts but also changing the very classification of insects. Despite being one of the most successful and diverse group of animals inhabiting planet Earth, they are poorly explored.

The word insect largely has strong negative connotations for most humans; though many of them are quite useful to human beings by yielding certain products directly used by the humans, others work as farmers' friends, being biocontrol agents, and still others are beneficial by providing various ecosystem services along with increasing crop productivity by facilitating and enhancing crop pollination.

Studies indicate that of the known 1 million species, hardly, 1% of them are harmful to human beings by the way of causing direct crop damages and lowering the yield, damaging the stored products and food produce, causing nuisance, or transferring disease-causing agents, besides causing nuisance and health hazards to our livestock. Such harmful insects are technically termed as pests and vectors. FAO estimates that annually between 20 and 40 percent of global crop production are lost to pests. It has been estimated that damages caused by these pests, vectors, and pathogens of crop plants are more than USD 13 billion per annum in India and around $250 billion globally. Possibly because of these facts, it has been emphasized that struggle between man and insects started long before the dawn of civilization, continued without break, and will probably continue as long as the human race exists.

It is these massive economic losses that are probably responsible for the global attention of entomologists towards curbing populations of harmful insects. This glaring monetary loss is probably the reason that most of the silently working beneficial insects providing ecosystem services are pushed to the back burner.

Through the ages, humans have been involved in finding ways and means to manage populations of insect pests. Cultural and chemical practices have been employed for the purpose since the tug of war between humans and insects started. Chemical practices have made their journey from initial crude options, such as ash, to more refined versions in the form of inorganic agrochemicals, synthetic organic chemicals to plant products. In addition to the above practices, farmers across the globe have also employed various physical, mechanical, cultural, legal, genetic, and ecological approaches. Of all these, chemical approach has by far been the most successful one till date. However, the use of chemicals, termed as pesticides, while providing an immediate remedy to overcome insect pest problems has resulted in severe long-term consequences, such as disruption of interspecific competition resulting in damage to farmers' friends, the biocontrol agents of these pests, resistance in pest species, resurgence of new pest species, and damage to the environment and the biodiversity along with the human health hazards. This has gradually also changed the very concept from pest eradication to pest control to pest management, including the concept of integrated pest management, with the basic objective to integrate various ecofriendly tools and techniques, such as cultural practices, biocontrol using pathogens, parasitoids, and predators (natural enemies) for the pest management, and minimizing the use of synthetic chemicals in modern agriculture.

In the last few decades, the humans have witnessed major advancements in life sciences; as a result, several new and powerful tools and techniques have evolved. This has led to great advancements in microbial nutrition, genetics, and their application in different fields. In modern era of biotechnology, the microbes have provided solutions to many of the human problems and necessities and thus serve as human and farmers' friends. The microbes have proved to be successful tools for the pest management. Similarly, there has been much advancement in the field of molecular biology, where many more techniques have evolved which can be helpful in the field of pest management too. Plant resistance, development of transgenic plants, and many more techniques are being considered the panacea to pest problems. On the other hand, there are widespread concerns of the safety of these microbial and biotechnological interventions with nontarget organisms, including humans. While the world stands divided on the ethical issues of these approaches and the many safety concerns, scientists believe that well thought of biotechnological interventions are probably the only safest ways possible for reducing pest attacks on crops.

Though several massive texts are available on insect pest management with exhaustive coverage of various means of insect pest management, my main objective to bring out a book entitled **Microbial Approaches to Insect Pest Management** is to bring precise but specialized information covering modern aspects of pest management. Also, through this publication, my idea is to present the Indian perspectives on this discipline before international readership, involving various specialists from microbiology.

I hope that the proposed book will not only present information on the modern and most effective means of pest management for postgraduate students and teachers and plant protection practitioners across the world but would also be quite useful to those in policy planning.

Lucknow, Uttar Pradesh, India Omkar
February 16, 2021

Acknowledgments

At the very outset, I express my gratitude to the contributors of different chapters contained in this book for sparing time from their routine to contribute their respective chapters to the level they could and I wanted.

I am also thankful to Dr. Geetanjali Mishra, associate professor, Department of Zoology, University of Lucknow, Lucknow, for her intellectual inputs. I also extend my special thanks to Dr. Swati Saxena, M/S Priya Singh, Apoorva Shandilya, Chandni Verma, Tripti Yadav, Shriza Rai, Lata Verma, Dipali Gupta, and Gaurvanvita Singh for their assistance and support in various ways.

I am also grateful to my wife Mrs. Kusum Upadhyay and nephew Mr. Mridul Pandey for their encouragements and support by sparing me from household activities for this work.

I also extend my thanks to Ms. Aakanksha Tyagi, senior editor, Springer Nature India, New Delhi, and Mr. Lenold Esithor, project manager at Springer Nature, for taking keen interest in this project and for their support in publishing this work in time.

Lucknow
February 16, 2021

Prof. Omkar

Contents

Editor and Contributors

About the Editor

Prof. Omkar, FNASc, He has been associated with teaching for about 35 years and research for more than 42 years. He is former head, Department of Zoology, University of Lucknow, Lucknow-226007, former coordinator, UGC-SAP (DRS-II), DST-FIST, DST-PURSE, and Centre of Excellence, Govt. of UP programs. He has worked on 10 more projects of state/central funding agencies. He specializes in environmental toxicology and entomology, with particular reference to the insect pest management. He is also an associate editor, International Journal of Tropical Insect Science, **Springer Nature**, and chief editor, Journal of Applied Bioscience, besides being president of International Society of Applied Biology.

He is a recipient of several awards, like Saraswati Samman by the Department of Higher Education, Govt. of UP, Prof. T N Ananthakrishanan Foundation Award, Rescholar Award of Excellence in Agricultural Entomology by the Association of Entomologists, Prof. G. S. Shukla Gold Medal by the Academy of Environmental Biology, Dr. S. Pradhan Memorial Lecture Award by The Division of Entomology, IARI & The Entomological Society of India, and ACCLAIM and Uddeepan awards from University of Lucknow. He is a fellow of The Entomological Society of India, Zoological Society of India, Society for Biocontrol Advancement, and fellow of The National Academy of Sciences, India (FNASc), besides many others. He has 16 books to his credit. He has published 237 research papers, 10 reviews, 44 reviews as book chapters, and 17 popular science articles. Prof. Omkar has guided more than two dozen Ph. D. students. He has visited the School of Biological Sciences, University of East Anglia, Norwich, UK (2006), for three months and Department of Zoology, University of South Bohemia, and Czech Academy of Sciences, Ceske Budejovice, Czech Republic (May 2016).

Contributors

Aasha Nematology Laboratory, Department of Zoology, Ch. Charan Singh University, Meerut, Uttar Pradesh, India

Karina Afzal Biocontrol Laboratory, Department of Botany, University of Lucknow, Lucknow, Uttar Pradesh, India

Abdoolnabi Bagheri Plant Protection Research Department, Hormozgan Agricultural and Natural Resources Research and Education Center, Agricultural Research Education and Extension Organization (AREEO), Bandar Abbas, Iran

N. Balakrishnan Department of Agricultural Entomology, Agricultural College and Research Institute, Tamil Nadu Agricultural University, Killikulam, Tamil Nadu, India

V. Balasubramani Department of Plant Biotechnology, CPMB&B, Tamil Nadu Agricultural University, Coimbatore, Tamil Nadu, India

Ashok K. Chaubey Nematology Laboratory, Department of Zoology, CCS University, Meerut, Uttar Pradesh, India

Mani Chellappan AINPVPM, Department of Agricultural Entomology, College of Agriculture, Vellanikkara, Kerala Agricultural University, Thrissur, Kerala, India

K. Elango Horticultural Research Station, Tamil Nadu Agricultural University, Kodaikanal, Tamil Nadu, India

Yaqoub Fathipour Department of Entomology, Faculty of Agriculture, Tarbiat Modares University, Tehran, Iran

Priyanka Gangwar Amity Institute of Biotechnology, Amity University, Lucknow, Uttar Pradesh, India

S. Harish Department of Plant Pathology, Tamil Nadu Agricultural University, Coimbatore, Tamil Nadu, India

Neeshma Jaiswal Department of Zoology, Baba Bhimrao Ambedkar University (A Central University), Lucknow, Uttar Pradesh, India

M. Kannan Department of Nano Science and Technology, Tamil Nadu Agricultural University, Coimbatore, Tamil Nadu, India

Marzieh Kashkouli Department of Entomology, Faculty of Agriculture, Tarbiat Modares University, Tehran, Iran

Sandeep K. Malhotra Parasitology Laboratory, Department of Zoology, University of Allahabad, Prayagraj, Uttar Pradesh, India

Shivji Malviya Department of Zoology, Govt. H.N.B. (P.G.) College, Naini, Prayagraj, Uttar Pradesh, India
U.P. Higher Education Service Commission, Prayagraj, Uttar Pradesh, India

Mohammad Mehrabadi Department of Entomology, Faculty of Agriculture, Tarbiat Modares University, Tehran, Iran

S. Mohankumar Department of Plant Biotechnology, CPMB&B, Tamil Nadu Agriculture University, Coimbatore, Tamil Nadu, India

M. Murugan Department of Agricultural Entomology, Tamil Nadu Agricultural University, Coimbatore, Tamil Nadu, India

S. Parthasarathy Department of Plant Pathology, Amrita School of Agricultural Sciences, Amrita Vishwa Vidyapeetham, Coimbatore, Tamil Nadu, India

S. R. Prabhukarthikeyan Crop Protection Division, ICAR—National Rice Research Institute, Cuttack, Odisha, India

T. P. Rajendran ICAR, New Delhi, India
ICAR-National Institute for Biotic Stress Management, Raipur, Chhattisgarh, India

M. T. Ranjith AINPVPM, Department of Agricultural Entomology, College of Agriculture, Vellanikkara, Kerala Agricultural University, Thrissur, Kerala, India

A. L. Reyaz Department of Plant Biotechnology, CPMB&B, Tamil Nadu Agricultural University, Coimbatore, Tamil Nadu, India

Majeed Askari Seyahooei Plant Protection Research Department, Hormozgan Agricultural and Natural Resources Research and Education Center, Agricultural Research Education and Extension Organization (AREEO), Bandar Abbas, Iran

Amritesh C. Shukla Biocontrol Laboratory, Department of Botany, University of Lucknow, Lucknow, Uttar Pradesh, India

Rajesh K. Tiwari Amity Institute of Biotechnology, Amity University, Lucknow, Uttar Pradesh, India

Mala Trivedi Amity Institute of Biotechnology, Amity University, Lucknow, Uttar Pradesh, India

Anita Yadav Department of Zoology, C.M.P.P.G. College, A Constituent College of the University of Allahabad, Prayagraj, Uttar Pradesh, India

Entomopathogenic Viruses 1

S. Harish, M. Murugan, M. Kannan, S. Parthasarathy, S. R. Prabhukarthikeyan, and K. Elango

Contents

S. Harish (✉)
Department of Plant Pathology, Tamil Nadu Agricultural University, Coimbatore, Tamil Nadu, India
e-mail: harish.s@tnau.ac.in

M. Murugan
Department of Agricultural Entomology, Tamil Nadu Agricultural University, Coimbatore, Tamil Nadu, India

M. Kannan
Department of NanoScience and Technology, Tamil Nadu Agricultural University, Coimbatore, Tamil Nadu, India

S. Parthasarathy
Department of Plant Pathology, Amrita School of Agricultural Sciences, Amrita Vishwa Vidyapeetham, Coimbatore, Tamil Nadu, India

S. R. Prabhukarthikeyan
Crop Protection Division, ICAR—National Rice Research Institute, Cuttack, Odisha, India

K. Elango
Horticultural Research Station, Tamil Nadu Agricultural University, Kodaikanal, Tamil Nadu, India

Omkar (ed.), *Microbial Approaches for Insect Pest Management*,
https://doi.org/10.1007/978-981-16-3595-3_1

Abstract

India has moved from an era of food shortages to a status of food self-sufficiency through improved modern agricultural technologies and sustainable agricultural practices. However, various constraints exist during the cultivation of the crop, which limits its production and productivity. In achieving the green revolution, chemicals have played a significant role in improving production and productivity. Nevertheless, continuous and indiscriminate use of insecticides pollute the environment and create health hazards to human beings. In this regard, biological control is an alternative strategy, which will be eco-friendly, cost-effective, restores soil fertility, and provides residue-free products. Of late, entomopathogenic bioagents have been exploited by the scientists for the management of various insect pests in modern agriculture. Of the various bioagents, viruses are being used as a promising tool for the management of economically important insect pests. Various viruses, *viz.* Nucleopolyhedrovirus (NPV), Granulosis viruses (GV) and Cytoplasmic Polyhedrosis viruses were used for the management of insect pests throughout the world. The host-specific viral particles are ingested by the insects and the virions infect the gut wall cells, fat body, and hemolymph, leading to death of the insects. The characteristics of the entomopathogenic viruses and the molecular mechanisms by which they infect and kill the insects needs to be explored in a detailed manner. Insect virus formulations have been developed by various research groups throughout the world and used for the management of insect pests. However, the mode of action, pathogenicity, time and duration of infection, specificity, persistence, etc. need to be considered for the development of stable and effective formulations. This review will highlight the characteristics of insect viruses, pathogenicity and mode of action, various formulations and their application in the management of insect pests.

Keywords

Entomopathogenic virus · Insect molecular taxonomy · Genomics · Biopesticides · Mass production

Learning Objectives

1. To explore various insect viruses and their potential to manage insect pests of crop plants from a global perspective.
2. To have a glimpse of the molecular taxonomy of insect viruses.
3. To understand the characteristics of the entomopathogenic viruses and the molecular mechanisms of virus infections and pathogenicity in insects.
4. To be acquainted with various available biopesticide formulations pertaining to insect viruses.
5. To know about the mass production, formulation, and commercialization strategies of entomopathogenic viruses.

1.1 Introduction

India has moved from an era of food shortages to a status of food self-sufficiency through improved modern agricultural technologies and sustainable agricultural practices. However, various constraints exist during the cultivation of the crops, which limits their production and productivity. In achieving the green revolution, chemicals have played a significant role in improving crop production, but the continuous and indiscriminate use of pesticides, vastly insecticides in crop production have brought pollution to the environment and created health hazards to human beings. A way forward, limiting such demerits from the use of pesticides, biocontrol agents serve as an alternative tool in pest management strategy, which is eco-friendly, cheaper, maintains soil fertility, and provides residue-free products. Considering the severity of infestation by insect pests on several crops and undisturbed ecosystems, such as forests, of late, entomopathogenic bioagents are exploited by the researchers for the control of various insects in crop husbandry and forest ecosystems.

Insects are the most diversified taxonomic group, appeared at least 400 million years ago and almost constitute every ecological niche, comprising over 58% of the identified global biodiversity (Takov et al. 2021). The entomofauna constitute around 80% of all animal species in the world. With the estimated population of insect species around 2.8–10 million on earth, only 1 million have been described. Obviously, insects are the dominant life forms on the earth, habituated ubiquitously, from humid tropical forests to icy polar regions. Among them, merely 10,000 are labeled as agricultural pests, which cause 20–40% yield losses annually at the field and postharvest levels (FAO 2019). Insect pests cause extensive losses in the production of agricultural crops, which ultimately distresses the food security and demand for global food production. Thus, the timely management of crop losses is a crucial part of global food security. During the 1940s, agricultural pest management relied mainly on chemical insecticides and still conquers the primary spot as a component for managing insect pests. Needless to state that, these chemical insecticides intensely improved the yield of important crops and anchored food security at all levels. Nevertheless, the continued use of synthetic chemical insecticides has led to negative effects on diversified nontarget organisms in the ecosystems due to high toxicity and has impaired the biodiversity, environment and the food chain of humans and animals (Epstein 2014). To evade such negative impacts on one side and to maintain the pest population below the economic threshold and obtain residue-free healthy foods, the natural mechanisms of insect regulations, such as the use of natural enemies and microorganisms have widely been adopted as biocontrol agents. Research on microbial biopesticide development and use are constantly being explored for the low probability of target insects to develop resistance, fewer environmental impacts, riskless human health, easy application and their substituting ability for chemical insecticides in the agricultural systems. In a natural ecosystem, insects are biologically regulated by epizootics of entomopathogenic organisms, like bacteria and viruses, through natural infections upon favorable conditions. Among the insect pathogens, the entomopathogenic viruses assume a critical function to build ecologically sound pest management systems (Prabhakar et al. 2017). Viruses are relatively abundant, small infectious noncellular forms that can bourgeon only inside the alive cells and are found associated with all types of organisms (Kalawate 2014), and many such viruses associated with insects are being explored for the past several years. The progress made with entomopathogenic viral insecticides has the prospective to reduce the existing dependence on synthetic chemical pesticides. The host selectivity of the most entomopathogenic viruses, especially baculoviruses, upholds the control of certain insect pests without harming beneficial insects and the natural ecosystem. While no discernible negative impacts of entomopathogenic viruses have been witnessed, the security of each entomopathogenic virus must be demonstrated before commercial exploitation, including harmfulness tests on warm-blooded animals, mammals, and different vertebrates. Moreover, it could be compulsory to examine through cytogenic tests, nucleic acid hybridization, and immunoassays, along with experiments for the conceivable perseverance of the entomopathogenic virus genome in nontarget organisms. Entomopathogenic viruses, potentially effective candidates for developing into commercial microbial viral bioinsecticides, underpin the view that measures of licensed versions of viral pesticides are needed to be pipelined for field applications. With this background, this chapter explores the possible role of entomopathogenic viruses in insect pest management.

1.2 Entomopathogenic Viruses

Insect and virus associations have long been reported and in several instances, they act as vectors of many important viruses that affect animals, humans, and plants, but insects themselves are also hosts to different viruses, termed as entomopathogenic viruses, which plays a pivotal role in environmental and agricultural sustainability by maneuvering their population. Entomopathogenic viruses are sub-microscopic, obligate, intracellular disease-causing pathogens and consist of either DNA or RNA in their genome encapsulated into a protein coat to form the nucleocapsids (Kalha et al. 2014) that could exclusively repeat inside a host insect. The evolutionary association of baculoviruses, bracoviruses, and nudiviruses with its hosts is long, maybe for in excess of 310 million years ago with the first insects observed during the Carboniferous period in the Paleozoic era. Moreover, the similar genomics of braconid wasps and bracoviruses suggest that broad diversification of these large-sized DNA viruses occurred alongside the broadening of insects during the Mesozoic era (Thézé et al. 2011). At present, there are in excess of 3000 reported virus infections that taint more than 20 different families of insects, the largest number depicted to date. Among the viruses, over 1690, are from the family Baculoviridae, which have been documented from over 1100 species of insects and mites (Eberle et al. 2012; Grzywacz and Moore 2017). The major families of viruses that cause pathogenic epizootics to insects are Baculoviridae, Densoviridae, Entomopoxvirinae and Reoviridae that have been mostly documented and developed commercially (Harrison et al. 2018). Among all these insect viruses, the family Baculoviridae is well studied at biological, ecological, molecular, and functional levels. Baculoviruses signify one of the most diverged assemblies among families of insect viruses and have probably evolved with holometabolous insect hosts. Baculoviruses have been isolated from above 600 insect hosts, with most of them belonging to nucleopolyhedrovirus (NPV); 456 isolates from Lepidoptera, 30 isolates from Hymenoptera, and 27 virus isolates from Diptera. Granuloviruses (GVs) are restricted to Lepidoptera with 148 isolates reported from China. Traditionally, baculoviruses are easily detected and isolated based on the visible symptoms on larvae, such as paralysis, whitening of the skin, tree-top hanging, and demise of the host, and in few insect viruses infections may be asymptomatic. Research on evolving non-baculoviruses for pest management has seen continued efforts, but only to a narrow extent. Members of different insect viruses are identified by further investigations using pathological and molecular diagnostics approaches.

1.3 Historical Perspectives of Entomopathogenic Viruses

One among the enigmatic things in the world, the developmental history of viruses is somewhat sooner than bacteria, which were first described only in the 1800s by porcelain filter experiments, yet have possibly co-existed with cellular life over nearly the whole of evolutionary history on this earth. Viruses of insects were reported over 2000 years back. Paleoentomology studies of preserved insects in a piece of amber between 15 and 200 million years ago had exposed the fossil of a grown-up sand fly contaminated with pathogens, such as fungi, trypanosomes, nematodes, cytoplasmic polyhedrosis viruses, and nucleopolyhedrovirus (Poinar and Poinar 2005). Invertebrate pathology and insect pathology may be the newly organized discipline, which includes insect virology, but its origins can be way back to 330 BC, where Aristotle emphasized the honey bee diseases on the basis of symptoms. During the sixteenth century, Girolamo Fracastoro (1478–1553), the

Italian Physician was the first person to propose that infectious entities might spread disease (Taylor 2014). Probably, the first insect viral disease called jaundice or grasserie in silkworm caterpillar (*Bombyx mori* L.) exhibited symptoms of swelling, shining, wilting, and melting caused by a baculovirus, which was described in the poem "De Bombyce" 1524 by Marco Girolamo Vida (Vida 1527; Ibarra and Del Rincón-Castro 2009). Similar symptoms were also observed by Merian (1679) and Nysten (1808) characterized the signs as "jaundice" of silkworms. Maestri (1856) revealed crystalline structures in the cells of jaundiced silkworm larvae, the first microscopic perception of nucleopolyhedrosis virus. In Germany, in the late 1800s, caterpillars of the nun moth, *Lymantria monacha*, climbed and died at the top of spruce trees, which exhibited the symptoms of virus infection. Bolle (1894) observed the disease-causing ability of crystal particles and their solvency in antacid gut juices of the caterpillar. Similarly, the pattern of wilting was observed in the gypsy moths by Glaser and Chapman (1913) and stated to be caused by microscopic particles, having capability of passing through bacteriological filters. In a while, White (1917) portrayed the first virus infections in honey bees. Glaser (1918) first proved sequential transmission of polyhedral disease using filtrates of gypsy moth diseased larvae. In Europe, a report of field application of polyhedrosis-like virus causing the disease *Wipfelkrankheit* in nun moth was documented in Europe (Ruzicka 1924). In 1926, Andre Paillot (1933) first discovered the GVs in the cabbage butterfly larvae and a few different kinds of viral infections as reported in his book "L Infection Chez les Insectes." Likewise, the first cytoplasmic polyhedrosis virus (cypovirus) was described by Ishimori (1934). Meantime, in 1936, a nucleopolyhedrovirus was first reported to attack European spruce sawfly larvae, *Gilpinia hercyniae*, brought unintentionally from Scandinavia to Canada, which later was hosted as a component of classical biological control program in Canada and followed was the sawfly epizootics caused by this virus in 1938. During the 1940s, the spruce sawfly incidentally introduced into North America was also effectively controlled by the subsequent introduction of a baculovirus (Kalawate 2014). The first electron micrographs of baculoviruses (NPVs) were got distributed, and new techniques to purify the viruses were established (Bergold 1947). In 1947, Furniss recorded an NPV in tussock moth caterpillars collected in Oregon (Martignoni 1999). Steinhaus (1949) published the first textbook "Principles of Insect Pathology," which depicted the scientific classification of viruses. Bird and Whalen (1954) obtained an NPV from Sweden in 1949, which became a widely used product in forestry that had existed long enough in Europe and North America. Constantin Vago (1963) first described the entomopoxviruses in the European cockchafer, *Melolontha melolontha*. The presence of GV was also described in Codling Moth, *Carpocapsa pomonella* (Tanada 1964). Huger (1966) discovered a beetle-specific virus named *Rhabdionvirus oryctes,* now known as *Oryctes* virus, befitted as an effective viral insecticide against pests of coconut and palm crops. Steinhaus and colleagues tested the first field application of NPV as a biological agent against the alfalfa caterpillar, *Colias eurytheme* in the field (Ibarra and Del Rincón-Castro 2009). Corn earworm, *Helicoverpa (Heliothis) zea* NPV (HezeSNPV) was commercially tested by the US Department of Agriculture

(USDA) for their control in the mid-1960s. It is a salient opening that the first commercial NPV bioinsecticide product was first developed in 1975 under the name Viron/H and later Elcar™ (*H. zea* NPV) by Sandoz Inc., to control many species of *Helicoverpa* complex (Lepidoptera: Noctuidae) especially *H. zea* in the United States (Ignoffo and Couch 1981). The *Dendrolimus spectailis* cypovirus type 1 (DsCPV-1) was another insect virus registered during the mid-1970s in Japan. Later in 1976, the Douglas fir tussock moth, *Orygia pseuedotsugata* multinucleocapsid nucleopolyhedrovirus (OrpsMNPV) was registered by the US Forest Service in the name of TM BioControl-1. Subsequently, Gypcheck™ was registered in US Environmental Protection Agency (US-EPA) against gypsy moth, *Lymantria dispar* during 1978, which remains to commercially exist. Velvet-bean caterpillar in soybean was controlled by NPV in the early 1980s in Brazil (Moscardi et al. 1981). In 1982, the manufacture of Elcar™ biopesticide was discontinued due to its fast degradation upon contact to direct sunlight. Therefore, research was directed to explore many adjuvants for supporting insecticidal property. Subsequently, researches were also conducted for genetic manipulation of insect viruses. In 1993, the first field trial of a transgenic viral bioinsecticide was established in the UK. *Heliothis armigera* NPV (HaSNPV) was formulated as an emulsifiable suspension and registered in China as a pesticide for cotton *Helicoverpa* sp. in 1993 (Zhang et al. 1995). Insect-hosted picorna-like virus has been identified in the brains of aggressive worker honey bees by Fujiyuki et al. (2004). In 2004, about two million ha of soybean crops were protected with *Anticarsia gemmatalis* Multiple Nucleopolyhedrovirus (AgMNPV) in Brazil (Moscardi et al. 2011). In China, around 200 insect virus isolates have been screened for their epizootics efficacy, and 32 insect viruses of them were found as potential bioinsecticides. There were 17 products made by 10 different companies that included HearNPV and SpliNPV (Sun 2015). Nowadays, several countries, viz., India, Brazil, China, and South Africa have increased the area of crops protected by insect viruses. Viruses have been used for long- and short-term insect pest control programs. In developing nations, the use of insect viruses for pest control program has been greatly successful. Large-scale commercial production of these viruses is done often as a small industry with supports from farmer groups. Globally, so far, numerous insect virus-based biopesticides are either registered with statuary bodies or in the investigation stage.

1.4 Taxonomic Classification of Entomopathogenic Viruses

Historically, several insect viruses were originally described and characterized by entomologists working on specific insect groups or pest species. The nature of mesobiotic viruses was mysterious before the 1900s but now viral metagenomics, high-throughput sequencing, and other molecular cum computational approaches have enlightened clearly about the abundance of virus particles in nature. In his book "Principles of Insect Pathology" (Steinhaus 1949), Steinhaus proposed the first scientific classification of insect viruses, giving scientific nomenclature to individual genera and type species. The viruses are classified by means of Latin Linnaean binomial names (LLBNs) in view of morphological attributes and afterward coordinated to take into genomic phylogeny. Viruses are divided into two broad nontaxonomic groups, viz., enveloped viruses and non-enveloped viruses. The fast-growing corpus of genomic data transforms the taxonomic approaches from traditional to genomics-based disputes. Virus taxonomy appeared as a discipline in the mid of the twentieth century. Viruses are grouped based on their nucleic acid composition, their genome structure, and the morphology of their external coats. The nucleic acid of insects may have either DNA or RNA genomes, never both. RNA viruses are composed of single- or double-stranded genomes. As per the polarity of their RNA, single-stranded RNA viruses can be again categorized as positive or negative-sense RNA viruses. Furthermore, RNA viruses can be portrayed by expressing a lipid envelope or not, and if they have genomes in a single (nonsegmented) or multiple (segmented) components, depending on the number of nucleocapsids packed within an occlusion-derived virus. Comparably, RNA viruses exhibit simple structures, condensed genome, and abundant replication, suiting them ideal for exploring their relationship with viruses of common origins, for constructing the taxonomy. The International Committee on Taxonomy of Viruses (ICTV) has consolidated the taxonomy of viruses up to date. The classification of viruses by taxonomists has been concentrated on the assemblage of closely related viruses, often in the background of an architype that highlights the rapid evolution of viruses, although the Baltimore viral classification system (Baltimore 1971) can be used to abode viruses into one of seven groups, which depends on their manner of mRNA synthesis (Kuhn 2020). The existing taxonomic classification of viruses is laid out in the Ninth Report of the International Committee on Virus Taxonomy (ICTV), just as the forward-thinking ICTV Taxonomy and Index to Virus Classification with its 15-rank classification system that closely line up with the Linnaean taxonomic system and may house the entire scale of genetic divergence in the virosphere, consists the Nomenclature Taxonomic Lists and Catalogue of Viruses that incorporates 6590 species and can be found on the site (https://talk.ictvonline.org/taxonomy/). The ICTV online index lists 22 virus families whose hosts include invertebrates for at least some members of the group. A great diversity of viruses representing nearly 39 virus families has been reported to be associated with arthropods to at least some amount as insect pathogens (Table 1.1). Of which seven families of insect viruses that have been used as an important entomopathogenic bioinsecticide, viz., Baculoviridae, Dicistroviridae, Iridoviridae, Nudiviridae, Parvoviridae, Picornaviridae, Poxviridae, Reoviridae, and Rhabdoviridae, cause sickness in insects. Be that as it may, the infections of viruses from families Baculoviridae and Reoviridae are considered the severest for their function as bioinsecticides (Kalawate 2014). The well-known genera of the family alpha-, beta-, delta-, and gamma-baculoviruses, the nucleopolyhedroviruses (NPV; *Alphabaculovirus* spp.), and granuloviruses (GV; *Betabaculovirus* spp.) have been commercially evolved as viral bioinsecticides against lepidopteran pests (Lacey et al. 2015). Recently, the International Committee on Taxonomy of Viruses (ICTV) approved three orders, 73 families, nine subfamilies, 287 genera, and 1950 species

Table 1.1 Major entomopathogenic virus families

Nucleic acid	Family
ssRNA (+)	Alphatetraviridae, Carmotetraviridae, Dicistroviridae, Iflaviridae, Nodaviridae, Permutotetraviridae, Solinviridae, Tetraviruses
ssRNA (−)	Rhabdoviridae
dsRNA	Birnaviridae, Reoviridae
ssDNA	Bidnaviridae, Parvoviridae, Circoviridae
dsDNA	Ascoviridae, Baculoviridae, Iridoviridae, Nudiviridae, Polydnaviridae, Poxviridae, Hytrosaviridae

of viruses, despite the fact that the last rundown incorporates an aggregate of 5450 viruses. Nomenclature of a few insect viruses follows a binomial standard, depending on the insect from which they were first isolated, given with signs, like GV or NPV in baculoviruses and individual segment numbers for cypoviruses and polydnaviruses. The detailed description of major virus families is as follows:

1.4.1 Ascovirus

Ascoviruses are enormous DNA viruses that usually infect noctuid larvae and are transmitted by Hymenopteran parasitoids. They were discovered in the larvae of *H. zea* in the 1970s (Adams et al. 1979) and named the ascovirus. It was isolated from the infected *S. frugiperda* in the early 1980s (Hamm et al. 1986). It infects insects of the Lepidopteran group, which consists of devastating agriculturally important insect pests. The infected host insects exhibit hypertrophy of cells and the nucleus ruptures in the lepidopteran larvae (Federici 1983). Ascoviruses are newer associates of the family *Ascoviridae*, with circular super-helix double-stranded DNA genomes of 156–186 kilobase pairs with 117–180 genes. Virions are large, oblong, enveloped with 130 nm in width by 200–400 nm long and contain 20 polypeptides, encoded for up to 180 open reading frames, with two tandems of inverted repeats. Virion can be reniform to bacilliform, ovoid, or allantoid in shape occluded inside vesicle-like impediment bodies made of minivesicles and protein. Ascoviruses are rare to transfer by oral ingestion, with infection rates around <15%, and field observations indicate that the virions are often mechanically transmitted during oviposition by female endoparasitic wasps of the Hymenopteran families Ichneumonidae and Braconidae (Bideshi et al. 2010) in a symbiotic manner. The virus particle replicates intracellularly and the infected insect larvae do not pupate. A larva infected with ascovirus is tougher to identify as being infected due to a lack of apparent symptoms in the field. It possesses unique developmental cytopathology in the infected larva, which is stunted with virus-instigated apoptosis leading to disintegration of the cells and producing a larger number of virions containing vesicles. These virions accumulate in the hemolymph of the infected caterpillars and are acquired later by the parasitic wasps that transmit the virus (Bideshi et al. 2005). Ascovirus infected larvae exhibit fragile delicate muscle versatility, diminished food consumption, diminished weight, hindered yellow body, shedding disappointment, neglected to pupate, and afterward die (Chen et al. 2020). The ascovirus genus contains five species, viz., *Heliothis virescens ascovirus 3a*, *Spodoptera frugiperda ascovirus 1a*, *Diadromus pulchellus 4a*, *Tricoplusia ni ascovirus 2a*, and *Tricoplusia ni ascovirus 6a* (Wei et al. 2014; Asgari et al. 2017). Basically, ascoviruses exhibit a low infection rate in the field ranging from 0.26% to 50%. On the other hand, parasitic wasp vectors, *Campoletis sonorensis*, *Cardiochiles nigriceps*, and *Microplitis croceipes* were able to transmit HvAV at a higher transmission rate (Tillman et al. 2004). *Heliothis virescens* ascovirus 3 h (HvAV-3 h) can infect *H. armigera*, *S. exigua*, and *S. litura* and makes it a conceivable biocontrol agent (Huang et al. 2012; Li et al. 2016). In ongoing analyses to improve per os infectivity of ascovirus, *Bacillus thuringiensis kurstaki* was engaged as a co-inoculant to harm the midgut of lepidopteran larvae, such as *H armigera*, *Mythimna separata*, *S. frugiperda*, and *S. litura* in formulations with isolates of *Heliothis virescens* ascovirus (HvAV-3 h and HvAV-3j). Nevertheless, the unique infection process and efficient transmission by hymenopteran endoparasitic wasps increase the probability of using an ascovirus as a biocontrol agent.

1.4.2 Baculovirus

The family Baculoviridae is the extensively studied group of entomopathogenic viruses, which consists of more than 600 viruses infecting 700 insect species globally, with applications in natural control of insect pests, gene therapy, vaccine production, expression vectors, and virology research. The discovery of an insect-baculovirus dates back thousands of years ago from a disease distressing silkworm (Rohrmann 2019) (Table 1.2). The word "baculovirus" is derivative from the Latin "baculum" which means a slimy liquid, poison, or stench. More recently, the definition attempted to convey two qualities of the virus: (1). possession of its own genetic material, which inside the host cell behaved as part of the cell, and (2). presence of a submicroscopic infective stage, the virion, which served as the vehicle for introducing the viral genome into a cell. Earlier, the stick shape of the virus denoted the rod-shaped nucleocapsids and later, the large greatly refractile polyhedron-shaped occlusion bodies followed by small, granular, and ellipsoidal occlusion bodies in cadavers of diseased insect were observed under the light microscope (Rohrmann 2019). Baculoviruses signify the biggest and most diverse family of DNA viruses. They are mainly classified into four different genera, viz., alpha-, beta-, gamma- and delta- baculovirus based on genome arrangement and the order of host insect. This replaces earlier classifications based on morphologically gathered groups of baculoviruses, viz., NPV, which infects 400 arthropod species belonging to seven orders, primarily in Coleoptera, Diptera, Hymenoptera, Neuroptera, Trichoptera, Decapoda (class Crustacea) and Lepidoptera (Murphy et al. 1995; Possee 1997; Herniou et al. 2012), and GV, infects more than 100 insect species, mostly the members of Lepidoptera and their host range is more narrow (Murphy et al. 1995). However, Martignoni and Iwai (1986) reported the

Table 1.2 Characteristics of Baculoviridae family

Given name	Genus	Targeted insect order	Occlusion body	Occlusion-derived viruses
Nucleopolyhedroviruses	Alphabaculovirus	Lepidoptera	Polyhedral (0.4–3)	Single or multiple nucleocapsids per envelope
Granuloviruses	Betabaculovirus	Lepidoptera	Granular (0.12–0.5)	Mostly single enveloped nucleocapsid
Nucleopolyhedroviruses	Gammabaculovirus	Hymenoptera	Polyhedral (0.4–1.1)	Single enveloped nucleocapsid
Nucleopolyhedroviruses	Deltabaculovirus	Diptera	Globular (0.4)	Single enveloped nucleocapsid

Siphonaptera (fleas) as a host. Viruses in three families, Baculoviridae, Entomopoxviridae, and Reoviridae, are unique because of the presence of occlusion bodies in which virions at a certain stage in their development are occluded at random. The occlusion bodies contribute to the stability and persistence of the viruses in the environment. Baculoviruses were established as biopesticides for the management of Lepidopteran insect pest species (Black et al. 1997).

Morphological Characteristics of Baculovirus

Francki et al. (1991) placed the two subgroups (NPV and GV) in the sub-family Eubaculovirinae and the third subgroup (nonoccluded viruses) in the sub-family Nudibaculovirinae. These two pathologically comparable, but morphologically and phylogenetically dissimilar baculoviruses were divided into two groups, viz., NPVs and GVs (Murphy et al. 1995). Studies on the morphological characteristics of polyhedral inclusion bodies by electron microscopy revealed two morphotypes: (1). single nucleocapsid nucleopolyhedroviruses (SNPVs) contain only a single nucleocapsid within a virion, and (2). multiple nucleocapsid nucleopolyhedroviruses (MNPVs) contain a few to many enveloped rod-shaped nucleocapsids structurally claw-like at base and ring-like at apex, encased by a lipid bilayer (Rohrmann 2019; Lei et al. 2020). NPV occlusion bodies have different size and shape in different species, viz., tetragonal, triangular, and pentagonal forms as in *Porthesia xanthocampa* (Ishikawa et al. 1966); multi-shaped and cuboidal as in *Euproctis similis* (Watanabe and Aratake 1974) and hexagonal as in *Eucalyptus similis* (Chu et al. 1975); spherical to irregular shape with size ranges from 0.5 to 2.5μm, 0.9 to 2.92μm, 1.0 to 2.0μm in *H. armigera, S. litura* and *Amsacta albistriga*, respectively (Rao et al. 2007); tetrahedral in shape in *Spilarctia obliqua* (Senthil Kumar et al. 2015); tetrahedral and a few were of hexagonal with 1.016–1.596μm size in *Euproctis chrysorrhoea* (Hussain et al. 2019); tetrahedral and triangular with the size ranged from 1.04 to 1.72μm in *S. obliqua* (Sivakumar et al. 2020a) (Figs. 1.1 and 1.2); tetrahedral shape and size of 1.64μm in *S. frugiperda* (Sivakumar et al. 2020b).

Baculovirus Replication

Viruses differ in their mode of replication, which involves the adsorption, uptake and uncoating, expression, replication of the viral genome and production of viral progeny. Enzymes that are present in the viral particles or in the host cells are required for replication. Viral replication involves three periods of development: (1) a latent or eclipse, (2) an exponential, and (3) a stationary period. During the eclipse period, the virus is undergoing uptake, uncoating and early stage of replication, and the virus is not infective. The exponential period is when the number of infectious virions increases exponentially until the number reaches a plateau at the stationary period (Rohrmann 2019).

Biochemical and Molecular Characteristics of Baculovirus

The viral particle (virus, virion, or vibrion) is composed of a protein shell (capsid) that surrounds the nucleic acid. The capsid provides the viruses with morphological

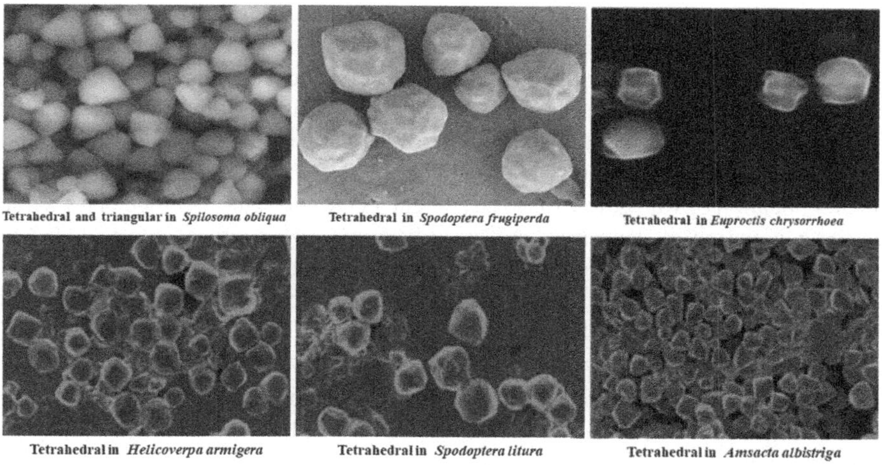

Fig. 1.1 Scanning electron microscopy micrograph of tetrahedral and triangular shape occlusion bodies of NPV. Source: Rao et al. (2007); Hussain et al. (2019); Sivakumar et al. (2020a) and Lei et al. (2020)

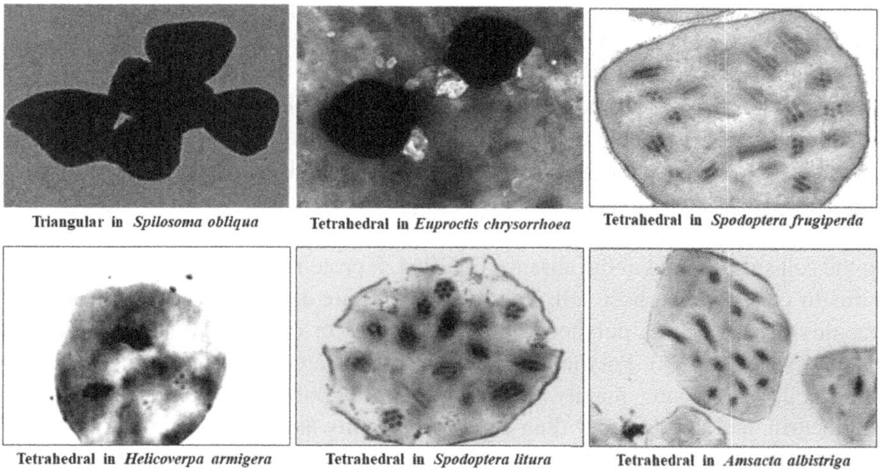

Fig. 1.2 Transmission electron microscopy micrograph of triangular and tetrahedral shape occlusion bodies of NPV. Source: Rao et al. (2007); Hussain et al. (2019); Sivakumar et al. (2020a) and Lei et al. (2020)

and functional properties, and the nucleic acid with the genetic constituent. Each virus has only one type of nucleic acid, either deoxyribonucleic acid (DNA) or ribonucleic acid (RNA). The nucleic acids may be single- or double- stranded. The nucleic acid together with the capsid forms the nucleocapsid. The simplest virus consists of nucleic acid and a capsid. Viroids have only nucleic acids and no capsids. The design of the capsid is of two major types: (1). helical assemblage (rod-shaped) and (2). Closed-shell (isometric, cubic, or quasi-spherical). In some helical and cubic viruses, the nucleocapsids are surrounded by envelopes that are lipid bilayer and may be related to components of the cell membrane. The envelope is acquired during viral replication or when the virus leaves or enters the cell. The envelope plays a role in the penetration of the virion into the cell. Some insect viruses are occluded in proteinaceous bodies that are referred to as viral occlusions, occlusions, or inclusion bodies. The occlusion body is more appropriate for the body containing virions, and the inclusion body should be a general term referring to a body with or without virions.

The members of the family *Baculoviridae* contain single large covalently closed circular supercoiled dsDNA with a size somewhere in the range of 80 and 180 kbp, encoding for 80 to 200 potential genes. They are enveloped within rod-shaped nucleocapsids (van Oers and Vlak 2007), 30–60 × 250–300 nm, containing >20 proteins with a large apical density of spikes (protein extensions) on the surface of the lipid envelope and are embedded in distinctive polyhedral or granular occlusion bodies, consisting of polyhedrin (NPV) or granulin (GV) with ovoid shape (granules) matrix protein. So far three protein families were identified in the baculovirus polyhedral, such as polyhedrins (Jarvis et al. 1991), glycoproteins, GP64, and membrane fusion (F) proteins (Monsma et al. 1996). The protein polyhedrin forms crystalline matrices around the virion for guarding against biochemical and physical degradation and help to maintain their biological activity (Hu et al. 1999). Once the polyhedra matrices are crumbled inside the host midgut, virions are free and activate the replication process. The fatty acid acylated protein GP64 is crucial in the propagation of the budded virus from cell to cell and binding to the cell surface, and in the absence of GP64, F proteins are indirectly assisting the virus to penetrate the host cell. Two phenotypes are expressed in the life cycle of baculoviruses, such as polyhedral/granular bodies or occlusion bodies (OBs) and budded virions (BVs). Polyhedral occlusion bodies are very much organized, proteinaceous structures mainly composed of a polyhedral envelope protein, polyhedrin, and p10 for providing complete stability and viability to the occlusion-derived virus. The matured OBs are a trimmed rhombic dodecahedron arrangement, which allows the assembly of the polyhedral unit in a rapid manner (Sajjan and Hinchigeri 2016). Another phenotype, budded virions (BVs) are responsible for the cell-to-cell appearance of the infection. BVs consistently contain envelope fusion proteins, viz., GP64 and F on its surface. The nucleocapsid has a cap and base plan that are joined in solid assembly with the bilayered envelope structure (Wang et al. 2016).

Interaction of Baculoviruses with Host Insects and Symptoms Development

The gross pathology of most lepidopteran larvae infected with NPVs shows no external signs of symptoms for 2–5 days after viral ingestion. The initial signs are the gradual changes in color and luster of the integument with an increase in opaqueness, milkiness, and glossiness. The hemolymph turns cloudy and milky (Fig. 1.3). The larva becomes less active and loses its appetite but may continue to feed up to a few days before death. The larva generally dies in 5–12 days, but virulent viral strains may kill very young larvae in 2–4 days. In some BV infections, the larval period is prolonged, even beyond the normal period of the larval stage. The prolonged life may be caused by the EGT gene (ecdysteroid UDP glucosyltransferase gene) located in the viral genomes. The product expressed by this gene decomposes ecdysone, the molting hormone, and thereby increases larval life. Such prolongation of larval life would benefit viral reproduction. Shortly before dying, the larva may move away from the food, disperse or climb an elevated location to hang from a branch or treetop by their abdominal, and caudal prolegs as in the case of "wipfelkrankheit" of the nun moth, *L. monacha*. Prior to death or shortly thereafter, the integument, if the hypodermal cells are infected, becomes fragile and easily torn when handled. Such a larva is in a wilted condition, typical of most nuclear polyhedroses. The larval body contents are a fluid mass. Although death usually occurs in the larval stage, some larvae may survive in the pupal or adult stages. The fecundity of the surviving, normal-appearing adult is unaffected, but the hatchability of the eggs may be reduced significantly (Tanada and Kaya 1993).

Mode of Action of Baculoviruses in Insect Body

The recent structural model of baculoviruses is shown in Fig. 1.4. The occluded form guards the virions against environmental deprivation and hostile alkaline conditions of the insect gut (Ji et al. 2010). Neonates, fourth and fifth instar larvae are ordinarily vulnerable to baculovirus. During the polyorganotrophic infection, the virions can

Fig. 1.3 Symptoms of nucleopolyhedrovirus and granulosis virus infection in pests. Source: Rao et al. (2007); Sivakumar et al. (2020a) and Lei et al. (2020)

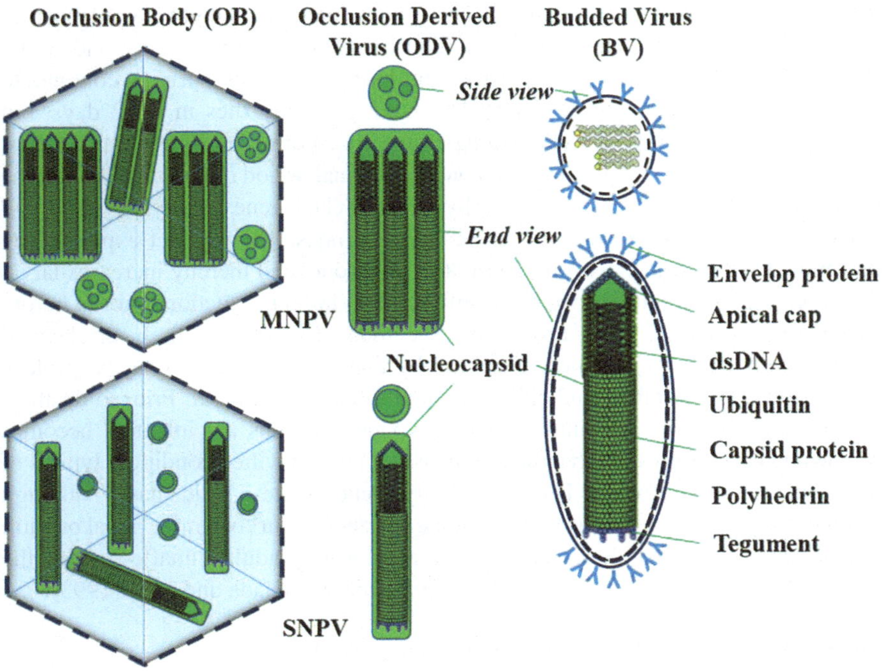

Fig. 1.4 Structural model of baculoviruses

penetrate the epidermal layer and move on fat bodies, hypodermis, hemocytes, tracheal lattices, silk organs, malpighian tubules, brain, corpus allatum, prothoracic organ, focal sensory system cells, pericardial cells, and midgut epithelium (Harrison and Hoover 2012), in spite of the fact that the replication of hymenopteran and dipteran NPVs is limited to the midgut epithelium. OBs are the reason for infection through the oral route. Inside the insect midgut alkaline conditions, occlusion-derived viruses are delivered and focus on the columnar epithelial cells of the midgut. Then, the replication begins, suppressing the resistant gene expression and producing BVs for secondary infection. The BVs are delivered into the host cells through ingestive endocytosis (Volkman and Goldsmith 1985) and penetrate the plasma membrane; nucleocapsids move toward the core and viral DNA gets out and begins its fast proliferation and transcription of all viral proteins. The symptomatic larval hemolymph befits turbid-milky instantly due to the replication of the virus particles (Chishti and Schaf 1990). The host defense related transcript and protein expression are ultimately shattered by baculovirus infection for 12–18 h (Katsuma et al. 2005). Immature larvae die within 2 days, and mature larvae die within 4–9 days (Afolami and Oladunmoye 2017). Occluded bodies of both NPV and GV, instituted within the nucleus of infected cells are essentially dormant structures that can survive in aggressive conditions upon their discharge from dead cadavers.

1.4.3 Cypovirus

The family of Reoviruses is portrayed as nonenveloped, icosahedral virions with genomes comprising 10–12 straight dsRNA molecules that replicate in arthropod vectors, for example, leafhoppers. The family comprises a genus Cypovirus, which is a cytoplasmic polyhedrosis virus (CPVs) in the subfamily Spinareovirinae that can only infect and be pathogenic for arthropods. They are mostly isolated from more than 250 species of Lepidoptera, and a few isolates are from Hymenoptera and Diptera insects of the suborder Nematocera (e.g., blackflies, midges, and mosquitoes) and rarely from Coleoptera or Neuroptera. Cypovirus was first portrayed in the cytoplasm of infected midgut cells in *B. mori* larvae in 1934 (Ishimori 1934). The cryoelectron microscopy upheld with a three-dimensional structure and has uncovered that the CPV capsid is made out of three significant structural proteins: the capsid shell protein, turret protein, and large protrusion protein (Yu et al. 2011). The cypovirus genus is a linear double-stranded RNA genome made out of 10 genome fragments that encode for 10–12 proteins. It has an expected size, which differs somewhere in the range of 1 and 4.2 kb, and the genome absolute size is 25 kb. The virus is distinctively occluded (either singly or multiply) inside the matrix of proteinaceous crystals called polyhedral inclusion bodies (Mori and Metcalf 2010), which appear to be engaged with structural auxiliary and transmission between individual insect hosts. The occlusions are icosahedral non-encompassed virions of 60–70 nm in width, having two shells encompassing the viral center, instead of single shells in different reoviruses. All 10 described genome segments (VP1, VP2 Pol, VP3, VP4, Nsp5, VP6, VP7, Nsp8, Nsp9, and polyhedrin) have been sequenced for some cypoviruses including *H. armigera* cypovirus 5 (Li et al. 2006), *L. dispar* cypovirus (Tan et al. 2008), *B. mori* cypovirus (Cao et al. 2012), and *Dendrolimus punctatus* cypovirus (Zhou et al. 2014). The positive strands contain a 5′-terminal type 1 cap structure (^{7m}GpppN$^{2'Om}$pNp. . .) in each duplex, which was first found in cypoviruses. The conserved terminal sequences of the dsRNA fragments are fluctuating in various CPV types, which are used to recognize the distinctive CPV species and associated with RNA replication and packaging. The minor capsid protein and major capsid protein are encoded by 1 and 3 genome segments, respectively (Chakrabarti et al. 2010); the RNA-dependent RNA polymerase is encoded by segment 2 (Ghorai et al. 2010), and polyhedrin is encoded by segment 10 (Sinha-Datta et al. 2005). ICTV assembles all the insect-specific reoviruses within the genus Cypovirus and recognizes and perceives 21 unique genera of cypoviruses (CPV-1 to CPV-21) with more than 74-member species. They are separated and depend on electrophoretic mobilities of the 10 dsRNA molecules, as well as nucleotide sequence and antigenic varieties. Cypoviruses spread through contact or fecal–oral courses, and upon ingestion the occlusion bodies (0.1–10μm) dissolve in the alkaline medium of the insect midgut. They usually replicate and form huge polyhedral bodies (no amino acid sequence) or release the occluded virions in the susceptible cytoplasm of larval midgut columnar epithelial cells. The production of large amounts of polyhedra gives the gut a symptomatically creamy-white appearance. The majority of infections exhibit chronic symptoms, often without wide larval mortality, which include reduced feeding due to infection in the gut cells, reduced absorptive capacity, reduced body size, frequent diarrhea, and malformed adults that have decreased life span and fecundity. The *B. mori* cytoplasmic polyhedrosis virus (BmCPV) is the type species and the most considered Cypovirus family as it causes mortality in silkworm (Jiang and Xia 2014). There are almost 250 reports of lepidopteran isolates and more than 25 from dipterans. Despite the fact that the Cypoviruses are incredibly irresistible and persist in insects, their use as viral insecticides is rare due to their generally chronic rather than acute intense symptomology. The only Cypovirus product registered for the control of pests in Japan, France, and China was Matsukemin of the pine caterpillar, *Dendrolimus spectabilis* (Kunimi 2007), but it lost its registration in 1995. *Norape argyrrhorea* cypovirus (NoarCPV) was used to control the natural population of devastative oil palm defoliator pest in Peru. Researchers are finding a framework for joining heterologous proteins into *B. mori* CPV occlusion bodies in an efficient form for use on protein chips (Ijiri et al. 2009). Likewise, Cypovirus disease is able to suppress chemical insecticide detoxification based on cytochrome P450 pathways. So, the other recommended use for Cypovirus is in the mix with synthetic insecticides for control of insect populaces safe.

1.4.4 Densovirus

Densoviruses (DNVs) (family: Parvoviridae, subfamily: Densovirinae) are habitually isolated from infected insects belonging to seven orders, viz., Hymenoptera, Hemiptera, Homoptera, Diptera, Blattodea, Orthoptera, and Lepidoptera and are relatively stable in the environment. Densoviruses are highly pathogenic, and infected tissues often show characteristic nuclear hypertrophy, caused by the accumulation of large virion particles, and cytoplasmic paracrystalline virion arrays. These virus particles are small, nonenveloped, icosahedral virions with a solitary linear 4–6 kbp, single-stranded ambisense DNA homotelomeric genome (Berns et al. 1995). It contains five major open reading frames (ORFs), encoding viral nonstructural replicative (NS1, NS2, and NS3) and structural coat proteins (VP1–4 and VP1) (Valles et al. 2013; Zhou et al. 2018), flanked by noncoding arrangements of variable length forming terminal palindromic organizations (Rhode and Iversen 1990) that can frame either a Y-shaped structure or a T-shaped structure or an unassuming two hairpin-like structures. Densoviruses undergo alternative splicing of mRNAs to yield the nonstructural protein 1 (NS1) endonuclease spending a rolling-hairpin mechanism to control replication (Cotmore and Tattersall 1996). Densovirus was first found in laboratory wax moth settlements (*Galleria mellonella*) and in insects reared commercially, for example, wax moth and silkworm. Densoviruses infect host larvae and exhibit a range of indications that include cuticular pigmentation, reformist loss of motion, and demise. The parasitism and harmfulness of Densoviruses differ significantly relying upon the infection species. Densovirus has a possible part in wing morph determination of the parasitized aphid (Ryabov et al. 2009). Some of the Densoviruses infecting important pests of agriculture are fairly virulent and host-specific and can be exploited as biopesticides. Studies recommended the introduction of *Galleria mellonella* densovirus into beehives against pervasions of wax moth. *Planococcus citri* densovirus infects the citrus mealybug (Thao et al. 2001). *Myzus persicae* densovirus infections are seen on the green peach aphid (van Munster et al. 2003). Biocontrol capability of *Junonia coenia* densovirus demonstrated to be lethal for *Spodoptera frugiperda*, the fall armyworm (Mutuel et al. 2010). The capability of a few densoviruses for the control of the oil palm pests, *Sibine fusca* and *Casphalia extranea,* was attempted in Africa, Egypt, and South America (Yu et al. 2012). Further, *Solenopsis invicta* densovirus (SiDNV) was evaluated against aggressive generalist forager fire ant in the United States for use as biopesticides, which is the first DNA virus in ants (Formicidae) and the first densovirus discovered in a hymenopteran insect (Valles et al. 2013). Most densoviruses cause grave diseases in their hosts and have been utilized for the biocontrol of significant insect pests owing to their high harmfulness and simplicity of transmission, and have been demonstrated to be easy going to progress as classical biopesticides providing a species-specific alternative to conventional insecticides (El-Far et al. 2012).

1.4.5 Dicistrovirus

Dicistroviruses (family Dicistroviridae) were formerly known as the "Cricket paralysis-like viruses" (Mayo 2002). The name Dicistrovirus refers to the characteristic monopartite RNA genome that includes two open reading frames (ORF) or dicistronic arrangement (Bonning 2009), first discovered with the *Acute bee paralysis virus* in 1963 infecting honey bees (Bailey et al. 1964). Introductory, and possibly incomprehensibly, a few dicistroviruses are existing in ordinary, sound arthropod occupants as persistent asymptomatic infections, which lead to death in due course. Dicistroviruses are predominantly dispersed among honey bee populations and coinfections can be regularly very high in different infections. Individuals from the family Dicistroviridae comprise 15 species, separated among three genera, in particular, Aparavirus, Cripavirus, and Triatovirus. All individuals from the family infect the gut tissues of many insect orders including Coleoptera, Lepidoptera, Orthoptera, Diptera, Hemiptera, and Hymenoptera, which serve as natural hosts, several being pathogenic to pests of agriculture, including vectors of plant viruses. Virions are small, nonenveloped, roughly spherical, isometric viruses at 30 nm in diameter with a monopartite, single-stranded, positive sense RNA genomes of 8–10 kb in length, encoding two long ORFs (Carrillo-Tripp et al. 2014), while recent confirmation shows that some members comprise the third ORF, termed ORFx and replicates in the host cytoplasm. The known ORFs are translated as polyproteins, then processed and matured into the individual viral proteins by the encoded protease. Dicistrovirus virions are stable under acidic conditions, whereas alkaline conditions cause uncoating by discharging the interactions between capsid proteins (Warsaba et al. 2020). Dicistroviruses are transmitted horizontally through a fecal–oral route (Chen and Siede 2007), often transmitted vertically by transovum. Although dicistroviruses use plants as vectors for spread into sap-sucking insects, knowingly *Rhopalosiphum padi virus* (RhPV) can be transmitted through plants, by circulating in the phloem vessels of the host plant from which it can be acquired by other aphids (Ban et al. 2007). The Big-Sioux River virus (BSRV) has also been detected in maize tissues (Wamonje et al. 2017), which is similar to the dicistrovirus. The potential host specificity and other desirable traits make several members of this group amenable for development as biopesticides to manage insects. One example is the use of the *Homalodisca coagulate* virus 1, a dicistrovirus as a biopesticide against a polyphagous insect glassy-winged sharp-shooter (GWSS) that voraciously feeds nearly 100 plant species, and the GWSS is a vector of devastating plant pathogenic bacteria, *Xylella fastidiosa*, which cause Pierce's disease of grapevines and citrus variegated chlorosis (Hunnicutt et al. 2008). There are different instances of the utilization of dicistrovirus for the control of pests that destructively affect crop plants including the use of the type member in the family, Cricket paralysis virus (CrPV; Cripavirus) for control of the *Dacus oleae* (olive fruit fly). It also has a wider host range and might target various hosts (Manousis et al. 1988); however, the early to mid-instars were the utmost susceptible stages than adults, which exhibit a point of resistance. Cricket paralysis virus on crickets and Aphid lethal paralysis virus (ALPV) infections of *Rhopalosiphum padi* exhibited, reduced feeding, paralysis of hind legs, followed by death. *Rhopalosiphum padi* virus (RhPV) was discovered in laboratory colonies of *R. padi* and *Schizaphis graminum,* which constantly appeared to diminish the fertility and life span. Three positive-strand RNA viruses, *Solenopsis invicta* virus (SINV-1), SINV-2, and SINV-3 seem to cause mortality on an intrusive red fire ant colony, which is reminiscent of honey bee colony collapse disorder (CCD) and all three earmarks of being a fantastic candidate as a natural control against fire ants in the United States (Valles et al. 2013). In *S. invicta* larvae and adults, infection of SINV-1 shows a strong tissue tropism for cells of the midgut (Hashimoto and Valles 2007). However, a recent study has shown that SINV-1 results in higher survival rates in the fields than chemical treatment (Tufts et al. 2014) in the spring and summer, which declines abruptly in the winter. The ALPV showed increased neurotropism and paralysis on the *R. padi* host during late stages of infection. Exploration on dicistroviruses that infect hymenopteran or hemipteran insects has been confronted by the lack of cell lines capable of supporting virus replication and challenging to proliferate in the amounts needed for biopesticide programs. As of late, *Helicoverpa armigera* stunt virus (HaSV) virion get together was perceived in plant protoplasts co-challenged with a plasmid expressing a capsid gene of HaSV and furthermore, plasmids conveying the cDNAs of the two HaSV genomic RNAs, implying a technique for dicistrovirus production. The RhPV had accumulated in the baculovirus-infected Sf21 cells, which magnificently expressed the recombinant RhPV clone in lepidopteran cells (Pal et al. 2007). The infection of Dicistrovirus exhibits pathological effects that include higher death rate, decreased fertility as well as the growth of the pest and to actuate cytopathic variances in back contaminated gut cells that elaborate loss of ribosomes and formation of intracellular vesicles. The dicistroviruses had the opportunity to switch to benign pest control methods. The nonappearance of a proper technique like cell lines for mass production of the virus has demonstrated to be a limitation in its utilization as a biopesticide. Attempts have been made to utilize the baculovirus expression system for in vitro production of dicistrovirus pesticides. The establishment of a dicistrovirus infectious clone (Kerr et al. 2015) opened up the opportunity of using Dicistroviruses as viral biopesticides. Additionally, Dicistroviruses offer an extraordinary model system for learning virus–host interactions, especially mechanisms of viral translational control and pathways of innate insect immunity.

1.4.6 Entomopoxvirus

Members of the family *Poxviridae* are divided into two sub-families based on a wide host range: The Entomopoxvirinae, which comprises insect poxviruses; and the Chordopoxvirinae, which involves vertebrate poxviruses (Goodwin et al. 1991). The perceived chicken-pox and little pox variola infections have a place with this family. The Entomopoxviruses (EPVs) were first found by Vago (1963) and are known to infect five orders of insects, viz., Coleoptera, Diptera, Hymenoptera, Orthoptera, and Lepidoptera (Murphy et al. 1995). There are 31 species of EPVs, divided among 3 assigned genera. Entomopoxviruses infected fat body cells of lepidopteran larvae, exhibit swelling and whitening, and extend to cell proliferation and hypertrophy. Eventually, slow mortality takes place at 3–10 weeks after infection and it may be a little slower in Coleoptera. They possess allantoid (ovoid) to brick-shaped (spindle) enveloped virions wrapped by the endoplasmic reticulum with 70–250 nm in width × 350–400 nm in length, containing single or monopartite linear dsDNA genome rich in AT residues, in size from 270 to 320 kbp, impeded inside spheroids. The EPVs are spread through insect feeding, and the virus replicates in the cytoplasm of susceptible insect hemocytes and adipose tissue cells (Lai-Fook and Dall 2000). Virus particles form into block molded intracellular mature virion (IMV), and it is lysed or acquires a second twofold layer from trans-Golgi and arises as outer encompassed extracellular enveloped virion (EEV). Mature virions are usually occluded in spheroids that comprise a major crystalline spheroidin protein host receptor approximately 109–115 kilo daltons (kDa) in size. The spheroids of some entomopoxviruses additionally contain a second spindle-shaped paracrystalline structure made out of fusolin protein. Earlier, the *Entomopoxvirinae* was divided into three genera, viz. *Entomopoxvirus A*, *Entomopoxvirus B*, and *Entomopoxvirus C* are based on their virion morphology, genome size, and host insects. Later, they were renamed as *Alphaentomopoxvirus,* which infects coleopteran beetles and comprises the type species *Melolontha melolontha entomopoxvirus* (MMEV) with other 13 species; *Betaentomopoxvirus* infects Lepidoptera and Oorthoptera, the type species infects *Amsacta moorei entomopoxvirus* (AMEV) with other 25 species, and the *Gammaentomopoxvirus* infects only Dipteran and the type species is *Chironomus luridus entomopoxvirus* (CLEV) with other 11 species. The unassigned fourth group comprises 4 species, which attacks hymenopterans. Entomopoxviruses have been used as potential bioinsecticides against orthopteran insects. *Melanoplus sanguinipes entomopoxvirus* (MSEV) and *Oedaleus senegalensis entomopoxvirus* (OSEV) infect major locust and grasshopper species. Among them, MSEV has a wide host range between locusts and grasshoppers. It enters upon ingestion and infects epithelium cells of the midgut and produces a systemic infection, such as suppression of pigments, delaying of development, and reduced food intake. Trypsin-like protease activity has been distinguished in relationship with MSEV occlusion bodies (Erlandson and Streett 1997). The full genomes of a few entomopoxviruses infecting Lepidoptera have been sequenced (Theze et al. 2013). Transgenic rice expressing the entomopoxvirus gene enhanced the susceptibility of armyworm larvae toward NPV (Hukuhara et al. 1999). An entomopoxvirus that causes chronic disease of the German cockroach (*Blattella germanica*) (Radek and Fabel 2000), and European spruce bark beetle (*Ips typographus*) can be utilized as biocontrol agents in the future. However, there are enduring expectations with the advancement of entomopoxvirus as expression vectors (Perera et al. 2010).

1.4.7 Iridovirus

Members of the family *Iridoviridae* are partitioned into two subfamilies: Alphairidovirinae and Betairidovirinae. The word "irido" is from Iris, the Greek goddess who manifests the rainbow. The trademark highlight of this family is to imitate a structure of a "rainbow-like" radiance of the vigorously infected insect tissues as mature virions accumulated as large paracrystalline arrays in the cytoplasm of infected cells. Members of the family are nonenveloped, nonoccluded, icosahedral viral particles of 120–300 nm in diameter with the core of nucleic acid and proteins. The genome is a linear-stranded DNA viral particles with a size of 200 kbp that infects a range of vertebrates and invertebrates, including insects (Williams 1996), whose viral particles fluctuate between 120 and 130 nm in size, which was isolated from the *Chilo suppressalis* (Asiatic rice borer). Iridovirus genome is a linear molecule ranging from 140 to 303 kb (Goorha and Murti 1982). The replication site of iridovirus is nuclear and cytoplasm of host insect, but virion assembly occurs completely in the cytoplasm. For the most part, individuals from the family Iridoviridae will be referenced as iridovirids (IV) to recognize them from the sensu stricto-invertebrate-iridoviruses (IIVs), which have a place with Betairidovirinae. The first invertebrate iridescent virus was reported in the mid-1950s from larval stage dipteran insects. This family of viruses has been isolated from Coleoptera, Diptera, Hemiptera, Hymenoptera, Lepidoptera, and Orthoptera (Boucias et al. 1987), in addition to few noninsect arthropods (Papp and Marschang 2019). The major capsid protein gene is generally used for the organization of Invertebrate-iridoviruses IIVs isolates. Most iridoviruses are transmitted by cannibalism or mechanical orally. Infections are chronic and tend to be systemic, the fat body and epidermis of transmission by parasitoids. Some iridoviruses prefer transovarial transmission but not easily transmitted hosts are mainly affected, nuclei of cells are damaged, causing the damage of hemocytes. The potential of iridoviruses in biocontrol is mainly prevalent by their diversity. However, it possesses low infectivity and chronic infection nature.

1.4.8 Nudiviruses

Nudiviruses (family: Nudiviridae) are recently perceived by the International Committee on Taxonomy of Viruses (ICTV) containing "nonoccluded baculoviruses." They show a more extensive host range than other arthropod viruses, infecting insects, and crustacean orders. It forms distinctive rod-shaped or ellipsoidal, enveloped, bacilliform nucleocapsids, super-coiled, circular double-stranded DNA viruses of 96 to 232 kbp, from which salient cases of endogenous nudivirus of the brown planthopper have been reported. These viruses have earlier been positioned in the family Baculoviridae because of their large, distinct lineage, close phylogenetic relationship, morphologically similar virions, replication in the nucleus, and hypertrophy of infected arthropod cells (Thézé et al. 2011). However, these Nudiviruses do not form occlusion bodies, which have imposed the foundation of the independent family. Six sequenced nudiviruses share 33 conserved genes, making a candidate core gene set, which results in these nudiviruses divided into two groups: the alpha- and betanudivirus genera (Bézier et al. 2015). Among the 33 core genes, 21 are shared with all baculoviruses (Rohrmann 2013), which denotes that the Baculoviridae and Nudiviridae are sister clades. Nudiviruses have been isolated from the orders Coleoptera, Diptera, Lepidoptera, and Orthoptera. Member species of the genus *Alpha nudivirus*, *Oryctes rhinoceros* Nudivirus Ma07 (OrNV), is a potential parasite for the rhinoceros beetle, *O. rhinoceros*, which devastatingly attacks coconut and oil palms (Bedford 2013) and has a 128-kbp genome encoding 139 open reading frames. Infections can occur systematically in both adult and immature stages of the host and cause mortality within 1–4 weeks. Virions assembled into occlusion bodies have been noted in nudivirus-infected hosts in some instances (Bézier et al. 2015). Transmission occurs horizontally and is initiated by oral ingestion of virions and infection of the host midgut epithelial tissues, from where it migrates to other cells. Biocontrol with OrNV has a prominent role in the control of the rhinoceros beetle for decades, especially because of high virulence on the insect larvae. However, recent evidence indicate that resistance could be evolving in some rhinoceros beetle populations in the Solomon islands (Etebari et al. 2020).

1.4.9 Polydnavirus

Polydnaviruses (PDVs) were discovered in-between 1960s and 1970s through electron microscopy but originally recognized in 1991 as a family of large double-stranded DNA viruses mutualistic with two endoparasitic Hymenopteran parasitoid wasp families, Braconidae and Ichneumonidae, respectively. The symbiotic relationship between polydnaviruses and wasps are about 100 million years old (Chen and van Achtenberg 2019), and "domesticated viruses" appear to elucidate how multidirectional wasp adapt to diverse host species. The mutualistic association between PDVs and Hymenopteran wasps was observed as the first instance of viruses progressing into beneficial symbionts (Edson et al. 1981). Members of the family *Polydnaviridae* include two endoparasitoid genera, Braconid-associated polydnaviruses termed as Bracovirus (BV) genus (32 species), whose type species infect the braconid wasp, *Cotesia melanoscela* (CmaBV) and Ichneumonid-associated polydnaviruses termed as Ichnovirus (IV) genus (21 species), whose type species infects the ichneumonid wasp, *Campoletis sonorensis* (CsIV). Successive reports specified that all Bracovirus-associating braconids are mutualistic with six subfamilies (Cardiochilinae, Cheloninae, Khoikholinae, Mendesellinae, Microgastrinae, and Miracinae) that form a monophyletic and hyperdiverse (50,000 species) assembly titled the microgastroid complex, while all Ichnovirus-associating ichneumonids mutualistic within two subfamilies named the Campopleginae and Banchinae (14,000 species) (Quicke et al. 2009). Paleovirology can exploit heritable horizontal gene transfer (HGT) of endogenous viral elements (EVEs) from viruses to hosts. The coevolution of viruses and their hosts and vectors together with dynamics involved in virus evolution is observed in Polydnavirus (Hull 2014). Polydnaviruses are observed as endogenous virus origins as their genomes are obtained from grown-up wasp to posterity as proviruses that get steadily coordinated into the parasitoid hymenopteran hosts (Dupuy et al. 2006) and support for the persistence of parasitoid wasps, as they paralyze larval stage Lepidoptera (Schmidt et al. 2001). Replication brings about the development of the encapsidated virus particles restricted to the nuclear region of specialized calyx cells in the ovaries of a female during wasp pupal-adult development. Members of this family produced nonoccluded, prolate ellipsoid virions (200 nm in length), containing a nucleocapsid; specifically, Bracovirus particles had single cylindrical capsid (40 $\times$ 25–100 nm) surrounded by an envelope, while in others multiple capsids were surrounded by a single envelope. Ichnovirus particles had fusiform or quasi-cylindrical or biconvex nucleocapsids ($\approx$85–330 nm) that are often with short tail-like appendages, which were individually enveloped by two unit membranes (Stoltz and Vinson 1979). The encapsidated large multipartite segments consist of circular supercoiled dsDNA that were nonequimolar abundance of segments, totaling 150–700 kbp when all the segments in particles (20 to more than 100) are aggregated; their individual sizes (2 to more than 30 kb) varied between wasp species. Those virions are structurally complex and contain at least 20–30 polypeptides, with sizes ranging from 10 to 200 kDa, possess uncharacterized lipids and carbohydrates (Webb and Strand 2005). Initial nucleotide sequences homologous to the DNA segments in particles were observed in the genomes of both male and female wasps through all lifecycle and cell types comprising the germlines (Fleming and Summers 1991). These particles replicate in pupal and adult stages of female wasps in calyx cells of the proximal ovaries and are assembled in their nuclei. Matured virions are gathered in the proximal ovary lumens and form a paste-like calyx fluid in the lateral oviducts with matured eggs of female wasps. Virions are introduced into the lepidopteran host by parasitized female wasps, which were shown to inject calyx fluid and that particles rapidly infect hemocytes, fat bodies, and other tissues. The nucleocapsids then travel to nuclear pores and encapsidated DNAs are released into the host nucleus and the genes are expressed (Strand and Burke 2019). In the interim, the wasp larvae develop from the hosts and grow into adults and are liberated from the hosts (Drezen et al. 2014). The wasps use the bracovirus particles containing virulent genes to ensure the survival of their offspring in the lepidopteran host, through changes in structure, repress the insusceptible reactions at oviposition, advancing wasp posterity advancement, and passing of the host (Strand and Burke 2015) to facilitate effective development of the parasitoid. A bracovirus symbiotic endoparasitoid, *Cotesia plutellae,* the wasp was used in combination with *B. thuringiensis* to control the insecticide-resistant populations of diamondback moth (*Plutella xylostella*). The incited immunosuppression by the bracovirus from the wasps in the parasitized larvae of the diamondback moth presented a more noteworthy susceptibility to *B. thuringiensis* infection (Park and Kim 2012) and thus can be exploited as biocontrol agents for the management of pest species in crops.

1.5 Genomics of Major Entomopathogenic Virus

Genomic studies on insect viruses are very treasured unveiling intrinsic assets crucial for classification, gene function, and insecticidal potential. The first entomopathogenic insect virus to be completely sequenced is the *Autographa californica* multiple nucleopolyhedrovirus, the type member of the NPVs (Ayres et al. 1994), and *Xestia c-nigrum* granulovirus was the first complete sequenced member in GV group,(Hayakawa et al. 1999) which contains 178,733 bp and 181 ORFs and is the biggest known baculovirus genome until now. Earlier, Sanger sequencing technique was utilized to sequence the viral genomes cloned in plasmids, and now with the existing advancements, next-generation sequencing (NGS) is employed at a nano-level for the huge scope of viral genomic sequencing. With the reformist advancement of omics approaches, the number of complete genomes has been sequenced quickly, offering a fortune of genomic information that adds to the comprehension of scientific classification and expected use of insect viruses. To date, more than 6 ascovirus, 115 baculovirus, 8 cypovirus, 19 densoviurs, 5 entomopoxvirus, 5 nudivirus, and 5 polydnavirus genomes have been sequenced completely, revealing more about baculovirus genomes (Herniou et al. 2012) (Table 1.3). Baculovirus gene content is generally separated into essential and nonessential genes (Rohrmann 2019), based on the direct or indirect involvement of virus infection establishment. Genome sizes range from 81,755 bp to 178,733 bp, eventually, the first sequenced AcMNPV genome has a size of 133,894 bp with 148 predicted nonoverlapping ORFs. The number of ORFs within the sequenced baculoviruses encoding more than 50 amino acids ranges from around 89 ORFs in *Neodiprion sertifer* GV to 181 ORFs in *Xestia-cnigrum* GV shows a lot of variation within the opposing ends. ORFs in the genomes of baculoviruses are present on both DNA strands and the ratio of ORFs with the clockwise orientation varies between 38 and 56%. Around 895 different genes were identified in the sequenced genomes

Table 1.3 Sequenced genomes of entomopathogenic viruses

Genus	Virus	Genbank Ref seq no.	Genome size (bp)
Ascovirus	*Heliothis virescens* ascovirus 3a	NC_009233	186,262
	Heliothis virescens ascovirus 3f	NC_044938	198,157
	Heliothis virescens ascovirus 3 g	NC_044939	199,721
	Spodoptera frugiperda ascovirus 1a	NC_008361	156,922
	Diadromus pulchellus ascovirus 4a	NC_011335	119,343
	Trichoplusia ni ascovirus 2c	NC_008518	174,059
Baculovirus	***Alphabaculovirus* (Group I)**		
	Anticarsia gemmatalis multiple nucleopolyhedrovirus-2D	NC_008520	132,239
	Anticarsia gemmatalis multicapsid nucleopolyhedrovirus-26–43	KR815455-KR815471	130,698-132,180
	Antheraea pernyi nucleopolyhedrovirus isolate Liaoning	NC_008035	126,629
	Autographa californica nucleopolyhedrovirus-C6	NC_001623	133,894
	Autographa californica nucleopolyhedrovirus-WP10	KM609482	133,926
	Bombyx mandarina nucleopolyhedrovirus-S1	NC_012672	126,770
	Bombyx mori nucleopolyhedrovirus-T3	NC_001962	128,413
	Catopsilia pomona nucleopolyhedrovirus-416	KU565883	128,058
	Choristoneura fumiferana DEF multiple nucleopolyhedrovirus	NC_005137	131,160
	Choristoneura fumiferana multiple nucleopolyhedrovirus isolate Ireland	NC_004778	129,593
	Choristoneura murinana alphabaculovirus strain Darmstadt	NC_023177	124,688
	Choristoneura occidentalis nucleopolyhedrovirus	NC_021925	128,446
	Choristoneura rosaceana nucleopolyhedrovirus-NB_1	NC_021924	129,052
	Condylorrhiza vestigialis multiple nucleopolyhedrovirus-PR.2002	NC_026430	125,767
	Dasychira pudibunda nucleopolyhedrovirus	KP747440	136,761
	Cyclophragma undans nucleopolyhedrovirus	KT957089	142,900
	Epiphyas postvittana multiple nucleopolyhedrovirus	NC_003083	118,584
	Hyphantria cunea nucleopolyhedrovirus-N9	NC_007767	132,959
	Lonomia obliqua multiple nucleopolyhedrovirus-SP2000	KP763670	120,023

(continued)

Table 1.3 (continued)

Genus	Virus	Genbank Ref seq no.	Genome size (bp)
	Orgyia pseudotsugata multiple nucleopolyhedrovirus	NC_001875	131,995
	Oxyplax ochracea nucleopolyhedrovirus isolate 435	NC_043529	113,971
	Philosamia cynthia ricini nucleopolyhedrovirus	JX404026	125,376
	Plutella xylostella multiple nucleopolyhedrovirus isolate CL3	DQ457003	134,417
	Rachiplusia ou multiple nucleopolyhedrovirus	NC_004323	131,526
	Thysanoplusia orichalcea nucleopolyhedrovirus-p2	NC_019945	132,978
	Alphabaculovirus (group II)		
	Adoxophyes honmai nucleopolyhedrovirus	NC_004690	113,220
	Adoxophyes orana nucleopolyhedrovirus	NC_011423	111,724
	Agrotis ipsilon multiple nucleopolyhedrovirus	NC_011345	155,122
	Agrotis segetum nucleopolyhedrovirus	NC_007921	147,544
	Agrotis segetum nucleopolyhedrovirus B isolate English	NC_025960	148,981
	Apocheima cinerarium nucleopolyhedrovirus	NC_018504	123,876
	Buzura suppressaria nucleopolyhedrovirus isolate Hubei	NC_023442	120,420
	Chrysodeixis chalcites nucleopolyhedrovirus	NC_007151	149,622
	Chrysodeixis chalcites single nucleopolyhedrovirus	JX535500, JX560539-JX560542	149,039-150,079
	Clanis bilineata nucleopolyhedrovirus-DZ1	NC_008293	135,454
	Cryptophlebia peltastica nucleopolyhedrovirus-SA	MH394321	115,728
	Ectropis obliqua nucleopolyhedrovirus-A1	NC_008586	131,204
	Euproctis pseudoconspersa nucleopolyhedrovirus isolate Hangzhou	NC_012639	141,291
	Helicoverpa armigera nucleopolyhedrovirus AC53	NC_024688	130,442
	Helicoverpa armigera multiple nucleopolyhedrovirus	NC_011615	154,196
	Helicoverpa armigera nucleopolyhedrovirus-C1	NC_003094	130,759
	Helicoverpa armigera nucleopolyhedrovirus-G4	NC_002654	131,405

(continued)

Table 1.3 (continued)

Genus	Virus	Genbank Ref seq no.	Genome size (bp)
	Helicoverpa armigera nucleopolyhedrovirus NNg1	NC_011354	132,425
	Helicoverpa zea single nucleopolyhedrovirus	NC_003349	130,869
	Hemileuca sp. nucleopolyhedrovirus-MEM	NC_021923	140,633
	Hyposidra talaca nucleopolyhedrovirus isolate India.001	MH261376	139,089
	Lambdina fiscellaria nucleopolyhedrovirus isolate GR15	NC_026922	157,977
	Leucania separata nucleopolyhedrovirus-AH1	NC_008348	168,041
	Lymantria dispar multiple nucleopolyhedrovirus-5-6	NC_001973	161,046
	Lymantria dispar multiple nucleopolyhedrovirus-27	KP027546	164,158
	Lymantria dispar multiple nucleopolyhedrovirus-BNP	KU377538	157,270
	Lymantria dispar multiple nucleopolyhedrovirus-2161	KF695050	163,138
	Lymantria dispar multiple nucleopolyhedrovirus-3029	KM386655	161,712
	Lymantria dispar multiple nucleopolyhedrovirus-49	KU862282	161,006
	Lymantria dispar multiple nucleopolyhedrovirus-3054	KT626570	164,478
	Lymantria dispar multiple nucleopolyhedrovirus-3041	KT626571	162,658
	Lymantria dispar multiple nucleopolyhedrovirus-ab-a624	KT626572	161,321
	Lymantria xylina nucleopolyhedrovirus-5	NC_013953	156,344
	Mamestra brassicae multiple nucleopolyhedrovirus strain K1	NC_023681	152,710
	Mamestra configurata nucleopolyhedrovirus A-90/2	NC_003529	155,060
	Mamestra configurata nucleopolyhedrovirus A-90/4	AF_539999	153,656
	Mamestra configurata nucleopolyhedrovirus B-96B	NC_004117	158,482
	Malacosoma neustria nucleopolyhedrovirus isolate ManeNPV-T2	KY968317	130,202
	Maruca vitrata multiple nucleopolyhedrovirus-MV8	NC_008725	111,953

(continued)

Table 1.3 (continued)

Genus	Virus	Genbank Ref seq no.	Genome size (bp)
	Mythimna unipuncta nucleopolyhedrovirus strain#7	NC_043530	148,482
	Operophtera brumata nucleopolyhedrovirus-MA	NC_040621	119,054
	Orgyia leucostigma nucleopolyhedrovirus-CFS77	NC_010276	156,179
	Peridroma alphabaculovirus isolate GR167	NC_024625	151,109
	Perigonia lusca single nucleopolyhedrovirus	NC_027923	132,831
	Pseudoplusia includens single nucleopolyhedrovirus-IE	NC_026268	139,132
	Spodoptera eridania nucleopolyhedrovirus-251	MH320559	149,090
	Spodoptera exempta nucleopolyhedrovirus-244.1	MH717816	129,528
	Spodoptera exigua multiple nucleopolyhedrovirus-US1	NC_002169	135,611
	Spodoptera frugiperda multiple nucleopolyhedrovirus-3AP2	NC_009011	131,331
	Spodoptera littoralis nucleopolyhedrovirus-Tun2	MG958660	137,099
	Spodoptera littoralis nucleopolyhedrovirus-AN1956	NC_038369	137,998
	Spodoptera litura multiple nucleopolyhedrovirus-G2	NC_003102	139,342
	Spodoptera litura nucleopolyhedrovirus II	JX454574	137,998
	Sucra jujuba nucleopolyhedrovirus isolate 473	NC_028636	135,952
	Trichoplusia ni single nucleopolyhedrovirus	NC_007383	134,394
	Troides aeacus nucleopolyhedrovirus	MH077961	125,477
	Urbanus proteus nucleopolyhedrovirus isolate southern Brazil	NC_029997	105,555
	Betabaculovirus		
	Adoxophyes orana granulovirus isolate English	NC_005038	99,657
	Agrotis segetum granulovirus-DA	NC_005839	131,680
	Artogeia rapae granulovirus isolate Wuhan	NC_013797	108,592
	Choristoneura occidentalis granulovirus (*Choristoneura fumiferana* granulovirus)	NC_008168	104,710
	Choristoneura diversana nucleopolyhedrovirus ChdiNPV-Hokkaido DNA	LC516821	122,827
	Clostera anachoreta granulovirus	NC_015398	101,487

(continued)

Table 1.3 (continued)

Genus	Virus	Genbank Ref seq no.	Genome size (bp)
	Clostera anastomosis granulovirus A-Henan	NC_022646	101,818
	Clostera anastomosis granulovirus B	NC_038371	107,439
	Cnaphalocrocis medinalis granulovirus-Enping	NC_029304	111,246
	Cryptophlebia leucotreta granulovirus-CV3	NC_005068	110,907
	Cydia pomonella granulovirus isolate Mexican 1	NC_002816	123,500
	Diatraea saccharalis granulovirus-Parana 2009	NC_028491	98,392
	Epinotia aporema granulovirus	NC_018875	119,082
	Erinnyis ello granulovirus-S86	NC_025257	102,759
	Helicoverpa armigera granulovirus	NC_010240	169,794
	Hyphantria cunea granulovirus isolate Hc1	MH923363	114,825
	Mocis latipes granulovirus isolate southern Brazil	NC_029996	134,272
	Mythimna unipuncta granulovirus B isolate MyunGV#8	NC_033780	144,673
	Phthorimaea operculella granulovirus-T	NC_004062	119,217
	Pieris rapae granulovirus	NC_013797	108,592
	Plodia interpunctella granulovirus isolate Cambridge	NC_032225	112,536
	Plutella xylostella granulovirus-K1	NC_002593	100,999
	Pseudalatia unipuncta granulovirus isolate Hawaiin	NC_013772	176,677
	Spodoptera frugiperda granulovirus isolate VG008	NC_026511	140,913
	Spodoptera litura granulovirus-K1	NC_009503	124,121
	Trichoplusia ni granulovirus LBIV-12	NC_038375	175,360
	Xestia c-nigrum granulovirus	NC_002331	178,733
	Gammabaculovirus		
	Neodiprion abietis nucleopolyhedrovirus	NC_008252	84,264
	Neodiprion lecontei nucleopolyhedrovirus	NC_005906	81,755
	Neodiprion sertifer nucleopolyhedrovirus	NC_005905	86,462
	Deltabaculovirus		
	Culex nigripalpus nucleopolyhedrovirus isolate Florida1997	NC_003084	108,252
Cypovirus	*Choristoneura occidentalis* cypovirus 16 segments 2–10	EU486988-EU201043	3768-1171
	Lymantria dispar cypovirus 1 isolate LdCPV1 segment 1–10	MN938831-MN938840	4146-920
	Lymantria dispar cypovirus 1 segment 1–9	AF389462-AF389470	4164-1187

(continued)

Table 1.3 (continued)

Genus	Virus	Genbank Ref seq no.	Genome size (bp)
	Cypovirus 14 RNA segment 1–10	NC_003006-NC_003015	4329-956
	Orgyia pseudotsugata cypovirus 5 segment 1–10	KC5883-KC588365	4126-883
	Heliothis armigera cypovirus 5 segment 1–10	NC_010670-NC_01066	4123-883
	Thyrinteina arnobia cypovirus 14 segment 1–10	MF161423-MF161431	4466-978
	Trichoplusia ni cypovirus 15 segment 1–11	NC_002557-NC_002566	4361-200
Densovirus	*Galleria mellonella* densovirus	NC_004286	6039
	Myzus persicae densovirus	NC_005040	5499
	Junonia coenia densovirus	NC_004284	5908
	Junonia coenia densovirus	KC883978	6032
	Bombus cryptarum densovirus isolate bcry3	NC_040626	3977
	Bombyx mori densovirus 1	NC_003346	5076
	Bombyx mori densovirus 3 isolate VD1	NC_020928	6543
	Bombyx mori densovirus 3 isolate VD2	NC_020927	6022
	Bombyx mori densovirus 5	NC_004287	5078
	Bombyx mori densovirus Zhenjiang segment VD1	EU623082	6543
	Bombyx mori densovirus Zhenjiang segment VD2	EU623083	6024
	Diaphorina citri densovirus	KX165268	5071
	Diatraea saccharalis densovirus	NC_001899	5941
	Helicoverpa armigera densovirus	NC_015718	4926
	Junonia coenia densovirus	NC_004284	5908
	Mythimna loreyi densovirus	NC_005341	6034
	Planococcus citri densovirus	NC_004289	5380
	Pseudoplusia includens densovirus	NC_019492	5990
	Pseudoplusia includens densovirus isolate IAF	JX645046	5990
Entomopoxvirus	*Amsacta moorei* entomopoxvirus "L"	NC_002520	232,392
	Melanoplus sanguinipes entomopoxvirus	NC_001993	236,120
	Choristoneura biennis entomopoxvirus "L" virophage	KJ683046	12,737
	Choristoneura biennis entomopoxvirus "L"	NC_021248	307,691
	Mythimna separata entomopoxvirus 'L'	NC_021246	281,182
Nudivirus	*Helicoverpa zea* nudivirus 2	JN418988	231,621
	Oryctes rhinoceros nudivirus strain LiboV	MT150137	125,846
	Oryctes rhinoceros nudivirus isolate Solomon islands	MN623374	125,917

(continued)

Table 1.3 (continued)

Genus	Virus	Genbank Ref seq no.	Genome size (bp)
	Oryctes rhinoceros virus	NC_011588	127,615
	Tipula oleracea nudivirus isolate 35	NC_026242	145,704
Polydnavirus	***Bracovirus***		
	Cotesia congregata bracovirus segment circle 1–36	NC_006633-NC_006662	27,346-17,477
	Microplitis demolitor bracovirus segment A-O	NC_007028-NC_007044	3611-34334
	Ichnovirus		
	Campoletis sonorensis ichnovirus	AF411011-NC_008007	6283-15812
	Glypta fumiferanae ichnovirus segment A1-E1	AB290007-NC_008837	5156-1533
	Hyposoter fugitivus ichnovirus segment A1-E1	NC_008947-NC_008998	2755-8851

of baculoviruses (Miele et al. 2011). The average G + C content in baculoviruses ranges from 32.4 to 57.5%; many baculoviruses have around 41% GC content. Although few hymenopteran baculoviruses, such as *Neodiprion sertifer* (Nese), *N. abietis* (Neab), and *N. lecontẹi* (Nele) NPV have lower GC contents just above 32%; several of them have significantly higher values (CfMNPV at 50.1%, AnpeNPV at 53.5%, OpMNPV at 55.1%, and LdMNPV at 57.5%). However, GC content is not correlated with the taxonomic classification of a baculoviruses due to large differences. About 115 species of baculovirus have been sequenced completely; out of which 83 belong to alphabaculoviruses, 27 betabaculoviruses, 3 gammabaculoviruses, and 1 deltabaculovirus. Genomic studies on betabaculoviruses (GVs) are limited due to the lack of a permissive cell lines. The *Alphabaculovirus* genus can be distributed into Groups I and II; consistent with the sequence and phylogeny of conserved genes for virus–cell fusion and receptor binding (IJkel et al. 2000; Jehle et al. 2006; Wang et al. 2016). Genes in Group I are described by their use of GP64 as their envelope fusion protein (EFP) mediate membrane fusion and are further divided into two clades: "a" and "b" based on phylogeny, and in Group II alphabaculoviruses, most of the betabaculoviruses (except *Diatraea saccharalis* granulovirus, which contains both *gp64* and *f*) and deltabaculoviruses exploit F protein as their EFP mediate membrane fusion (Ardisson-Araújo et al. 2016). The baculovirus GP64 proteins and F proteins are activated at acidic pH but differ from an amino acid sequence, biochemical, and structural properties (IJkel et al. 2000). Baculovirus genome organization consists of circular DNA, becomes infectious after cellular entry and uncoating, and contains no virion-associated proteins. DNA replication, late gene transcription, and virion structure are governed by a set of 38 gene homologs conserved across their genomes, have therefore been so-called baculovirus core genes (conserved proteins), and are shared among alpha-, beta-, gamma-, and deltabaculoviruses (Herniou et al. 2003; Miele et al. 2011; Javed et al. 2017; Wang et al. 2019). These core genes encode for the life cycle and comprise proteins responsible for viral DNA replication, gene transcription, virion architecture, DNA packaging, virion assembly, and interaction with host proteins. Moreover, 28 conserved genes are commonly identified in all sequenced lepidopteran alpha- and betabaculoviruses and among them, 11 genes (including *gp64*) are prominent in Group I alphabaculoviruses. In addition, numerous genes have been identified that are unique to a few or even a single NPV or GV species of baculoviruses. Some subsets of noncore genes are united by respective lineages of baculoviruses. These noncore genes have greatly affected the enormous diversification of baculoviruses. Baculoviruses commonly have a homologous region (hr), which contains a palindrome that is normally flanked by short direct repeats situated elsewhere in the genome, so the development of complete a genome map is essential for confirmation. Several researchers have proposed various bioinformatic approaches to detect genes based on genomic sequences, gene content, or genome modifications. Also, a combination of these approaches was pragmatic to re-assess baculoviridae classification. Still, there may be more orthologous sequences that may not be recognized due to the mutations during the course of evolution. Evolution in baculoviruses is always exponential along with its host by genetic flow, gene loss, gene gain by horizontal transfer, gene duplications, genetic recombination, and transposition by mobile elements. Baculovirus evolution relies on variation in content, size, the architecture of gene, G + C content, codon utilizations, and intergenic space content.

1.6 Biopesticide Strategies for Entomopathogenic Viruses

Sustainable management of natural resources implicates the employment of trials, which are environmentally safe in all domains of life. The adverse effects of pesticide application practiced in agriculture, and allied environment has directed toward its restricted use, accompanied by their re-assessment for food security, which has diverted the attention of entomologists to biological control of insect pests. The overall biopesticide market in 2013 was estimated at around $3 billion of all out yield insurance market and is extended to develop more than $4.5 billion by 2023 (Olson 2015). The North American Free Trade Agreement (NAFTA) countries use up to just about 45% of overall sold biopesticides, while the European Union uses 20%. There are more than 990 biopesticides organizations formally enlisted with Central Insecticides Board and Registration Committee (CIBRC), India, which advances potential biopesticides for agriculture. Also, the use and regulation of viral biopesticide products were listed for use on various crops in Argentina, Australia, Africa, Brazil, Canada, China, India, Europe, and United States (Kabaluk et al. 2010).

The entomopathogenic virus–based biopesticide products infecting pests have been known over the past 40 years (Buerger et al. 2007). Several IPM programs involving the use of entomopathogenic viruses have been developed worldwide.

Baculoviruses play an important role in controlling world significant lepidopteran pest species, such as *Helicoverpa* spp. (Rowley et al. 2011), *Spodoptera* spp., and *Plutella xylostella* (Lacey et al. 2015), which implies its significance as an alternative to the conventional system. Baculovirus products signify $50–70 million per annum in a global biopesticides market appraised to be a value of $2.8 billion a year (Wilson et al. 2020). Almost 60% of the recognized insect viruses belong to the family Baculoviridae, and it is assessed that more than 100 baculovirus biopesticides are commercially available on the global market. These viruses could be used in Integrated Pest Management (IPM) programs against 30% of major insect pests of agricultural significance (Table 1.4). In 1892, baculovirus was attempted against an eruptive forest defoliator. NPV was applied in an augmentative manner against natural populations of Nun moth (*Lymantria monacha* L.) in Europe (Huber 1986). Similarly, the natural population of the gypsy moth (*Lymantria dispar* L.) was moderately minimized with the introduction of LdMNPV in 1913 in the United States. The first efficacious control of an eruptive forest defoliator was by using a baculovirus in Canada during the 1930s. During the late 1940s, the alfalfa caterpillar (*Colias philodice eurytheme*) was controlled through the artificial aerial spray of polyhedrosis virus in the United States (Steinhaus and Thompson 1949). In India, the first report on the presence of NPV was from gram pod borer in Gujarat, India (Patel et al. 1968). The successful cases in the application of nucleopolyhedrosis viruses identified were *Autographa californica* MNPV (Ayres et al. 1994), *Bombyx mori* NPV (Gomi et al. 1997; Gomi et al. 1999), *Lymantria dispar* MNPV (Kuzio et al. 1999), and *Orgyia pseudotsugata* MNPV (Ahrens and Rohrmann 1995). Crop-defoliating insect population hindered similar destruction following outbreaks, again checked by baculovirus, which produced mortality (Fuxa 1982; Fuxa 2004). Brazilian Ministry of Agriculture, Livestock, and Food Supply has begun the work on baculovirus to control *Spodoptera frugiperda* in Maize and Sorghum at 1984. During the 1990s, *Lymantria dispar* MNPV formulations were registered under names Gypchek, Disparivirus, and Virin-ENSH; *Orgyia pseudotsugata* MNPV under trade names TM BioControl-1 and Virtuss were used against lepidopteran larvae in forest ecosystem (Reardon et al. 1996). Numerous viral insecticides were registered with a wider host range, such as *Anagrapha falcifera* (AnfaNPV), *Autographa californica* (AucaMNPV), and *Mamestra brassicae* (MabrMNPV). MabrMNPV was registered against *H. armigera*, *M. brassicae*, *P. xylostella*, and *P. operculella*. The AucaMNPV type baculovirus was mostly infective to distantly related hosts, and AnfaNPV and AucaMNPV were reported to infect nearly 30 lepidopteran species (Adams and McClintock 1991). The SPOD-X™ containing particles of *Spodoptera exigua* NPV to control vegetable crop pests and Spodopterin™ containing *Spodoptera littoralis* NPV were used to control devastating pests of cotton, corn, chilli, tomatoes, and maize. Lepidopteran pests were controlled successfully with *Spodoptera frugiperda* NPV in Brazil (Moscardi 1999). The *Heliothis zea* NPV was re-registered under the name Gemstar™ as viral bioinsecticide of *Helicoverpa armigera* (Mettenmeyer 2002). Turfgrass pest control employed *Agrotis ipsilon* NPV (AgipMNPV), which gave good control of early instars. However, its persistence was restricted by frequent mowing and exposure to

Table 1.4 Commercialized entomopathogenic virus–based products

Virus	Products	Insect pest	Host plant
Baculoviruses			
Lymantria dispar multiple nucleopolyhedrovirus	Gypcheck, Dispavirus, Disparivirus, and Virin-ENSH	Gypsy moth	Hardwood trees and conifers
Orgyia pseudotsugata multiple nucleopolyhedrovirus	TM Biocontrol-1 and Virtuss	Douglas fir tussock moth	Douglas fir, spruce fir, and ornamental trees
Neodiprion sertifer nucleopolyhedrovirus	Neocheck-S and Virox	European spruce sawfly	Forest trees
Neodiprion lecontei nucleopolyhedrovirus	Lecontivirus	Redheaded pine sawfly	Pine speciesand Norway spruce
Neodiprion abietis nucleopolyhedrovirus	Abietiv	Balsam fir sawfly	Balsam fir, spruce, and larch
Autographa californica multiple nucleopolyhedrovirus	VPN 80, Lepigen, Lepigen CCAB, and Loopex	Multiple pests	Vegetables, ornamentals, maize, and cotton
Anagrapha falcifera multiple nucleopolyhedrovirus	Certis	Lepidopteran larva including alfalfa looper, cotton bollworm, and tobacco budworm	Vegetables, fruit, ornamentals, maize, and cotton
Autographa californica multiple nucleopolyhedrovirus + Spodoptera albula nucleopolyhedrovirus	VPN-ULTRA	Multiple pests	Vegetables and alfalfa
Anticarsia gemmatalis multiple nucleopolyhedrovirus	Baculovirus soja WP, GRAP Baculovirus, Verpavex, Baculo-Soja, Baculovirus AEE, Protege Polygen, multigene, Baculovirus Nitra, and Cooper virus SC	Velvet bean caterpillar	Soybean and legumes
Heliocoverpa armigera single nucleopolyhedrovirus	Hanpv	Helicoverpa complex (cotton bollworm, podborer, Old	Corn, cotton, legumes, soybean, tobacco, and vegetables

(continued)

Table 1.4 (continued)

Virus	Products	Insect pest	Host plant
		World bollworm)	
Helicoverpa zea single nucleopolyhedrovirus	HzNPV-CCAB	Helicoverpa complex	Corn
Helicoverpa zea multiple nucleopolyhedrovirus	Gemstar LC, Armigen, and Heligen	Helicoverpa complex (corn earworm, tomato fruitworm, and tobacco budworm)	Corn, cotton, soybean, tomato, and tobacco
Mamestra brassicae multiple nucleopolyhedrovirus	Mamestrin	Multiple pests	
Mamestra configurata multiple nucleopolyhedrovirus		Bertha armyworm	
Spodoptera exigua multiple nucleopolyhedrovirus	Spodex, SPOD-X, Spodopterin, Spexit, Vir-ex, and Senpv	Beet armyworm	Asparagus, beets, cotton, cereals, celery, lettuce, tomato, and oilseeds
Spodoptera exempta multiple nucleopolyhedrovirus	–	African armyworm	Barley, maize, rice, pasture grass, and wheat
Spodoptera frugiperda nucleopolyhedrovirus	Cartugen, CartuchoVit, and Vircontrol S.f	Fall armyworm	Corn
Spodoptera sunia nucleopolyhedrovirus	VPN 82	Costa Rican armyworm	Vegetables
Chrysodeixis includens nucleopolyhedrovirus	Chrysogen, Loopavir, Chrysogen, and CCAB	Soybean looper	Soybean
Condylorrhiza vestigialis multiple nucleopolyhedrovirus	Baculovirus Alamo	Alamo moth	Poplar
Helicoverpa armigera nucleopolyhedrovirus	Diplomata–K and owner	Cotton bollworm	Corn, cotton, soybean, and tobacco
Helicoverpa zea multiple nucleopolyhedrovirus + *Chrysodeixis includens nucleopolyhedrovirus* (Heliothinae)	Surtivo Soja and Surtivo CCAB	Tobacco budworm and Helicoverpa complex	Corn, cotton, soybean, and tobacco
Spodoptera litura nucleopolyhedrovirus	Spodocide, Spodoterin, Spodicide, Senpv,	Tobacco cutworm, African cotton	Corn, cotton, soybean, and tobacco

(continued)

Table 1.4 (continued)

Virus	Products	Insect pest	Host plant
	biovirus–S, and Somstar-SL	leafworm, and fall armyworm	
Spodoptera littoralis nucleopolyhedrovirus	Littovir, Spodopterin, and Spodo-Cide	Mediterranean brocade moth	Cotton, tomatoes, soybean, sugarcane, sugar beet, grapes, and ornamentals
Helicoverpa armigera multiple nucleopolyhedrovirus	Helicide, Virin-H, Helocide, biovirus-H, Helicop, Heligard, Helicovex, Heliokill, Helitec, and Somstar-Ha	Cotton bollworm and corn ear worm	Corn, cotton, soybean, and tobacco
Adoxophyes orana granulovirus Swiss isolate (BV-001)	Capex	Summer fruit tortrix moth	Apple and pear
Adoxophyes honmai granulovirus	–	Smaller tea tortrix	Tea, ornamental trees, and shrubs
Cydia pomonella granulovirus	Madex 3, CYD-X, C-X HP, Granusal, Granupom, Madex top, Madex max, Madex plus, Madex twin, Carposin, Virin-CyAP, Carpovirus plus, Carpovirusine, and Virosoft^{CP4}	Codling moth and oriental fruit moth	Pome fruit and walnut
Agrotis segetum granulosis virus	–	Common cutworm, turnip moth, black cutworm, and mole cricket	Maize, potatoes, sugarbeet, field crops, ornamentals, and others
Cryptophlebia leucotreta granulovirus	Cryptogran, Cryptex, and Gratham	False codling moth	Citrus, cotton, macadamia nuts, avocadoes, and grapes

(continued)

Table 1.4 (continued)

Virus	Products	Insect pest	Host plant
Erinnyis ello granulovirus	Baculovirus erinnyis	Ello sphinx	Cassava
Homona magnanima granuloviruses	–	Oriental tea tortrix	Tea, flowers, pome fruit
Plodia interpunctella Granulovirus	NutGuard-V	Indian meal moth	Dried fruits and nuts
Plutella xylostella granulovirus	Plutellavex	Diamond back moth	Cabbage
Phthorimaea operculella granulovirus	Tutavir, Baculovirus Corpoica, Matapol, Metapol-plus, Bacu-Turin, and PTM baculovirus	Leafminer and potato tuber moth	Tomato and potato
Cypoviruses			
Dendrolimus spectabilis cypovirus 1	Matsukemin	Pine caterpillar	Pines

UV (Bixby-Brosi and Potter 2010). Remarkably, the combination of two granuloviruses in a single formulation that consisted of *Adoxophyes orana* GV and *Homona magnanima* GV was developed for controlling two tortrix pests of tea in Japan (Kunimi 2007). The invasive introduction and outbreak of forest defoliators, such as balsam fir sawfly and pine false webworm, were effectively terminated by NeabNPV in America (Moreau and Lucarotti 2007). In another landmark, NPVs were used against a major polyphagous pest, beet armyworm that had advanced resistance to chemical pesticides due to their extensive use. Isolates of *Spodoptera exigua* multicapsid nucleopolyhedrovirus (SeMNPV) successfully controlled beet armyworm on vegetable crops under greenhouse and field conditions (Lasa et al. 2007) because of very specific host range. The first commercially available baculovirus in South Africa was *Cryptophlebia leucotreta* GV against false codling moth. Also, the NPV of *Spodoptera exempta* (SpexNPV) was developed for control of the African armyworm, a major migrant pest in Africa (Grzywacz et al. 2008). Insecticidal activity of two geographically distinct strains of SeMNPV, which were lively constituents of Vir-ex (Spain) and Spexit (Switzerland), was measured on the geographically irrelevant host colonies (Elvira et al. 2013). The Iberian isolate HearSNPV-SP1 exhibited the fastest killing potential and was recommended for use against the cotton bollworm in Spain (Arrizubieta et al. 2014). Annually, CpGV-based product Madex™ had been applied up to 2.5 million ha of pome orchards in American and European countries, and HaSNPV was applied to 1 million ha of cotton in China. The Brazilian research institute EMBRAPA devised the program for soya IPM in the 1970s and identified the potential of AgMNPV with a low application rate of 1.5×10^{11} OBs/ha. Subsequently, it was used over almost two million hectares to control the soybean velvet bean caterpillar and sugarcane borer in Brazil (Moscardi et al. 2011) and expanded to Paraguay and Mexico. But in most of

NPVs, application rates were much higher at 10^{12} OBs/ha, for HearNPV, HzSNPV, SpexNPV, and SeMNPV (Lacey et al. 2015). Similarly, Granulovirus products with a single virion per OB required higher application rates reported at 2–3 × 10^{13} OBs/ha for field applications of CrleGV and PlxyGV (Moore et al. 2015); though on some plantation crops, the use of 10^{13}OBs/ha seems to be the optimum rate for CpGV and AdorGV (Lacey et al. 2015). The new viral mixture developed with the inductive ability of the *S. frugiperda granulovirus* (SpfrGV) isolate VG008 was used to augment the infectivity of *S. frugiperda* multiple nucleopolyhedrovirus (SpfrMNPV) isolate SfCOL tested against *S. frugiperda* larvae in the field at 97.5%: 2.5% (8 × 10^{11} OBs/ha) (Cuartas-Otálora et al. 2019). This enhanced effect was achieved by the virulence proteins, such as chitinases and enhancins, with the affinity toward intestinal chitin and mucin proteins associated with the peritrophic membrane (PM) (Slavicek 2012). In India, attempts were made to develop the *Hyblaea puera* NPV against teak defoliator, *Amsacta albistriga* NPV against red hairy caterpillar on groundnut; *Oryctus* baculovirus and cypovirus against rhinoceros beetle and red palm weevil in coconut, respectively. In the Kashmir region, field application of LyobMNPV at 2.5 × 10^{12} OBs/ha reduced the larval population density by 25–63% (Gupta et al. 2016) and followed with LydMNPV tested in vitro comparatively with LyobMNPV (Gani et al. 2017). The intensive use of Granulovirus CpGV applied to control the codling moth in pome fruits led to the development of virus-resistant insects detected in several orchards in Europe (Sauer et al. 2017). It had been overcome by incorporating the new isolates into a product that has now been brought to market. Recombinant baculoviruses, especially of AcNPV, BmNPV, and HearSNPV, have also been employed to attain viral strains with notable insecticidal activities as well as to improve the rapidity. Recombinant AcNPV virions expressing scorpion neurotoxin AaHIT from the *Androctonus australis* caused a reduction of 25% in the LT_{50}. When introduced the same recombinant into the genome of *Helicoverpa armigera* SNPV, the ST_{50} was reduced to 17–34%. Genes expressing toxins of *Bacillus thuringiensis* δ-endotoxins and Corn Protein URF13 have also been integrated into the AcNPV genome (Merryweather et al. 1990; Korth and Levings 1993), and URF13 showed a 40% reduction in the LT_{50}. An AcMNPV-enMP2 recombinant, expressing the MacoNPV enhancin gene, which degrades the peritrophic membrane of the insects under control of its native promoter was developed and characterized (Li et al. 2003). These recombinant expressions of exogenous toxins and enzymes with the insecticidal potential enhanced the efficacy of the virus-mediated insect lethality.

1.7 Mass Production and Application of Insect Viruses

Biological control of agricultural pests has gained importance in recent years due to increased stress to reduce agrochemicals and their residues in the environment and food. Currently, insect viruses of a few families have been efficiently used as biopesticides. They are safe for people and wildlife, and their specificity is very narrow. Rather than conventional pesticides, insect viruses unveil insect specificity, eco-friendly, low cost, and safety performance. Their application as bioinsecticides is meager because of its limitations, like slow killing action (4–5 days post-application) and technical difficulties for in vitro commercial production. In world-wide, the degree of biopesticides production is just 4.8% of the total pesticides, and in the United States, the production scenario is 6%, while in India, biopesticides production is only 3–5%, with an estimated annual growth rate of 2.5%. It is required to raise up to half of the general pesticide market by 2050 (Parker and Sander 2017). Under these conditions, economically feasible and rapid methodologies are anticipated. Research on the field efficacy studies of insect virus formulations against agricultural pests begins with nucleopolyhedrovirus in the mid-1950s. In 1972, the nucleopolyhedrovirus commercial formulation Virön/H in the United States was the first of its kind in the world. Then, the AgMNPV has been formulated as a wettable powder in 1986 (Moscardi et al. 2011). Since then, insect viruses' commercialization has been continuously encouraged, given a better understanding of the genome sequence, in vitro and in vivo culturing and multiplication, and formulation (Table 1.5).

Currently, in vivo produced baculoviruses of economically important insects, like heliothine and Spodoptera pests in Australia, China, Europe, North America, and India; *Anitcarsia gemmatalis* in Argentina and Brazil; *Cydia pomonella* in Argentina, Canada, Europe, New Zealand and the United States have met the real demand and are widely successful at field level. Nearly 10 more products of HearNPVs strains, viz., Biovirus-H, Helicide, Helocide, Heliguard, Helicop, and Virin-H for controlling *Helicoverpa armigera* attacking Bengal gram, cotton, chillies, groundnut, maize, okra, red gram, sunflower, sorghum, tomato, etc., were registered. Similarly, the products of SlNPVs strains, viz., Biovirus-S, Biokill-S, Spodo-Cide, Spodi-Cide, and Spodoterin for controlling *Spodoptera litura* in beetroot, cabbage, cauliflower, cotton, groundnut, soybean, sunflower, tobacco, etc., with a recommended dose of 500 ml/ha containing 100 and 250 larval equivalent (LE) (one LE $= 6 \times 10^9$ OBs) of NPV are available. Globally, China is highly demanded for viral biopesticides, which have registered 45 viral biopesticides. It includes nine baculovirus products, which are commercially existing with the Chinese manufacturers, viz., *Helicoverpa armigera* NPV (HearNPV), *Autographa californica* NPV (AucaMNPV), *Buzura suppressaria* NPV (BuzuNPV), *Euproctis pseudoconspersa* NPV (EupsNPV), *Extropic obliqua* NPV (EcobNPV), *Leucania (Mythimna) separata* NPV (LeseNPV) *Plutella xylostella* GV (PlxyGV), *Pieris rapae* GV (PiraGV), *Spodoptera exigua* NPV (SeMNPV), *Spodoptera littoralis* NPV (SpliNPV), and *Spodoptera litura* NPV (SpltNPV). Similarly, other viruses, viz., *Dendrolimus punctatus* CPV, *Periplaneta fuliginosa* DNV, and *Oryctes rhinoceros* nudivirus have also been developed (Lacey et al. 2015). In Thailand and Vietnam, HearNPV, SpltNPV, and SeMNPV were registered for the control of lepidopteran pests. In Brazil, Guatemala, Mexico, and South America, there is a well-established program for the production and use of *Anticarsia gemmatalis* NPV (AngeMNPV) against velvet bean caterpillar (*Anticarsia gemmatalis* Hübner) on soybean. In addition, Brazil also devised promotional program on *Spodoptera frugiperda* NPV (SpfrMNPV) for pest control in maize, soybean, cotton, and

Table 1.5 Effective entomopathogenic viruses as biological control of insect pests of crops and their commercial production

Sl. No.	Name of the virus	Name of the pest	Commercial producers
I. Baculovirus			
1.	Corn earworm NPV (HezeSNPV)	*Helicoverpa zea*: corn earworm, tomato fruitworm, tobacco budworm, *Heliothis virescens*	Certis (USA)
2.	Cotton bollworm NPV (HearNPV)	Cotton bollworm, podborer, *Helicoverpa armigera*	Andermatt (Switzerland), AgBioTech (Australia), Jiyuan Baiyun Industry Company Ltd. (China), Bio-Control Research Labs (India), Kenya Biologics (Kenya), plus other producers in India
3.	Unbarred *Spodoptera* moth (army worm NPV (SdalNPV)	*Spodoptera albula*	Agricola el Sol (Guatamala)
4.	Beet armyworm NPV (SpexMNPV)	*Spodoptera exigua*	Andermatt, (Switzerland), Certis (USA), Jiyuan Baiyun Industry Company Ltd.,(China) BioTech (Thailand)
5.	Egyptian cotton leafworm NPV (SpliNPV)	*Spodoptera littoralis*	Andermatt (Switzerland)
6.	Tobacco armyworm NPV (SpltNPV)	*Spodoptera litura*	Biocontrol Research Lab, Ajay Biotech, Bassarass Biocontrol, Biotech International, BioControl Research Labs (India) Jiyuan Baiyun Industry Company Ltd. (China)
7.	Gypsy moth, NPV (LydiMNPV)	*Lymantria dispar*	USDA (USA), Sylvar Technology (Canada), Andermatt (Switzerland)
8.	Velvetbean caterpillar, NPV (AngeMNPV)	*Anticarsia gemmatalis*	Coodetec. CNP So, Nova Era Biotechnologica Agricola, Nitral Urbana Laboratorios, Coop Central Milenio Agro Ciencias (Brazil)
9.	Red headed pine sawfly NPV (NeleNPV)	*Neodiprion lecontei*	Sylvar Technology (Canada)
10.	Douglas fir tussock moth NPV (OrpsNPV)	*Orygia pseudotsugata*	Canadian Forest Service
11.	Balsam fir sawfly NPV, (NeabNPV)	*Neodiprion abietis*	Sylvar Technology (Canada)

(continued)

Table 1.5 (continued)

Sl. No.	Name of the virus	Name of the pest	Commercial producers
12.	Alfalfa looper NPV (AucaMNPV)	*Autographa californica*	Agricola el Sol (Guatamala)
13.	Cabbage looper (TrniSNPV)	*Trichoplusia ni*	Andermatt (Switzerland)
14.	Tea geomotrid EcobNPV	*Extropic obliqua*	Small scale commercial production China
15.	Tea tussock moth (Eups NPV)	*Euproctis pseudoconspersa*	Small scale commercial production China
16.	Tea moth (BuzuNPV)	*Buzura suppressaria*	Small scale commercial production China
17.	Teak defoliator (HypeNPV)	*Hyblea peura*	Kerala Forest Research Institute (India
18.	Imported cabbageworm (PiraGV)	*Artogeia (Pieris) rapae*	Small scale commercial production China
19.	Oriental armyworm, (LeseNPV)	*Leucania (Mythimna) separata*	Small scale commercial production China
20.	Fall armyworm (MNPV), (*Spfr*NPV)	*Spodoptera frugiperda*	Certis (Europe) produces for Africa, Laboratory scale production by ICAR-NBAIR, India
GV			
1.	Diamond back moth GV (PlxyGV)	*Plutella xylostella*	Jiyuan Baiyun Industry Company Ltd. (China)
2.	Sugarcane early shoot borer (GV)	*Chilo infuscatellus*	Laboratory scale production by ICAR-Sugarcane Breeding Institute, Coimbatore, India
3.	Sugarcane early internode borer (GV)	*Chilo sacchariphagus indicus*	
4.	Codling moth GV (CpGV)	*Cydia pomonella*	Certis (USA), BioTepp (Canada), Arysta Lifescience (France), Andermatt (Switzerland), Hoerst (Germany), BioBest (Belgium), Arysta Life Science (France), Agro Roca (Argentine)
5.	False codling moth GV (CrleGV)	*Cryptophlebia leucotreta*	Andermatt (Switzerland), River Bioscience (South Africa)
6.	Potato tubermoth GV (PhopGV)	*Phthorimaea operculella*	Centro Internacional de la Papa (Peru), Proinpa (Bolivia)
7.	Summer fruit totrix GV (AdorGV)	*Adoxophyes orana*	Andermatt (Switzerland)

(continued)

Table 1.5 (continued)

Sl. No.	Name of the virus	Name of the pest	Commercial producers
8.	Tea tortrix (HomaGV)	*Homona magnanima*	Arysta life science (Japan)
9.	Smaller tea tortrix GV (AdhoGV)	*Adoxophyes honmai*	Arysta life science (Japan)
II. Reoviridae (CPV)			
1.	Masson pine moth cypovirus (CPV)	*Dendrolimus punctatus*	Small scale commercial production China
III. Parvoviridae			
1	Cockroach densonucleosis virus (DNV)	*Periplaneta fuliginosa*	Small scale commercial production China
IV	Nudiviruses	*Oryctes rhinoceros*	Locally produced for autodissemination

Source: Lacey et al. (2015)

beans; *Condylorrhiza vestigialis* NPV (CoveNPV) on poplar trees and *Erinnyis ello* GV for cassava pest control. One of the most broadly developed viruses is the Mexican codling moth, *Cydia pomonella* granulovirus (CpGV) used for more than 20 years in organic pome fruit orchards in Europe, North, and South America. Several products were registered based on the Mexican GV isolate (CpGV-M) (Eberle et al. 2012), which are now used worldwide. However, mass rearing of host insects to produce virus insecticides has not yet been effectively mounted up to meet commercial satisfactoriness with capable of continuous high proficiency production. Earlier, the production of commercial viral insecticides under in vivo systems on specially reared or wild living insects, compared to their chemical pesticide counterparts remains a constraint (Grzywacz et al. 2014).

As of late, in vitro cell culture systems have been exploited to produce insect viruses (van Oers et al. 2015). Presently, the commercial production of insect viruses is done in vivo by inoculating the virus particles on its target insect in the field or laboratory on an artificial diet. Under some circumstances, an artificial diet is unavailable for few host insects, which may not support under artificial conditions. In such conditions, in vitro multiplication of virus under insect cell lines with fermentors is utilized, which offers static, sterile, and controllable yields. The protein GP64 is mandatory for budding, transmission, and replication of budded virus (BV) particles in cell lines under various culture conditions (Blissard 1996). Crucial factors that should be rectified at in vivo production are microbial contaminants, insect proteins, and cuticles. Arguably, in vitro production is less potent than insect-infected methods. The major constraint of in vitro production is the continuous inoculation cycle of virus particles into the cell lines, which leads to a mutation in relative genetic stability at defective interfering particles and few polyhedra, which tends to loss of efficacy of the polyhedral (Rhodes 1996; Souza et al. 2001). Nonhomologous origin of DNA replication is a contrivance of defective interfering particle mutants, and thereby deletion of nonhomologous origin improved constancy of virus upon continuous inoculation in cell cultures (Pijlman et al. 2002), and a few polyhedral mutants are ascribed by a defect of a 25-kDa protein. Crucial screening of suitable cell lines and media is another challenge that has to be addressed under in vitro production. Then, a plague screening assay was employed to in vitro production of several polyhedral variants. In vitro production is the main preference on commercial production for baculoviruses, due to concise methodology. The lepidopteran cell line from the embryonic tissue of *Helicoverpa armigera* was highly susceptible to HaNPV (6.3×10^6 NPV/ml). The cell lines from larval and pupal ovaries of *Spodoptera litura* were found highly susceptible to SlNPV ($5–6 \times 10^6$ NPV/ml), which can able to cause complete mortality to the second instar larvae (Pant et al. 2002). The most exploited lepidopteran cell lines for baculovirus production are BCIRL/AMCYAfO(/T)V-CLG, BCIRL-Cc-AM, NIV-HA-197, RIRI-PX1, Se6FHA, Sf9, Sf21, and High five cells (Arunkarthick et al. 2017). Cell lines, BTI-TN-5B1–4 and IPLB-Sf21, are most potent in culturing under serum-free media and in a bioreactor.

There are several insect cell lines capable of supporting viral replication on a large scale, such as the *Heliothes zea* (HzAMI) cell line to produce HearNPV, the *Spodoptera frugiperda* SF9 cell line to produce SfMNPV and the *Anticarsia gemmatalis* cell line to produce AgMNPV. These insect cells are relatively friable compared to the bacterial and yeast based on the large scale cell culture systems. A cell line from the glassy-winged sharpshooter appeared to be indulgent to infection by transfection with RhPV RNA (Boyapalle et al. 2007) and a baculovirus-mediated clone expression of dicistrovirus in lepidopteran cells lead to infections in aphids (Pal et al. 2007). Almost 200 cell lines have been recognized from around 70 species of insects with the ultimate goal of research in the production of viral biopesticides in large quantity on a commercial scale. The slower killing nature and low concentration of BV titers and technical difficulties for in vitro commercial production remain as a hindrance to their more extensive use (Szewczyk et al. 2006). Likewise, a large portion of the viral pesticides ought to be showered during night hours due to deadly impacts by UV. Additives in the form of spreaders, surfactants (molasses), phagostimulants (monosodium glutamate and trans-ABCD), UV protectants (Blankophor P167, Tinopal, Stilbene subsidiaries with and without titanium dioxide, specialized dyes, chemicals, and natural substances), wetting agents, etc., had been used in the formulations to protect from the detrimental effects on the virus and enhance the exhibition and persistence under field conditions. An ongoing strategy uncovers the upgrade in the UV resistance by exposing nano-zinc oxide-binding peptides on the surfaces of their OBs, particularly in baculoviruses. Before the commercialization of viral biopesticides to agricultural environs, their biosafety, toxicity, persistence behavior, and influence on ecosystems have to be evaluated. Baculoviruses have been approved as low-risk biocontrol agents by the Organization for Economic Co-operation and Development (OECD) based on the Assessment of its Environmental Applications and published biosafety studies confirming its use and application in agriculture. At present, large-scale application of baculovirus can be done at a comparatively low cost. To combat future demands, in vitro cell culture remains a reliable approach to overcome supply and cost constraints (Moscardi et al. 2011). The use of insect viruses as biopesticides in developed and developing countries achieved them numerous efficient, environmental, and social advantages. The incredible preferred position of insect viruses, which are natural control agents, that they do not cause any harm to the health of farmers; do not kill natural predators and parasitoids of insect pests; do not disrupt the environment; do not pollute woodlands, springs; and waterways; and do not contaminate products *in natura* to be sold in the racks of grocery stores, leaving no residues in blossoms, organic products, fruits, and vegetables. Every one of these components joined with the specificity and ease of handling of insect viruses in relation to chemical pesticides and the target insects make it one of the best biological control agents.

1.7.1 Mass Production of Insect Viruses

In India, *Helicoverpa armigera* and *Spodoptera litura* are major important polyphagous pests damaging a wide variety of food, fiber, oilseed, fodder, and horticultural crops. The nucleopolyhedrovirus of *H. armigera* (HaNPV) and *S. litura* (SlNPV) are used for the management of *H. armigera* and *S. litura* on chickpea, cotton, pigeonpea, tomato, castor, groundnut, cauliflower, and sunflower. Mass production of nucleopolyhedrovirus (NPV) on a commercial scale is restricted to in vivo procedures in host larvae, which are obtained in field collection from the above host plants and also mass cultured in the laboratory using semisynthetic diet. Rearing of larvae in the natural host plant will involve a frequent change of food at least once a day during the incubation period of 5–9 days, increasing the handling time and cost. In order to reduce the cost, field-collected larvae are released into semi-synthetic diet treated with virus inoculum.

The NPV of *H. armigera* and *S. litura* are propagated in early fifth instar larvae. The virus is multiplied in a facility away from the host culture laboratory. The dose of the inoculum used is 6×10^9 polyhedral occlusion bodies (POB) in 10 ml suspension. The virus is applied onto the semisynthetic diet (lacking formaldehyde) dispensed previously in 5 ml glass vials. A blunt end polished glass rod (6 mm) is used to distribute the suspension containing the virus uniformly over the diet surface. Early fifth instar larvae are released singly into the glass vials after inoculation and plugged with cotton and incubated at a constant temperature of 25 °C in a laboratory incubator. When the larvae exhausted the feed, a fresh untreated diet is provided. The larvae are observed for the development of virosis and the cadavers collected carefully from individual bottles starting from the fifth day. Approximately, 200 cadavers were collected per sterile cheese cup (300 ml) and the contents were frozen immediately. Depending upon the need, cadavers were removed from the refrigerator and thawed very rapidly by agitation in water.

1.7.2 Processing and Formulation of NPV

The method of processing of NPV requires greater care to avoid losses during processing. The cadavers are brought to normal room temperature by repeatedly thawing the container with the cadaver under running tap water. The cadavers are homogenized in sterile ice-cold distilled water at the ratio of 1:2.5 (w/v) in a blender or precooled all-glass pestle and mortar. The homogenate is filtered through double-layered muslin and repeatedly washed with distilled water. The ratio of water to be used for this purpose is 1:7.5–12.5 (w/v) for the original weight of the cadaver processed. The leftover mat on the muslin is discarded and the filtrate can be semi-purified by differential centrifugation. The filtrate is centrifuged for 30–60 s at 500 rpm to remove debris. The supernatant is next centrifuged for 20 min at 5000 rpm. Then the pellet containing the polyhedral occlusion bodies (POB) is suspended in sterile distilled water and washed three times by centrifuging the pellet in distilled water at low rpm followed by centrifugation at high rpm. The pellet finally collected is suspended in distilled water and made up to a known volume, which is necessary to calculate the strength of the POB in the purified suspension. Cherry et al. (2000) stated that many attempts have been made to improve the performance of NPV. A wide range of formulations has been employed making use of locally available adjuvants as stickers, UV protectants, and phagostimulants. The different types of formulations are crude aqueous suspensions, Wettable powder, oil-based emulsifiable concentrate, lyophilized POBs powder, and microencapsulated formulations, etc. During field spraying, along with NPV formulations 10% crude sugar or jaggery, 0.1% teepol and 0.1% tinopal were mixed well to enhance the efficiency of NPV for effective pest management (Fig. 1.5).

1.8 Future Perspectives

The perceptions for developing insect virus-based biopesticides targeting cropping and forest zones are necessary for the establishment of environment-friendly, specific, safe, low-cost insect pest management tool in the near future. Major attention should be focused on understanding the nature that underlies host specificity of viral species, which can lead to the identification of complex multispecies-virus products that can infect several pests simultaneously in a crop at a time. It is necessary to understand the role of exogenous toxins, enzymes, and other infectivity factors to alleviate specificity and improving the efficacy of the virus. Biopesticide formulations with UV-resistance to increase or improve the field level persistence and long shelf-life comparable to chemical pesticides need to be developed. Research should also focus on the phyto-inactivation of viruses, development of large-scale in vitro production systems, targeted significance with fast and earlier cessation strains, lowering the production cost, and improving the virus-based product quality. Enhancing the shelf-life and field persistence through

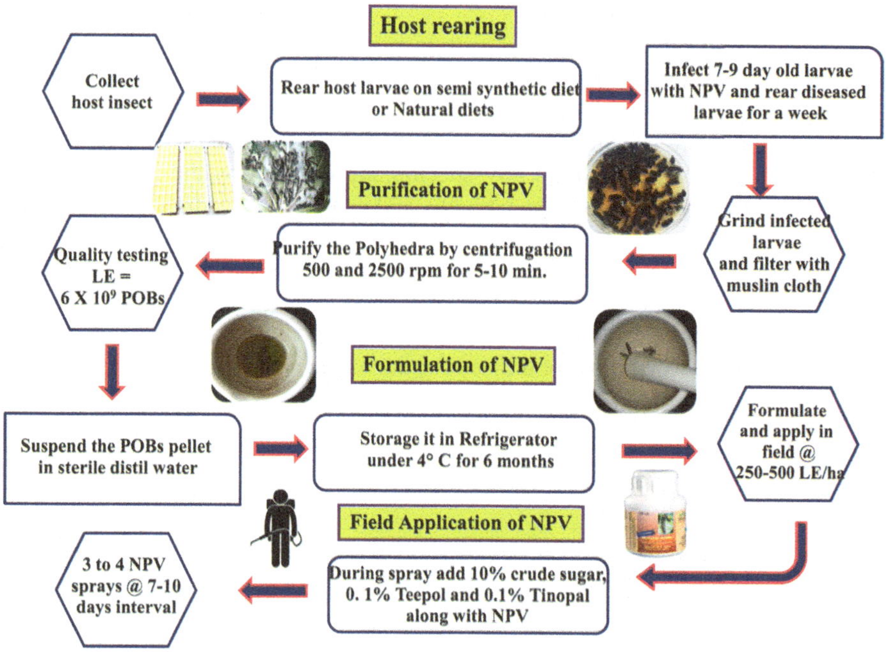

Fig. 1.5 Mass production and formulation of insect virus

nanoencapsulation will pave the way for the commercialization of insect viruses/ products in pest management.

1.9 Conclusion

Of the various bioagents, viruses are being used as a promising tool for the management of economically important insect pests of agricultural and horticultural crops. Various insect viruses, viz. NPV, Granulosis viruses (GV), and Cytoplasmic Polyhedrosis viruses were used for the management of insect pests throughout the world. The genomic characteristics of the entomopathogenic viruses and the molecular mechanisms by which the viruses infect and kill the insects need to be recognized for the successful application of the viruses under field conditions. Simultaneously, mass production and development of various stable formulations under on-farm and laboratory conditions have to be standardized. Thus, exploring and characterization of new entomopathogenic viruses with higher pathogenicity will play a major role in the sustainable management of insect pests in futuristic agriculture.

1.10 Points to Remember

1. Various insect viruses exist in nature, which are pathogenic to pests of agricultural and horticultural crops.
2. Insect viruses are DNA/RNA viruses with varied genome size.
3. The viral particles, which are host-specific, are ingested by the insects and the virions infect the gut wall cells, fat body, and hemolymph leading to death of the insects.
4. Among the nine known entomopathogenic viruses, baculoviruses are the major viruses exploited against insect pests.
5. Various commercialized entomopathogenic virus–based products are available throughout the world and are utilized for the management of insect pests.

References

Adams JR, McClintock JT (1991) Baculoviridae, nuclear polyhedrosis viruses part 1. Nuclear polyhedrosis viruses of insects. In: Adams JR, Bonami JR (eds) Atlas of invertebrate viruses. CRC Press, Boca Raton, FL, pp 87–204

Adams J, Stadelbacher E, Tompkins G (1979) A new virus-like particle isolated from the cotton bollworm, *Heliothis zea* biological control. Ann Meet Electron Microsc Soc Am 30:238–249

Afolami OI, Oladunmoye MK (2017) Baculoviruses: emerging frontiers for viral biocontrol of insect pests of agricultural importance. J Adv Microbiol 5(4):1–7

Ahrens CH, Rohrmann GF (1995) Replication of *Orgyia pseudotsugata* baculovirus DNA: lef-2 and ie-1 are essential and ie-2, p34, and Op-iap are stimulatory genes. Virology 212:650–662

Ardisson-Araújo DM, Melo FL, Clem RJ, Wolff JL, Ribeiro BM (2016) A betabaculovirus-encoded gp64 homolog codes for a functional envelope fusion protein. J Virol 90(3):1668–1672

Arrizubieta M, Williams T, Caballero P, Simon O (2014) Selection of a nucleopolyhedrovirus isolate from Helicoverpa armigera as the basis for a biological insecticide. Pest Manag Sci 70 (6):967–976

Arunkarthick S, Asokan R, Aravintharaj R, Niveditha M, Krishna Kumar NK (2017) A review of insect cell culture: establishment, maintenance and applications in entomological research. J Entomol Sci 52(3):261–273

Asgari S, Bideshi DK, Bigot Y, Federici BA, Cheng X, ICTV (2017) Report consortium, 2017, ICTV virus taxonomy profile: *Ascoviridae*. J Gen Virol 98:4–5

Ayres MD, Howard SC, Kuzio J, Lopez-Ferber M, Possee RD (1994) The complete DNA sequence of *Autographa californica* nuclear polyhedrosis virus. Virology 202:586–605

Bailey L, Gibbs AJ, Woods RD (1964) Sacbrood virus of the larval honey bee (*Apis mellifera* linnaeus). Virology 23:425–429

Baltimore D (1971) Expression of animal virus genomes. Bacteriol Rev 35(3):235–241

Ban L, Didon A, Jonsson LM, Glinwood R, Delp G (2007) An improved detection method for the *Rhopalosiphum padi* virus (RhPV) allows monitoring of its presence in aphids and movement within plants. J Virol Methods 142:136–142

Bedford GO (2013) Biology and management of palm dynastid beetles: recent advances. Annu Rev Entomol 58:353–372

Bergold G (1947) Die Isolierung des Polyeder-Virus und die Natur der Polyeder. Z Naturforsch 2b:122–143

Berns KI, Bergoin M, Bloom M, Lederman M, Muzyczka N, Siegl G, Tal J, Tattersall P (1995) Parvoviridae. Arch Virol Suppl 10:169–178

Bézier A, Thézé J, Gavory F, Gaillard J, Poulain J, Drezen JM, Herniou EA (2015) The genome of the nucleopolyhedrosis-causing virus from *Tipula oleracea* sheds new light on the Nudiviridae family. J Virol 89(6):3008–3025
Bideshi DK, Tan Y, Bigot YA, Federici BA (2005) Viral caspase contributes to modified apoptosis for virus transmission. Genes Dev 19:1416–1421
Bideshi DK, Bigot Y, Federici BA, Spears T (2010) Ascoviruses. In: Asgari S, Johnson KN (eds) Insect virology. Caister Academic Press, Norfolk, pp 3–34
Bird FT, Whalen MM (1954) Stages in the development of two insect viruses. Can J Microbiol 1:170–174
Bixby-Brosi AJ, Potter DA (2010) Evaluating a naturally occurring baculovirus for extended biological control of the black cutworm (Lepidoptera: Noctuidae) in golf course habitats. J Econ Entomol 103:1555–1563
Black BC, Brennan LA, Dierks PM, Gard IE (1997) Commercialization of baculoviral insecticides. In: Lois KM (ed) The baculoviruses. Springer, Berlin, pp 341–387
Blissard GW (1996) Baculovirus–insect cell interactions. Cytotechnology 20:73–93
Bolle G (1894) Il giallume od il mal del grasso del baco da seta. Communicazione preliminare. Atti. E Mem. Dell' I.R. Soc Agric Gorizia 34:133–136
Bonning BC (2009) The dicistroviridae: an emerging family of invertebrate viruses. Virol Sin 24:415
Boucias DG, Maruniak JE, Pendland JC (1987) Characterization of an iridovirus isolated from the sourthern mole cricket, Scapteriscus vicinus. J Invertebr Pathol 50:238–245
Boyapalle S, Pal N, Miller WA, Bonning BC (2007) A glassywinged sharpshooter cell line supports replication of *Rhopalosiphum padi* virus (Dicistroviridae). J Invertebr Pathol 94:130–139
Buerger P, Hauxwell C, Murray D (2007) Nucleopolydrovirus introduction in Australia. Virol Sin 22:173–179
Cao G, Meng X, Xue R, Zhu Y, Zhang X, Pan Z, Zheng X, Gong C (2012) Characterization of the complete genome segments from BmCPV-SZ, a novel *Bombyx mori* cypovirus 1 isolate. Can J Microbiol 58:872–883
Carrillo-Tripp J, Bonning BC, Miller WA (2014) Challenges associated with research on RNA viruses of insects. Curr Opin Insect Sci 6:1–7
Chakrabarti M, Ghorai S, Mani SKK, Ghosh AK (2010) Molecular characterization of genome segments 1 and 3 encoding two capsid proteins of *Antheraea mylitta* cytoplasmic polyhedrosis virus. Virol J 7:181–191
Chen YP, Siede R (2007) Honey bee viruses. Adv Virus Res 70:33–80
Chen XX, van Achtenberg C (2019) Systematics, phylogeny & evolution of braconid wasps: 30 years of progress. Annu Rev Entomol 64:335–358
Chen G, Liu H, Mo B-C, Hu J, Liu S-Q, Bustos-Segura C, Xue J, Wang X (2020) Growth and development of *Helicoverpa armigera* (Lepidoptera: Noctuidae) larvae infected by *Heliothis virescens ascovirus* 3i (HvAV-3i). Front Physiol 11:93
Cherry AJ, Rabindra RJ, Parnell MA, Geetha N, Kennedy JS, Grzywacz D (2000) Field evaluation of *Helicoverpa armigera* nucleopolyhedrovirus formulations for control of the chickpea pod-borer, *H. armigera* (Hubn.), on chickpea (Cicer arietinum var. Shoba) in southern India. Crop Prot 19:51–60
Chishti MZ, Schaf KA (1990) Studies on the polyhedral disease of silkworm, Bombyx mori in Jammu & Kashmir State. Indian J Seric 29(1):155–157
Chu GK, Xie RD, Zhang HJ, Yao YE, Fang ZJ (1975) On a nuclear polyhedrosis virus of mulberry tussock moth, *Euproctis similis* Fuessly (Lepidoptera: Lymantridae) and field test for the moth control. Acta Microbiol Sin 15:93–100. (in Chinese)
Cotmore SF, Tattersall P (1996) Parvovirus DNA replication. Cold Spring Harbor Monograph Arch 31:799–813
Cuartas-Otálora PE, Gómez-Valderrama JA, Ramos AE, Barrera-Cubillos GP, Villamizar-Rivero LF (2019) Bio-insecticidal potential of nucleopolyhedrovirus and granulovirus mixtures to

control the fall armyworm *Spodoptera frugiperda* (J.E. Smith, 1797) (Lepidoptera: Noctuidae). Viruses 11:684

Drezen J-M, Chevignon G, Louis F, Huguet E (2014) Origin and evolution of symbiotic viruses associated with parasitoid wasps. Curr Opin Insect Sci 6:35–43

Dupuy C, Huguet E, Drezen J-M (2006) Unfolding the evolutionary history of polydnaviruses. Virus Res 117:81–89

Eberle KE, Jehle JA, Hüber J (2012) Microbial control of crop pests using insect viruses. In: Abrol DP, Shankar U (eds) Integrated pest management: principles and practice. CABI Publishing, Wallingford, pp 281–298

Edson KM, Vinson SB, Stoltz DB, Summers MD (1981) Virus in a parasitoid wasp: suppression of the cellular immune response in the parasitoid's host. Science 211:582–583

El-Far M, Szelei J, Yu Q, Fediere G, Bergoin M, Tijssen P (2012) Organization of the ambisense genome of the *Helicoverpa armigera* densovirus. J Virol 86:7024

Elvira S, Ibargutxi MA, Gorria N, Munoz D, Caballero P, Williams T (2013) Insecticidal characteristics of two commercial *Spodoptera exigua* nucleopolyhedrovirus strains produced on different host colonies. Biol Microbial Control 106:50–56

Epstein L (2014) Fifty years since 'silent spring'. Annu Rev Phytopathol 52:377–402

Erlandson MA, Streett DA (1997) Entomopoxviruses associated with grasshoppers and locusts: biochemical characterization. Memoirs Entomol Soc Canada 129(S171):131–146

Etebari K, Filipović I, Rašić G, Devine GJ, Tsatsia H, Furlong MJ (2020) Complete genome sequence of *Oryctes rhinoceros* nudivirus isolated from the coconut rhinoceros beetle in Solomon Islands. Virus Res 278:197864

FAO (2019) The state of food and agriculture. Moving forward on food loss and waste reduction. FAO, Rome

Federici BA (1983) Enveloped double-stranded DNA insect virus with novel structure and cytopathology. PNAS 1 80(24):7664–7668

Fleming JG, Summers MD (1991) PDV DNA is integrated in the DNA of its parasitoid wasp host. Proc Natl Acad Sci USA 88:9770–9774

Francki RIB, Fauquet CM, Knudson D, Brown F (1991) Classification and nomenclature of viruses. In: Fifth Report of the International Committee on Taxonomy of Viruses. Springer-Verlag, Wien, New York, p 452

Fujiyuki T, Takeuchi H, Ono M, Ohka S, Sasaki T, Nomoto A, Kubo T (2004) Novel insect picorna-like virus identified in the brains of aggressive worker honeybees. J Virol 78:1093–1100

Fuxa JR (1982) Prevalence of viral infections in populations of fall armyworm,*Spodoptera frugiperda*, in southeastern Louisiana. Environ Entomol 11:239–242

Fuxa JR (2004) Ecology of nucleopolyhedroviruses. Agric Ecosyst Environ 103:27–43

Gani M, Gupta RK, Zargar SM, Kour G, Monobrullah M, Kandasamy T, Mohanasundaram A (2017) Molecular identification and phylogenetic analyses of multiple nucleopolyhedrovirus isolated from Lymantria obfuscata (Lepidoptera: Lymantriidae) in India. Appl Entomol Zool 52:389–399

Ghorai S, Chakrabarti M, Roy S, Chavali VRM, Bagchi A, Ghosh AK (2010) Molecular characterization of genome segment 2 encoding RNA dependent RNA polymerase of *Antheraea mylitta* cytoplasmic polyhedrosis virus. Virology 404:21–31

Glaser RW (1918) The polyhedral virus of insects with a theoretical consideration of filterable viruses generally. Science 48:301–302

Glaser RW, Chapman JW (1913) The wilt disease of gypsy moth caterpillars. J Econ Entomol 6:479–488

Gomi S, Zhou CE, Yih W, Majima K, Maeda S (1997) Deletion analysis of four of eighteen late gene expression factor gene homologues of the baculovirus, BmNPV. Virology 230:35–47

Gomi S, Majima K, Maeda S (1999) Sequence analysis of the genome of *Bombyx mori* nucleopolyhedrovirus. J Gen Virol 80(5):1323–1337

Goodwin RH, Milner RJ, Beaton CD (1991) Entomopoxvirinae. In: Adams JR et al (eds) Atlas of invertebrate viruses. CRC Press, Boca Ratón, pp 259–285

Goorha R, Murti KG (1982) The genome of frog virus 3, an animal DNA virus, is circularly permuted and terminally redundant. Proc Natil Acad Sci USA 79:248–252

Grzywacz D, Moore S (2017) Production, formulation & bioassay of baculoviruses for pest control. In: Lacey LA (ed) Microbial control of insect and mite pests (Chapter 7). Academic Press, Amsterdam, pp 109–124

Grzywacz D, Mushobozi W, Parnell M, Jolliffe F, Wilson K (2008) The evaluation of Spodoptera exempta nucleopolyhedrovirus (SpexNPV) for the field control of African armyworm (*Spodoptera exempta*) in Tanzania. Crop Prot 27:17–24

Grzywacz D, Stevenson PC, Mushobozi WL, Belmain S, Wilson K (2014) The use of indigenous ecological resources for pest control in Africa. Food Sec 6:71–86

Gupta RK, Gani M, Kaul V, Bhagat RM, Bali K, Samnotra RK (2016) Field evaluation of *Lymantria obfuscata* multiple nucleopolyhedrovirus for the management of Indian gypsy moth in Jammu and Kashmir, India. Crop Prot 80:149–158

Hamm JJ, Pair SD, Marti OG (1986) Incidence and host range of a new ascovirus isolated from fall armyworm, *Spodoptera frugiperda* (Lepidoptera: Noctuidae). Fla Entomol 69:524–531

Harrison R, Hoover K (2012) Baculoviruses and other occluded insect viruses. In: Vega F, Kaya H (eds) Insect pathology. Elsevier, Amsterdam, pp 73–131

Harrison R, Herniou E, Jehle J, Theilmann D, Burand J, Becnel J, Krell P, van Oers M, Mowery J, Bauchan G (2018) ICTV report consortium. ICTV virus taxonomy profile: baculoviridae. J Gen Virol 99:1185–1186

Hashimoto Y, Valles SM (2007) *Solenopsis invicta* virus-1 tissue tropism and intra-colony infection rate in the red imported fire ant: a quantitative PCR-based study. J Invertebr Pathol 96:156–161

Hayakawa T, Ko R, Okano K, Seong S, Goto C, Maeda S (1999) Sequence analysis of the Xestia C-Nigrum granulovirus genome. Virology 262:277–297

Herniou EA, Olszewski JA, Cory JS, O'Reilly (2003) The genome sequence and evolution of baculoviruses. Annu Rev Entomol 48:211–234

Herniou EA, Arif BM, Becnel JJ, Blissard GW, Bonning B, Harrison R, Jehle JA, Theilmann DA, Vlak JM (2012) Family baculoviridae. In: Andrew MQ et al (eds) Virus taxonomy, classification and nomenclature of viruses, ninth report of the international committee on taxonomy of viruses. Elsevier Academic Press, Amsterdam, pp 163–173

Hu Z, Luijckx T, Dinten LC, Oers MM, Hajós JP, Bianchi FJ, Lent JW, Zuidema D, Vlak JM (1999) Specificity of polyhedrin in the generation of baculovirus occlusion bodies. J Gen Virol 80:1045–1053

Huang GH, Garretson TA, Cheng XH, Holztrager MS, Li SJ, Wang X, Cheng XW (2012) Phylogenetic position and replication kinetics of Heliothis virescens ascovirus 3h (HvAV-3h) isolated from *Spodoptera exigua*. PLoS One 7(7):e40225

Huber J (1986) Use of baculovirus in pest management programs. In: Granados RR, Federici BA (eds) The biology of baculoviruses, Vol. 2, Practical applications for insect control. CRC Press, Boca Raton, FL, pp 181–202

Huger AM (1966) A virus disease of the Indian Rhinoceros beetle, *Oryctes Rhinoceros* (Linnaeus), caused by a new type of insect virus, *Rhabdinovirus Oryctes* gen. N., sp. n. J Invertebr Pathol 8:38–51

Hukuhara T, Hayakawa T, Wijonarko A (1999) Increased baculovirus susceptibility of armyworm larvae feeding on transgenic rice plants expressing an entomopoxvirus gene. Nat Biotechnol 17:1122–1124

Hull R (2014) Plant virology, 5th edn. Academic Press, Waltham, MA, p 1104

Hunnicutt LE, Mozoruk J, Hunter WB, Crosslin JM, Cave RD, Powell CA (2008) Prevalence and natural host range of *Homalodisca coagulate* virus-1 (HoCV-1). Arch Virol 153:61–67

Hussain B, Sivakumar G, Kannan M, War AR, Ballal CR (2019) First record of a nucleopolyhedrovirus infecting brown-tail moth larvae, *Euproctis chrysorrhoea* (L.) (Lepidoptera: Lymantriidae) in India. Egyptian J Biol Pest Control 29:11

Ibarra JE, Del Rincón-Castro CM (2009) Insect viruses diversity, biology and use as bioinsecticides. In: Del Claro K et al (eds) Tropical biology and conservation management, Volume 7: Phytopathology and entomology. EOLSS Publications, p 304

Ignoffo CM, Couch TL (1981) The nucleopolyhedrosis virus of Heliothis species as a microbial pesticide. In: Burges HD (ed) Microbial control of pests and plant diseases. Academic, London, pp 329–362

Ijiri H, Coulibaly F, Nishimura G, Nakai D, Chiu E, Takenaka C, Ikeda K, Nakazawa H, Hamada N, Kotani E, Metcalf P, Kawamata S, Mori H (2009) Structure-based targeting of bioactive proteins into cypovirus polyhedra and application to immobilized cytokines for mammalian cell culture. Biomaterials 30:4297–4308

IJkel WF, Westenberg M, Goldbach RW, Blissard GW, Vlak JM, Zuidema D (2000) A novel baculovirus envelope fusion protein with a proprotein convertase cleavage site. Virology 275 (1):30–41

Ishikawa Y, Tanabe H, Nakayama C, Asayama T (1966) Studies on the nuclear polyhedrosis of *Porthesia xanthocampa*. J Sericult Sci Japan 35:174–180

Ishimori N (1934) Contribution a l'e'tude de la grasserie du ver a´ soie (*Bombyx mori*). C.R. Seances Soc. Biol. Soc Franco Japonaise Biol 116:1169–1170

Jarvis DL, Bohlmeyer DA, Garcia A (1991) Requirements for nuclear localization and supramolecular assembly of a baculovirus polyhedrin protein. Virology 185:795–810

Javed MA, Biswas S, Willis LG, Harris S, Pritchard C, Van Oers MM, Donly BC, Erlandson MA, Hegedus DD, Theilmann DA (2017) *Autographa californica* multiple nucleopolyhedrovirus AC83 is a per os infectivity factor (PIF) protein required for occlusion-derived virus (ODV) and budded virus nucleocapsid assembly as well as assembly of the PIF complex in ODV envelopes. J Virol 91:02115–02116

Jehle JA, Lange M, Wang HL, Hu ZH, Wang YJ, Hauschild R (2006) Molecular identification and phylogenetic analysis of baculoviruses from Lepidoptera. Virology 346(1):180–193

Ji, X., Sutton,G., Evans, G., Axford, D., Owen, R., & Stuart, D.I (2010) How baculovirus polyhedra fit square pegs into round holes to robustly package viruses. EMBO J, 29:505–514

Jiang L, Xia QY (2014) The progress and future of enhancing antiviral capacity by transgenic technology in the silkworm *Bombyx mori*. Insect Biochem Mol Biol 48:1–7

Kabaluk JT, Svircev AM, Goette MS, Woo SG (2010) The use and regulation of microbial pesticides in representative jurisdictions worldwide. IOBC Global 99

Kalawate A (2014) Microbial viral insecticides. In: Sahayaraj K (ed) Basic and applied aspects of biopesticides. Springer, New Delhi, pp 47–68

Kalha C, Singh P, Kang S, Hunjan M, Gupta V, Sharma R (2014) Entomopathogenic viruses and bacteria for insect–pest control. In: Abrol DP (ed) Integrated pest management: current concepts and ecological perspective. Academic Press, San Diego, CA, pp 225–244

Katsuma S, Tanaka S, Omuro N, Takabuchi L, Daimon T, Imanishi S, Yamashita S, Iwanaga M, Mita K, Maeda S, Kobayashi M, Shimada T (2005) Novel macula-like virus identified in *Bombyx mori* cultured cells. J Virol 79:5577–5584

Kerr CH, Wang QS, Keatings K, Khong A, Allan D, Yip CK, Foster LJ, Jan E (2015) The 5′ untranslated region of a novel infectious molecular clone of the dicistrovirus cricket paralysis virus modulates infection. J Virol 89:5919–5934

Korth KL, Levings CS (1993) Baculovirus expression of maize mitochondrial protein URF13 confers insecticidal activity in cell cultures and larvae. Proc Natl Acad Sci U S A 90:3388–3392

Kuhn JH (2020) Virus taxonomy. Reference module in life sciences, B978-0-12-809633-8.21231-4.

Kunimi Y (2007) Current status and prospects on microbial control in Japan. J Invertebr Pathol 95:181–186

Kuzio J, Pearson MN, Harwood SH, Funk CJ, Evans JT, Slavicek JM, Rohrmann GF (1999) Sequence and analysis of the genome of a baculovirus pathogenic for *Lymantria dispar*. Virology 253:17–34

Lacey LA, Grzywacz D, Shapiro-Ilan DI, Frutos R, Brownbridge M, Goettel MS (2015) Insect pathogens as biological control agents: back to the future. J Invertebr Pathol 132:1–41

Lai-Fook J, Dall DJ (2000) Spindle bodies of *Heliothis armigera* entomopoxvirus develop in structures associated with host cell endoplasmic reticulum. J Invertebr Pathol 75:183–192

Lasa R, Pagola I, Ibanex I, Belda JE, Williams T, Caballero P (2007) Efficacy of *Spodoptera exigua* multiple nucleopolyhedrovirus as a biological insecticide for beet armyworm control in greenhouse of southern Spain. Biocontrol Sci Tech 17:221–232

Lei C, Yang J, Wang J, Hu J, Sun X (2020) Molecular and biological characterization of *Spodoptera frugiperda* multiple Nucleopolyhedrovirus field isolate and genotypes from China. Insects 11:777

Li Q, Li L, Moore K, Donly C, Theilmann DA, Erlandson M (2003) Characterization of *Mamestra configurata* nucleopolyhedrovirus enhancin and its functional analysis via expression in an *Autographa californica* M nucleopolyhedrovirus recombinant. J Gen Virol 84:123–132

Li Y, Tan L, Li Y, Chen W, Zhang J, Hu Y (2006) Identification and genome characterization of *Heliothis armigera* cypovirus types 5 and 14 and *Heliothis assulta* cypovirus type 14. J Gen Virol 87:387–394

Li SJ, Hopkins RJ, Zhao YP, Zhang YX, Hu J, Chen XY, Xu Z, Huang GH (2016) Imperfection works: survival, transmission and persistence in the system of *Heliothis virescens* ascovirus 3h (HvAV-3h), *Microplitis similis* and *Spodoptera exigua*. Sci Rep 6:21296

Maestri A (1856) Frammenti anatomici, fisiologici e patologici sul baco da seta. Fratelli Fusi, Pavia

Manousis T, Arnold MK, Moore NF (1988) Electron microscopical examination of tissues and organs of *Dacus oleae* flies infected with cricket paralysis virus. J Invertebr Pathol 51:119–125

Martignoni ME (1999) History of TM BioControl-1: the first registered virus-based product for control of a forest insect. Am Entomol 45:30–37

Martignoni ME, Iwai PJ (1986) A catalogue of viral diseases of insects, mites, and ticks. USFS Pacific Northwest Research Station General Technical Report PNW-195

Mayo MA (2002) A summary of taxonomic changes recently approved by ICTV. Arch Virol 147:1655–1663

Merian MS (1679) Del' Raupen Wunderbare Verwandelung, und Sonderbare Blumennahrung. J. A. Graff, Niirnberg, p 155

Merryweather AT, Weyer U, Harris MPG, Hirst M, Booth T, Possee RD (1990) Construction of genetically engineered baculovirus inscticides containing the *Bacillus thuringiensis* subsp. *kurstaki* HD-73 delta endotoxin. J Gen Virol 71:1535–1544

Mettenmeyer A (2002) Viral insecticides hold promise for bio-control. Farming Ahead 124:50–51

Miele SAB, Garavaglia MJ, Belaich MN, Ghiringhelli PD (2011) Baculovirus: molecular insights on their diversity and conservation. Int J Evol Biol 15:379424

Monsma SA, Oomens AGP, Blissard GW (1996) The GP64 envelope fusion protein is an essential baculovirus protein required for cell-to-cell transmission of infection. J Virol 70:4607–4616

Moore SD, Kirkman W, Richards GI, Stephen P (2015) The *Cryptophlebia leucotreta* granulovirus - 10 years of commercial field use. Viruses 7:1284–1312

Moreau G, Lucarotti CJ (2007) A brief review of the past use of baculoviruses for the management of eruptive forest defoliators and recent developments on a sawfly virus in Canada. For Chron 83:105–112

Mori H, Metcalf P (2010) Cypoviruses. In: Asgari S, Johnson KN (eds) Insect virology. Caister Academic Press, Great Britain, p 436

Moscardi F (1999) Assessment of the application of baculoviruses for control of Lepidoptera. Annu Rev Entomol 44:257–289

Moscardi F, Allen GE, Greene GL (1981) Control of the velvetbean caterpillar by nuclear polyhedrosis virus and insecticides and impact of treatments on the natural incidence of the entomopathogenic fungus *nomuraea rileyi*. J Econ Entomol 74(4):480–485

Moscardi F, Souza ML, Castro MEB, Moscardi ML, Szewczyk B (2011) Baculovirus pesticides: present state and future perspectives. In: Ahmad I et al (eds) Microbes and microbial technology. Springer, New York, pp 415–445

Murphy FA, Fauquet CM, Bishop DHL, Ghabrial SA, Jarvis AW, Martelli GP, Mayo MA, Summers MD (eds) (1995) Virus taxonomy - the classification and nomenclature of viruses: sixth report of the international committee on taxonomy of viruses. Springer-Verlag, NewYork

Mutuel D, Ravellac M, Chabi B, Multeau C, Salmon J-M, Fournier P, Ogliastro M (2010) Pathogenesis of *Junonia coenia* densovirus in *Spodoptera frugiperda*: a route of infection that leads to hypoxia. Virology 403:137–144

Nysten PH (1808) Recherches sur les maladies des vers a soie et les moyens de les pre'venir. De L'Imprimerie Impe'riale, Paris

Olson S (2015) An analysis of the biopesticide market now and where is going. Outlooks Pest Manag 26:203–206

Paillot A (1933) *L'infection Chez Les Insectes*. G. Patissier, 'I'revoux, p 535

Pal N, Boyapalle S, Beckett R, Miller WA, Bonning BC (2007) A baculovirus-expressed dicistrovirus that is infectious to aphids. J Virol 81:9339–9345

Pant U, Sudeep AB, Athawale SS, Vipat VC (2002) Baculovirus studies in new indigenous lepidopteran cell lines. Indian J Exp Biol 40(1):63–68

Papp T, Marschang RE (2019) Detection and characterization of invertebrate Iridoviruses found in reptiles and prey insects in Europe over the past two decades. Viruses 11:600

Park B, Kim Y (2012) Immunosuppression induced by expression of a viral RNase enhances susceptibility of *Plutella xylostella* to microbial pesticides. Insect Sci 19:47–54

Parker KM, Sander M (2017) Environmental fate of insecticidal plant-incorporated protectants from genetically modified crops: knowledge gaps and research opportunities. Environ Sci Technol 51:12049–12057

Patel RC, Patel JK, Patel PB, Singh RB (1968) Nuclear polyhedrosis of gram pod borer, Heliothis armigera. J Econ Entomol 61:191–193

Perera S, Li Z, Pavlik L, Arif B (2010) Entomopoxviruses. In: Asgari S, Johnson K (eds) Insect virology. Caister Academic Press, Norfolk, pp 83–102

Pijlman GP, Dortmans JCFM, Vermeeschm AMG, Yang K, Martens RW, Goldbach RW, Vlak JM (2002) Pivotal role of the non-hr origin of DNA replication in the genesis of defective interfering baculoviruses. J Virol 76:5605–5611

Poinar G, Poinar R (2005) Fossil evidence of insect pathogens. J Invertebr Pathol 89:243–250

Possee RD (1997) Baculoviruses as expression vectors. Curr Opin Biotechnol 8:569–572

Prabhakar CS, Choudhary A, Choudhary JS, Sood P, Mehta PK (2017) Role of insect viruses in the management of insect pests. In: Anwer A (ed) Biopesticides and bioagents: novel tools for pest management. Apple Academic Press Inc., Waretown, NJ, pp 209–238

Quicke DLJ, Laurenne NM, Fiton MG, Broad GR (2009) A thousand and one wasps: a 28S rDNA and morphological phylogeny of the Ichneumonidae (Insecta: Hymenoptera) with an investigation into alignment parameter space and elision. J Nat Hist 43:1305–1321

Radek R, Fabel P (2000) A new entomopoxvirus from a cockroach: light and electron microscopy. J Invertebr Pathol 75:19–27

Rao GVR, Rupela OP, Rameshwar Rao V, Reddy YVR (2007) Role of biopesticides in crop protection: present status and future prospects. Indian J Plant Protect 35:1–9

Reardon R, Podgwaite JP, Zerillo RT (1996) GYPCHEK - the gypsy moth nucleopolyhedrosis virus product. USDA Forest Service Publication FHTET-96–16

Rhode SL, Iversen P (1990) Parvovirus genomes - DNA sequences. In: Tijssen P (ed) Handbook of parvoviruses, vol 1. CRC Press, Boca Raton, FL, pp 31–57

Rhodes DJ (1996) Economics of baculovirus-insect cell production systems. Cytotechnology 20:291–297

Rohrmann GF (2013) Baculovirus molecular biology. In: National center for biotechnology 1434 Information (US), 3rd edn, Bethesda, MD

Rohrmann GF (2019) Baculovirus molecular biology, 4th edn. National Center for Biotechnology Information (US), Bethesda, MD. https://www.ncbi.nlm.nih.gov/books/NBK543458/

Rowley DL, Pophamb HJR, Harrison RL (2011) Genetic variation and virulence of nucleopolyhedroviruses isolated worldwide from the heliothine pests *Helicoverpa armigera*, *Helicoverpa zea*, and *Heliothis virescens*. J Invertebr Pathol 107:112–126

Ruzicka J (1924) Die Neusten Erfahrungen Uber Die Nonne in Bohmen. Cent Ges Forsta 50:33–68

Ryabov EV, Keane G, Naish N, Evered C, Winstanley D (2009) Densovirus induces winged morphs in asexual clones of the rosy apple aphid, *Dysaphis plantaginea*. Proc Natl Acad Sci U S A 106:8465–8470

Sajjan DB, Hinchigeri SB (2016) Structural organization of baculovirus occlusion bodies and protective role of multilayered polyhedron envelope protein. Food Environ Virol 8:86–100

Sauer AJ, Schulze-Bopp S, Fritsch E, Undorf-Spahn K, Jehle JA (2017) A third type of resistance to *Cydia pomonella* granulovirus in codling moths shows a mixed Z-linked and autosomal inheritance pattern. Appl Environ Microbiol 83:e01036–e01017

Schmidt O, Theopold U, Strand M (2001) Innate immunity and its evasion and suppression by hymenopteran endoparasitoids. Bioessays 23:344–351

Senthil Kumar CM, Jacob TK, Devasahayam S, Silva SD, Jinsha J, Rajna S (2015) Occurrence and characterization of a tetrahedral nucleopolyhedrovirus from *Spilarctia obliqua* (Walker). J Invertibr Patholol 132:135–142

Sinha-Datta U, Chavali VRM, Ghosh AK (2005) Molecular cloning and characterization of *Antheraea mylitta* cytoplasmic polyhedrosis virus polyhedrin gene and its variant forms. Biochem Biophys Res Commun 332:710–718

Sivakumar G, Kannan M, Ramesh Babu V, Mohan M, Kumari S, Rangeshwaran R, Venkatesan T, Ballall CR (2020a) Characterization and field evaluation of tetrahedral and triangular nucleopolyhedrovirus of *Spilosoma obliqua* (SpobNPV) strain NBAIR1 against jute hairy caterpillar. Egyptian J Biol Pest Control 30:1–7

Sivakumar G, Kannan M, Ramesh Babu V, Mohan M, Sampath Kumar M, Raveendran P, Venkatesan T, Rangeshwaran R, Ballall CR, Ram Kumar P (2020b) Isolation and characterization of indigenous nucleopolyhedrovirus infecting new invasive fall armyworm *Spodoptera frugiperda* (J.E. Smith) (Lepidoptera: Noctuidae) in India. Curr Sci 119:860–864

Slavicek J (2012) Baculovirus enhancins and their role in viral pathogenicity. In: Adoga MP (ed) Molecular biology. IntechOpen, Rijeka, Croatia, pp 147–168

Souza ML, Castro MEB, Barros AML, Sihler W, Moscardi F (2001) Ana'lise, de DNA de isolados de nucleopolyhedrovirus de Anticarsia gemmatalis utilizados no controle da lagarta da soja no Brasil. Boletim de Pesquisa e Desenvolvimento Embrapa Recursos Gene'ticos e Biotecnologia

Steinhaus EA (1949) Principles of insect pathology. McGraw-Hill Book Co., Inc., New York, p 757

Steinhaus EA, Thompson CG (1949) Granulosis disease in the buckeye caterpillar *Junonia coenia* Hubner. Science 110:276–278

Stoltz DB, Vinson SB (1979) Viruses and parasitism in insects. Adv Virus Res 24:125–171

Strand MR, Burke GR (2015) Polydnaviruses: from discovery to current insights. Virology 479–480:393–402

Strand MR, Burke GR (2019) Polydnaviruses: evolution and function. In: Bonning BC (ed) Insect molecular virology. Advances and emerging trends. Caister Academic Press, Poole, Dorset, pp 163–181

Sun X (2015) History and current status of development and use of viral insecticides in China. Viruses 7:306–319

Szewczyk B, Hoyos-Carvajal L, Paluszek M, Skrzecz W, de Souza ML (2006) Baculoviruses-re-emerging biopesticides. Biotechnol Adv 24:143–160

Takov D, Pilarska D, Linde A, Barta M (2021) Infectious and parasitic diseases of phytophagous insect pests in the context of extreme environmental conditions. Central Eur Forestry J 67

Tan L, Zhang J, Li Y, Jiang H, Cao X, Hu Y (2008) The complete nucleotide sequence of the type 5 *Helicoverpa armigera* cytoplasmic polyhedrosis virus genome. Virus Genes 36:587–593

Tanada Y (1964) A Granulosis virus of the codling moth, *Carpocapsa pomonella* (Linnaeus) (Olethreutidae, Lepidoptera). J Insect Pathol 8:378–380

Tanada Y, Kaya HK (1993) Insect pathology. Academic Press, Inc., San Diego, CA

Taylor MW (2014) Introduction: a short history of virology. In: Taylor MW (ed) Viruses and man: a history of interactions. Springer, Swizerland, pp 1–22

Thao M, Wineriter S, Buckingham G, Baumann B (2001) Genetic characterization of a putative Densovirus from the mealybug, *Planococcus citri*. Curr Microbiol 43:457–458

Thézé J, Bézier A, Periquet G, Drezen J-M, Herniou EA (2011) Paleozoic origin of insect large dsDNA viruses. Proc Natl Acad Sci U S A 108(38):15931–15935

Theze J, Takatsuka J, Li Z, Gallais J, Doucet D, Arif B, Nakai M, Herniou EA (2013) New insights into the evolution of Entomopoxvirinae from the complete genome sequences of four entomopoxviruses infecting *Adoxophyes honmai*, *Choristoneura biennis*, *Choristoneura rosaceana* & *Mythimna separata*. J Virol 87:7992–8003

Tillman PG, Styer EL, Hamm JJ (2004) Transmission of ascovirus from *Heliothis virescens* (Lepidoptera: Noctuidae) by three parasitoids and effects of virus on survival of parasitoid *Cardiochiles nigriceps* (Hymenoptera: Braconidae). Environ Entomol 33:633–643

Tufts DM, Hunter WB, Bextine B (2014) *Solenopsis invicta* virus (Sinv-1) infection and insecticide interactions in the red imported fire ant (Hymenoptera: Formicidae). Fla Entomol 97:1251–1254

Vago C (1963) A new type of insect virus. J Insect Pathol 5:275–276

Valles SM, Porter SD, Choi MY, Oi DH (2013) Successful transmission of *Solenopsis invicta* virus 3 to *Solenopsis invicta* fire ant colonies in oil, sugar and cricket bait formulations. J Invertebr Pathol 113:198–204

van Munster M, Dullemans AM, Verbeek M, van den Heuvel JF, Reinbold C, Brault V, Clérivet A, van der Wilk F (2003) Characterization of a new densovirus infecting the green peach aphid *Myzus persicae*. J Invertebr Pathol 84(1):6–14

van Oers MM, Vlak JM (2007) Baculovirus genomics. Curr Drug Targ 8(10):1051–1068. Bentham Science Publishers

van Oers MM, Pijlman GP, Vlak JM (2015) Thirty years of baculovirus insect cell protein expression: from dark horse to mainstream technology. J Gen Virol 96:6–23

Vida, M.H (1527) The Silkworm (Bombycum): a poem in two books. Translated into English by S. Pullein (1750). S. Powell, Dublin

Volkman LE, Goldsmith PA (1985) Mechanism of neutralization of budded *Autographa californica* nuclear polyhedrosis virus by monoclonal antibody: inhibition of entry by adsorptive endocytosis. Virology 143:185–195

Wamonje FO, Michuki GN, Braidwood LA, Njuguna JN, Musembi Mutuku J, Djikeng A, Harvey JJW, Carr JP (2017) Viral metagenomics of aphids present in bean and maize plots on mixed-use farms in Kenya reveals the presence of three dicistroviruses including a novel Big Sioux River virus-like dicistrovirus. Virol J 14:188

Wang Q, Bosch BJ, Vlak JM, van Oers MM, Rottier PJ, van Lent JWM (2016) Budded baculovirus particle structure revisited. J Invertebr Pathol 134:15–22

Wang X, Shang Y, Chen C, Liu S, Chang M, Zhang N, Hu H, Zhang F, Zhang T, Wang Z, Liu X, Lin Z, Deng F, Wang H, Zou Z, Vlak JM, Wang M, Hu Z (2019) Baculovirus *per os* infectivity factor complex: components and assembly. J Virol 93:e02053–e02018

Warsaba R, Sadasivan J, Jan E (2020) Dicistroviruses: host molecular interactions. Curr Issues Mol Biol 34:83–112

Watanabe H, Aratake Y (1974) Two isolates of a nuclear-Polyhedrosis virus of the Brown-tail moth, *Euproctis similis*, exhibiting different inclusion body shapes. J Invertibr Patholol 24:383–386

Webb BA, Strand MR (2005) The biology and genomics of polydnaviruses. In: Gilbert LI et al (eds) Comprehensive molecular insect science, vol 6. Elsevier, San Diego, CA, pp 323–360

Wei YL, Hu J, Li SJ, Chen ZS, Cheng XW, Huang GH (2014) Genome sequence and organization analysis of *Heliothis virescens* ascovirus 3f isolated from a *Helicoverpa zea* larva. J Invertebr Pathol 122:40–43

White GF (1917) Sacbrood. U.S. Dep Agric Bull 431:1–55

Williams T (1996) The Iridoviruses. Adv Virus Res 46:345–412

Wilson K, Grzywacz D, Curcic I, Scoates F, Harper K, Rice A, Paul N, Dillon A (2020) A novel formulation technology for baculoviruses protects biopesticide from degradation by ultraviolet radiation. Sci Rep 10:13301

Yu XK, Ge P, Jiang JS, Atanasov I, Zhou ZH (2011) Atomic model of CPV reveals the mechanism used by this single-shelled virus to economically carry out functions conserved in multishelled Reoviruses. Structure 19:652–661

Yu Q, Fediere G, Abd-Alla A, Bergoin M, Tijssen P (2012) Iteravirus-like genome organization of a densovirus from *Sibine fusca* Stoll. J Virol 86:8897–8898

Zhang GY, Sun XL, Zhang ZX, Zhang ZF, Wan FF (1995) Production and effectiveness of the new formulation of Helicoverpa virus pesticide-emulsifiable suspension. Virol Sin 10:242–247

Zhou Y, Qin T, Xiao Y, Qin F, Lei C, Sun X (2014) Genomic and biological characterization of a new cypovirus isolated from Dendrolimus punctatus. PLoS One 9:e113201

Zhou C, Liu S, Song W, Luo S, Meng G, Yang C, Yang H, Ma J, Wang L, Gao S, Wang J, Yang H, Zhao Y, Wang H, Zhou Z (2018) Characterization of viral RNA splicing using whole-transcriptome datasets from host species. Sci Rep 8:3273

Entomopathogenic Bacteria

2

Priyanka Gangwar, Mala Trivedi, and Rajesh K. Tiwari

Contents

Abstract

Bacteria are well known and extensive in the environment. They have developed a variety of interactions with pests, including essential symbiosis. The term biocontrol applies to the process where living organisms are used to check growth in the population density of specific pest to the extent that ecological balance is maintained and without making them extinct. Bacteria have been used as biocontrol agents for the biocontrol of pests for over a century. The scientific communities have also looked at them as an important component of integrated

P. Gangwar · M. Trivedi · R. K. Tiwari (✉)
Amity Institute of Biotechnology, Amity University, Lucknow, Uttar Pradesh, India
e-mail: rktiwari@lko.amity.edu

Omkar (ed.), *Microbial Approaches for Insect Pest Management*,
https://doi.org/10.1007/978-981-16-3595-3_2

pest management to develop ecofriendly pest management system for crop protection and improvement. It became more pertinent in view of problems posed with synthetic chemicals as they induced harmful changes in nontarget insects and pests.

Microbial pesticides, as living organism or their products or byproducts, offer an environment-friendly alternative with target specificity. They control pests through their nontoxic mechanisms, which are pathogenic to them. The most studied bacterium belonging to the family *Bacillaceae* is *Bacillus thuringiensis* (BT). Besides that, some more bacteria have been reported as potent biocontrol agents, viz. *B. sphaericus, Bacillus popilliae, Bacillus lentimorbus, Paenibacillus popilliae, Serratia entomophila, Brevibacillus laterosporus, Chromobacterium subtsugae,* and *Yersinia entomophaga.*

Bacterium, as the active ingredient in living form, their products or byproducts can control many different kinds of pests. *Aedes aegypti* can easily be controlled by *Bacillus thuringiensis.* On the other hand, to check the population of *Culex quinquefasciatus* and other mosquitoes, *B. sphaericus* showed promising results. The host range of *Bacillus lentimorbus* is larvae (grubs) of Japanese beetle. However, specificity is also a limiting factor, as single application can be effective to a single pest species. Some physical factors, namely heat, desiccation, and radiation, and storage procedures reduce the effectiveness of these bioinsecticides. Despite that, bioinsecticides present great promise to develop better and environmentally friendly pest control programs. The aim of this chapter is to give a holistic and concise picture of insect pathogenic bacteria and their use as bioinsecticides in pest management programs.

Keywords

Bioinsecticides · Biocontrol · Bacteria, · *Bacillus* · Integrated pest management

Learning Objectives

The chapter has the following learning outcomes:

1. The majority of entomopathogenic bacteria belong to the families Bacillaceae, Pseudomonadaceae, Enterobacteriaceae, Streptococcaceae, and Micrococcaceae.
2. The entomopathogenic bacteria enter in insects by ingestion, where they produce toxins, disrupting epithelium of midgut that later cause death of insects.
3. Use of microbes in pest management is an ecofriendly approach to control pests in agricultural crops and to reduce the use of synthetic pesticides.

2.1 Introduction

In the agricultural fields, the farmers have always encountered huge crop damages as well as losses due to pests and diseases. Pesticides that include insecticides, herbicides, and fungicides are employed in modern agriculture to control pests and diseases, and to increase the crop yield. In both developed and developing countries, the use of synthetic pesticides has increased dramatically during the last few decades, and the control of pests with synthetic chemicals results in several problems (Gamliel et al. 1997). The residues of these synthetic insecticides cause toxic effects on humans and wildlife (e.g., birds, beneficial insects like honeybees). Another environmental concern is the contamination of groundwater (Lacey and Siegel 2000).

Despite many years of effective control of pests by insecticides, the continuous use of these chemicals has threatened their effectiveness. It includes the development of insecticide resistance in the target pest species and the deregistration of insecticides due to human health and environmental issues (Nicholson 2007). Therefore, environment-friendly options are the demand of the present time. Improvement in pest control strategies and efficacies leads to higher quality and a greater quantity of agricultural produce. Therefore, there is a great need to develop effective, biodegradable, and environment-friendly biopesticides.

Thus, because of the hazardous effects of insecticides, one of the environment-friendly methods is developed to protect the plant from plant pathogens recently, that is, the use of antagonistic microorganisms called biological control agents or referred to as "biopesticides" (Mazid et al. 2011). Various researches and facts have proved that biological control is a powerful plant disease management tool that can bring huge benefits. Microorganisms that can cause disease in an insect pest and, to some extent, in other arthropods are known as entomopathogens (Tanzini et al. 2001). Many beneficial microorganisms, such as *Pseudomonas fluorescens*, *Trichoderma* spp., *Bacillus subtilis*, and *Fusarium* nonpathogen, have been studied and tested for their efficacy against various plant pathogens. Some of them have been released and marketed as biopesticides (EPA 2011).

2.2 Biological Pesticides

Biopesticides are microorganisms or their products used in pest management and are not related to synthetic pesticides. They are derived from natural enemies, such as animals, plants, bacteria, and certain minerals (EPA 2015), and can be cultured and applied. For example, rapeseed oil and baking soda have insecticidal effects and are considered biopesticides. However, Sudakin (2003) and Gupta and Dikshit (2010) pointed out a broader definition of biopesticides, that is, biopesticides are biochemical pesticides, naturally occurring substances that can control pests through their nontoxic mechanisms. They are living organisms, like viruses, bacteria, fungi, protozoans, and nematodes or their byproducts (phytochemical products and microbial products) that can be used to manage harmful pests of crops.

The unique mode of action of biopesticides makes them different from synthetic pesticides. Biopesticides are usually effective in small amounts and degrade quickly, thereby reducing environmental exposure and avoiding pollution problems. Therefore, they are safer for the environment and human health; and have not found any residual impact on humans (Gupta and Dikshit 2010). According to the US Environmental Protection Agency (USEPA 2010), to classify a substance or a mixture of substances as a biopesticide active ingredient, three conditions are essential: (i) it must be naturally occurring, (ii) must have a nontoxic mode of action against the target pest, and (iii) must have a history of nontoxic exposure to humans and the environment.

2.3 Microbial Pesticides

Biopesticides are further divided into three major categories: microbial pesticides, plant-incorporated protectants (PIPs), and biochemical pesticides (Pathak et al. 2017). Microbial pesticides belong to naturally occurring bacteria, fungi, viruses, etc. Plant-incorporated protective agents (PIP) are the pesticides produced inside the plants from their genetic material. For example, scientists have incorporated *Bt* insecticidal protein genes into plant genomes. However, biochemical pesticides are natural substances, such as plant extracts and fatty acids, from which pests can be controlled through nontoxic mechanisms.

Microbial pesticides can be effectively used as a substitute for chemical pesticides. They are biological pesticides derived from microorganisms, like bacteria or fungi (Pathak et al. 2017). The pathogenic effects of these microorganisms on target pests are host-specific. The mode of action of microbial insect pathogens is by invasion through the outer skin or intestinal tract of insects, which leads to the reproduction of pathogens inside the host. Further, these pathogens produce insecticidal toxins and cause the death of the host. These toxins are identified as peptides, but their structure, toxicity, and specificity greatly differ (Burges 1981).

Moreover, microbial pesticides can be used alone or in combination with other pest management tools (Rathod et al. 2014). Therefore, these microbial pesticides can effectively replace synthetic pesticides with target specificity. A comprehensive pest management plan, also termed as integrated pest management (IPM), evaluates various control measures, cultural operations, weather, potential interactions between pests and crops, and considers all available pest control measures (Flint and Van den Bosch 1981). They leave little or no residue on food crops, thus are efficient and safe for humans and nontarget animals. Since microbial pesticides are species-specific, so other natural enemies are not threatened. They are, therefore, ecologically safe and capable of maintaining biodiversity in the ecosystem.

Microbial pesticides also promote the survival of beneficial insects in treated crops (Hangay et al. 2008). For example, the bacterium, *Paenibacillus popilliae,* is used in biological control. It causes milky spore disease and is useful for controlling

"Japanese beetles." It is very specific to the host species and harmless to other natural enemies. That is why microbial insecticides have been widely used as biocontrol agents in the past three decades (Lacey et al. 2001).

2.4 Bacterial Insect Pathogenesis

Microorganisms that are pathogenic to arthropods, especially to insects, are called entomopathogens. Several naturally occurring bacteria, nematodes, fungi, and viruses infect various insects, mites, and ticks and play an important role to manage them (Kalha et al. 2014). Entomopathogenic bacteria are single-celled prokaryotes, whose size ranges from less than 1µm to several µm. The mode of action of pathogenesis by different bacteria is mentioned below.

2.4.1 Mode of Action

To infect any host, the first task of a pathogen is to enter the host's cells and cavities of its body. This can take place through three main mechanisms: through a lesion, attack of the body cavity through the nematode vector, and the consumption of infected food by the host (Fig. 2.1, Vallet-Gely et al. 2008). On infection of the host, the pathogen enters the hemolymph (Evans et al. 2006). An example of infection by nematode vector is *Photorhabdus luminescens* and *Xenorhabdus nematophila*,

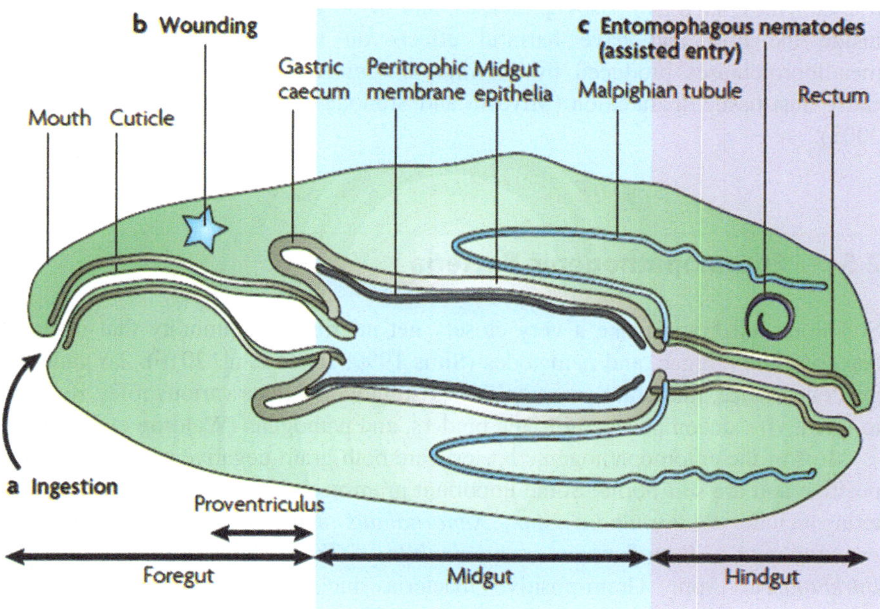

Fig. 2.1 Main routes of bacterial infection in the host. (Vallet-Gely et al. 2008)

which first colonize the gut of the nematode in a symbiotic manner. The nematode then infects the insect and allows the bacteria to colonize inside the insect. Finally, the bacteria infect and kill the insects (Forst et al. 1997).

After entry into the insect's body, the next step is to maintain and colonize the insect. A variety of enzymes are secreted for this process, which work against the gut products of the insect, start producing biofilm, and modify the host gut (Jarrett et al. 2004; Vallet-Gely et al. 2008). AFP (antifading propagase) gene present in *Serratia* spp. is a good example of this. It has toxic effects on host gut epithelial cells and favors bacterial colonization (Hurst et al. 2007).

The bacteria have to escape from the host immune system while living in the gut of the insect. They do this either by avoiding recognition by the immune system or by suppressing the immune response (Vallet-Gely et al. 2008). Host insects use a mechanism to prevent itself from infection by producing reactive oxygen species (ROS) and antimicrobial peptides (AMPs) (Hinnebusch et al. 2002; Parsek and Singh 2003) (Fig. 2.2). On the contrary, bacteria produce proteins that either protect them from the effects of these compounds or degrade these compounds. For example, *Pseudomonas entomophila* produces zinc metalloprotease AprA that acts against host AMP's (Liehl et al. 2006). After evading the host's defense mechanism, the bacteria start killing the host.

Although several factors involved in the pathogenesis of bacteria have been described, it is still unclear what exactly causes the insect's death. It seems that either excessive bacterial proliferation or the production of toxic factors cause damage to the host insect and eventually causes its death (Fig. 2.3). The different stages of infection and bacterial response are outlined in Fig. 2.4. Various degradative enzymes, such as lipases, proteases, and hemolysin, are produced by bacteria inside the host and have harmful effects on the host. For example, the metalloproteinases produced by *Serratia marcescens* inactivate host AMP and cause host tissue degradation (Miyoshi and Shinoda 2000; Ffrench-Constant et al. 2003).

2.5 Entomopathogenic Bacteria

Soil flora and fauna make a very closely net microbial community that includes bacteria, fungi, algae, and nematodes (Sims 1990; Jones et al. 2010). To maintain general balance, microbial communities existing in soil have various roles assigned to them, viz., decomposers, nitrogen binders, and pathogens (Waldrop et al. 2000).

Most of the entomopathogenic bacteria are both gram-negative as well as gram-positive and are soil-borne. Some important gram-negative entomopathogenic bacteria include *Photorhabdus* spp., *Xenorhabdus* spp., *Serratia* spp., *Yersinia entomophaga, Pseudomonas entomophila, Chromobacterium* spp., and *Burkholderia* spp. Gram-positive bacteria include *Bacillus thuringiensis, Lysinibacillus sphaericus, Paenibacillus* spp., *Brevibacillus laterosporus, Clostridium bifermentans, Saccharopolyspora spinosa,* and *Streptomyces* spp..

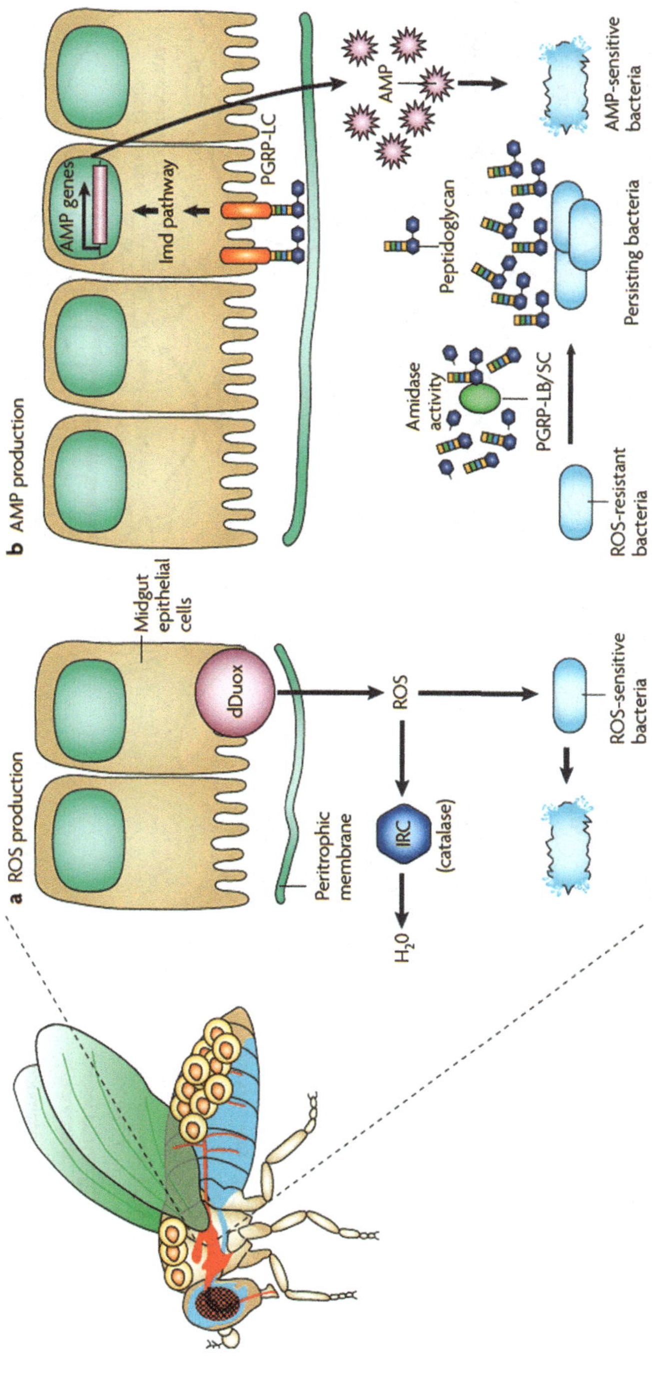

Fig. 2.2 Overview of the local immune response in Drosophila. (Vallet-Gely et al. 2008)

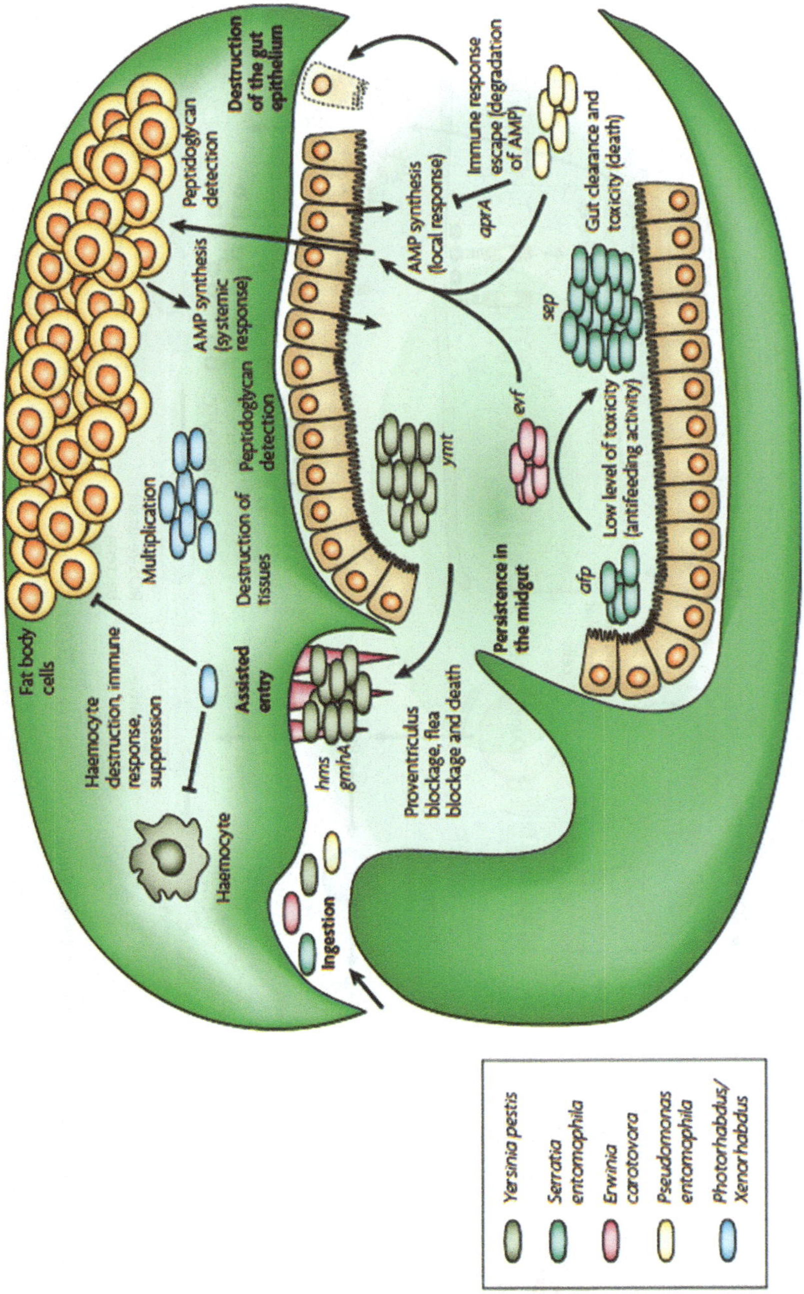

Fig. 2.3 Schematic diagram of different bacterial infections in the host. (Vallet-Gely et al. 2008)

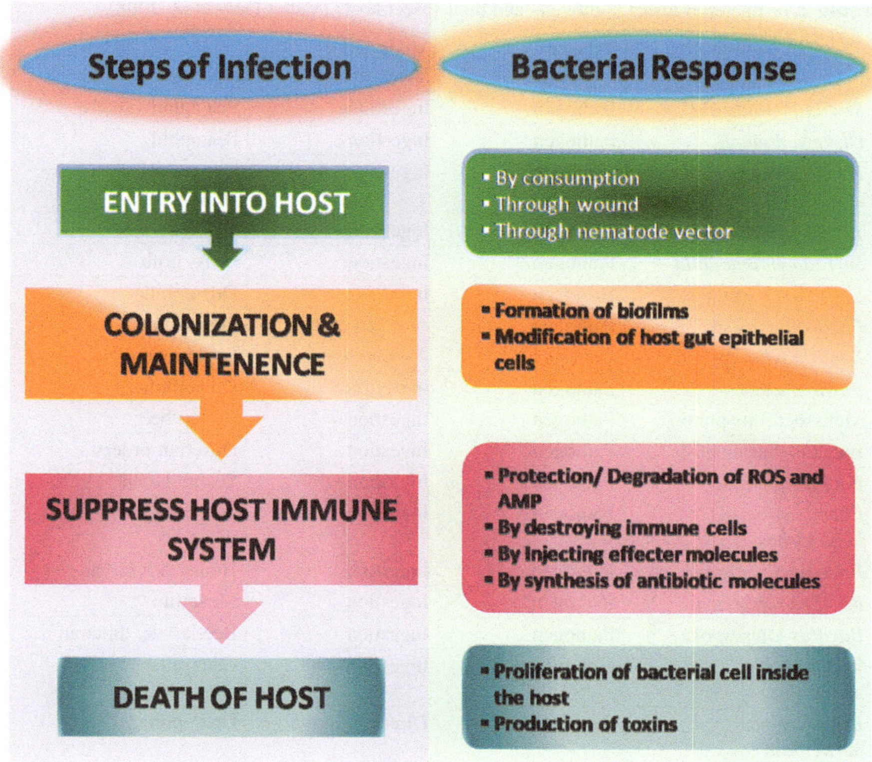

Fig. 2.4 Different stages of bacterial infection and responses in the host

2.5.1 Types of Entomopathogenic Bacteria Other than BT

Several arthropod pathogens have been identified by the bacterial families Bacillaceae, Enterobacteriaceae, Pseudomonadaceae, Micrococcaceae, and Streptococcaceae so far. Usually, these families contain epiphytes, yet some pathogens have been shown to have very toxic effects on their hosts. The two genera (*Photorhabdus* and *Xenorhabdus*) carried by nematodes have been used for agricultural purposes in various ways (Toh et al. 2006; Shigenobu et al. 2000; Akman et al. 2002). They live symbiotically in the alimentary canal of entomopathogenic nematodes and produce a series of toxins (Forst et al. 1997; Ffrench-Constant et al. 2007). These toxins can be applied over leaves of crp plants as extracts or cell suspensions to control insects (Munson et al. 1991). Also, genes that encode toxins can be used to develop transgenic plants to protect crops (Ffrench-Constant and Bowen 2000).

Among entomopathogenic bacteria, the family Bacillaceae has attracted a lot of attention. *Bacillus thuringiensis* (*Bt*) is a soil-borne bacterial species and a series of many deadly pathogens. It is most widely used as a biocontrol agent for insect pests (Pigott and Ellar 2007; Bravo et al. 2007; Kalha et al. 2014). Some other bacterial

Table 2.1 Bacterial insect pathogens and their insect hosts (Vallet-Gely et al. 2008)

Bacteria	Type of interaction	Mode of interaction	Host
Erwinia aphidicola	Pathogen	Ingestion	Pea aphid
Dickeya dadantii	Pathogen	Ingestion	Pea aphid
Pseudomonas entomophila	Pathogen	Ingestion	Drosophila, Bombyx, galleria
Yersinia pestis	Pathogen	Ingestion	Rat flea
Serratia entomophila	Pathogen	Ingestion	Grass grub
Serratia marcescens	Pathogen	Ingestion	Drosophila
Photorhabdus sp.	Pathogen	Assisted entry	Lepidopteran
Xenorhabdus sp.	Pathogen	Assisted entry	Lepidopteran
Vibrio cholera	Pathogen	Ingestion	Drosophila
Melissococcus pluton	Pathogen	Ingestion	Honey bee
Bacillus thuringiensis	Pathogen	Ingestion	Different orders
Bacillus papillae	Pathogen	Ingestion	Scarab larvae
Paenibacillus lentimorbus	Pathogen	Ingestion	Scarab larvae
Paenibacillus larvae	Pathogen	Ingestion	Honey bee larvae
Bacillus sphaericus	Pathogen	Ingestion	Mosquito
Bacillus laterosporus	Pathogen	Ingestion	Bee larvae, dipteran
Pseudomonas aeruginosa	Opportunistic	Ingestion	Caterpillar
Pseudomonas aeruginosa	Opportunistic	Direct injection	Drosophila
Bacillus cereus	Opportunistic	Ingestion	Galleria mellonella
Erwinia carotovora	Infectious	Ingestion	Drosophila larvae
Shigella spp.	Passive	Ingestion	Vector house fly
Rickettsia spp.	Vector	Ingestion	Cat flea
Bartonella spp.	Vector	Ingestion	Cat flea

species of *Bacillus* and *Paenibacillus* are also found pathogenic for coleopteran, dipteran, and lepidopteran insects. For example, *B. sphaericus* is highly toxic to mosquitoes, whereas *Bacillus popilliae* and *Paenibacillus popilliae* cause milk spore disease and are used against Japanese beetle larvae (Baumann et al. 1991; Charles et al. 1997; Davidson et al. 1975; Zhang et al. 1997).

Till now more than 100 species of *Clostridium* spp., *Paenibacillus* spp., and *Bacillus* spp. have been identified as biocontrol agents and are highly pathogenic to arthropods (Table 2.1 and Fig. 2.5). Besides, it is reported that biopesticides based on heat-inactivated *Burkholderia rinojensis* and *Chromobacterium subtsugae* can target different types of mites, as they have multiple modes of action (Burkhead et al. 1994; Janisiewicz and Roitman 1988; Martin et al. 2007). Some nonspore-forming ones that belong to the genus *Pseudomonas, Xenorhabdu, Photorhabdus, Yersinia,* and *Serratia* have also received great attention as microbial agents (Waterfield et al. 2001; Zhang et al. 2009; Marshall et al. 2012; Vodovar et al. 2005).

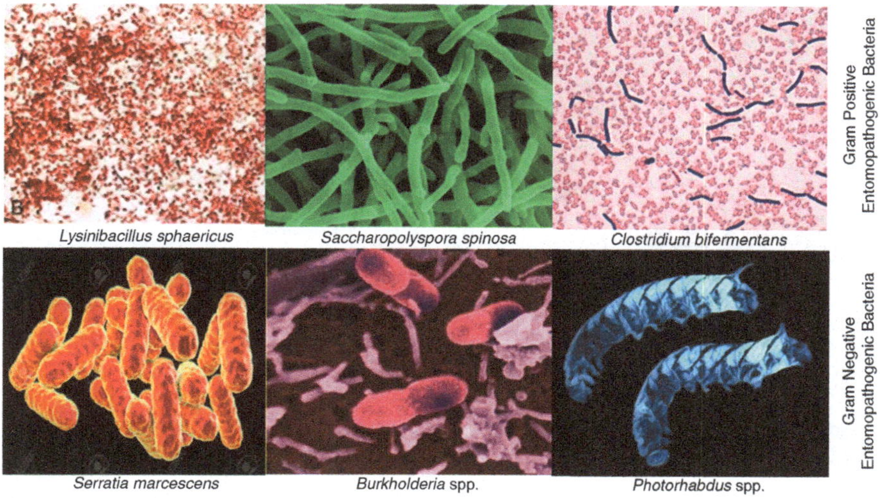

Fig. 2.5 Diagrams of different gram-positive and gram-negative entomopathogenic bacteria

2.5.2 Insect Pathogenic Bacteria Belong to Different Groups/Classes

The four bacterial phyla to which entomopathogens belong are Firmicutes, Actinobacteria, Proteobacteria, and Tenericutes (Fig. 2.6). And family includes Neisseriaceae, Enterobacteriaceae, Paenibacillaceae, and Bacillaceae. Plenty of data on entomopathogens are related to the genus *Bacillus*.

2.5.3 Gram-Positive Entomopathogenic Bacteria

Bacillaceae

Bacillus thuringiensis

Bacillus thuringiensis (BT) is a soil-borne naturally occurring bacterium that has been widely studied and used for pest control naturally. During the last decades, some of its crystal-producing strains have been popularized as the main active substances used in microbial pest management programs (Vega et al. 2012). The pathogenic action of this bacterium is generally followed by spores and crystalline inclusions containing insecticidal δ-endotoxins. The δ-endotoxins interact with receptors present in the midgut epithelial cells of insects and cause cell lysis followed by gut paralysis and death (Pigott and Ellar 2007; Bravo et al. 2007).

B. thuringiensis subsp. *kurstaki* (*Btk*) is commonly used to control lepidopteran larvae. Different strains of *Btk* with important commercial value are ABTS-351, EG2348, HD-1, PB 54, SA-11, and SA-12. Another strain of *B. thuringiensis* subsp. *aizawai* (*Bta*) (ABTS-1857) is also used to suppress the populations of armyworms

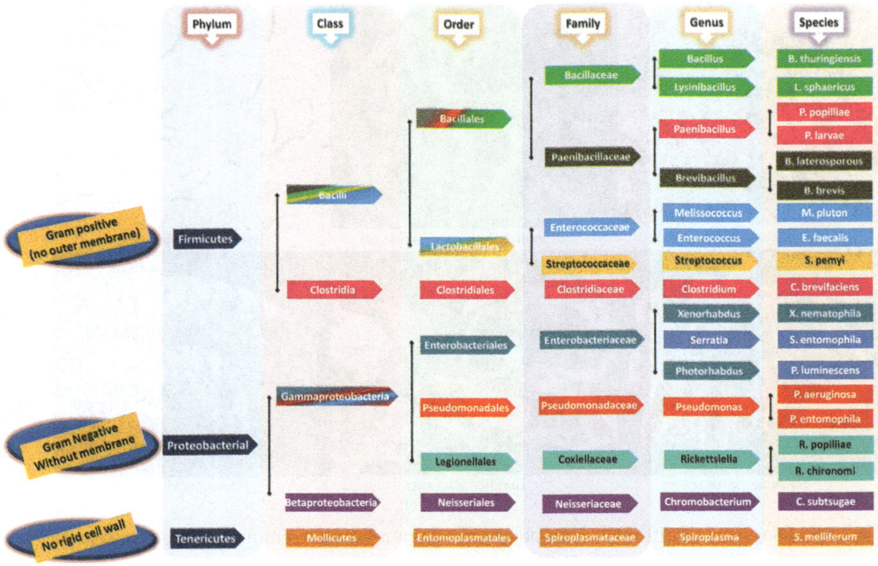

Fig. 2.6 Broad classification of entomopathogenic bacteria

and diamondback moth larvae. Besides, strains belonging to subspecies *israelensis* (*Bti*) have been used to manage mosquitoes and simulids, and of tenebrionids (*Btt*) to fight against Coleoptera (Glare and O'Callaghan 2000).

Lysinibacillus Sphaericus

Entomopathogenic strains belonging to the group of *Lysinibacillus sphaericus* species are characterized by the production of spherical endospores. The endospores are closely related to the parasporal crystals with equivalent proportions of binary protein toxins (BinA and BinB) (Baumann et al. 1991). The insecticidal mode of action involves the lysis of the microvilli epithelial cells of the insect midgut (Charles et al. 1997). Also, the vegetative cells of certain strains produce toxins (Mtx proteins) that kill mosquitoes. Therefore, the main targets of commercially prepared *L. sphaericus* strains are mosquitoes, blackflies, and nonbiting midges.

Paenibacillaceae

Paenibacillus Spp.

Various species of the genus *Paenibacillus* cause pathogenicity to pests, such as American Foulbrood disease in honeybees caused by the spore-forming bacterium *P. larvae* subsp. *larvae* (Davidson 1973). Similarly, *P. popilliae* and *P. lentimorbus* cause milky diseases in phytophagous coleopteran larvae. Even though *P. popilliae* is not directly responsible for the insecticidal effect, the production of parasporal inclusions within the sporangial cells has been reported. After entering the host, the spores begin to germinate in the host's midgut. The pathogenicity of this bacterium can be found as septicemia.

Brevibacillus laterosporus

Brevibacillus laterosporus is a pathogen of invertebrates with a wide range of antimicrobial effects (Ruiu 2013). During the sporulation process, it produces the typical canoe-like parasporal body, which is tightly combined with the sporecoat and gives the species unique morphological characteristics. The insecticidal effects of different strains of *B. laterosporus* on different classes of insects including Coleoptera, Diptera, and Lepidoptera, as well as phytopathogenic bacteria, nematodes, fungi, and molluscs have been reported.

Recently, the whole genome of *B. laterosporus* has been deciphered, revealing its ability to produce various toxins (Djukic et al. 2011; Sharma et al. 2012). Some strains that are toxic to corn rootworms (*Diabrotrica* spp.) and other coleopteran larvae produce insecticidal secreted proteins (ISPs). The protein acts as a binary toxin in the insect midgut and has a high degree of homology with the plant insecticidal protein produced by *B. thuringiensis* (Warren 1997).

Clostridiaceae

Clostridium bifermentans

Clostridium bifermentans strains are highly toxic to mosquitoes and black flies and produce three main proteins involved in insecticidal action (Nicolas et al. 1990). Among these mosquitocidal proteins, Cbm71 is found similar to *B. thuringiensis* delta endotoxins (Barloy et al. 1996).

Actinobacteria

Saccharopolyspora spinosa

During the screening procedure, *Saccharopolyspora spinosa,* a species of bacteria was found from a sugar mill rum and later revealed the insecticidal activity of its isolate "A83543" (Mertz and Yao 1990). The subsequent analysis emphasized the widespread toxicity of specific compounds isolated from the fermentation broth, which was later renamed "spinosyns." It is a new insecticide that contains a structurally unique glycosylated macrolactone that has selective activity against a wide variety of insect pests. Spinosyns have a unique mode of action, including the postsynaptic nicotinic acetylcholine and gamma-aminobutyric acid (GABA) receptors (Bond et al. 2004; Watson 2001).

The interaction of these receptors ultimately leads to the destruction of neuronal activity and the resulting paralysis and death of insects (Perry et al. 2007). Despite its widespread pathogenic activity, it has a lower risk with nontarget species than other pesticides (Sparks et al. 2001).

Streptomyces Spp.

Different *Streptomyces* spp. are related to herbivorous insects that use their cellulose decomposing properties for pathogenicity (Book et al. 2014). Other species and strains of this genus produce various metabolites as effective toxins that have high insecticidal activity (Copping and Menn 2000). Some of the insecticidal compounds produced by *Streptomyces* species are Antimycin A, Macrotetralides, Piericidins, Prasinons, and Flavensomycin (Ruiu 2015). Another substance "Avermectins" produced by *Streptomyces avermitilis* was discovered that had insecticidal and repellent activity (Turner and Schaeffer 1989). These compounds target GABA receptors present in the peripheral nervous system of the insect. The GABA binding to macrocyclic lactone derivatives produces a cascade of events that cause inhibition of neurotransmission and finally paralysis of the neuromuscular system (Bloomquist 1996).

2.5.4 Gram-Negative Entomopathogenic Bacteria

Gammaproteobacteria

Photorhabdus Spp. and *Xenorhabdus* Spp.

Entomopathogenic members of the genus *Photorhabdus* and *Xenorhabdus* are endosymbionts and are usually related to the genus *Heterorhabditis* and *Steinernema* species, respectively (Ffrench-Constant et al. 2007). Once the nematode actively enters the insect body, it releases symbiotic bacteria into the insect hemocoel under pathological action. Here, released bacteria produce and spread various antimicrobial compounds to combat the growth of other microorganisms. They also release various enzymes that contribute to the degradation of hemocoel and make an ideal environment for the development of the nematode population (Tailliez et al. 2006).

Serratia Spp.

The relationship of *Serratia* spp. with insects or insect-pathogenic nematodes has been fully proven (Zhang et al. 2009; Abebe et al. 2011; Torres-Barragan et al. 2011). The pathogen of the grass grub *Costelytra zealandica* produces a group of insecticidal toxins named Sep protein (SepA, SepB, SepC) that show similarities to the insecticidal toxins of *P. luminescens* (Jackson et al. 1992; Hurst et al. 2000). Different species of this genus produce multiple virulence factors. On the other hand, recent genome sequencing of *S. nematodiphila* has also highlighted other pathogenic factors of *Serratia* species (Kwak et al. 2015). It has recently been demonstrated that the pathogenicity of *Serratia marcescens* is increased by the action of serralysin metalloproteinase. It allows bacteria to suppress cellular immunity by reducing the adhesion properties of immune surveillance cells in the insect hosts (Ishii et al. 2014).

Yersinia entomophaga

Yersinia entomophaga is a nonspore-forming pathogenic bacterium with the characteristic of producing insecticidal toxin complex (Yen-Tc). It shows the similarity with *Photorabdus* spp. products (Hurst et al. 2011). These complexes include three Y protein families, A, B, and C, and two chitinases (Chi1 and Chi2) (Landsberg et al. 2011). Various studies on its wide range of toxins and its post-ingestion histopathological effects in the midgut epithelium of insects have been reported (Marshall et al. 2012). Promising research has been carried out under field conditions in which insecticides containing *Y. entomophaga* are used to combat the pasture pest porina (Ferguson et al. 2012).

Pseudomonas entomophila

A ubiquitous bacterium, *P. entomophila,* infects insect larvae orally in a different order and determines extensive intestinal cell damage. The host–pathogen interaction has been studied in *Drosophila melanogaster* that highlights specific immune responses after ingestion (Vodovar et al. 2005). The complete genome sequencing of *P. entomophila* recently highlighted a specific secretion system and related toxins, which may be responsible for its pathogenicity (Vodovar et al. 2006).

Betaproteobacteria

Chromobacterium Spp.

The *Chromobacterium subtsugae* strain named PRAA4-1 T was isolated from soil samples in Maryland (USA). This strain has high insecticidal activity against different species of insects, viz., the Western corn rootworm, *Diabrotica virgifer*, the Diamondback moth, *Plutella xylostella* L., the Southern corn rootworm, and Diabrotica undecimpunctata (Martin et al. 2007). This strain has broad-spectrum activity by multiple pathogenic actions, possibly due to the production of different chemical compounds. Among the bacterial metabolites, *C. subtsugae* specifically synthesizes the violacein, a derivative of tryptophan, which gives its colonies of the characteristic purple color. Besides, various compounds produced by this species have been characterized and found its insecticidal effects (Asolkar et al. 2019). It is reported that the biologically active compound is related to the stationary growth phase and the thermal stability of the insecticidal toxin has also been confirmed (Koivunen et al. 2009).

Burkholderia Spp.

Different insect species carry symbiotic bacteria, of the genus *Burkholderia*, most of which are related to specific intestinal areas (Kim et al. 2013; Martinson et al. 2011). In addition to their mutual interactions with insects, it has recently been reported that *Burkholderia* spp. affect the oviposition and fecundity of bean bug, *Riptortus pedestris* (Kil et al. 2014). The ability of *Burkholderia* species against different plant pathogens as a biocontrol agent has also been reported (Burkhead et al. 1994; Janisiewicz and Roitman 1988). The broth culture of the bacterial strain, named A396, showed oral toxicity and contact effects on *Spodoptera exigua* and

Tetranychus urticae. Its insecticidal and acaricidal properties persist even after heat treatment. Therefore, its commercial formulations are mostly based on heat-killed cells against a variety of chewing and sucking insects and mites.

2.6 Advantages

Biocontrol agents are generally acknowledged as low-risk substances compared to traditional chemical pesticides. Various benefits are associated with the use of entomopathogenic bacteria as biocontrol agents. Like other natural enemies, insect pathogens, such as bacteria and viruses, can achieve considerable control over target populations. For example, their method of action is generally more complex than that of conventional chemicals. They target a wide variety of active sites, thereby reducing the chances of developing resistant pests. Although entomopathogenic bacteria are efficient enough for pest management in organic farming; nevertheless their rotational use and combination of chemicals are strongly promoted to achieve full efficacy and ecostability. Several studies have reported the mutual compatibility and synergistic effects of entomopathogenic bacteria and chemical substances (Morris 1972; Seleena et al. 1999; Musser et al. 2009).

The use of entomopathogenic bacteria in pest management programs has many more advantages. Organic pesticides are nontoxic and nonpathogenic during their handling, dilution, mixing, and application. This property makes them safe for the environment as well as for agricultural workers. The aggregates of these entomopathogenic bacteria and viruses spoil easily, so it leaves fewer residues on crops. Therefore, these can be used even when the crops are almost ready for the harvest. However, they are easy to incorporate into organic farming agreements. Nevertheless, the effectiveness of biopesticides is usually achieved through its correct application in the field. In the long term, organic pesticides are more effective than chemical pesticides. Therefore, it proves to be quite helpful in reducing the total load of chemical pesticides on food, feed, or fiber crops.

2.7 Disadvantages

Although these biopesticides have different benefits from chemical pesticides, their mass production is still a daunting challenge because they require a special substrate and even live host insects to grow. The special formulation and storage procedures increase its production cost and time. Thus, it makes these biopesticides more expensive and less readily available than conventional insecticides. Due to which farmers with large cropping areas may find it difficult to use biopesticides continuously.

One of the advantages of biological pesticides is its high specificity. However, this biggest advantage is also its biggest disadvantage because if the crops are attacked by nontargeted pests, they will have immunity. This means that multiple types of biological pesticides may be required to control all pests at one time.

Therefore, the potential market for these products becomes restricted. Microbial pesticides are dependent on environmental biotic and abiotic factors. It has a limited lifespan. The effectiveness of many microbial pesticides is reduced by exposure to heat, solution, or ultraviolet radiation. Thus, their efficacy shows variability against the population of target pests of different areas. Consequently, the correct timing and proper application of biopesticides are particularly important. Additionally, frequent exposure to toxins puts evolutionary pressure on pests to resist that toxin. Due to this, organisms develop and increase their resistance against control treatments.

2.8 Conclusions

Currently, there is a need to feed the growing human population and limit losses in the production of major crops. On the other hand, the land available for farming and cultivation on the Earth is also limited. It, therefore, requires the development of new technologies to support productivity improvement and its continuous progress in pest management systems. To reduce the health risks caused by the increased use of chemical pesticides in agriculture, ecofriendly pest and disease management methods are being developed and evaluated globally. In view of this, entomopathogenic bacteria and viruses have a wide range of biocontrol agents.

The targeted toxicity and their nontoxic effects on nontarget organisms make biological pesticides ideal tools for integrated pest management (IPM) programs. More attention should be paid to its use in combination with chemical pesticides and the integration of biological agents into production systems. At the same time, it is also necessary to encourage public funding schemes, pesticide companies, and commercial investors to establish biological pesticide companies. Thus, biopesticides can work effectively by combining performance along with minimal application and safety benefits to humans and the environment. It may be considered that their role in agriculture and horticulture can be enriched by encouraging their use at the government and nongovernment levels shortly.

2.9 Points to Remember

1. The synthetic insecticides and pesticides used in pest management cause hazardous effects on the environment and human health.
2. Use of broad-spectrum pesticides destroys natural enemies of pests.
3. Biological pesticides provide an ecofriendly measure for pest management.
4. Entomopathogenic bacteria other than *Bt* can also be used in insect pest management.
5. They are a valuable source of insect toxins for the biocontrol of pests.
6. Discovery of new biopesticides is critical in tackling environmental degradation and pest resistance development.

References

Abebe E, Akele FA, Morrison J et al (2011) An insect pathogenic symbiosis between a *Caenorhabditis* and *Serratia*. Virulence 2:158–161

Akman L, Yamashita A, Watanabe H et al (2002) Genome sequence of the endocellular obligate symbiont of tsetse flies, Wigglesworthia glossinidia. Nat Genet 32:402–407

Asolkar R, Huang H, Koivunen M, Marrone P (2019) *Chromobacterium* bioactive compositions and metabolites. US Patent 10,405,555

Barloy F, Delécluse A, Nicolas L, Lecadet MM (1996) Cloning and expression of the first anaerobic toxin gene from *Clostridium bifermentans subsp. Malaysia*., encoding a new mosquitocidal protein with homologies to *Bacillus thuringiensis* delta-endotoxins. J Bacteriol 178:3099–3105

Baumann P, Clark MA, Baumann L, Broadwell AH (1991) *Bacillus sphaericus* as a mosquito pathogen: properties of the organism and its toxins. Microbiol Rev 55:425–436

Bloomquist JR (1996) Ion channels as targets for insecticides. Annu Rev Entomol 41:163–190

Bond JG, Marina CF, Williams T (2004) The naturally derived insecticide spinosad is highly toxic to Aedes and Anopheles mosquito larvae. Med Vet Entomol 18:50–56

Book AJ, Lewin GR, McDonald BR et al (2014) Cellulolytic *Streptomyces* strains associated with herbivorous insects share a phylogenetically linked capacity to degrade lignocellulose. Appl Environ Microbiol 80:4692–4701

Bravo A, Gill SS, Soberón M (2007) Mode of action of *Bacillus thuringiensis* cry and Cyt toxins and their potential for insect control. Toxicon 49:423–435

Burges HD (1981) Safety, safety testing and quality control of microbial pesticides. In: Burges HD (ed) Microbial control of pests and plant diseases 1970–1980. Academic Press, London, pp 739, 949 pp–767

Burkhead KD, Schisler DA, Slininger PJ (1994) Pyrrolnitrin production by biological control agent *Pseudomonas cepacia* B37w in culture and in colonized wounds of potatoes. Appl Environ Microbiol 60:2031–2039

Charles JF, Silva-Filha MH, Nielsen-LeRoux C, Humphreys MJ, Berry C (1997) Binding of the 51- and 42-kDa individual components from the Bacillus sphaericus crystal toxin to mosquito larval midgut membranes from Culex and Anopheles sp. (Diptera: Culicidae). FEMS Microbiol. Lett 156:153–159

Copping LG, Menn JJ (2000) Biopesticides: a review of their action, applications and efficacy. Pest Manag Sci 56:651–676

Davidson EW (1973) Ultrastructure of American foulbrood disease pathogenesis in larvae of the worker honey bee., Apis mellifera. J Invertebr Pathol 21:53–61

Davidson EW, Singer S, Briggs JD (1975) Pathogenesis of *Bacillus sphaericus* strain SSII-1 infections in *Culex pipiens quinquefasciatus* (= C. pipiens fatigans) larvae. J Invertebr Pathol 25(2):179–184

Djukic M, Poehlein A, Thürmer A, Daniel R (2011) Genome sequence of *Brevibacillus laterosporus* LMG 15441, a pathogen of invertebrates. J Bacteriol 193:5535–5536

E.P.A (2011) Exposure Factors Handbook 2011 Edition (Final Report). U.S. Environmental Protection Agency, Washington, DC, EPA/600/R-09/052F

E.P.A (2015) What are biopesticides?

Evans JD, Aronstein K, Chen YP et al (2006) Immune pathways and defence mechanisms in honey bees *Apis mellifera*. Insect Mol Biol 15:645–656

Ferguson CM, Barton DM, Harper LA et al (2012) Survival of *Yersinia entomophaga* MH96 in a pasture ecosystem and effects on pest and non-target invertebrate populations. N Z Plant Prot 65:166–173

Ffrench-Constant RH, Bowen DJ (2000) Novel insecticidal toxins from nematode-symbiotic bacteria. Cell Mol Life Sci 57:828–833

Ffrench-Constant RH, Dowling A, Waterfield NR (2007) Insecticidal toxins from *Photorhabdus* bacteria and their potential use in agriculture. Toxicon 49:436–451

Ffrench-Constant R, Waterfield N, Daborn P et al (2003) *Photorhabdus*: towards a functional genomic analysis of a symbiont and pathogen. FEMS Microbiol Rev 26:433–456

Flint ML, Van den Bosch R (1981) Introduction to integrated pest management. Plenum Press, New York, p 240

Forst S, Dowds B, Boemare N, Stackebrandt E (1997) *Xenorhabdus* and *Photorhabdus* spp.: bugs that kill bugs. Annu Rev Microbiol 51:47–72

Gamliel A, Grinstein A, Peretz Y et al (1997) Reduced dosage of methyl bromide for controlling verticillium wilt of potato in experimental and commercial plots. Plant Dis 81:469–474

Glare TR, O'Callaghan M (2000) *Bacillus thuringiensis*: biology, ecology and safety. John Wiley, USA, Chichester, New York, pp 2–80

Gupta S, Dikshit AK (2010) Biopesticides: an ecofriendly approach for pest control. J Biopest 3: 186–188

Hangay G, Gruner SV, Howard FW et al (2008) Microbial control of insects. In: Encyclopedia of entomology. Springer, The Netherlands, pp 2367–2373

Hinnebusch BJ, Rudolph AE, Cherepanov P et al (2002) Role of *Yersinia* murine toxin in survival of *Yersinia pestis* in the midgut of the flea vector. Science 296:733–735

Hurst MRH, Beard SS, Jackson TA, Jones SM (2007) Isolation and characterization of the *Serratia entomophila* antifeeding prophage. FEMS Microbiol Lett 270:42–48

Hurst MRH, Glare TR, Jackson TA, Ronson CW (2000) Plasmid-located pathogenicity determinants of *Serratia entomophila*, the causal agent of amber disease of grass grub, show similarity to the insecticidal toxins of *Photorhabdus luminescens*. J Bacteriol 182:5127–5138

Hurst MRH, Jones SA, Binglin T et al (2011) The main virulence determinant of *Yersinia entomophaga* MH96 is a broad-host-range toxin complex active against insects. J Bacteriol 193:1966–1980

Ishii K, Adachi T, Hamamoto H, Sekimizu K (2014) *Serratia marcescens* suppresses host cellular immunity via the production of an adhesion-inhibitory factor against immunosurveillance cells. J Biol Chem 289:5876–5888

Jackson TA, Pearson JF, O'Callaghan M, Mahanty HK, Willocks M (1992) Pathogen to product - development of Serratia entomophila (Enterobacteriaceae) as a commercial biological control agent for the New Zealand grass grub (Costelytra zealandica). In: Jackson TA, Glare TR (eds) Use of pathogens in scarab pest management. Intercept, Andover, pp 191–198

Janisiewicz WJ, Roitman J (1988) Biological control of blue mold and gray Mold on apple and pear with *Pseudomonas cepacia*. Phytopathology 78:1697

Jarrett CO, Deak E, Isherwood KE et al (2004) Transmission of *Yersinia pestis* from an infectious biofilm in the flea vector. J Infect Dis 190:783–792

Jones A, Jeffery S, Gardi C, Jones A, Montanarella L, Marmo L, Miko L, Ritz K, Peres G, Römbke J, van der Putten WH (2010) European atlas of soil biodiversity. European Commission: European Soil Data Centre (ESDAC), Luxembourg

Kalha CS, Singh RP, Kang SS, Hunjan MS, Gupta V, Sharma R (2014) Entomopathogenic viruses and bacteria for insect-pest control. In: Abrol DP (ed) Integrated pest management: current concepts and ecological perspectives. Academic, San Diego, pp 225–244

Kil YJ, Seo MJ, Kang DK et al (2014) Effects of Enterobacteria (*Burkholderia sp.*) on development of *Riptortus pedestris*. J Fac Agric Kyush Univ 59:77–84

Kim JK, Kim NH, Jang HA et al (2013) Specific midgut region controlling the symbiont population in an insect-microbe gut symbiotic association. Appl Environ Microbiol 79:7229–7233

Koivunen M, Chanbusarakum L, Fernández L et al (2009) Development of a new microbial insecticide based on *Chromobacterium subtsugae*. IOBC/WPRS Bull 45:183–186

Kwak Y, Khan AR, Shin JH (2015) Genome sequence of *Serratia nematodiphila* DSM 21420T, a symbiotic bacterium from entomopathogenic nematode. J Biotechnol 193:1–2

Lacey LA, Frutos R, Kaya HK, Vail P (2001) Insect pathogens as biological control agents: do they have a future? Biol Control 21:230–248

Lacey, L.A. & Siegel, J.P (2000) Safety and ecotoxicology of entomopathogenic bacteria. In: Entomopathogenic bacteria: from laboratory to field application Kluwer Academic Publishers, Dordrecht, pp 253–273

Landsberg MJ, Jones SA, Rothnagel R et al (2011) 3D structure of the *Yersinia entomophaga* toxin complex and implications for insecticidal activity. Proc Natl Acad Sci U S A 108:20544–20549

Liehl P, Blight M, Vodovar N et al (2006) Prevalence of local immune response against oral infection in a *Drosophila/Pseudomonas* infection model. PLoS Pathog 2:0551–0561

Marshall SDG, Hares MC, Jones SA et al (2012) Histopathological effects of the *Yen-Tc* toxin complex from *Yersinia entomophaga* MH96 (Enterobacteriaceae) on the *Costelytra zealandica* (Coleoptera: *Scarabaeidae*) larval midgut. Appl Environ Microbiol 78:4835–4847

Martin PAW, Gundersen-Rindal D, Blackburn M, Buyer J (2007) *Chromobacterium subtsugae sp. nov.*, a betaproteobacterium toxic to Colorado potato beetle and other insect pests. Int J Syst Evol Microbiol 57:993–999

Martinson VG, Danforth BN, Minckley RL et al (2011) A simple and distinctive microbiota associated with honey bees and bumble bees. Mol Ecol 20:619–628

Mazid S, Kalita JC, Rajkhowa RC (2011) A review on the use of biopesticides in insect pest management. Int J Sci Adv Technol 1:169–178

Mertz FR, Yao RC (1990) *Saccharopolyspora spinosa sp. nov.* isolated from soil collected in a sugar mill rum still. Int J Syst Bacteriol 40:34–39

Miyoshi S, Shinoda S (2000) Microbial metalloproteases and pathogenesis. Microbes Infect 2:91–98

Morris ON (1972) Susceptibility of some forest insects to mixtures of commercial *Bacillus thuringiensis* and chemical insecticides, and sensitivity of the pathogen to the insecticides. Can Entomol 104:1419–1425

Munson MA, Baumann P, Kinsey MG (1991) Buchnera gen. nov. and *Buchnera aphidicola sp. nov.*, a taxon consisting of the mycetocyte associated, primary endosymbionts of aphids. Int J Syst Bacteriol 41:566–568

Musser FR, Nyrop JP, Shelton AM (2009) Integrating biological and chemical controls in decision making: European corn borer (Lepidoptera: *Crambidae*) control in sweet corn as an example. J Econ Entomol 99:1538–1549

Nicholson GM (2007) Fighting the global pest problem: preface to the special Toxicon issue on insecticidal toxins and their potential for insect pest control. Toxicon 49:413–422

Nicolas L, Hamon S, Frachon E et al (1990) Partial inactivation of the mosquitocidal activity of *Clostridium bifermentans serovar Malaysia* by extracellular proteinases. Appl Microbiol Biotechnol 34:36–41

Parsek MR, Singh PK (2003) Bacterial biofilms: an emerging link to disease pathogenesis. Annu Rev Microbiol 57:677–701

Pathak DV, Yadav R, Kumar M (2017) Microbial pesticides: development, prospects and popularization in India. In: Plant-microbe interactions in agro-ecological perspectives. Springer, Singapore, pp 455–471

Perry T, McKenzie JA, Batterham P (2007) A D α 6 knockout strain of *Drosophila melanogaster* confers a high level of resistance to spinosad. Insect Biochem Mol Biol 37:184–188

Pigott CR, Ellar DJ (2007) Role of receptors in *Bacillus thuringiensis* crystal toxin activity. Microbiol Mol Biol Rev 71:255–281

Rathod NP, Vala GS, Dudhat AS, Kachhadiya NM (2014) Field efficacy of bio-pesticides alone and in combination with newer insecticides against *Helicoverpa armigera* of pigeonpea. Int J Plant Prot 7:128–131

Ruiu L (2013) *Brevibacillus laterosporus*, a pathogen of invertebrates and a broad-spectrum antimicrobial species. Insects 4:476–492

Ruiu L (2015) Insect pathogenic bacteria in integrated pest management. Insects 6:352–367

Seleena P, Lee HL, Chiang YF (1999) Compatibility of *Bacillus thuringiensis* serovar israelensis and chemical insecticides for the control of Aedes mosquitoes. J Vector Ecol 24:216–223

Sharma V, Singh PK, Midha S et al (2012) Genome sequence of *Brevibacillus laterosporus* strain *GI-9*. J Bacteriol 194:1279

Shigenobu S, Watanabe H, Hattori M et al (2000) Genome sequence of the endocellular bacterial symbiont of aphids *Buchnera sp.* APS. Nature 407:81–86

Sims GK (1990) Biological degradation of soil. In: Lal R, Stewart BA (eds) Advances in soil science: soil degradation. Springer, Berlin, pp 289–330

Sparks TC, Crouse GD, Durst G (2001) Natural products as insecticides: the biology., biochemistry and quantitative structure-activity relationships of spinosyns and spinosoids. Pest Manag Sci 57: 896–905

Sudakin DL (2003) Biopesticides. Toxicol Rev 22:83–90

Tailliez P, Pagès S, Ginibre N, Boemare N (2006) New insight into diversity in the genus *Xenorhabdus*, including the description of ten novel species. Int J Syst Evol Microbiol 56: 2805–2818

Tanzini M, Alves S, Setten A, Augusto N (2001) Compatibilidad de agent estensoactivos com *Beauveria bassiana* y *Metarhizium anisopliae*. Manejo Integrado de Plagas 59:15–18

Toh H, Toh H, Weiss BL et al (2006) Massive genome erosion and functional adaptations provide insights into the symbiotic lifestyle of *Sodalis glossinidius* in the tsetse host. Genome Res 16: 149–156

Torres-Barragan A, Suazo A, Buhler WG, Cardoza YJ (2011) Studies on the entomopathogenicity and bacterial associates of the nematode *Oscheius carolinensis*. Biol Control 59:123–129

Turner MJ, Schaeffer JM (1989) Mode of action of Ivermectin. In: Ivermectin and Abamectin. Springer, New York, pp 73–88

U.S.E.P.A (2010) Pesticide registration manual: chapter 3 - additional considerations for biopesticide, products. http://www.epa.gov/pesticides/bluebook/chapter3.html#pheromone. Accessed 18 Aug 2020

Vallet-Gely I, Lemaitre B, Boccard F (2008) Bacterial strategies to overcome insect defences. Nat Rev Microbiol 6:302–313

Vega FE, Meyling NV, Luangsa-ard JJ, Blackwell M (2012) Fungal entomopathogens. In: Vega FE, Kaya HK (eds) Insect pathology, 2nd edn. Academic Press, San Diego, California, pp 171–220

Vodovar N, Vallenet D, Cruveiller S et al (2006) Complete genome sequence of the entomopathogenic and metabolically versatile soil bacterium *Pseudomonas entomophila*. Nat Biotechnol 24:673–679

Vodovar N, Vinals M, Liehl P et al (2005) Drosophila host defense after oral infection by an entomopathogenic Pseudomonas species. Proc Natl Acad Sci U S A 102:11414–11419

Waldrop MP, Balser TC, Firestone MK (2000) Linking microbial community composition to function in a tropical soil. Soil Biol Biochem 32:1837–1846

Warren G (1997) Vegetative insecticidal proteins: novel proteins for control of corn pests. In: Advances in insect control: the role of transgenic plants. Taylor & Francis, London, pp 109–121

Waterfield NR, Bowen DJ, Fetherston JD et al (2001) The *tc* genes of *Photorhabdus*: a growing family. Trends Microbiol 9:185–191

Watson GB (2001) Actions of insecticidal spinosyns on γ-aminobutyric acid responses from small-diameter cockroach neurons. Pestic Biochem Physiol 71:20–28

Zhang J, Hodgman TC, Krieger L et al (1997) Cloning and analysis of the first *cry* gene from *Bacillus popilliae*. J Bacteriol 179:4336–4341

Zhang CX, Yang SY, Xu MX et al (2009) *Serratia nematodiphila sp. nov.*, associated symbiotically with the entomopathogenic nematode *Heterorhabditidoides chongmingensis* (Rhabditida: *Rhabditidae*). Int J Syst Evol Microbiol 59:1603–1608

Bacillus thuringiensis 3

A. L. Reyaz, N. Balakrishnan, V. Balasubramani, and S. Mohankumar

Contents

A. L. Reyaz · V. Balasubramani · S. Mohankumar
Department of Plant Biotechnology, CPMB&B, Tamil Nadu Agricultural University, Coimbatore, Tamil Nadu, India

N. Balakrishnan (✉)
Department of Agricultural Entomology, Agricultural College and Research Institute, Tamil Nadu Agricultural University, Killikulam, Tamil Nadu, India

Omkar (ed.), *Microbial Approaches for Insect Pest Management*,
https://doi.org/10.1007/978-981-16-3595-3_3

Abstract

Bacillus thuringiensis (*Bt*) is an omnipresent, spore-forming soil bacterium that is distinguished by the production of proteins, which are active on a wide range of insect pests. The potential of *Bt* to be utilized as an insecticidal spray was perceived in the mid-twentieth century and since that time numerous *Bt*-based biopesticides have been marketed. The approach of molecular biology made it possible to make plants to express the genes coding for proteins, which protect the plants from the pests. The first *Bt* crop was commercialized in 1996, and today all the biotech crops including *Bt* crops are planted in an area of 190.4 million hectares in 29 countries. Development of *Bt* crops is a lengthy and costly process that may take about 13 years with an expense of $136 million. The procedure of getting endorsements by government administrative organizations is among the most basic in the later phases of the advancement procedure and takes about 25% of all the expenses in bringing a *Bt* crop to the market. In spite of the cutting edges over the chemical insecticides, insects are able to develop resistance against *Bt* crops. Insect resistance management is another challenge that has to be addressed for sustaining the benefits of *Bt* crops. The genome editing tool, CRISPR-Cas9 and Next Generation Sequencing (NGS) technologies are much useful for understanding resistance mechanisms and also to improve the properties of *Bt* through genetic manipulation and aid in their biocontrol applications.

Keywords

Bacillus thuringiensis (*Bt*) · Biopesticide · Mode of action · *Bt* transgenic crops · Mechanism of resistance · Management · Genome editing · CRISPR-Cas9

Learning Objectives

This chapter is focused to learn the following objectives:

1. Bio-pesticides and types.
2. Brief history of *Bacillus thuringiensis*: discovery to commercialization.

(continued)

3. *Bt* based products: registered and commercialized.
4. *Bt* formulations in India: types, manufacturers and hurdles.
5. Genetic constituent, structure and mode of action.
6. *Bt* based GE crops: transformation to commercialization.
7. Commercialized transgenic *Bt* cotton traits across the globe.
8. Insect resistance development and its mechanism against *Bt* crops.
9. Resistance management.
10. Use of genome editing tool for studying insect resistance.
11. Safety of *Bt* crops.

3.1 Introduction

Bacillus thuringiensis Berliner (*Bt*) is an ubiquitous, aerobic, Gram-positive, rod shaped, spore-forming bacterium. A large number of *Bt* isolates have been found and grouped into subspecies, such as *Bt* subsp. *thuringiensis*, *Bt* subsp. *kurstaki*, *Bt* subsp. *aizawai*, *Bt* subsp. *israelensis and Bt* subsp. *tenebrionis*. *Bt* differs from other bacilli by producing crystal proteins, which are toxic to various insect pests and cancer cells, but not harmful to other vertebrates and human beings. This characteristic feature led to the development of *Bt* as biopesticide. Insecticidal crystal proteins (Cry) and vegetative insecticidal proteins (Vip) are produced during the sporulation and vegetative stages of the growth, respectively, besides the additional toxins, such as cytolytic (Cyt) and secretory insecticidal proteins (Sip). The proteins are classified based on the homology of their amino acid sequence. Since the first *cry* gene (*cry1Aa*) was cloned from *Bt* subsp. *kurstaki* (HD-1) in 1981, the search for new *cry* genes has been going on continuously worldwide as an effort to find out the *cry* gene content/novel gene(s) in the *Bt* isolates and till date, 78 *cry* gene, 4 *vip* gene, and 3 *cyt* gene families have been reported. Various PCR-based methods are used for detection of the genes in *Bt* isolates.

The parasporal crystalline inclusions show wide spectrum toxicity against lepidopteran, coleopteran, dipteran, hemipteran, hymenopteran insects, and also to some nematodes. Toxin(s) of *Bt* acts as stomach poison. When the larval stage of the insect pests feed upon the crystal proteins, midgut protease enzymes solubilize the protoxin and convert it into an activated toxin under alkaline pH condition. This activated toxin recognizes specific receptor present on the surface of midgut epithelial membrane and binds to it, which leads to membrane insertion, pore formation, ionic imbalance (osmolysis), cell lysis, paralysis of intestine, cessation of feeding, and finally death of the insect. Based on the insecticidal nature of the crystal (Cry) proteins, several commercial formulations of microbial *Bt* insecticides were developed for the management of the insect pests. Though the Cry proteins and Vip proteins cause the mortality in similar fashion, they are structurally dissimilar and have different binding sites.

Currently, biopesticides share about 5% of the total crop protection market globally, with a value of about $3 billion. However, it is projected to grow at a CAGR of 14.7% from an estimated value of USD 4.3 billion in 2020 to reach USD 8.5 billion by 2025. Almost 90% of the microbial biopesticides currently available in the market are derived from *Bt* only. The major disadvantage with *Bt* biopesticide is its rapid inactivation by environmental factors. Scientists attempted to overcome this problem by late 1980 by developing insect resistant transgenic plants expressing *Bt* Cry/Vip protein(s) by introducing codon optimized *Bt cry/vip* genes into the plant system through genetic engineering techniques. Genetically modified (GM) crops/ Biotech crops were approved for cultivation in 1996 at global level, and biotech crop production is in an increasing trend every year. However, there have been reports of field evolved resistance in insect pests against *Bt* Cry proteins in transgenic *Bt* crops. Hence, there is a need for designing strategies to delay the development of resistance in target insect populations. In this chapter, different aspects of *Bt* right from its discovery, biopesticide to development of transgenic crops, insect resistance, and its management are discussed.

3.2 Biopesticides

As per Environmental Protection Agency—EPA (www.epa.gov; accessed on 26.8.2020) biopesticides are certain types of pesticides derived from natural materials, such as animals, plants, bacteria, and certain minerals. These fall into three classes majorly.

1. Biochemical pesticides: These are naturally occurring substances that control pests by nontoxic mechanisms, e.g., insect sex pheromones.
2. Microbial pesticides: These contain microorganisms (bacterium, fungus, protozoan, or virus) as an active ingredient. The most widely used microbial pesticides are strains and subspecies of bacterium, *B. thuringiensis*.
3. Plant incorporated protectants (PIPs): These are pesticidal substances that plants produce from genetic material that has been added to the plant.

Biopesticides have long been endorsed as potential options to synthetic pesticides. However, they have not yet achieved the desired level of usage. But they could shift the supremacy of chemical pesticides. At present, biopesticides involve a small share of the total crop protection market worldwide, with a value of about $3 billion globally, accounting for just 5% of the total crop protection market (Damalas and Koutroubas 2018). The status of biopesticides has been reviewed in many countries across the globe, such as India (Kumar et al. 2019), Iran (Karimi et al. 2019), China (Huang et al. 2007), South Africa (Hatting et al. 2019), Canada (Brownbridge and Buitenhuis 2019), Brazil (Mascarin et al. 2019), United States (Arthurs and Dara 2019), and New Zealand (Glare and O'Callaghan 2019). Approximately 90% of the microbial biopesticides presently in the market are from *B. thuringiensis* (Kumar and Singh 2015).

3.2.1 *Bt*-Based Biopesticides

Bt has been reported worldwide from diverse habitats, such as soil, water, dust from grain storage, dead insects, spider web, leaves from deciduous trees, and diverse conifers (Martin and Travers 1989; Ramalakshmi and Udayasuriyan 2010; Shishir et al. 2014; Unalmis et al. 2015; Reyaz et al. 2017). It produces crystalline inclusions, which consist of one or more insecticidal proteins known as δ-endotoxins (Schnepf et al. 1998; Reyaz and Arulselvi 2016). Different *Bt* strains contain different protein crystals with different shapes, such as bipyramidal, spherical, cubiodal (Ramalakshmi and Udayasuriyan 2010; Reyaz et al. 2017; Ganesh et al. 2018). For the benefit of readers, we have provided the microphotograph (scanning electron microscope) showing different protein crystals with different shapes (Fig. 3.1). This phenotypic characteristic feature of *Bt* is utilized to isolate it from different *Bacillus* species (Vilas-Bôas et al. 2007). When insecticidal crystal proteins are ingested by a susceptible insect, these are solubilized by high alkaline pH of the insect midgut and proteolytically activated by midgut proteases. After that, these activated toxins bind to specific receptors located in the insect cell membrane leading to the destruction of the epithelial cells lining of the insect gut. It is generally believed that these toxins act by creating pores in the cell membrane (Bravo et al. 2007). Even though the bacterium itself contributes in the fatality of the insect, the δ-endotoxins are competent enough to kill some species on their own if produced at sufficient high doses (Raymond et al. 2010). According to the current nomenclature (Crickmore et al. 2020) the insecticidal proteins produced by *B. thuringiensis* should be preferably called as pesticidal proteins instead of Cry toxins, *Bt* toxins, etc.

Extracellular proteases produced by *B. thuringiensis* act on the pesticidal proteins and reduce their yield. The deletion of extracellular protease genes in *B. thuringiensis* can lead to enhanced yield of pesticidal proteins. Tan and Donovan et al. (2000) showed that deletion of the extracellular proteases genes: neutral protease A (*nprA*) gene and alkaline protease A (*aprA*) gene, increases the yield of the Cry1Bb protein in *B. thuringiensis*. Recently, Soonsanga et al. (2020) developed

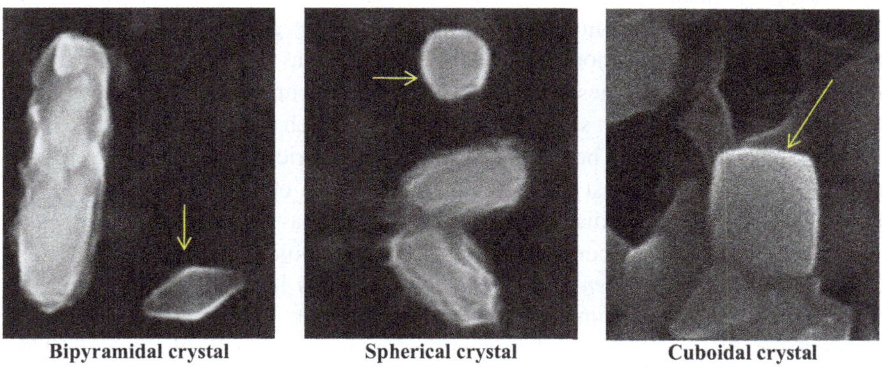

Fig. 3.1 Different shapes of *Bt* crystal proteins

a CRISPR-Cas9 [Clustered Regularly Interspaced Short Palindromic Repeats–CRISPR-Associated protein 9 (Cas9)] system that provided efficient *B. thuringiensis* gene editing in both chromosomal genes and plasmid gene as compared to CRISPR-Cas9 from *Streptococcus pyogenes*, which was previously reported capable of facilitating gene editing in *E. coli* (Jiang et al. 2013).

Thus, there are possibilities that the genome editing tool could be used to enhance the efficacy of the *Bt* based biopesticides. Wide toxicity studies of *Bt* isolates have shown these are devoid of beta-exotoxin and are not toxic or pathogenic to mammals (McClintock et al. 1995). Moreover, *Bt* strains producing beta-exotoxin cannot be registered for their use in plant protection because of their toxicity to mammals (Sebesta et al. 1981). Thus, CRISPR-Cas9 genome editing tool could also be useful to eliminate mammalian toxic protein(s), if any, from a potential *Bt* isolate candidate, which could facilitate their way towards commercialization.

3.2.2 Brief History and Development of *B. thuringiensis*

The history of *B. thuringiensis* discovery is more than a century old. It was first described in Japanese writing by Ishiwata (1901) during investigation of bacterial ailment of silkworms. Later, in 1905 he found the potential pathogenic characters of the *Bt* and named it as sottoin-Bacillus (sudden death-bacillus) (Ishiwata 1905). Afterwards, Berliner (1915) reported an analogous *Bacillus* bacterium that killed flour moths and named it *B. thuringiensis* for the Thuringia locale in Germany, where the bacterial disease was found. This followed investigations into the utility of *Bt* as an insecticide spray (Mattes 1927) and field trials against *Lymantria dispar* (L.) and *Ostrinia nubilalis* (Hubner) in the United States (Metalnikov and Chorine 1929; Husz 1930). Vouk (1930) found that plants sprayed with *Bt* remained protected from European corn borer than unsprayed plants. Later, in 1938, Libec, a France laboratory produced the first *Bt* based bioformulation and named it as Sporeine (Entwistle et al. 1993).

During 1950s, large-scale culture of *Bt* by fermentation technique at low cost, proved by Steinhaus in the United States, enabled to utilize it as a commercial insecticide. Primarily, formulations of the spores were generated by fermentation (Andrews et al. 1987). Although *Bt* product Sporeine was used from 1938 in France, it was not registered as a pesticide in the United States until 1961. Dulmage (1970) discovered a more active *Bt* subsp. *kurstaki* (HD1), which was commercialized in the United States as Dipel. Then other *Bt* products Thuricide and Biobit were also developed and used against lepidopteran pests (Baum et al. 1999). In late 1970s, another new *Bt* strain was discovered from Israel and was termed as *B. thuringiensis* subsp. *israelensis*, which controls mosquito species (Goldberg and Margalit 1977). Subsequently, *B. thuringiensis* subsp. *tenebrionis* was identified in Germany and found to be effective against coleopteran insect pests (Krieg et al. 1983). The demand for *Bt* based insecticides in agricultural sector declined, in mid-1970s, because of more effective chemical pesticides. In the 1980s, *Bt* research got momentum due to progress in biotechnology. First, Schnepf and Whiteley (1981) cloned a crystal toxin gene from *Bt* subsp. *kurstaki* into *Escherichia coli*, since then much research has been performed to improve target spectra and to find out more infectious strains of *Bt* (Osman et al. 2015).

Bt was produced in large-scale level by bioreactors and marketed in low price (Andrews et al. 1987). It was predicted that market share growth of biopesticide will outpace that of chemical ones, with an annual growth rate of 15% (Marrone 2014). These were far and wide available in North America and embody 55% of the bioinsecticide market. However, they were less popular in European Union (EU), and represent only 8% of this market. In the EU, low level usage of *Bt* products is primarily owing to the greater complexity of EU-based biopesticide regulations (Balog et al. 2017). Strains of *B. thuringiensis* for Lepidoptera remain the most popular products (Arthurs and Dara 2019).

The first transgenic tobacco and tomato plants were generated using *Bt* crystal (Cry) proteins. Initially, these *Bt* expressing transgenic plants were commercialized in the United States (Frutos et al. 1999). In 1996, transgenic cotton and maize plants expressing *Bt* crystal proteins were successfully introduced in the United States. In India, *Bt* cotton was commercialized in 2002 and the area under genetically modified crops is in increasing trend, world over. Owing to the specific characteristics of *Bt* proteins like safety to nontarget organisms and environmental friendly nature, usage of chemical insecticides has been reduced in insect resistant transgenic crops (O'Callaghan et al. 2004).

3.2.3 Serotyping and Servovars

In the search of novel *B. thuringiensis* strains harbouring new pesticidal proteins and biological activities several screening programs have been established across the globe. As a result, by 1996, 50,000 *B. thuringiensis* strains were deposited in different collections worldwide (Sanchis et al. 1996). The immunological reaction to the bacterial flagellar antigen, H serotyping, has been established as the method for the classification of the wide diversity of *B. thuringiensis* strains (De Barjac 1981) and is still the most widely accepted subspecific classification technique for *B. thuringiensis*. At present, there are 71 H serotypes (Table 3.1) and 83 serovars of *B. thuringiensis*. However, there are some limitations of this classification, such as it is unable to process the self-agglutinated and immobile (non-flagellar) strains. Strains from the same serovar may not necessarily be sharing the biochemical, toxicological, or genetic characteristics, e.g., serovar *morrisoni* (H8a,8a), which includes some strains with mosquitocidal activity (Padua et al. 1984), others active against coleopteran larvae (Höfte et al. 1987), and some others are active against lepidopteran larvae (de Barjac and Frachon 1990). The different subspecies of *B. thuringiensis* and their host spectrum are given in Table 3.2.

Table 3.1 *Bacillus thuringiensis* serovars based upon H serotyping

H serotype	Serovars	H serotype	Serovars	H serotype	Serovars
H1	*thuringiensis*	H19	*tochigiensis*	H45	*roskildiensis*
H2	*finitimus*	H20a,20b	*yunnanensis*	H46	*chanpaisis*
H3a,3c	*alesti*	H20a,20c	*pondicheriensis*	H47	*wratislaviensis*
H3a,3b,3c	*kurstaki*	H21	*colmeri*	H48	*balearica*
H3a, 3d	*sumiyoshiensis*	H22	*shandongiensis*	H49	*muju*
H3a,3d,3e	*fukuokaensis*	H23	*japonensis*	H50	*navarrensis*
H4a, 4b	*sotto*	H24a,24b	*neoleonensis*	H51	*xiaguangiensis*
H4a,4c	*kenyae*	H24a,24c	*novosibirsk*	H52	*kim*
H5a,5b	*galleriae*	H25	*coreanensis*	H53	*asturiensis*
H5a,5c	*canadensis*	H26	*silo*	H54	*poloniensis*
H6	*entomocidus*	H27	*mexicanensis*	H55	*palmanyolensis*
H7	*aizawai*	H28a,28b	*monterrey*	H56	*rongseni*
H8a,8b	*morrisoni*	H28a,28c	*jagathesan*	H57	*pirenaica*
H8a,8c	*ostriniae*	H30	*medellin*	H58	*argentinensis*
H8b,8d	*nigeriensis*	H 31	*toguchini*	H59	*iberica*
H9	*tolworthi*	H32	*cameroun*	H60	*pingluonsis*
H10a, 10b	*darmstadiensis*	H33	*leesis*	H61	*sylvestriensis*
H10a,10c	*londrina*	H34	*konkukian*	H62	*zhaodongensis*
H11a, 11b	*toumanoffi*	H35	*seoulensis*	H63	*bolivia*
H11a, 11c	*kyushuensis*	H36	*malaysiensis*	H64	*azorensis*
H 12	*thompsoni*	H37	*andaluciensis*	H65	*pulsiensis*
H13	*pakistani*	H38	*oswaldocruzi*	H66	*graciosensis*
H14	*israelensis*	H39	*brasiliensis*	H67	*vazensis*
H15	*dakota*	H40	*huazhongensis*	H68	*thailandensis*
H16	*indiana*	H41	*sooncheon*	H69	*pahangi*
H17	*tohokuensis*	H42	*jinghongiensis*	H70	*sinensis*
H18a,18b	*kumamotoensis*	H43	*guiyangiensis*	H71	*jordanica*
H18a,18c	*yosoo*	H44			

Source: International Entomopathogenic *Bacillus* Center (IEBC), Pasteur Institute, France; website accessed on 4.10.20
http://www.wfcc.info/ccinfo/index.php/strain/display/590/bacteria/90

Table 3.2 *B. thuringiensis* subspecies and their host spectrum

Bacillus thuringiensis subsp.	Susceptible insects
kurstaki, kurstaki HD2, *thunngiensis, aizawai, darmstadiensis, entomocidus*	Lepidopterans
israelensis, kyushuensis	Dipterans
kurstaki HD1, *aizawai* IC1, HD249	Dipterans and lepidopterans
tenebrionis, san diego	Coleopterans

3.2.4 Commercially Available *Bt* Bioformulations

The majority of *Bt* products (bioformulations) in market possess viable spores, pesticidal crystal proteins, proteases, chitinases, phospholipases, vegetative insecticidal proteins, and various unidentified virulent factors besides inerts/adjuvants (Priest 1992). In order to bring a *Bt* based formulation into global market, a number of core areas are required to be attended, which include their activity spectrum, persistence and recycling, and enhancement of formulations by means of using conventional and simple adjuvants/additives, which are cost effective (Priest 1992). Previously, lactose-acetone technique was used as a method to recover *Bt* spores. However, now advanced techniques, like ultracentrifugation, microfiltration and vacuum filtration, to separate active ingredients (insoluble solids) from soluble liquid (inert) fraction are employed, which resulted in efficient recovery of the active ingredient (Brar et al. 2006). A number of factors have to be considered while choosing an adjuvant or additive in the preparation of bioformulation, such as type of formulation, *viz.*, solid, liquid, or encapsulated, type of insect species, developmental stage of the insect like neonate, instars, or adults, way of application, *viz.*, aerial/terrestrial, foliar boom and nozzle spray hydraulic, timing of spray, volume application rate and spray droplet size spectrum, hurdles of penetration, such as waxy, hairy, or thick leaves or sediments, place of application, *viz.*, aquatic or terrestrial or in sensitive areas, water chemistry like hard or soft, low or high pH (Brar et al. 2006).

Some researchers reported the connection between the particle sizes of *Bt* powder and their control efficacy; *e.g.*, Kim and Je (2012) investigated the relationship between the particle size of spray-dried *B. thuringiensis* NT0423 technical powder and its insecticidal activity against diamondback moth. Their results suggested that *B. thuringiensis* NT0423 technical powder with smaller particles is better in controlling diamondback moths, but excessively small particles (<10μm) possibly reduce the insecticidal activity. In a similar fashion, Vimala Devi and Vineela (2015) also reported the particle size in suspension concentrate formulation of *Bt* effects its efficacy against *Helicoverpa armigera.* They observed the feeding cessation in larvae within 24 h after treatment with *Bt* particles 70μm and lower (a local strain DOR Bt-1) while low larval feeding with *Bt* particles of 105μm and greater larval feeding with *Bt* particles of 210μm. Many *Bt* based formulations have been commercialized across the globe in many countries. We have presented some of the products approved for usage in various countries (Table 3.3).

B. thuringiensis has been traditionally multiplied through submerged fermentation. The high cost of raw materials in the production of *Bt* formulations has hindered their commercial application to a great level. Various researchers reported production of *Bt* using cost-effective agro-industrial materials/wastes such as cotton seed meal, corn steep liquor, beans, peanuts, linseed meal, kitchen wastes, and wastewater sludge (Table 3.4).

Table 3.3 Commercially available *Bt*-based bioformulations in the global market

Bt subspecies/strain used	Trade name of the product	Target insects
The United States		
B. thuringiensis subsp. *kurstaki*	BMP 123 (2X)	Lepidopterans
B. thuringiensis subsp. *kurstaki* strain SA	Bonide, Costar, SAN 420I, Thuricide HPC, Javelin WG, Javelin WG2	Lepidopterans
B. thuringiensis subsp. *kurstaki* EG2348	Condor WP, Cutlass	Lepidopterans
B. thuringiensis subsp. *kurstaki* EG7826	Crymax WP, Lepinox WDG	Lepidopterans
B. thuringiensis subsp. *kurstaki* ABTS-351	Biobit HP, Dipel DF	Lepidopterans
B. thuringiensis subsp. *israelensis*	BMP 144 (2X WP)	Mosquitoes and black flies
B. thuringiensis subsp. *israelensis* EG2215	Gnatrol, Aquabac	Mosquitoes and flies
B. thuringiensis subsp. *aizawai*	M-Trak	Colorado potato beetle
B. thuringiensis subsp. *aizawai* GC-91	Agree 50 WP	Lepidopterans
B. thuringiensis subsp. *aizawai* NB200	Florbac 15,000 WG	Moth larvae
B. thuringiensis subsp. *aizawai* ABTS-1857	XenTari (Valent USA), Agree WG	Caterpillars
B. thuringiensis subsp. *tenebrionis*	Novodor	Colorado potato beetle
Europe		
B. thuringiensis subsp. *kurstaki*	BMP 123	Lepidopterans
B. thuringiensis subsp. *kurstaki* ABTS	Batik	Lepidopterans
B. thuringiensis subsp. *kurstaki* strain ABTS-351	Dipel	Lepidopterans
B. thuringiensis subsp. *aizawai* GC-91	Turex	Lepidopterans
B. thuringiensis subsp. *israelensis* AM65	VectoBac	Sciarids
B. thuringiensis subsp. *tenebrionis*	Novodor	Coleopterans
Japan		
B. thuringiensis subsp. *kurstaki*	Toarowaa Esmark Guardjet, Dipol, Tuneup Fivestar BioMax DF	Lepidopterans
B. thuringiensis subsp. *aizawai*	Quark Xen Tari, Florbac Sabrina	Lepidopterans
B. thuringiensis aizawai + *kurstaki*	Bacilex	Lepidopterans
B. thuringiensis japonensis	Bui Hunter	Cockchafers and white grubs

India		
B. thuringiensis subsp. *kurstaki*	Bio-Dart-M, Abtec Btk, Deflin WG, Dipel, Dipole, Agni Gold Btk, Halt, Bioasp, Biolep, Biobit, BT Killer, Caterpilin, Cezar, Taciobio-Btk, VBT, RB Bt, Neelstaki, Minchu, Mahastra, Lipel SP, Krishi BioPrasar, Kavach Bt, Jas BT	Lepidopterans
B. thuringiensis subsp. *israelensis*	Tacibio, Technar, BacticideWP/DT/AS, Bacto Power-Bti, Biodart-M, Larvect 50, Deltafix, MoskitosMlb, RK Biovecta, Spicbio-Bti, Vectobar, Vectobac AS/G	Mosquitoes and blackflies
Australia		
B. thuringiensis subsp. *kurstaki*	Biocrystal, Caterpillar Killer, Dipel, Costar, Delfin, Full-Bac WDG	Lepidopterans
B. thuringiensis subsp. *israelensis*	Aquabac, BTI, Teknar, Vectobac	Mosquito larvae
B. thuringiensis subsp. *aizawai*	Agree, Bacchus, XenTari	Lepidopterans
Africa		
B. thuringiensis subsp. *kurstaki*	Agree	Lepidopterans
B. thuringiensis subsp. *israelensis*	VectoBac	Mosquito
B. thuringiensis subsp. *aizawai* and *kurstaki*	Dipel, Rokur, Thuricide	Lepidopterans
Kenya		
B. thuringiensis	Vectonil 50 WP	Lepidopterans
B. thuringiensis subsp. *kurstaki*	Biokil WP	
B. thuringiensis subsp. *kurstaki* strain SA-11	Delfin 6.4 WG	
B. thuringiensis subsp. *kurstaki* strain ABTS-351	Dipel DF	
B. thuringiensis subsp. *kurstaki* strain ABTS-351	Dipel 2X	
B. thuringiensis subsp. *aizawai*	Xen Tari	
B. thuringiensis subsp. *kurstaki*	Halt 50% WP, Thuricide HP, H7 Florbac WG	
Canada		
B. thuringiensis subsp. *kurstaki*	Bioprotec 3P, Bioprotec CAF, Dipel 2X DF, Thuricide HPC	Lepidopterans
B. thuringiensis subsp. *aizawai*	XenTari WG	

(continued)

Table 3.3 (continued)

Bt subspecies/strain used	Trade name of the product	Target insects
Austria		
B. thuringiensis subsp. *kurstaki*, strain ABTS 351, PB 54, SA 11, SA12 and EG 2348	Foray 48B, Delfin WG, Bactospeine DF, Lepinox Plus	Lepidopterans
B. thuringiensis subsp. *aizawai*, strain ABTS-1857 and GC-91	XenTari, Florbac, Raupenfrei, Buchsbaumzünslerfrei, Zünsler und Raupenfrei	Lepidopterans
B. thuringiensis subsp. *israelensis* (Serotype H-14), strain AM65-52	Gnatrol SC	Fungus gnats
Croatia		
B. thuringiensis subsp. *kurstaki*, SOJ: PB54, Zbirka kulture: br. CECT 7209	Baturad WP	*Lymantria dispar, Hyphantria cunea*
B. thuringiensis subsp. *kurstaki*, SOJ:ABTS 351, Zbirka kulture br. ATCC SD-1275	Biobit WP	*Hyphantria cunea, Pieris brassicae, Lobesia botrana, Eupoecilia ambiguella, Prays oleae*
	Foray 48 B	*Tortrix viridana* *Operophtera brumata* *Lymantria dispar* *Thaumetopoea pityocampa*
B. thuringiensis subsp. *tenebrionis*, SOJ:NB 176 (TM 14 1), Zbirka kulture: br. SD-5428	Novodor 3%	Colorado beetle
Egypt		
B. thuringiensis subsp. *kurstaki*	Protecto 9.4%	Lepidopterans
B. thuringiensis	Dipel 2X 6.4% WP, Agerine 6.5% WP, Agree 50%WG, Bay 8.8% SC, Biotect 9.4% WP, Delfin 85% WG, Dipel DF 6.4% DF, W-Bus 8% WP	
Greece		
B. thuringiensis subsp. *kurstaki* strain EG 2348	Lepinox plus, Cordalene, Rapax	Lepidopterans

B. thuringiensis subsp. *kurstaki* strain ABTS 351	Bactecin DP, Bactecin ER, Bathikur DP, Bactospeine WG, Foray 48B, Dipel 2X	
B. thuringiensis subsp. *kurstaki* strain PB 54	Belthirul 32,000 WP, Lepiback, Bactoil SC, Amcobac, Bacillus Chemia DP	
B. thuringiensis subsp. *kurstaki* strain SA 11	Delfin WG	
B. thuringiensis subsp. *kurstaki* strain SA12	Costar WG	
B. thuringiensis (serotype h-14)	Aquabac XT, Aquabac 200G	
B. thuringiensis subsp. *aizawai* strain ABTS-1857	Xen Tari WG	
B. thuringiensis subsp. *aizawai* strain GC-91	Agree WP	
B. thuringiensis subsp. *israelensis* serotype H-14	Vectobac G	Mosquitoes
Hungary		
B. thuringiensis subsp. *kurstaki*	Dipel, Bactospeine, Foray 76B, Lepinox Plus, Novodor FC	Lepidopterans
Ireland		
B. thuringiensis subsp. *kurstaki*	Lepinox Plus	Lepidopterans
B. thuringiensis subsp. *aizawai*	Agree 50 WG	Lepidopterans
Lithuanian		
B. thuringiensis var. *kurstaki* ABTS- 351	Dipel DF, Foray 76B	Lepidopterans
Maltese islands		
B. thuringiensis subsp. *kurstaki* strain EG 2348-375	Lepinox Plus	Lepidopterans
Slovenia		
B. thuringiensis subsp. *aizawai* strains ABTS-1857 and GC-91	Agree WG	Lepidopterans
B. thuringiensis subsp. *kurstaki* strains ABTS 351, PB 54, SA 11, SA12 and EG 2348	Delfin WG	Lepidopterans
Switzerland		
B. thuringiensis subsp. *kurstaki*	Bio Buxus, Bio Garden, Delfin, Gegen Buchsbaumzunsler, Bio Raupen Stop, BioHop, Dipel DF, Sanoplant Dipel	Lepidopterans
B. thuringiensis subsp. *israelensis*	Bio Garden, Trauermucken-stop, Solbac	Mosquitoes

(continued)

Table 3.3 (continued)

Bt subspecies/strain used	Trade name of the product	Target insects
B. thuringiensis subsp. *tenebrionis*	Novodor 3FC	Colorado potato beetle
B. thuringiensis subsp. *aizawai*	Agree WP, Xen Tari WG	Lepidopterans
Turkey		
B. thuringiensis	Agree 50WG	*Helicoverpa armigera, Tuta absoluta, Lobesia botrana*
B. thuringiensis subsp. *kurstaki*	Biobit WP (16,000 IU/mg)	*L. botrana,Cydia pomonella, Ectomyelois ceratoniae, H. armigera, T. absoluta, Spodoptera littoralis*
	Delfin WG	*T. absoluta, C. pomonella, E. ceratoniae, H. armigera, L. botrana*
B. thuringiensis subsp. *kurstaki* ABTS-351	Dipel DF BT 54%	*Cryptoblabes gnidiella, Archips rosanu, S. littoralis, E. ceratoniae, L. botrana*
B. thuringiensis subsp. *aizawai* strains ABTS-1857	Florbac WG	*Archips rosanus, Archips xylosteanus, Tuta absoluta, L. botrana, S. littoralis*

Table 3.4 Different kinds of cost-effective raw materials used for commercial production of *Bt*

Cost-effective raw material	Reference
Cotton seed meal, fish meal, beef blood, slaughterhouse residues, corn steep liquor, sorter liquor, fodder yeasts, horse beans, kidney beans, lima beans, soybeans, chick peas, lentils and peanuts	Salama et al. (1983)
Spent brewer's yeast and waste cassava starch	Ejiofor (1991)
Wheat bran, rice bran and rice husk	Suyanandana et al. (1996)
Lentil meal, great northern white bean, defatted soy flour, wheat germ shoots, feather meal, blood meal, meat meal, cottonseed meal, sunflower meal, canola meal, linseed meal, and casamino acid	Morris et al. (1997)
Wastewater sludge and Sewage sludge	Lachhab et al. (2001), Tirado-Montiel et al. (2001), Brar et al. (2004), Yezza et al. (2004), Yezza et al. (2005), Vidyarthi et al. (2002), Zhuang et al. (2011)
Fish meal, gruel	Zouari et al. (2002)
Potato, common sugar, and Bengal gram	Poopathi and Kumar (2003)
Wheat bran	Devi et al. (2005)
Soybean flour, groundnut cake powder, and wheat bran extract	Prabakaran and Balaraman (2006)
Barley, groundnut and soyabean	Shojaaddini et al. (2010)
Cotton seed meal, soya meal, sunflower meal	Dhingra and Chaudhary (2011)
Kitchen waste, Food waste, Home compost	Zhang et al. (2013), Zou et al. (2016), Ballardo et al. (2020)
Sucrose, Cane molasses, Soybean flour and Milk whey	Salazar-Magallon et al. (2015)
Mustard meal, sesame meal, linen meal, jojoba meal, sugar beet pulp, cotton seed meal, olive meal, soybean meal, wheat germ meal, fodder yeast, beans meal, corn meal, banana peels, beans peels, pea peels and potato peels	El-Bendary et al. (2016)
Soy fibre residue and Wood sticks	Ballardo et al. (2016)
Wheat bran, rabbit feed, cabbage leaves, potato tubercles, and cactus	Hasanain (2017)
Wheat flour, corn flour, soyabean seed powder, cotton seed powder	Monika and Govind (2017)
Nonsterile organic fraction of municipal solid waste	Ballardo et al. (2017)
Fine sand, sugar beet pulp and sesame meal	El-Bendary et al. (2017)
Gruel, starch, molasses, chickpea, soybean meal, fish meal and potato	Zghal et al. (2018)

3.2.5 Biopesticides in India

At the international level, biopesticides are gaining wide popularity because of their safe nature and target-specific. But their usage, in developing countries, like India, is still in diminutive level in contrast to chemical pesticides. In order to encourage the usage of biopesticides, the Indian government has placed them into several of the agricultural schemes (Mishra et al. 2020). The governing body that regulates biopesticides in India is the Central Insecticides Board and Registration Committee (CIBRC). This regulates the biopesticides under the Insecticides Act of 1968 and Insecticides Rules of 1971. The board counsels both the central and state governments on technical matters associated to the production, sale, delivery, and utilization of insecticides including biopesticides to make sure safety to humans and animals. Once the importer or manufacturer submits the formulation and data on efficacy, toxicity, and packaging, CIBRC confirms it and then awards license to public and private firms for large-scale production, distribution, and sale of biopesticides. New products by the manufacturers can be registered under section 9(3B) (provisional registration for a new active ingredient used in India) or 9 (3) (regular registration) section of the Insecticides Act. Presently Indian pesticide market contains 5% biopesticides, which comprises of at least 15 microbial species and 970 microbial formulations registered by CIBRC. As of 2017, over 30 products based on *B. thuringiensis* (*Bt*) subsp. *kurstaki* and 12 based on *Bt* subsp. *israelensis* were registered (Kumar et al. 2019). Both public and private institutions are involved in identification and development of entomopathogenic organisms, as microbial biopesticides in India.

3.2.6 *Bt* Formulations Against the Insect Pests

According to the Directorate of Plant Protection, Quarantine and Storage (DPPQS) in 2017, there were 361 governmental and private biocontrol laboratories working in India (DPPQS 2017). In public sector, genetic profiling and comparative bioassays of indigenous *Bt* strains against lepidopteran pests have been conducted by institutes, like ICAR-National Bureau of Agricultural Insect Resources (NBAIR) (Ramanujam et al. 2014), ICAR-Indian Agricultural Research Institute (IARI) (Gupta et al. 2020), ICAR-Indian Institute of Horticultural Research (IIHR) (Ramasamy et al. 2020), Tamil Nadu Agricultural University (TNAU) (Ramalakshmi and Udayasuriyan 2010; Reyaz et al. 2019; Kaviyapriya et al. 2019), Periyar University (Reyaz et al. 2017), University of Agricultural Sciences, Dharwad (Nethravathi et al. 2010; Goudar et al. 2012). The ICAR-Indian Institute of Oil Research (IIOR) has developed solid-state fermentation technology and commercialized the production of indigenous *Btk* isolates (DOR*Bt*-1 and DOR*Bt*-5). This technology has been licensed to 37 biopesticide entrepreneurs for distribution and sale, which resulted in substantial resource generation to the institution. Some of the companies that produce *Bt* formulations are listed in Table 3.5. The ICAR-NBAIR has also developed liquid

Table 3.5 List of companies involved in the manufacturing of *Bt* bioinsecticides in India

S. no.	Active ingredient in formulation	Name of the producing unit	Formulation
1.	*B. thuringiensis* subsp. *kurstaki*	Ajay Biotech Ltd., India	7.5% WP
2.		Amit Biotech Pvt. Ltd., Kolkata, India	0.5% WP
3.		M/s Indore Biotech Inputs & Research P. Ltd., India	0.5% WP
4.		M/s Neelagriva Biosciences Pvt. Ltd., Hyderabad, India	0.5% WP
5.		M/s Krishi bio Product & Research Pvt. Ltd., Indore, India	0.5% WP
6.	*B. thuringiensis* subsp. *kurstaki* H 3a, 3b, 3c	M/s Kan Biosys Pvt. Ltd., Pune, India	0.5% WP
7.	*Bacillus thuringiensis* var. *kurstaki*	M/s Biotech International Ltd., New Delhi	Formulation and Technical
8.	*B. thuringiensis* subsp. *israelensis*	Bacto Power India Pvt. Ltd., India	5 AS
9.		M/s Biotech International Ltd., India	Tech (I) and 1.15% WP
10.		M/s. Amit Biotech (P) Ltd., Kolkota, India	5% AS
11.	*B. thuringiensis* subsp. *israelensis*	M/s Kilpest India Ltd., Bhopal, India	5% AS
12.	*B. thuringiensis* subsp. *israelensis*	Ajay Biotech (India) Ltd., Pune, India	5% WP
13.	*Bacillus thuringiensis* subsp. *israelensis* (H-14)	Aventis Crop Sci. Ltd., India	Vectobac 12 AS
14.	*B. thuringiensis* var. *kurstaki;* Serotype-3a, 3b, 3c	Directorate of Oilseeds Research, Hyderabad, India	0.5% WP
15.		Neesa Agritech Pvt. Ltd., Ahmedabad, India	0.5% WP
16.		Gujarat Eco Microbial Technologies, India	0.5% WP
17.		Prathibha Biotech, Hyderabad, India	0.5% WP
18.		Varsha Bioscience & Technology, Hyderabad, India	0.5% WP
19.		Shri Ram Solvent Extractions Pvt. Ltd., Jaspur, India	0.5% WP
20.		Sri Biotech, Hyderabad, India	0.5% WP
21.		Surya Bio Products, Eluru, Andhra Pradesh, India	0.5% WP
22.		Nitapol Industries, Kolkata, India	0.5% WP
23.		Poshak Bio research (P) Ltd., Gujarat, India	0.5% WP
24.		International Panacea Ltd., New Delhi, India	0.5% WP
25.		Agro Biotech Research Centre, Kottayam, India	0.5% WP
26.		Ajay Bio-Tech (India) Ltd., Pune, India	0.5% WP
27.		Hindustan Bioenergy Ltd., Lucknow, India	0.5% WP
28.		Bharat Biocon Pvt. Ltd., New Delhi, India	0.5% WP
29.		Bio-Control Laboratory, Varanasi, India	0.5% WP

fermentation technology for the indigenous *Btk* isolates PDBCBT1 and NBAIIBTG4 (Ramanujam et al. 2014).

There are many reports of such *Btk* formulations for use against bollworms, loopers, and other lepidopteran pests in India. For example, liquid *Btk* formulations have been evaluated against *H. armigera* and *Mauruca vitrata* in pigeon pea in the states of Andhra Pradesh (Kumar et al. 2016), Telangana (Vimala Devi and Vineela 2015), and Punjab (Kumar and Kaur 2017) in India. Kesavan et al. (2003) evaluated different commercial formulations of *B. thuringiensis* (Delfin, Biobit, Dipel and Halt) against sugarcane early shoot borer, *Chilo infuscatellus* in Tamil Nadu, India. They reported the order of efficacy of *Bt* formulations as Delfins>Biobit>Dipel>Halt. They also observed that insecticide Sevidol was less effective compared to Delfin, Biobit, and Dipel. *Btk* has provided excellent control of several citrus pests. Rao et al. (2015) reported up to 100% mortality of citrus leaf miner, *Phyllocnistis citrella*, for up to 10 days after spraying in a study conducted in Andhra Pradesh. Application of Dipel (1 kg/ha) also resulted in >90% reduction of citrus butterfly, *Papilio demoleus* larvae in sweet orange (Gopalakrishnan and Gangavisalakshy 2005). In sorghum, Delfin (1 g/l) reduced the leaf damage caused by the stem borer, *Chilo partellus* by 67% compared to control in Karnataka (Jose et al. 2008). In cabbage, Bioasp @ 1 kg/ha reduced the leaf damage of diamondback moth, *Plutella xylostella,* by more than 80% (Singh et al. 2015).

The future market for *Bt* products may be improved by commercializing additional indigenous strains for endemic and emerging pests. The Indian subcontinent contains a diversity of *Bt* strains harbouring new putative genes (Rangeshwaran et al. 2014). Reyaz et al. (2017) identified 68 *Bt* strains with four crystalline inclusion types from the Kashmir Valley. Subbanna et al. (2019) reported 80 *Bt* strains from a novel ecological niche of Uttarakhand Himalayas, several of which (UKBt3, UKBt11, UKBt13, and UKBt18) showed good insecticidal activity against *H. armigera*, *Pieris brassicae*, *P. xylostella*, and *Spodoptera litura* under laboratory conditions and are good candidates for commercialization.

B. thuringiensis subsp. *israelensis* (*Bti*) was the very first strain described for having insecticidal activity outside Lepidoptera. It was isolated from a water pond in the Negev desert with activity against dipteran larvae (Goldberg and Margalit 1977). Its discovery opened up the possibility for using this as biolarvicide in mosquito control programs. In India, various reports have been published using *Bti* for mosquito control (Batra et al. 2000; Mittal 2003; Poopathi et al. 2003). The effectiveness of such products differs among strains and targets; however, under definite conditions, they are very much effective. For instance, Bactoculicide, an imported powder formulation of *Bti* from Russia when applied at 0.5 gm^{-2} (5 kg/ha), resulted in >90% reduction of *Aedes aegypti* and *A. albopictus* mosquito larvae for 4–5 weeks in small breeding habitats in industrial scraps, and *Culex quinquefasciatus* approximately 2 weeks in drains (Mittal 2003).

3.2.7 Limitations for Growth of Microbial Biopesticides Market in India

The growth of Indian microbial pesticide market is limited by a number of factors. The primary limitations cited are shortage of large-scale production unit facilities, quality control problems, which includes low microbial count that results in poor performance in the field, availability of products that are not registered (Arora et al. 2010; Gupta and Dikshit 2010; Mishra et al. 2015; Mishra et al. 2020). A study conducted by the National Pesticide Manufacturers Association showed that several products marketed in India have no address or identifiable company registration number and no active ingredients listed on the label. In the analysis, it was evident that the issues of fake or unregistered microbial pesticides were mainly widespread in the states of Tamil Nadu, Andhra Pradesh, Karnataka, Maharashtra, Madhya Pradesh, and Gujarat (FICCI 2015). Another study on the quality of Microbial biopesticide-based products carried out in India, showed that 50–70% of these had problems like too much moisture content in solid formulations, smaller number of colony propagules than listed on the label or contaminants, and therefore did not meet the required CIBRC standards (Ramanujam et al. 2014). Shelf life is another constraint for some microbial biopesticides in rural areas, where there is a dearth to the availability of fresh products and refrigerated storage (Mishra et al. 2015). Some formulations take time to kill, lack persistence at field level due to high UV radiations, and are less soluble in water, which are additional challenges in developing commercially viable microbial pesticides (Aneja et al. 2016). Finally, the strict, costly and lengthy registration process in India hampers the development of microbial biopesticides. At present, the time taken between granting patents and registering biopesticide formulations in India exceeds 5 years (Venkatesan and Pattar 2017). This may be one of the causes for aggravating the sale of unregistered products, which may be of underprivileged quality.

3.3 Genetic Constituent of *Bt*

The genome size of *Bt* is about 5 to 6.7 Mbp (Bravo and Soberon 2008; Barbosa et al. 2015; Hollensteiner et al. 2017; Li et al. 2017; Reyaz et al. 2019; Reyaz et al. 2020). Initially, in 1980s, plasmid curing method was used and found that mega plasmids present inside the *Bt,* carrying genes are responsible for insecticidal activity (González Jr et al. 1981). The *Bt* genes producing parasporal crystals were found to be in extra chromosomal plasmid (Beard et al. 2001), which exists in circular and other linear forms (Bravo and Soberon 2008). Reyaz et al. (2019) reported that the genome size of *Bt* strain T414 as 5.5 Mbp that includes *cry1, cry2, vip3,* and *cyt2* genes present in plasmids. Self-transmissible properties of these *cry* genes is responsible for conjugal transfer (Wilcks et al. 1998; Makart et al. 2017). Diversity of *cry* genes in *Bt* is due to the transposable elements (Lereclus et al. 1986).

3.4 Three Domain Structure of *Bt* Cry Proteins

Most of the Cry proteins have different amino acid sequence and they require the conserved three domain structure for their insecticidal activity (de Maagd et al. 2003; Bravo et al. 2007; Pardo-Lopez et al. 2013). Domain I consists of 7 to 8 α-helix structure, essential for proteolytic activation of all three domains and responsible for pore formation and membrane insertion (Schnepf et al. 1998; Ben-Dov 2014; Xu et al. 2014). Domain II is known as central domain and has 3 β-sheets, which is involved in reorganization and binding of receptor, oligomerization and membrane insertion (Xu et al. 2014). Domain III involved in β sandwich of two antiparallel β sheets regulating insect specificity and ion channel interactions (Xu et al. 2014). X-ray crystallographic structure of Cry7Ca1 protein has 2.3 Å global shape with three domains, domain I consists of seven helix cluster involve in pore formation, domain II resembles prism-shaped with three β sheets, and domain III has β sandwich that interacts with domain I and domain II (Jing et al. 2019). The putative receptor binding sites (β5-β6 and β7-β8) were identified in domain II by chimeric scanning and domain swapping mutagenesis in Cry2Aa protein that are exhibiting toxicity against dipterans and lepidopterans (Morse et al. 2001).

3.5 Mode of Action of *Bt* Toxins

Bt infection in insects leads to changes in nutrient absorption, degenerative transformation, loss of appetite, physiological disorders, total paralysis, and finally death. The infected larvae become blackened. *Bt* acts as a stomach poison. It produces crystal protein(s) during sporulation stage. It has to be ingested by the insects to cause mortality in insects. When the insect ingests the crystal protein, which is in protoxin form, is solubilized by alkaline environment in the midgut, followed by the action of midgut proteases to cleave the protoxin to activated toxin, recognition of specific receptors present on the surface of midgut epithelial membrane by the activated toxin fragment and leading to formation of pores (Carroll and Ellar 1993; Kirouac et al. 2002), resulting in ionic imbalance (osmolysis) in cell (Knowles and Ellar 1987), disruption of midgut epithelial membrane cells, paralysis, and finally death of the intoxicated larvae. In classical model, pore formation and membrane insertion were not clear, hence the mechanisms behind the receptor binding and their role against toxicity were further investigated (Gómez et al. 2002; Pardo-López et al. 2006; Likitvivatanavong et al. 2011). New models were proposed to understand the mechanism of pore formation by sequential binding model (Bravo et al. 2002; Gómez et al. 2002; Pacheco et al. 2009) and the signaling pathway model (Zhang et al. 2005; Zhang et al. 2006).

The sequential binding model (Soberón et al. 2007; Bravo and Soberon 2008; Jimenez-Juarez et al. 2008; Bravo et al. 2011; Pardo-Lopez et al. 2013) explains that when the crystal proteins are ingested by the larvae, midgut enzymes activate the monomeric toxin that binds to GPI anchored and cadherin receptors in the midgut epithelial membrane. Further proteolytic action eliminates the α1 helical in domain I,

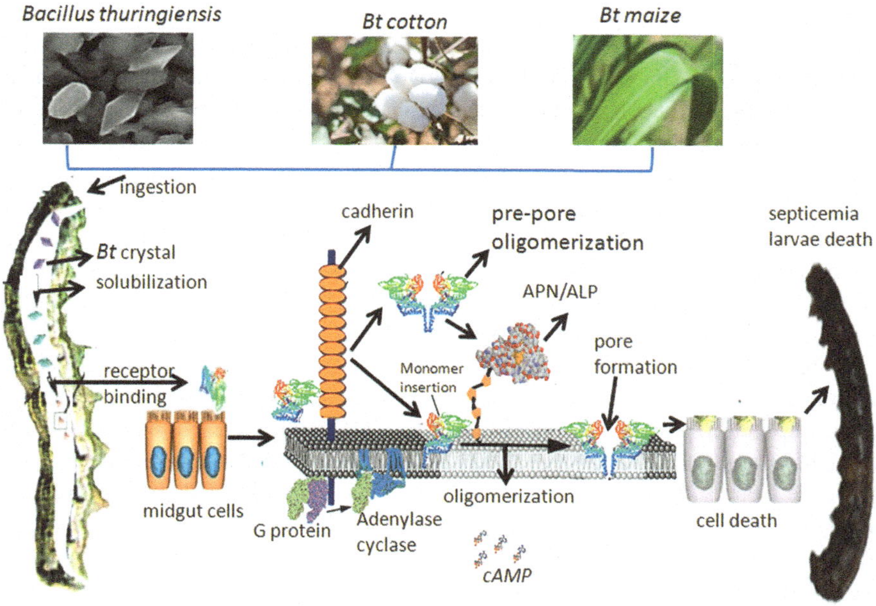

Fig. 3.2 Mode of action of *Bt*. (Modified from Adang et al. 2014)

leading to oligomerization of toxin and formation of pore, which results in the rupturing of midgut epithelial cells and finally leads to the death of larvae (Fig. 3.2).

The pore formation mechanism was demonstrated in *Manduca sexta* with Cry1Ab protein and also studied in Cry1Aa and Cry1Ac proteins, which were found to be involved in the recognition of cadherin-like protein and a glycosyl-phosphatidylinositol (GPI)-anchored aminopeptidase-N receptors present in the midgut epithelial membrane of lepidopteran larvae (Gómez et al. 2007; Pigott and Ellar 2007; Likitvivatanavong et al. 2011). Binding of Cry1Ab protein in the specific cadherin receptor lead to conformational changes and proteolytic cleavage resulted in dimer and tetrameric oligomer formation (pre pore formation) (Gómez et al. 2002). Bravo et al. (2004) demonstrated oligomeric form of toxin showed higher affinity to bind the amino peptidase receptor than monomeric form of toxin. Vachon et al. (2012) discussed on several reports for sequential binding model requires α1 helical removal and formation of pre pore forming structure by oligomeric toxin.

3.6 Classification of *Bt* Proteins

The first gene encoding a *B. thuringiensis* crystal protein was cloned in 1981 (Schnepf and Whiteley 1981). Due to the continuous isolation and screening, numerous genes were cloned subsequently. Höfte and Whiteley (1989) proposed a nomenclature for the classification of these proteins according to which proteins were classified based on their insecticidal activities. Proteins toxic to lepidopteran insects were classified as CryI proteins, proteins toxic to both lepidopterans and dipterans were classified as CryII and proteins active against coleopterans and dipterans were classified as CryIII and CryIV, respectively. Although, this nomenclature systematically classified proteins that had been formerly been given arbitrary names, it soon became evident that there were major limitations of this nomenclature. For instance some proteins that shared sequence homology often had different insecticidal specificities, thus requiring them to be put into different primary classification groups. The need to obtain comprehensive bioassay data before a protein could be classified was another major limitation of this classification.

To overcome these limitations, Crickmore et al. (1998) introduced a revised nomenclature, which classified the proteins solely by amino acid identity. A four-level naming system was adopted in which proteins that shared at least 45% sequence identity were placed in the primary classification group (Cry1, Cry2, Cry3, Cry4, etc.), proteins that shared less than 78% identity were assigned different secondary ranks (Cry1A, Cry1B, Cry1C, etc.), proteins that shared less than 95% sequence identity (Cry1Aa, Cry1Ab, Cry1Ac, etc.) were allocated tertiary rank and finally a fourth level was used for the proteins that shared greater than 95% identity (Cry1Aa1, Cry1Aa2, Cry1Aa3, etc.). The efforts to find other proteins with improved or more diverse activities led to the characterization of proteins from a wider range of bacteria and with a variety of different protein folds.

It became clear that this revised nomenclature had limitations in representing the diverse range of proteins that had been identified. Therefore again a new nomenclature was introduced, which retained the basic principles of the 1998 version. This system of classification provides specific mnemonics to represent different structural groups. For the intention of consistency, the vast majorities of the proteins have either retained their name or have a new name that clearly refers the previous one. Other pesticidal proteins not previously included in the nomenclature have been incorporated into this version of nomenclature (Crickmore et al. 2020). This nomenclature has categorized pesticidal proteins of various microbial origins including *B. thuringiensis* into the 16 classes (Table 3.6).

3.6.1 Primary Rank Proteins

Primary rank proteins share 45% or less than 45% sequence identity with each other and have always been a challenging job to discover as compared to other protein ranks, such as secondary rank proteins, tertiary rank or quaternary rank. These proteins were discovered by both public institutions (e.g., University of Washington, Cambridge University, University of California, Hokkaido University, Huazhong Agricultural University) and private institutions (e.g., Mycogen, Monsanto, Novartis Agribusiness Biotechnology Research North Carolina, Ecogen) across the globe, such as United States, Mexico, China, India, Russia, Japan, Belgium, and France. A list of primary rank proteins is provided in Table 3.7. According to the different proteins (delta endotoxins) produced by the *Bt* isolates

Table 3.6 Different classes of pesticidal proteins according to latest nomenclature

S. No.	Class	Description (PDB codes)
1.	App	Pesticidal proteins with a predominantly Alpha helical structure
2.	Cry	Proteins in which the active form normally consists of three domains
3.	Cyt	Cytolytic, normally single domain, proteins such as Cyt2Aa (1CBY)
4.	Gpp	Pesticidal proteins with homology to the aegerolysins
5.	Mcf	Pesticidal proteins related to Mcf proteins
6.	Mpf	Pesticidal proteins that have homology to the membrane attack complex/ Perforin superfamily
7.	Mpp	Pesticidal proteins with homology to the Etx_Mtx2 family
8.	Mtx	Pesticidal proteins related to Mtx1 proteins
9.	Pra	Proteins related to the *Photorhabdus* insect-related toxin A component
10.	Prb	Proteins related to the *Photorhabdus* insect-related toxin B component
11.	Spp	Pesticidal proteins related to sphaericolysin
12.	Tpp	Pesticidal proteins with homology to the Toxin_10/Bin family
13.	Vip	Mnemonic retained for Vip3 related proteins
14.	Vpa	Pesticidal proteins related to Vip2 catalytic component
15.	Vpb	Pesticidal proteins related to the binding partner of the Vip1/2 binary toxin, previously known as Vip1 and to the structurally-related proteins previously known as Vip4
16.	Xpp	A holding name for pesticidal proteins with currently unclassified homology groups

Source: Crickmore et al. (2020)

and their binding ability to the midgut receptors in the insects, mortality happens in the insects (Table 3.8).

3.6.2 *Bt* Proteins Deployed in Commercialized Genetically Engineered (GE) Crops

Only a small number of the known *Bt* insecticidal proteins have been exploited for the development of GE crops. The Cry proteins used for control of lepidopteran insects (foliage feeding) include very well characterized 3 domains Cry proteins or their altered three forms. These are Cry1Ab, Cry1Ac, Cry1Fa2, Cry2Ab, Cry2Ae, and Cry1A.105, and Vip3Aa. Similarly, very few numbers have been utilized for controlling the coleopteran insects, which include Cry3A, Cry3Bb1, Cry34Ab1 and Cry35Ab1, and there is only one protein Cry51Aa2, which is used for controlling hemipteran insects. The list of commercialized *Bt* genes is given in Table 3.9.

Cry1Ab1

The *cry1Ab1* gene was first cloned and described by Wabiko et al. (1986) and is one of the most studied 3-domain Cry proteins. It is produced as a protoxin that is activated by proteases enzymes in the midgut of an insect and active toxin binds to the receptors on the midgut cells and makes a pore on it, which leads to the leakage

Table 3.7 List of primary rank proteins from *Bt* discovered in various countries

Primary rank	Country	Reference
App		
App6Aa1 (Cry6Aa1)	Mycogen Corporation, San Diego, California, USA	Narva et al. (1991)
Cry		
Cry1Aa1	University of Washington, Seattle, Washington, USA	Schnepf et al. (1985)
Cry2Aa1	Ecogen, Inc., Langhorne, Pennsylvania, USA	Donovan et al. (1988a)
Cry3Aa1	Mycogen Corporation, San Diego, California, USA	Herrnstadt et al. (1987)
Cry4Aa1	Cambridge University, UK.	Ward and Ellar (1987)
Cry5Aa1	Mycogen Corporation, San Diego, California, USA	Narva et al. (1991)
Cry7Aa1	Plant Genetic Systems, Ghent, Belgium	Lambert et al. (1992)
Cry8Aa1	Mycogen Corporation, San Diego, California, USA	Foncerrada et al. (1992)
Cry9Aa1	Institute of Microbial Genetics, Moscow, Russia	Smulevitch et al. (1991)
Cry11Aa1	Ecogen Inc., Langhorne, Pennsylvania, USA	Donovan et al. (1988b)
Cry12Aa1	Mycogen Corporation, San Diego, California, USA	Narva et al. (1991)
Cry13Aa1	Mycogen Corporation, San Diego, California, USA	Narva et al. (1995)
Cry14Aa1	Mycogen Corporation, San Diego, California, USA	Payne et al. (1996)
Cry19Aa1	Unité des Bactéries Entomopathogènes, Institut Pasteur, Paris, France	Rosso and Delécluse (1997)
Cry20Aa1	Department of Entomology, University of California, USA	Lee and Gill (1997)
Cry21Aa1	Mycogen Corporation, San Diego, California, USA	Payne et al. (1996)
Cry24Aa1	Department of Entomology, University of California, USA	Kawalek (1998)
Cry25Aa1	Department of Entomology, University of California, USA	Kawalek (1998)
Cry26Aa1	Laboratory of Protein Chemistry, Institute for Microbial Genetics, Moscow, Russia	Wojciechowska et al. (1999)
Cry27Aa1	Biotechnology & Food Research Institute, Fukuoka Industrial Technology Center, Fukuoka, Japan	GenBank: BAD82796.1
Cry28Aa1	Laboratory of Protein Chemistry, Institute for Microbial Genetics, Moscow, Russia	Wojciechowska et al. (1999)
Cry29Aa1	Laboratoire des Bactéries et Champignons Entomopathogènes, Institut Pasteur, Paris, France	Juárez-Pérez et al. (2003)
Cry30Aa1	Laboratoire des Bactéries et Champignons Entomopathogènes, Institut Pasteur, Paris, France	Juárez-Pérez et al. (2003)

Cry31Aa1	Biotechnology & Food Research Institute, Fukuoka Industrial Technology Center, Fukuoka, Japan	Mizuki et al. (2000)
Cry32Aa1	Department of Molecular Microbiology, School of Biotechnology, Madurai Kamaraj University, Tamil Nadu, India	Balasubramanian et al. (2002)
Cry39Aa1	Department of Applied Bioscience, Hokkaido University, Sapporo, Japan	Ito et al. (2002)
Cry40Aa1	Department of Applied Bioscience, Hokkaido University, Sapporo, Japan	Ito et al. (2005)
Cry41Aa1	Biotechnology & Food Research Institute, Fukuoka Industrial Technology Center, Fukuoka, Japan	Yamashita et al. (2005)
Cry42Aa1	Biotechnology & Food Research Institute, Fukuoka Industrial Technology Center, Fukuoka, Japan	GenBank: BAD35166.1
Cry43Aa1	Chiba Prefectural Agricultural Research Center, Chiba, Japan	Yokoyama et al. (2004)
Cry44Aa1	Department of Applied Bioscience, Hokkaido University, Sapporo, Japan	Ito et al. (2006)
Cry47Aa1	CSIRO Livestock Industries, Queensland, Australia	Kongsuwan et al. (2005)
Cry50Aa1	Kyushu University, Fukuoka, Japan	GenBank: BAE86999.1
Cry52Aa1	China Academy of Agricultural Sciences, Beijing, China	GenBank: ABU96490.1
Cry53Aa1	China Academy of Agricultural Sciences, Beijing, China	GenBank: ABV55105.1
Cry54Aa1	Rice Research Institute, Sichuan Agricultural University, Yaan, China	Tan et al. (2009)
Cry56Aa1	Rice Research Institute, Sichuan Agricultural University, Sichuan, China	Zhu et al. (2010)
Cry57Aa1	Departamento de Biotecnologia y Bioquimica, Cinvestav, Irapuato, Mexico	Noguera and Ibarra (2010)
Cry58Aa1	Departamento de Biotecnología y Bioquímica, Cinvestav, Irapuato, México	Noguera and Ibarra (2010)
Cry59Aa1	Departamento de Biotecnología y Bioquímica, Cinvestav, Irapuato, México	Noguera and Ibarra (2010))
Cry61Aa1	Huazhong Agricultural University, Wuhan, China	Guan et al. (2014)
Cry62Aa1	Rice Research Institute, Sichuan Agricultural University, Sichuan, China	GenBank: AEA92302.1
Cry63Aa1	Graduate School of Biosphere Science, Hiroshima University, Hiroshima, Japan	Nagamatsu et al. (2010)
Cry65Aa1	Huazhong Agricultural University, Wuhan, Hubei, China	Peng et al. (2015)
Cry66Aa1	Huazhong Agricultural University, Wuhan, Hubei, China	GenBank: AEB52311.1
Cry67Aa1	Huazhong Agricultural University, Wuhan, Hubei, China	GenBank: AEJ35086.1
Cry68Aa1	Rice Research Institute, Sichuan Agricultural University, Sichuan, China	Guan et al. (2012)

(continued)

Table 3.7 (continued)

Primary rank	Country	Reference
Cry69Aa1	Rice Research Institute, Sichuan Agricultural University, Sichuan, China	Guan et al. (2014)
Cry70Aa1	Rice Research Institute, Sichuan Agricultural University, Sichuan, China	Li et al. (2015)
Cry71Aa1	Rice Research Institute, Sichuan Agricultural University, Sichuan, China	Li et al. (2015)
Cry72Aa1	Rice Research Institute, Sichuan Agricultural University, Sichuan, China	Li et al. (2015)
Cry73Aa1	College of Life Science and Technology, Huazhong Agricultural University, Wuhan, China	Ye et al. (2012)
Cry79Aa1	College of Life Sciences, Northeast Agriculture University, Harbin, Heilongjing, China	Haitao et al. (2019)
Cyt		
Cyt1Aa1	Research Institute ITAL, Wageningen, The Netherlands	Waalwijk et al. (1985)
Cyt2Aa1	Department of Biochemistry, University of Cambridge, England	Koni and Ellar (1993)
Cyt3Aa1	Rice Research Institute, Sichuan Agricultural University, China	Zhu et al. (2020)
Gpp		
Gpp34Aa1	Dow AgroSciences, San Diego, California, USA	Ellis et al. (2002)
Mpp		
Mpp5Aa1 (Sip1Aa1)	Monsanto Company, St. Louis, MO, USA Ecogen Inc., Langhorne, PA, USA	Donovan et al. (2006)
Mpp5Aa1 (Cry15Aa1)	Department of Microbiology, University of Washington, Seattle, Washington, USA	Brown and Whiteley (1992)
Mpp23Aa1 (Cry23Aa1)	Monsanto Company, St. Louis, MO, USA	Donovan et al. (2000)
Mpp33Aa1 (Cry33Aa1)	Biotechnology & Food Research Institute, Fukuoka Industrial Technology Center, Fukuoka, Japan	San Kim et al. (2003)
Mpp38Aa1 (Cry38Aa1)	Monsanto Company, Chesterfield, Missouri and Ecogen Inc., Belmar, New Jersey, USA	Baum et al. (2004)
Mpp45Aa1 (Cry45Aa1)	Fukuoka Industrial Technology Centre, Fukuoka, Japan	Okumura et al. (2005)
Mpp46Aa1 (Cry46Aa1)	Department of Chemistry, Kyushu University, Fukuok, Japan	Ito et al. (2004)

Mpp51Aa1 (Cry51Aa1)	Biotechnology Research Institute, Chinese Academy of Agricultural Sciences, Beijing, China	Huang et al. (2007)
Mpp60Aa1 (Cry60Aa1)	Hunan Normal University, China; University of California, USA	Sun et al. (2013)
Mpp64Aa1 (Cry64Aa1)	Biotechnology and Life Science, Sojo University, Kumamoto, Japan	Ekino et al. (2014)
Tpp		
Tpp35Aa1 (Cry35Aa1)	Dow AgroSciences, San Diego, California, USA	Ellis et al. (2002)
Tpp36Aa1 (Cry36Aa1)	Monsanto Company St. Louis, USA	Rupar et al. (2003)
Tpp78Aa1 (Cry78Aa1)	School of Life Science, Northeast Agricultural University, China; Institute of Plant Protection, Chinese Academy of Agricultural Sciences, Beijing, China	Wang et al. (2018b)
Tpp80Aa1	Guangxi University, China, Jiyang College of Zhejiang A&F University, Zhuji China; Hainan Institute of Tropical Agricultural Resources, China Cuixi Academy of Biotechnology, China; Biotechnology Research Institute of Chinese Academy of Agricultural Sciences, Beijing, China	Zhou et al. (2020c)
Vip		
Vip3Aa1	Chiba Agricultural Biotechnology, NC, USA	Estruch et al. (1996)
Vpa2Aa1		
Vpa2Aa1 (Vip2Aa1)	Scripps Research Institute, La Jolla, California, USA, Novartis Agribusiness Biotechnology Research, North Carolina, USA	Han et al. (1999)

Note: Primary rank proteins such as Cry16Aa1, Cry17Aa1, Mpp74Aa1 (old name Cry74Aa1), (*Clostridium bifermentans*), Mpp75Aa1 (old name Cry49Aa1) (*Brevibacillus laterosporus*); Cry18Aa1 (*Bacillus popilliae*); Cry48Aa1, Tpp49Aa1 (old name Cry49Aa1) (*Bacillus sphaericus*); Cyt4Aa1, Cyt5Aa1, Cyt6Aa1, and Cyt7Aa1 (*Dickeya dadantii*) have not been included in the list as these were reported from other microbes

Table 3.8 Host spectrum of *Bt* delta-endotoxins (Cry and Cyt)

Target insects	*Bt* delta-endotoxins
Lepidopterans	Cry1A, Cry1B, Cry1C, Cry1D, Cry1E, Cry1F, Cry1G, Cry1H, Cry1I, Cry1J, Cry1J, Cry1K, Cry2a, Cry7B, Cry8D, Cry9A, Cry9B, Cry9C, Cry9E, Cry15 A, Cry22A, Cry32A, Cry51A, Cry54A,Cry56A
Dipterans	Cry1A, Cry1B, Cry1C, Cry2A, Cry4A, Cry4B, Cry10, Cry11A, Cry11B, Cry16A, Cry19A, Cry19B, Cry20A, Cry24C, Cry27A, Cry32B, Cry32C, Cry32D, Cry39A, Cry44A, Cry47A, Cry56A, Cry60A, Cry60B, Cry69A, Cry80Ab1, Cyt1A, Cyt1B, Cyt2A, Cyt2B
Coleopterans	Cry1B, Cry3A, Cry3B, Cry3C, Cry7A, Cry8A, Cry8B, Cry8C, Cry8D, Cry8E, Cry8F, Cry8G, Cry9D, Cry14A, Cry18A, Cry22A, Cry22B, Cry23A, Cry34A, Cry34B, Cry35A, Cry35B, Cry36A, Cry37A, Cry43A, Cry43B, Cry55A, Cyt1A, Cyt2C
Hemipterans	Cry2A, Cry3A, Cry11A, Cry51A, Cry78A
Nematodes	Cry5A, Cry5B, Cry6A, Cry6B, Cry1I, Cry12A, Cry13A, Cry14A, Cry21A, Cry55A

Modified from Palma et al. (2014)

Table 3.9 List of *Bt* sub species and their genes used in commercial products

Gene	Source
Active against lepidopteran insects	
cry1A.105	*B. thuringiensis* subsp. *kumamotoensis*
cry1Ab	*B. thuringiensis* subsp. *kurstaki*
cry1Ab (truncated)	Synthetic form of Cry1Ab from *B. thuringiensis* subsp. *kumamotoensis*
cry1Ab-Ac	Synthetic fusion gene derived from *B. thuringiensis*
cry1Ac	*B. thuringiensis* subsp. *kurstaki* (strain HD73)
cry1C	Synthetic gene derived from *B. thuringiensis*
cry1F	*B. thuringiensis* var. *aizawai*
cry1Fa2	Synthetic form of *cry1F* gene derived from *B. thuringiensis* var. *aizawai*
cry2Ab2	*B. thuringiensis* subsp. *kumamotoensis*
cry2Ae	*B. thuringiensis* subsp. *dakota*
cry9C	*B. thuringiensis* subsp. *tolworthi* (strain BTS02618A)
cry1F (modified)	Synthetic form of *cry1F* gene from *B. thuringiensis* var. *aizawai*
vip3Aa	*B. thuringiensis* (strain AB88)
vip3Aa20	*B. thuringiensis* (strain AB88)
Active against coleopteran insects particularly against corn root worm	
cry34Ab1	*B. thuringiensis* (strain PS149B1)
cry35Ab1	*B. thuringiensis* (strain PS149B1)
cry3A	*B. thuringiensis* subsp. *tenebrionis*
cry3Bb1	*B. thuringiensis* subsp. *kumamotoensis*
cry3A (modified)	Synthetic form of *cry3A* gene from *B. thuringiensis* subsp. *tenebrionis*
Active against hemipteran insects *Lygus hesperus* and *L. lineolaris*	
cry51Aa2 (modified)	*B. thuringiensis*

Source: www.isaaa.org

of the fluids. Insect stops feeding and eventually dies. It is used for a wide range lepidopteran, which are economically significant for example, *Ostrinia* spp., *Diatraea* spp. (van Frankenhuyzen 2009).

Cry1Ac

Cry1Ac is one of the most potent 3-domain Cry proteins, which is active against the lepidopterans, particularly the noctuid members. This lead to its widespread deployment in GE crops, especially cotton. This 3-domain Cry insecticidal protein has a long history of study (Schnepf et al. 1998). It was first cloned and described by Adang et al. (1985) from *B. thuringiensis* subsp. *kurstaki* (strain HD73).

Cry1Fa2

Cry1Fa2 is 3-domain protein, which shows high potency against *Spodoptera* spp. (van Frankenhuyzen 2009). It was discovered by Payne and Sick (1993). Because of its potency against *Spodoptera* spp. it is used in combination with other Cry1A proteins, which are low toxic against these pests.

Cry1A.105

This is a chimeric insecticidal protein, which was developed by Monsanto Co. It has four parts from four different domains. Domains 1 and 2 amino acid sequences of this protein are identical to respective domains of Cry1Ab and Cry1Ac proteins, domain 3 is very much identical to the Cry1F protein and C-terminal domain is identical to the Cry1Ac protein. Therefore, this chimeric protein combines most of the insecticidal properties displayed by the Cry1A and Cry1F proteins (Biosafety Clearing House 2014).

Cry2Ab

Cry2Ab is a three domain protein that has activity against both lepidopteran and dipteran insects (Widner and Whiteley 1989). This protein does not share midgut binding site with Cry1 proteins (Hernández-Rodríguez et al. 2008), thus are good candidates for pyramiding with Cry1 proteins in insect resistant crops. It is often targeted to chloroplast for the better expression levels and reduced negative plant phenotypes in GE plants (Corbin and Romano 2006).

Cry2Ae

Its sequence is 89.7% identical to Cry2Ab. Its mechanism of action is also different from the Cry1 proteins as it doesn't share binding sites with Cry1 proteins (Caccia et al. 2010; Gouffon et al. 2011). Like Cry2Ab, Cry2Ae is also a good candidate for combining with Cry1 protein in pyramiding for insect resistance management.

Vip3Aa

Vip3Aa is a secretory insecticidal protein, which is expressed during the vegetative state of *Bt* growth, before sporulation (Estruch et al. 1996). Vip3Aa is mostly active against lepidopteran pests of cotton and corn. Sequence of Vip3Aa protein and its mode of action are different when compared to Cry1 or Cry2A insecticidal proteins

(Lee et al. 2006). So, Vip3Aa is valuable tool to slow down the development of insect resistance in combination with Cry1 or Cry2 pesticidal proteins in transgenic plants (Narva et al. 2013; Chakroun et al. 2016b).

Cry3Aa and mCry3Aa (Modified Cry3Aa)

Western corn rootworm (*Diabrotica* spp.) (WCR) is one of the most important pests of corn. The first reported coleopteran-active *Bt* insecticidal protein (Herrnstadt et al. 1987) has not been useful for the control *Diabrotica* spp. (Herrnstadt et al. 1987; Slaney et al. 1992). In order to improve the activity against the WCR, Walters et al. (2008) engineered Cry3Aa that had a chymotrypsin/cathepsin G protease site at in the loop between α-helix 3 and α-helix 4 of domain I. This change in native Cry3Aa improved insecticidal activity against WCR larvae.

3.7 *Bt* Transgenic Crops

Bt formulations are environmentally safe. However, weakness of this technology includes the requirement for repetitive application in the due course, reasonably short timeframe of efficacy in environmental conditions, the incapability to reach the insects with specific feeding behavior; for instance, those that nourish plant sap or else underneath the ground on plant roots and timing of applications.

The advancement in plant transformation technology (biolistic and *Agrobacterium*-mediated) fashioned the opportunities to deliver and express genes in the plants. These genetically engineered / modified (GE/GM) plants have one or more useful traits like insect resistance, herbicide tolerance, disease resistance, abiotic stress tolerance, and nutritional improvement (Kumar et al. 2020). In 32 different crops, various transgenic events (525 Nos.) have been developed in many countries across the globe. Among these, maize accounts for the maximum number of events (238), followed by cotton (67), potato (49), Argentine canola (42), soybean (41), carnation (19) and others (https://www.isaaa.org/ accessed on January 1, 2021). At present, ten insect resistant transgenic crops have been commercialized for cultivation, majority of which are based on insecticidal genes (different types of *cry* gene(s) or *vip* gene) from *B. thuringiensis* (Kereša et al. 2008; https://www.isaaa.org/ accessed on January 1, 2021).

In the initial plant transformation attempts, expression of *Bt* Cry proteins was very low; however, plant tolerance to insect feeding was achieved. Fischhoff et al. (1987) developed transgenic tomato plants resistant to *Helicoverpa zea*, with truncated Cry1Ac. Vaeck et al. (1987) performed the transformation of tobacco with Cry1A, which resulted in *Manduca sexta* (L.) resistant GE plants. In order to increase the expression levels of the Cry proteins in transgenic plants, efforts were put to know the divergence in gene structure and codon usage between *Bt* and the particular host plants. Perlak et al. (1991) were successful in increasing the expression of Cry protein levels up to 100-fold by creating synthetic Cry-encoding genes with a codon usage specific to target plants towards that preferred by plants and deficient in mRNA destabilizing sequences. Partly modified or completely modified transgenes coding for Cry1Ab or Cry1Ac resulted in elevated expression and a more proportion of GE tomato and tobacco tolerant to *M. sexta* damage. Similarly, Adang et al. (1993) also attained success with a modified *cry3Aa* gene expressed in GE potato that was resistant to *Leptinotarsa decemlineata*. These early achievements in producing GE plants expressing *Bt* proteins laid down the stage for an industry-wide inclination among seed producers to produce GE *Bt* crops.

3.7.1 Transformation Technologies

A number of factors play a role for an effective transformation of a plant. They are (i) availability of competent target tissues that are capable for propagation or regeneration, (ii) proficient means for delivery of target DNA, (iii) ability to sort out transformed cells, and (iv) competence to recover productive GE plants (Hansen and Wright 1999). Several plant tissue types are amenable to transformation, which include leaf tissue, immature embryos, embryogenic shoot tips, embryogenic suspension cultures, immature cotyledonary-nodes, and hypocotyls (Lee et al. 2013). The choice of a tissue type intended for transformation depends on various aspects including simplicity and accessibility (e.g., free from patent restrictions), but finally it is important that fertile, GE plants are developed.

The two most commonly methods employed for DNA delivery are *Agrobacterium*-mediated transformation and particle bombardment. The former uses the gene-transfer machinery of the bacterium to deliver or introduce a desired piece of DNA (T-DNA) into the host cell, which eventually is incorporated into the genome. It can be employed to introduce DNA to both dicots and monocots, can deliver reasonably large pieces of DNA, and characteristically a small number of T-DNA copies are integrated into the genome of the host at a single location. In the particle bombardment and other physical delivery approach particles of different materials (gold or tungsten silicon fiber "whiskers") are layered with DNA and physically delivered into cells (Smith and Hood 1995; Hansen and Wright 1999; Petolino and Arnold 2009). Many examples of using this method for generation of commercialized biotech crops are available. Some examples of the commercialized events, which have been produced/developed by *Agrobacterium*-mediated transformation, are given in Table 3.10.

As compared to *Agrobacterium*-mediated transformation, particle bombardment frequently produces complex events harbouring multiple copies and/or fragments of DNA and integration of the DNA into multiple genomic regions (Finer and Dhillon 2007). Various events of crops have been produced using particle bombardment method, which has been commercialized, are given in Table 3.11.

Genes that allow transformed cells, tissues, or plants to get differentiated from non-transformed ones are called selectable marker genes. These make an important element of plant transformation systems. The selectable marker genes include antibiotic resistant genes (e.g., *nptII* gene that confers resistance to the antibiotics kanamycin and neomycin), herbicide tolerant genes (e.g., *pat* gene that confers tolerance to the herbicide glufosinate) or other genes (e.g., *pmi* gene that enables

Table 3.10 Commercialized *Bt* crops developed through *Agrobacterium*-mediated transformation

Event name and code	Trade name	Developer
Cotton (*Gossypium hirsutum L.*)		
Name: 31707 Code: Not available	BXN Plus Bollgard Cotton	Monsanto Company (including fully and partly owned companies)
Name: BXN10224 (10224) Code: BXN-1Ø224-4	BXN Cotton	Monsanto Company (including fully and partly owned companies)
Name: COT102 (IR102) Code: SYN-IR1Ø2-7	VIPCOT Cotton	Syngenta
Name: MON1076 Code: MON-89924-2	Bollgard Cotton	Monsanto Company (including fully and partly owned companies)
Name: MON1698 Code: MON-887Ø2-4	Roundup Ready Cotton	Monsanto Company (including fully and partly owned companies)
Name: MON88913 Code: MON-88913-8	Roundup Ready Flex Cotton	Monsanto Company (including fully and partly owned companies)
Cowpea (*Vigna unguiculata*)		
Name: AAT709A Code: AAT-7Ø9AA-4	Not available	African Agricultural Technology Foundation (AATF)
Eggplant (*Solanum melongena*)		
Name: Bt Brinjal Event EE1 Code: Bt Brinjal Event EE1	BARI Bt Begun-1, -2, -3 and -4	Maharashtra Hybrid Seed Company (MAHYCO)
Maize (*Zea mays L.*)		
Name: 5307 Code: SYN-Ø53Ø7-1	Agrisure Duracade	Syngenta
Name: 59122 Code: DAS-59122-7	Herculex RW	Dow AgroSciences LLC and DuPont (Pioneer Hi-Bred International Inc.)
Name: MIR162 Code: SYN-IR162-4	Agrisure Viptera	Syngenta
Name: MON88017 Code: MON-88Ø17-3	YieldGard VT Rootworm RR2	Monsanto Company (including fully and partly owned companies)
Poplar (*Populus* sp.)		
Name: *Bt* poplar, poplar 12 (Populus nigra) Code: Not available	Not available	Research Institute of Forestry (China)
Potato (*Solanum tuberosum L.*)		
Name: 1210 amk Code: Not available	Lugovskoi plus	Centre Bioengineering, Russian Academy of Sciences
Name: 2904/1 kg Code: Not available	Elizaveta plus	Centre Bioengineering, Russian Academy of Sciences
Name: ATBT04-6 Code: NMK-89761-6	Atlantic NewLeaf potato	Monsanto Company (including fully and partly owned companies)
Name: BT23 Code: NMK-89675-1	New Leaf Russet Burbank potato	Monsanto Company (including fully and partly owned companies)
Name: RBMT22-262 Code: Not available	New Leaf Plus Russet Burbank potato	Monsanto Company (including fully and partly owned companies)
Name: SEMT15-15 Code: NMK-8993Ø-4	Shepody NewLeaf Y potato	Monsanto Company (including fully and partly owned companies)

Source: www.isaaa.org

Table 3.11 List of commercialized events developed through the particle bombardment method

Event name and code	Trade name	Developer
Cotton (*Gossypium hirsutum L.*)		
Name: Event1 Code: JKCH-1947 Bt	JK 1	JK Agri Genetics Ltd. (India)
Name: MON15985 Code: MON-15985-7	Bollgard II Cotton	Monsanto Company (including fully and partly owned companies)
Maize (*Zea mays L.*)		
Name: Bt176 (176) Code: SYN-EV176-9	NaturGard KnockOut, Maximizer	Syngenta
Name: CBH-351 Code: ACS-ZMØØ4-3	Starlink Maize	Bayer Crop Science (including fully and partly owned companies)
Name: DBT418 Code: DKB-89614-9	Bt Xtra Maize	Monsanto Company (including fully and partly owned companies)
Name: MON810 Code: MON-ØØ81Ø-6	YieldGard, MaizeGard	Monsanto Company (including fully and partly owned companies)
Name: MON863 Event Code: MON-ØØ863-5	YieldGard Rootworm RW, MaxGard	Monsanto Company (including fully and partly owned companies)
Name: TC1507 Code: DAS-Ø15Ø7-1	Herculex I, Herculex CB	Dow AgroSciences LLC and DuPont (Pioneer Hi-Bred International Inc.)
Rice (*Oryza sativa L.*)		
Name: Tarom molaii + *cry1Ab* Code: Not available	Not available	Agricultural Biotech Research Institute, Iran

Source: www.isaaa.org

plants to use mannose as a carbon source in tissue culture systems) (Rosellini 2012). In some cases, genes of interest and the selectable marker genes are incorporated in the host genome as two-independent events, which allow removing selectable marker genes from the commercial product *via* conventional breeding processes. However, in majority of the commercialized events, the selectable marker gene is incorporated along with the genes of interest. Table 3.12 summarizes the marker genes used by different companies and their functions.

3.7.2 Commercialized *Bt* Crops

Bt crops are the plants genetically engineered (modified) to contain the insecticidal protein(s) from the bacterium, *B. thuringiensis*, to be resistant to certain insect pests. The first *Bt* crop that was commercialized in the United States back in 1995 was *Bt* potato. This was followed by *Bt* corn and cotton in 1996 (Betz et al. 2000). These crops are rapidly accepted by farmers across the globe and 29 countries have approved the cultivation of biotech crops. Information on the status of deregulated biotech traits is maintained by several organizations, such as the Biotechnology

Table 3.12 Different kinds of marker gene/selectable marker genes used in commercialized crop events

Crop	Marker gene/selectable marker gene	Gene source	Product	Function
JK 1	*nptII*	*E. coli* Tn5 transposon	Neomycin phosphotransferase II enzyme	Allows transformed plants to metabolize neomycin and kanamycin antibiotics during selection
Bollgard II Cotton	*nptII*	*E. coli* Tn5 transposon	Neomycin phosphotransferase II enzyme	Allows transformed plants to metabolize neomycin and kanamycin antibiotics during selection
	aad	*E. coli*	3″(9)-O-aminoglycoside adenylyltransferase enzyme	Allows selection for resistance to aminoglycoside antibiotics such as spectinomycin and streptomycin
	uidA	*E. coli*	Beta-D-glucuronidase (GUS) enzyme	Produces blue stain on treated transformed tissue, which allows visual selection
Starlink Maize	*bla*	*E. coli*	Beta lactamase enzyme	Detoxifies beta lactam antibiotics such as ampicillin
	bar	*Streptomyces hygroscopicus*	Phosphinothricin N-acetyltransferase (PAT) enzyme	Eliminates herbicidal activity of glufosinate (phosphinothricin) herbicides by acetylation
VIPCOT cotton	*aph4 (hpt)*	*E. coli*	Hygromycin-B phosphotransferase (hph) enzyme	Allows selection for resistance to the antibiotic hygromycin B

Source: www.isaaa.org

Industry Organization (BIO, http://www.biotradestatus.com/), the Center for Environmental Risk Assessment (CERA 2012) and the International Service for the Acquisition of Agri-Biotech Applications (ISAAA, http://www.isaaa.org) maintain the information on the status of deregulated biotech traits. In each and every case, the information provided in these online resources is a synopsis of information provided in public by various governmental regulatory authorities.

3.7.3 *Bt* Cotton: Commercialized Events

Cotton plant is a dicot shrub, which is native to tropical and subtropical regions across the world, including USA, Africa, Egypt, and India and the greatest diversity

Table 3.13 Trade names of cotton crop events approved for commercialization in various countries

Trade name	Commercial traits
Bollgard cotton, Ingard (event name: Mon531)	*cry1Ac*
Bollgard II cotton	*cry1ac, cry2ab2*
Bollgard III	*vip3A(a), cry1Ac, cry2Ab2*
TwinLink cotton	*cry1Ab, cry2Ae*
WideStrike cotton	*cry1Ac, cry1AF, (pat) (syn)*
VIPCOT cotton	*vip3A(a)*
Roundup Ready × Bollgard cotton	*cp4 epsps (aroA:CP4), cry1Ac*
Bollgard III × Roundup Ready Flex	*cp4 epsps (aroA:CP4)*
Roundup Ready Bollgard II Cotton	*cp4 epsps (aroA:CP4), cry1Ac, cry2Ab2*
Glytol × Twinlink × VIPCOT cotton	*cry1Ab; Cry2Ae, vip3A (a), 2mepsps (bar)*
BXN plus Bollgard cotton	*cry1Ac, bxn*

Source: www.isaaa.org

of wild cotton species is found in Mexico, followed by Australia and Africa. As mentioned above, transgenic *Bt* cotton was first introduced in the United States in 1996 for commercial purpose. At that time *Heliothis virescens*, a notorious pest, had developed resistance to majority of the insecticides used for its management (Wolfenbarger 1981; Sparks 1981; Sparks et al. 1993; Leonard et al. 1988; Subbaratnam and Radhika 2005). However, it was susceptible to proteins in *Bt* cotton. Thus, *Bt* cotton was adopted largely because of its effectiveness in controlling this pest. At the same time, it was also introduced in Australian 1996 where the main target pest was *H. armigera* (Downes and Mahon 2012). In 2020, *Bt* cotton was approved for commercialization (food, feed, or cultivation) in 27 countries (Argentina, Australia, Brazil, Burkina Faso, Canada, China, Colombia, Costa Rica, Eswatini, Ethiopia, European Union, India, Japan, Malaysia, Mexico, New Zealand, Nigeria, Pakistan, Paraguay, Philippines, Singapore, South Africa, South Korea, Sudan, Taiwan, United States) across the globe (https://www.isaaa.org/ December, 2020). The trade names of some commercialized events in cotton crop are given in Table 3.13.

3.8 Insect Resistance to *Bt* Toxins

The widespread acceptance of *Bt* sprays and *Bt* crops inevitably imposes intense selective pressure for insect pests to develop resistance to *Bt* proteins, thus thinning the benefits provided by *Bt* biotechnology (Tabashnik and Carrière 2017). Resistance is a genetically based reduction in the susceptibility of a population toward an insecticide (National Research Council 1986). By definition, it is an evolutionary phenomenon because it involves changes in allele frequencies in a population (Georghiou 1972; Hartl 1988). In order to document resistance, LC_{50} studies have to be conducted.

3.8.1 Laboratory Selection to *Bt* Toxins

In the past it was expected that insects would not become resistant to *Bt* toxins given that insects and *Bt* have coevolved. Starting in the mid-1980s a number of insect species with diverse levels of resistance to *Bt* crystal proteins were reported by laboratory selection experiments, using either laboratory-adapted insects or insects collected from wild populations (Tabashnik 1994). The laboratory-selected *B. thuringiensis*-resistant colonies offer an opportunity to study the resistance inheritance (Liu and Tabashnik 1997), find out the mechanism of resistance (Wang et al. 2007; Song et al. 2015; Guo et al. 2015; Zhou et al. 2020a, b), estimation of resistance allele frequency (Walsh et al. 2014), evaluation of fitness cost (Bird and Akhurst 2004), development of molecular diagnostic tools for detection (Flagel et al. 2015), testing models to predict resistance evolution and improve the ongoing resistance management strategies (Zhao et al. 2003). Examples of laboratory-selected crop pests for resistance to Cry toxins include *Plodia interpunctella* (McGaughey 1985; McGaughey and Beeman 1988; Gomis-Cebolla et al. 2018), *P. xylostella* (Kirsch and Schmutterer 1988; Liu and Tabashnik 1997; Zhao et al. 2003; Guo et al. 2015; Zhu et al. 2015; Guo et al. 2019; Zhou et al. 2020b), *L. decemlineata* Say (Whalon et al. 1993), *Cadra cautella* (McGaughey and Beeman 1988), *Trichoplusia ni* (Estada and Ferre´ 1994; Wang et al. 2007; Song et al. 2015), *Spodoptera littoralis* (Mu¨ller-Cohn et al. 1996), *Spodoptera exigua* (Moar et al. 1995), *H. virescens* (Stone et al. 1989; Gould et al. 1992; Lee et al. 1995; Pickett 2009; Gulzar et al. 2012), *Culex quinquefasciatus* (Georghiou and Wirth 1997), *Spodoptera frugiperda* (Bernardi et al. 2016; Chandrasena et al. 2018), *H. armigera* (Kranthi et al. 2000; Akhurst et al. 2003; Walsh et al. 2014; Gomis-Cebolla et al. 2018), *Helicoverpa punctigera* (Walsh et al. 2014; Wang et al. 2016), *H. zea* (Anilkumar et al. 2008; Welch et al. 2015; Yang et al. 2020b), *S. litura* (Barkhade and Thakare 2010), *Pectinophora gossypiella* (Patin et al. 1999; Tabashnik et al. 2000; Liu et al. 2001a, b), *Ostrinia furnacalis* (Alves et al. 2006; Han et al. 2009; Xu et al. 2010; Wang et al. 2016; Gomis-Cebolla et al. 2018; Shabbir et al. 2018; Wang et al. 2019), *P. xylostella, O. nubilalis* (Huang et al. 1997; Siqueira et al. 2004; Pereira et al. 2008), *Diabrotica virgifera virgifera* (Lefko et al. 2008; Meihls et al. 2008, 2011, 2012; Oswald et al. 2011; Frank et al. 2013).

3.8.2 Field Evolved Resistance

Globally, transgenic plants producing *Bt* insecticidal proteins have been planted to manage some important crop pests. Majority of the target pests of *Bt* crops have been sustainably and effectively controlled by proactive resistance management strategies, such as the refuge strategy and the pyramid strategy in many countries (Wu 2014; Tabashnik and Carrière 2017). An analysis of global monitoring data (44 cases) by Tabashnik and Carrière (2019) showed 19 cases where no significant decrease in susceptibility had occurred after 21 years of exposure to *Bt* crops. These 19 cases comprised data from seven countries (Brazil, United States, South Africa,

Australia, Mexico, Spain, China) indicating susceptibility to five toxins (Cry1Ac, Cry1Ab, Cry2Ab, Cry1Fa, Vip3Aa) in *Bt* crops for populations of eleven species of lepidopteran pests: *Chrysodeixis includens, Earias biplaga, Diatraea grandiosella, H. armigera, H. punctigera, H. virescens, O. nubilalis, H. zea, P. gossypiella, S. frugiperda*, and *Sesamia nonagroides.*

Although most of the insect populations remained susceptible to *Bt* toxins in plants, but there have been incidences of field evolved resistance to one or more than one *Bt* toxins in *Bt* crops in many countries across the globe. In some cases, field-evolved resistance has been reported with only statistically significant decrease in susceptibility but no reduced efficacy of the *Bt* crop, e.g., *Diatraea saccharalis* in the United States to Cry1Ab (Huang et al. 2012), *O. furnacalis* in the Philippines to Cry1Ab (Alcantara et al. 2011) and *H. armigera* in China to Cry1Ac (Jin et al. 2015), *H. armigera* in Pakistan to Cry1Ac (Jamil et al. 2020). Whereas there also have been reports of field evolved resistance where more than 50% of individuals in a population were resistant and the efficacy of the *Bt* crop was reduced in the field (practical field evolved resistance) (Table 3.14). In *Bt* cotton, practical field resistance has been reported for populations of one pest species (*P. gossypiella*), in *Bt* corn for five species (*Busseola fusca, D. saccharalis, D. virgifera virgifera, S. frugiperda*, and *Striacosta albicosta*), and in *Bt* corn and *Bt* cotton for *H. zea.* Recently practical resistance to *Bt* corn (Cry1F) in *S. frugiperda* and *Bt* cotton (Cry1Ac + Cry2Ab) in *P. gossypiella* has been documented in Argentina and India, respectively (Table 3.14). The fastest case of field evolved resistance to a *Bt* toxin (Cry1F) in *Bt* crop with reduced efficacy is that of *S. frugiperda* in the US territory of Puerto Rico in just three years. This is also the first case of resistance leading to the pulling out a *Bt* crop from the marketplace (Storer et al. 2010, 2012).

In India, the pink bollworm was reported to feed on both *Bt* cotton producing Cry1Ac (BG) (Dhurua and Gujar 2011) and *Bt* cotton with Cry1Ac + Cry2Ab (BGII) (Naik et al. 2018; Reyaz et al. 2018) in fields. Naik et al. (2018) reported that in central and southern India, the annual average PBW larval recovery from *Bt* cotton (Cry1Ac + Cry2Ab) had been high (28.85–72.49%) during 2014–2017 and the 50% lethal concentration (LC_{50}) of Cry1Ac increased from a mean of 0.330μg mL^{-1} in 2013 to a mean of 6.938μg mL^{-1} in 2017. The resistance ratio (RR) increased from a mean of 47.12 in 2013 to a mean of 1387 in 2017. The LC_{50} value for Cry2Ab increased from a mean of 0.014μg mL^{-1} in 2013 to a mean of 12.51μg mL^{-1} in 2017 and the RR increased from a mean of 5.4 in 2013 to a mean of 4196 in 2017. The resistance ratios (RRs) to Cry1Ac were 26–262 and those to Cry2Ab were 1–108 in northern India (Naik et al. 2018).

Fand et al. (2019) conducted extensive roving surveys in 83 locations covering 16 major cotton growing districts of Maharashtra state, India and reported a widespread infestation of pink bollworm on *Bt* cotton in the surveyed sites, which ranged from 40 to 95% and accounted for an estimated yield losses between 20 and 30%. Recently, Agrawal et al. (2020) performed midgut transcriptome analysis of the BGII resistant pink bollworm population from India, which revealed 1741 unigenes were differentially expressing than susceptible population. Out of those, 1024 unigenes were down-regulated and 717 were up-regulated. They found that genes,

Table 3.14 List of practical field evolved resistance in insects to pesticidal proteins in *Bt* crops

S. no.	Insect	Crop	Toxin	Country	Reference
1.	*B. fusca*	Corn	Cry1Ab	S. Africa	Van Rensburg (2007)
2.	*D. saccharalis*	Corn	Cry1A.105	Argentina	Grimi et al. (2015), Grimi et al. (2018)
3.	*D. saccharalis*	Corn	Cry1Fa	Argentina	Grimi et al. (2018)
4.	*D. v. virgifera*	Corn	Cry3Bb	USA	Gassmann et al. (2011), Andow et al. (2016), Shrestha et al. (2018), Shrestha and Gassmann (2019)
5.	*D. v. virgifera*	Corn	Cry34/35Ab	USA	Gassmann et al. (2016), Ludwick et al. (2017), Gassmann et al. (2020)
6.	*D. v. virgifera*	Corn	mCry3A	USA	Gassmann et al. (2014), Andow et al. (2016)
7.	*D. v. virgifera*	Corn	eCry3.1Ab	USA	Jakka et al. (2016), Zukoff et al. (2016), Andow et al. (2016)
8.	*H. zea*	Corn	Cry1Ab	USA	Dively et al. (2016), Storer et al. (2001)
9.	*H. zea*	Corn	Cry1A.105	USA	Dively et al. (2016), Kaur et al. (2019)
10.	*H. zea*	Cotton	Cry1Ac	USA	Ali et al. (2006)
11.	*H. zea*	Cotton	Cry2Ab	USA	Ali and Luttrell (2007)
12.	*H. zea*	Corn	Cry1A.105+ Cry2Ab	Louisiana, United States	Kaur et al. (2019)
13.	*S. albicosta*	Corn	Cry1Fa	USA	Ostrem et al. (2016), Difonzo et al. (2016), Smith et al. (2017), Eichenseer et al. (2008)
14.	*S. albicosta*	Corn	Cry1Fa	Canada	Smith et al. (2017)
15.	*S. frugiperda*	Corn	Cry1Ab	Brazil	Omoto et al. (2016)
16.	*S. frugiperda*	Corn	Cry1F	Brazil	Farias et al. (2016)
17.	*S. frugiperda*	Corn	Cry1F	Argentina	Chandrasena et al. (2018), Vassallo et al. (2019)
18.	*S. frugiperda*	Corn	Cry1F	USA	Storer et al. (2010), Huang et al. (2014)
19.	*P. gossypiella*	Cotton	Cry1Ac	India	Dhurua and Gujar (2011), Mohan et al. (2015), Naik et al. (2018)
20.	*P. gossypiella*	Cotton	Cry2Ab	India	Naik et al. (2018)

Modified from Tabashnik and Carrière (2017) and Tabashnik and Carrière (2019)

which had been reported earlier to be associated with *Bt* resistance (APN, ABCA, ABCG8, and cadherin) in other insects, were found down-regulated in BGII resistant population and on the other hand, unigenes related to metabolic resistance, such as cytochrome P450, glutathione s-transferase and carboxylesterase, were up-regulated in resistant pink bollworm compared to susceptible pink bollworm.

In China, the concentration of Cry1Ac in *Bt* cotton differed over time, which permitted survival of susceptible larvae of both pests during some of the growing season in China; thus, increasing the risk of resistance (Wu and Guo 2005). Zhang et al. (2011) collected and tested cotton bollworm from 13 sites in five provinces of northern China and from Shawan and Shache in northwestern China. Their study revealed a decreased susceptibility to Cry1Ac in northern China, where *Bt* cotton has been planted intensively in contrast with Shawan and Shache of northwestern China, where *Bt* cotton planting has been restricted. Further, the resistance monitoring data of Wan et al. (2012) showed that susceptibility of pink bollworm to Cry1Ac decreased significantly in 2008–2010 compared with 2005–2007 in six provinces of the Yangtze river valley, China. Their laboratory diet bioassays revealed that the mean LC_{50} for Cry1Ac was twice as high in 2008–2010 compared to the data of 2005–2007. It indicates that the frequency of resistance to Cry1Ac increased in the field populations of pink bollworm tested. However, recent analysis of 11 years of field monitoring data from six provinces of this Yangtze river valley of China, indicated that *P. gossypiella* resistance to single-toxin *Bt* cotton has delayed or even reversed. This is attributed to a novel seed mixture strategy serendipitously adopted by millions of growers, which involved planting of second-generation seeds from crosses between *Bt* and non-*Bt* cotton, that yielded a refuge of 25% non-*Bt* plants randomly interspersed within fields of *Bt* cotton (Wan et al. 2017).

Bt corn has not been approved in China yet. However, *Bt* maize transgenic varieties are undergoing regulatory trials and two varieties of *Bt* maize were recently issued safety certificates by the Ministry of Agriculture and Rural Affairs of the People's Republic of China, which indicates that the commercialization of *Bt* maize will come soon in China (Li et al. 2020). In this connection, prior to the commercialization of a transgenic corn crop, knowledge of diagnostic concentrations and resistance allele frequencies in pests of interests could help in monitoring resistance development and evaluating resistance management strategies in the future. Li et al. (2020) investigated baseline susceptibility to *Bt* toxins and *Bt* toxin resistance allele frequencies in *O. furnacalis* populations, a major pest of corn, collected from Huanghuaihai, summer corn region in China during 2015 to 2018. The median lethal concentration (LC_{50}) values of the *Bt* toxins Cry1Ab, Cry1Ac, and Cry1F for 15 different populations of *O. furnacalis* ranged from 0.887 to 1.617, 1.251 to 2.594, and 4.146 to 6.465 ng cm^{-2}, respectively. Concentrations of 93, 45, and 197 ng cm^{-2} of Cry1Ab, Cry1Ac, and Cry1F, respectively, which killed >99% of individuals of eight *O. furnacalis* populations collected in 2017, were identified as diagnostic concentrations for monitoring susceptibility in *O. furnacalis* populations in this region. The F_2 screening method with these diagnostic concentrations showed the resistance allele frequencies related to Cry1Ab, Cry1Ac and Cry1F as 0.002, 0.001 and 0.001, respectively, in 2018.

3.8.3 Vip Proteins in Transgenic Crops and Resistance Scenario

Bt cells produce insecticidal proteins apart from parasporal crystals proteins (Cry and Cyt) during vegetative state and secretes into the culture medium. These are referred to as Vegetative Insecticidal Proteins (Vip) and are the next generation insect pest killers. Vip1 and Vip2 heterodimer toxins have an insecticidal activity against many coleopteran and hemipteran pests. Vip3, the most extensively studied family of Vip toxins, is effective against lepidopterans (Chakroun et al. 2016a, b; Syed et al. 2020; Chakrabarty et al. 2020). However, recently, for the first time, Wang et al. (2020a) have reported that Vip3Aa protein has also activity against larvae of *A. aegypti.* As Vip proteins are genetically different from the Cry proteins and show very low sequence and structural homology with Cry proteins. These may have unique binding sites in target host cells (Adang et al. 2014; Chakroun et al. 2016a). The studies of Gomis-Cebolla et al. (2018) revealed that the colonies resistant to Cry1A proteins, Dipel (*H. armigera, T. ni, O. furnacalis*, and *P. interpunctella*) or Cry2Ab (*H. armigera* and *T. ni*) were not cross-resistant to Vip3 proteins. Vip3A proteins are currently used in combination with Cry1/Cry2/(Cry34/35Ab1) proteins in *Bt* corn and *Bt* cotton products (Ludwick et al. 2017; DiFonzo et al. 2018). Reports have shown that Vip3Aa is highly effective in fields (Burkness et al. 2010; Yang et al. 2015). As mentioned earlier, laboratory selections of insects to *Bt* toxins are helpful to study the resistance inheritance, mechanism, allele frequency, fitness cost, development of molecular diagnostic tools for detection, testing models to predict resistance evolution and improve the ongoing resistance management strategies. Many laboratory selections with Vip3Aa also have been studied, which yielded 285 to >3000-fold resistance to Vip3A toxin in five major lepidopteran pests (*H. armigera, H. punctigera, H. virescens, S. frugiperda*, and *S. litura*). These studies demonstrated the genetic potential of the pests for field-evolved resistance to Vip3 proteins (Mahon et al. 2012; Barkhade and Thakare 2010; Bernardi et al. 2016; Chakroun et al. 2016b; Pickett et al. 2017).

In Australia, it was observed that prior to the introduction of Vip3 based crops, the estimated frequency of alleles conferring resistance to Vip3Aa was higher than expected: 0.034 for *H. armigera* (Chakroun et al. 2016b) and 0.010 for *H. punctigera* (Downes et al. 2016). A similar kind of study in Brazil during 2013–2014 revealed the presence of resistant alleles in *S. frugiperda.* The resistance allele frequency to Vip3Aa20 overlaid on diet or *Bt* corn leaves was similar, 0.0012 and 0.0011, respectively, with an overall frequency of 0.0009 in 2013 to 2014 (Bernardi et al. 2015). Recently, Yang et al. (2020a) have documented presence of major Vip3Aa resistance alleles conferring high resistance to Vip3Aa protein in field populations of *H. zea* in Texas, United States. This is an alarming situation because there have been already the incidence of field resistance to Cry1 and Cry2 proteins in *H. zea* populations in the United States (Ali et al. 2006; Ali and Luttrell 2007; Dively et al. 2016; Storer et al. 2001; Kaur et al. 2019).

3.9 Mechanism of Resistance

Bt toxin mode of action has been intensively studied and frequently reviewed. Despite that many details of mode of action of *Bt* Cry toxins are far from understood, the key steps in mode of action, i.e., crystal solubilization, proteolytic activation, receptor binding, membrane insertion, and pore formation are agreed by and large (Knowles 1994; Schnepf et al. 1998). Insects could develop resistance to Cry toxins due to alteration at any step of the sequential process of intoxication (Heckel 1994). A variety of proteins have been identified and characterized as receptors or putative receptors for Cry binding of *Bt* toxins. These include cadherins, amino peptidases (APNs), alkaline phosphatases (ALPs), ATP-binding cassette (ABC) transporters (Wu 2014; Heckel 2012).

The modification in the binding characteristics of the receptors either in cadherin or amino peptidase showed significant level of resistance to *Bt* proteins (Gahan et al. 2001; Bravo et al. 2004). For example, deletion mutation in cadherin receptors and (Xu et al. 2005) amino peptidase receptors showed resistance in *H. armigera* (Zhang et al. 2009) to Cry1Ac protein. Similarly modification in receptors altered the resistance (Baxter et al. 2005; Higuchi et al. 2007; Gahan et al. 2010a, b; Khajuria et al. 2011). Sivakumar et al. (2007) demonstrated susceptibility of *H. armigera* against Cry1Ac protein by silencing the HaAPN1. Likewise reduction in the susceptibility against *M. sexta* was achieved by silencing cadherin receptor gene (Fabrick et al. 2009). Fabrick et al. (2014) reported alternative splicing and highly variable cadherin transcripts were associated with field-evolved resistance of pink bollworm to *Bt* cotton in India. Similarly, Morin et al. (2003), Wang et al. (2018a), Wang et al. (2019) Wang et al. (2020a, b, c), and Fabrick et al. (2020) linked alteration in a cadherin receptor to the resistance of pink bollworm to *Bt* toxin. Mutation in cadherin conferred resistance to *Bt* toxin in other insects also such as in *Chilo suppressalis* larvae (Zhou et al. 2020a), *H. armigera* (Xu et al. 2005; Yang et al. 2007; Zhao et al. 2010; Zhang et al. 2012).

Many studies have also revealed that ABC transporter gene is involved in Cry1Ac intoxication in insects (Gahan et al. 2001; Gahan et al. 2010a, b; Atsumi et al. 2012; Xiao et al. 2014). Guo et al. (2015) showed that down-regulation of ABC transporter gene (PxWhite) is linked with Cry1Ac resistance in *P. xylostella*. Ocelotl et al. (2017) demonstrated that *ABCC2* gene is associated with *Bt* Cry1Ac toxin oligomerization and membrane insertion in diamondback moth. Further, Guo et al. (2019) performed the CRISPR/Cas9-mediated knockout of both the *PxABCC2* and *PxABCC3* genes, which resulted in high-level resistance to *B. thuringiensis* Cry1Ac toxin in the diamondback moth; thus validated their role in resistance. Reduced expression of the P-glycoprotein gene, *PxABCB1* was also found to be resistance to *Bt* Cry1Ac toxin in *P. xylostella* (Zhou et al. 2020b).

Midgut cadherins and ATP-binding cassette transporter proteins (ABCs) are among the receptors of *B. thuringiensis* Cry1A toxins in several insects (Heckel 2012; Wu 2014; Ocelotl et al. 2017). Disruption of these genes has been identified as genetically linked to resistance to Cry1A toxins in several insects, such as *H. armigera* (Xu et al. 2005; Yang et al. 2007; Zhao et al. 2010; Zhang et al.

2012; Xiao et al. 2014), *P. gossypiella* (Morin et al. 2003; Fabrick et al. 2014; Wang et al. 2018a; Wang et al. 2019; Wang et al. 2020b; Fabrick et al. 2020) and *H. virescens* (Gahan et al. 2001; Gahan et al. 2010a, b).

3.10 Validation of Insect Resistance with the Genome Editing Tool CRISPR-Cas9

The genome editing tool, CRISPR-Cas9 is a powerful genetic manipulation tool and represents an invaluable system for the precise editing of genes in diverse species (Jinek et al. 2012; Hsu et al. 2014). This technology has been used for the validation of genes related to *Bt* toxin resistance in insects. As described above in the section of "Mechanism of Resistance" many genes have been identified whose disruption or mutation is linked to resistance. Many scientists validated the genes related to insect resistance using CRISPR-Cas9.

Wang et al. (2016) used the CRISPR/Cas9 genome editing system to knock out cadherin gene (*HaCad*) from the Cry1Ac-susceptible SCD strain of *H. armigera*. The results of Western blotting experiment confirmed that HaCad was no longer expressed in the edited line while an intact HaCad of 210 kDa was present in the parental SCD strain. The bioassays showed that edited line *H. armigera* exhibited 549-fold resistance to Cry1Ac compared with susceptible strain, but no significant change in susceptibility to Cry2Ab. It provided strong evidence for HaCad as a functional receptor of Cry1Ac. Similarly, Guo et al. (2019) did functional validation of ABC transporter genes in *Bt* resistance by the CRISPR/Cas9 system. They also utilized the novel CRISPR/Cas9 genome engineering system to successfully construct two knockout strains from Cry1Ac susceptible *P. xylostella*: the ABCC2KO strain that was homozygous for a 4-bp deletion in exon 3 of the *PxABCC2* gene, and the ABCC3KO strain, which was homozygous for a 5-bp deletion in exon 3 of the *PxABCC3* gene. Both of these strains produced truncated ABCC proteins. The bioassay results with these strains indicated high levels of resistance to the Cry1Ac protoxin. This indicated a causal link between alterations in these functional candidate genes (*PxABCC2* and *PxABCC3*) and Cry1Ac resistance in *P. xylostella*. Similarly, Huang et al. (2020) used CRISPR-mediated knockouts to evaluate the role of five genes (*SeAPN1, SeCad1, SeABCC1, SeABCC2, or SeABCC3*) encoding *Bt* toxin receptors in *S. exigua* and compared susceptibility to *Bt* toxins Cry1Ac, Cry1Fa, and Cry1Ca between the parent susceptible strain and each of five strains homozygous for the knockout of one of the candidate genes. Their results revealed that *SeABCC2* has a major role and *SeCad1* a minor role in mediating toxicity of Cry1Ac and Cry1Fa. *SeABCC2* also has a minor role in toxicity of Cry1Ca. In addition, the results entailed a little or no role for the other three candidate receptors in toxicity of Cry1Ac or Cry1Fa; or for the four candidate receptors other than *SeABCC2* in toxicity of Cry1Ca. In *O. furnacalis* CRISPR-mediated knockout of the *ABCC2* gene resulted in high-level resistance to the *Bt* toxin Cry1Fa indicating its role as a receptor (Wang et al. 2020c). Similarly, CRISPR/Cas9-mediated genome editing of *H. armigera* with mutations of an ABC transporter gene *HaABCA2* conferred resistance to *Bt* Cry2A toxins (Wang et al. 2017). Many researchers have demonstrated the role of aminopeptidases (Rajagopal et al. 2009; Sivakumar et al. 2007; Sun et al. 2020) and alkaline phosphatases (Hua et al. 2009; Jiménez et al. 2012; Martins et al. 2010; Lee et al. 2014) as binding proteins for *Bt* toxins in the midgut of insects. Guo et al. (2020) recently identified and characterized aminopeptidase as binding protein for *Bt* toxin Cry3Aa in the midgut of *Monochamus alternatus* (Coleoptera: Cerambycidae).

The other studies of functional validation for receptors in insects involved in *Bt* protein resistance using CRISPR-mediated gene knockouts of *PxABCC2* and *PxABCC3* in *P. xylostella* (Liu et al. 2020); knockout of three aminopeptidase N genes in *H. armigera* (Wang et al. 2020b); knockout ABC transporter gene *HaABCA2* in *H. armigera* (Wang et al. 2017); knockout of the *ABCC2* gene in *O. furnacalis* (Wang et al. 2020c); knockout of the cadherin gene in *S. frugiperda* (Zhang et al. 2020); knockout of cadherin gene in *C. suppressalis* (Zhou et al. 2020a), have been carried out. Thus, CRISPR/Cas9 technique can act as a powerful and efficient genome editing tool to study gene function in agricultural pests.

3.11 Insect Resistance Management

When the *Bt* crops were commercialized for the first time in the nineties, strategies for delaying pest resistance were totally dependent on theoretical projections from modeling. From that time onwards, global monitoring has documented both significant successes and failures in terms of managing pest resistance to *Bt* crops. These successes include sustained susceptibility of *H. armigera* and *H. punctigera* in Australia; *H. virescens*, *O. nubilalis*, and *P. gossypiella* in the United States; and *P. gossypiella* in China (Tabashnik and Carrière 2017).

The refuge strategy has been the primary approach to delay evolution of insect resistance to *Bt* crops, wherein host plants that do not produce *Bt* toxins (refuges) was expected to boost survival of susceptible pests (Shelton et al. 1998; Shelton et al. 2000; Bates et al. 2005). Laboratory and green house experiments, large-scale studies, retrospective evaluations of global resistance-monitoring data of field-evolved resistance showed that refuges can delay resistance (Tabashnik et al. 2009; Tabashnik et al. 2013). Greenhouse trials of Shelton et al. (1998) revealed that pure stands of *Bt*-expressing plants (0% refuge) resulted in fast development of highly resistant diamondback moth populations, and increased size of the refuge delayed the development of resistance. Their finding also showed that the position of the refuge plants significantly affected the development of resistance. When both plant types were mixed in a random spatial arrangement (mixed seedling model), larvae were able to move between plant types. As they moved from refuge plants to *Bt*-expressing plants, they died and caused an overall decline in the number of susceptible alleles (Shelton et al. 1998; Zhao et al. 2005). In refuge strategy rare survivors from *Bt* crops mate with the relatively abundant susceptible pests that flourish in refuges. If inheritance of resistance is recessive the resulting heterozygous progeny from such mating will be killed by *Bt* crops, thus greatly delaying the evolution of the resistance. This is sometimes called as the high-dose refuge strategy. The refuge strategy works best if *Bt* crops produce constantly a dose of *Bt* proteins high enough to kill all, or almost all, of the heterozygous insects that feed on *Bt* plants. On the contrary, if resistance is non-recessive or partial recessive and *Bt* crops do not produce the high-dose standard for the indented pest and the usefulness of the refuge strategy will be mainly compromised (Bates et al. 2005; Wu 2014; Tabashnik et al. 2009, 2013). If the desired high dose is not achieved, then resistance can be delayed using more abundant refuges, which compensates for the existence of the heterozygous progeny on *Bt* plants by decreasing the proportion of the population selection for resistance (Tabashnik et al. 2009).

The other two strategies being used are: refuges in combinations with pyramided *Bt* crops and planting random mixtures of *Bt* and non-*Bt* seeds. In gene pyramiding (gene stacking) strategy for *Bt* resistance management, crops produce two or more dissimilar *Bt* pesticidal proteins (e.g., Cry1Ac + Cry2Ab in Cotton Bollgard II) in order to delay resistance, improve efficacy against some pests, and broaden the spectrum of pest controlled. The toxins used for pyramiding should have no cross resistance. Pyramided crops were first commercialized in 2003 and are being used currently in many countries (Carrière et al. 2016; Wu 2014).

In Arizona, pink bollworm control program was started in 2006, which involved the mass releases of irradiated, partially sterile pink bollworm moths. After using this approach, pink bollworm populations declined heavily and the percentage of pink bollworm infested cotton bolls collected from non-Bt cotton fields dropped to >90 with the elimination of insecticide sprays against this pest (Tabashnik et al. 2010; Liesner et al. 2011, 2018, 2019). The control of this pest saved $192 million from 2014 to 2019 in the United States (Tabashnik et al. 2020).

In India, transgenic cotton producing Cry1Ac and Cry1Ac + Cry2Ab proteins were commercialized in 2002 and 2006, respectively (Choudhary and Gaur 2010, 2015). Keeping in view of the insect resistance management, Genetic Engineering Approval Committee (GEAC), Ministry of Environment, Forest and Climate Change (Government of India), approved commercialization of *Bt* cotton (Cry1Ac) with the mandate of planting refuge crop of non-*Bt* in five perimeter rows or 20% of the sown area (structured refuge) whichever is more (Mohan 2018). Following the GEAC requirement, seed companies provided 120 g of non-*Bt* cotton seeds as a separate packet with every packet of 450 g of *Bt* cotton seeds (Kranthi et al. 2017; Mohan 2018). *Bt* cotton managed American bollworm *H. armigera* (Hubner), spotted bollworm, *Earias vitella* (Fabricious) and pink bollworm, *P. gossypiella*. Apart from these bollworms, it also provided protection against other minor lepidopteran insect pests, such as leaf eating caterpillars, hairy caterpillars and semiloopers. In the beginning, the *Bt* cotton technology in India performed well. However, toward the end of the first decade of the commercialization, *Bt* cotton expressing Cry1Ac, lost its battle against the pink bollworm (Choudhary and Gaur 2010; Dhurua and Gujar 2011).

In the second decade of *Bt* cotton cultivation, pink bollworm has become a major pest on the *Bt* cotton with single (Cry1Ac) or dual proteins (Cry1Ac + Cry2Ab) in central India and southern India (Naik et al. 2018; Mohan 2017; Fand et al. 2019;

Tabashnik and Carrière 2019). Many reasons can be attributed to the pink bollworm resistance to *Bt* proteins in cotton in India but certain factors that appear likely are: (i) Non-compliance of cotton farmers to refuge strategy, (ii) Supply of fraudulent refuge seeds by the seed companies, (iii) Extensive cultivation of long duration *Bt* cotton hybrids with diverse flowering and fruiting windows, and (iv) Extension of normal crop season through ratooning or providing supplementary irrigation and fertilizer applications (Kranthi 2015; Kranthi et al. 2017). Recently, Naik et al. (2020) performed the genetic diversity analysis of 214 Indian pink bollworm populations collected from nine cotton growing states comprising of forty four major cotton growing districts using mitochondrial cytochrome oxidase I (COI) gene and concluded that there is pink bollworm population expansion in India.

Low compliance with refuge planting could be due to misunderstanding the purpose of refuges and a misconception that there will be reduction in yield as 20% area is meant for refuge. Recognizing the problems with resistance and the lack of farmer compliance with planting separate blocks of non-*Bt* cotton as structured refuges, in December 2016, Ministry of Agriculture, Cooperation and Farmers Welfare, Government of India (GoI) endorsed the implementation of "refuge-in-bag" (RIB) strategy. A succeeding notification directed the *Bt* seed companies to implement RIB with isogenic refuge, wherever available, and employ a complete shift to isogenic refuge by December 2019. This recent notification directs that the *Bt* trait purity of the blend should be between 90% and 95% and that of the isogenic non-*Bt* refuge seeds of the corresponding *Bt* hybrids should be between 5% and 10% (Mohan 2018). In addition to RIB, Fand et al. (2019) have proposed that refuge planting should be made a mandatory requirement and added some other points for promoting implementation of refuge strategy, that: (i) growers should be provided incentives only if they fulfill the strict requirements of refuge planting, and (ii) to receive compensation from the government, in case of eventual damage and yield loss due to PBW. They also suggested that pre-monsoon sowing in the months of April–May in irrigated pockets should be avoided in order to lessen the PBW menace in cotton.

The shift from structured refuge to RIB leads the farmers to have refuge crops in their field invariably and there is no need to maintain separate refuge. However, as pink bollworm has got high levels of resistance to both Cry1Ac and Cry2Ab expressed in cotton in India, this refuge is too little and too late to substantially remedy the high level of resistance (Tabashnik and Carrière 2019). Furthermore, when the US Environmental Protection Agency (EPA) approved the reduction of the non-*Bt* crop refuge size to 5% for pyramided *Bt* corn, scientists in the United States have challenged it citing evidence is lacking that such a low refuge percentage can delay pest resistance (Alyokhin 2011; Tabashnik and Gould 2012; Yang et al. 2014; Carrière et al. 2016). Results from modeling and small-scale experiments showed that seed mixtures may significantly accelerate resistance relative to block refuges when larvae move extensively between *Bt* and non *Bt* plants (Carrière et al. 2016). It is pertinent to mention that analysis of 11 years of field monitoring data from Yangtze river valley of China, where millions of farmers serendipitously employed a novel seed mixture strategy by planting second-generation seeds from crosses between *Bt* and non-*Bt* cotton, indicating that this approach delayed or even reversed *P. gossypiella* resistance to single-toxin *Bt* cotton while sustaining pest suppression. This strategy produced a refuge of 25% non-*Bt* plants randomly interspersed within fields of *Bt* cotton (Wan et al. 2017).

Currently, there is no *Bt* cotton crop available or in pipeline for commercialization in India, which could be used to control the resistant pink bollworm populations. Thus IPM is the best option to deal with this resistance menace and achieve sustainable pest suppression. Various kinds of IPM strategies recommended for pink bollworm control in India include: (i) growing early to medium maturing cotton hybrids, (ii) judicious use of insecticides based on scouting and thresholds, (iii) biological control with natural enemies, (iv) termination of the crop by December, (v) strict avoidance of ratoon cotton after harvest, (vi) destruction or removal of crop residues after harvest, (vii) deep summer ploughings, crop rotation, and (viii) pheromones for mass trapping and mating disruption (Kranthi 2015; Mohan 2017; Tabashnik and Carrière 2019).

Insects are outstandingly adaptable and are anticipated to evolve resistance to any management strategy. However, discovery of new proteins or genetically modification of existing *Bt* toxins that can kill resistant pest populations to native *Bt* toxins and utilization of insecticidal proteins from bacteria other than *Bt*, insects, animals, plants that act as inhibitors of insect digestive enzymes (e.g., protease inhibitors, a-amylase inhibitors, and cholesterol oxidase) will continue to provide new tools for insect pest management. RNAi technology, in which a small double-stranded RNA (dsRNA) causes a sequence-specific suppression of target gene expression (Agrawal et al. 2003; Rana and Mohankumar 2016; Rana et al. 2020), has a great potential to act as a substitute, or complement, to *Bt* pesticidal proteins in transgenic crops for managing insect pests (Kim et al. 2015; Head et al. 2017; Zhang et al. 2010; Zhang et al. 2020; Ma and Zhang 2019; Zhu and Palli 2020). The regular assessment of global patterns of field-evolved resistance to transgenic crops will provide experimental support for a framework to effectively manage pest resistance in current and future transgenic crops (Shelton et al. 2000; Tabashnik and Carrière 2017; Tabashnik and Carrière 2019).

3.12 Safety of *Bt* Crops

The toxins of the *Bt* crops did not show any negative effect on soil bacteria, actinomycetes, fungi, protozoa, algae, nematodes, or earthworm. *Bt* corn or *Bt* cotton was found to have no significant effect on populations of beneficial insects (Abbas 2018). The population of target pests is suppressed in *Bt* crops. However, the secondary insect pests attain the major pest status at times. Lu (2010) reported that high infestation of mirid bugs in *Bt* cotton in China and attained the status of key pest on *Bt* cotton. Similarly, the continuous cultivation of *Bt* cotton leads to the infestation by aphids and mealybugs in India (Losey et al. 1999).

Hilbeck et al. (1998, 1999) found significant mortality on larvae of *Chrysoperla carnea,* which were fed on artificial diet mixed with Cry1Ab toxin or fed on

S. littoralis reared on artificial diet mixed with the Cry1Ab and Cry2A toxins. However, Mendelshon et al. (2003) conducted laboratory studies and found that pollen containing Cry toxins was not toxic to coccinellids, green lacewings (*Chrysoperla* spp.), or honeybees. They also found that beneficial arthropods were substantially more abundant in *Bt* crops than in crops treated with chemical pesticides. Dahi (2013) reported that the populations of the predators were not affected in *Bt* cotton ecosystem.

There were no effects on weight and survival of honey bees when they were fed with Cry1Ab sweet corn pollen. In field studies also, the honey bee colonies foraging in *Bt* corn plots and the honey bees fed with *Bt* pollen cakes showed no adverse effects on bee-weight, foraging activity, and colony performance (Rose et al. 2007). There were studies indicating that the mice/rat/sheep fed with *Bt* crop produces were not affected (Wang et al. 2002; deVendomois et al. 2009; Anilkumar et al. 2010).

3.13 Conclusions

1. Biopesticides pose potentially less risk to humans and the environment as compared to chemical pesticides and have been attracting global attention as a safer strategy than chemical pest control.
2. More cooperation between the public and private sectors is required in order to facilitate the development, manufacturing and sale of these ecofriendly alternatives.
3. Research on characterization of *Bt* isolates and development of effective formulations would boost the commercialization and use of biopesticides.
4. Assuring the availability of the *Bt* biopesticides to the farmers at affordable cost, especially in developing countries, is also important.
5. The efficacy of *Bt* formulations in the field has to be improved by nanotechnology methods.
6. Continuous research efforts for identification and characterization of alternate pesticidal proteins of *Bt* or from any other sources is also important for counteracting the insect resistance development.
7. Making awareness among the farmers of developing countries to adopt resistance management strategies to delay the development of resistance in insects.

3.14 Future Perspectives

1. The viability and longevity of *Bt* based formulations under field conditions can be increased with the help of nanotechnology.
2. Novel *Bt* proteins against various pests could be discovered and employed in integrated pest management, using next generation sequencing technologies.

3.15 Points to Remember

- *B. thuringiensis* is a ubiquitous Gram positive bacterium, having wide spectrum of pesticidal activity against lepidopterans, dipterans, coleopterans, hemipterans, hymenopteran insects, and nematodes.
- At present, there are 71 H serotypes and 83 serovars of *B. thuringiensis.*
- Currently, biopesticides share 5% of the total crop protection market globally, with a value of about $3 billion, of which 90% is from *Bt* based microbial biopesticides.
- *Bt* is a stomach poison and it causes mortality in insects by a series of steps, *viz.* activation of gut proteases, fragmentation of protoxin in to toxin, binding with the midgut receptors, pore formation, osmolysis, cell lysis, cessation of feeding and finally death of the insect.
- *Bt* produces parasporal crystal (Cry) proteins majorly contain delta-endotoxins, which lead to the mortality in insects. Besides, the cytolytic proteins (Cyt) and vegetative insecticidal proteins (Vip) of *Bt* also cause mortality.
- The limitation factors for *Bt* biopesticides are shortage or non-availability of large-scale production unit facilities and quality control issues and high cost of formulations.
- The genome size of *Bt* is about 5 to 6.7 Mbp and presence of transposable element is responsible for diversity of *cry* genes.
- Cry proteins have different amino acid sequence and require three domain structures for their insecticidal activity. Domain I is essential for proteolytic activation, pore formation, and membrane insertion. Domain II is essential for reorganization and binding of receptor, oligomerization, and membrane insertion and domain III is for insect specificity and ion channel modification.
- Based on amino acid sequence homology, a four-level naming system was adopted in which proteins that shared less than 45% sequence identity were placed in the primary rank (Cry1, Cry2, Cry3, Cry4, etc.), proteins that shared 45–78% identity were assigned different secondary ranks (Cry1A, Cry1B, Cry1C, etc.), proteins that shared 78–95% sequence identity (Cry1Aa, Cry1Ab, Cry1Ac, etc.) were allocated tertiary rank and finally a fourth level was used for the proteins that shared greater than 95% identity (Cry1Aa1, Cry1Aa2, Cry1Aa3, etc.).
- Recent classification of bacterial pesticidal proteins (BPP) has 16 classes of proteins from various microbial origins including *Bt.*
- The *Bt* proteins, *viz.* Cry1Ab, Cry1Ac, Cry1Fa2, Cry2Ab, Cry2Ae, Cry1A.105, Cry3A, Cry3Bb1, Cry34Aa1, Cry35Ab1, Cry51Aa2, and Vip3Aa have been deployed in commercialized genetically engineered crops.
- Practical field evolved resistance has been reported for *P. gossypiella* in *Bt* cotton (single toxin Cry1Ac and dual toxins Cry1Ac + Cry2Ab), *B. fusca, D. saccharalis, D. virgifera virgifera*, *S. frugiperda*, and *S. albicosta* in *Bt* corn, and *H. zea* in both *Bt* corn and cotton. The fastest case of field evolved resistance in *S. frugiperda* to *Bt* maize (Cry1F) has been recorded in the US territory of Puerto Rico in just 3 years.

- Genome editing technologies will be helpful for understanding the mechanism of resistance in insects.
- The resistance management strategies (especially refugia) have to be followed by the farmers scrupulously to delay the development of insect resistance.

Suggested Websites for Readers

http://www.omafra.gov.on.ca
https://www.aatf-africa.org
https://www.infonet-biovision.org
http://cr.biosafetyclearinghouse.net
http://www.geacindia.gov.in
http://www.isaaa.org
https://www.bpprc.org
http://www.bgsc.org
http://www.wdcm.org
http://www.wfcc.info

References

Abbas MST (2018) Genetically engineered (modified) crops (*Bacillus thuringiensis* crops) and the world controversy on their safety. Egypt J Biol Pest Control 28:52

Adang MJ, Staver MJ, Rocheleau TA, Leighton J, Barker RF, Thompson DV (1985) Characterized full-length and truncated plasmid clones of the crystal protein of *Bacillus thuringiensis* subsp. *kurstaki* HD-73 and their toxicity to *Manduca sexta*. Gene 36(3):289–300

Adang MJ, Brody MS, Cardineau G, Eagan N, Roush RT, Shewmaker CK, Jones A, Oakes JV, McBride KE (1993) The reconstruction and expression of a *Bacillus thuringiensis cryIIIA* gene in protoplasts and potato plants. Plant Mol Biol 21(6):1131–1145

Adang MJ, Crickmore N, Jurat-Fuentes JL (2014) Diversity of *Bacillus thuringiensis* crystal toxins and mechanism of action. In: Advances in insect physiology, vol 47. Academic Press, New York, pp 39–87

Agrawal N, Dasaradhi PVN, Mohmmed A, Malhotra P, Bhatnagar RK, Mukherjee SK (2003) RNA interference: Biology, mechanism, and applications. Microbiol Mol Biol Rev 67(4):657–685

Agrawal A, Venkatesan T, Ramasamy GG, Syamala R, Muthugounder M, Rai A (2020) Transcriptome alterations of field-evolved resistance in *Pectinophora gossypiella* against *Bt* Bollgard II cotton in India. J Appl Entomol 144(10):929–940. https://doi.org/10.1111/jen.12805

Akhurst RJ, James W, Bird LJ, Beard C (2003) Resistance to the Cry1Ac δ-endotoxin of *Bacillus thuringiensis* in the cotton bollworm, *Helicoverpa armigera* (Lepidoptera: Noctuidae). J Econ Entomol 96(4):1290–1299

Alcantara E, Estrada A, Alpuerto V, Head G (2011) Monitoring Cry1Ab susceptibility in Asian corn borer (Lepidoptera: Crambidae) on Bt corn in the Philippines. Crop Prot 30(5):554–559

Ali MI, Luttrell RG (2007) Susceptibility of bollworm and tobacco budworm (Lepidoptera: Noctuidae) to Cry2Ab2 insecticidal protein. J Econ Entomol 100(3):921–931

Ali MI, Luttrell RG, Young SY III (2006) Susceptibilities of *Helicoverpa zea* and *Heliothis virescens* (Lepidoptera: Noctuidae) populations to Cry1Ac insecticidal protein. J Econ Entomol 99(1):164–175

Alves AP, Spencer TA, Tabashnik BE, Siegfried BD (2006) Inheritance of resistance to the Cry1Ab *Bacillus thuringiensis* toxin in *Ostrinia nubilalis* (Lepidoptera: Crambidae). J Econ Entomol 99 (2):494–501

Alyokhin A (2011) Scant evidence supports EPA's pyramided *Bt* corn refuge size of 5%. Nat Biotechnol 29(7):577–578

Andow DA, Pueppke SG, Schaafsma AW, Gassmann AJ, Sappington TW, Meinke LJ, Mitchell PD, Hurley TM, Hellmich RL, Porter RP (2016) Early detection and mitigation of resistance to *Bt* maize by western corn rootworm (Coleoptera: Chrysomelidae). J Econ Entomol 109(1):1–12

Andrews RE, Faust RM, Wabiko H, Raymond KC, Bulla LA (1987) The biotechnology of *Bacillus thuringiensis*. Crit Rev Biotechnol 6(2):163–232

Aneja KR, Khan SA, Aneja A (2016) Biopesticides an eco-friendly pest management approach in agriculture: status and prospects. Kavaka 47:145–154

Anilkumar KJ, Rodrigo-Simón A, Ferré J, Pusztai-Carey M, Sivasupramaniam S, Moar WJ (2008) Production and characterization of *Bacillus thuringiensis* Cry1Ac-resistant cotton bollworm *Helicoverpa zea* (Boddie). Appl Environ Microbiol 74(2):462–469

Anilkumar B, Gopala A, Kalukumar B, Usha M, Anjaneyulu Y, Gopi KS (2010) Serobiochemical studies in sheep fed on *Bt* cotton plants. Toxicol Int 17(2):99–101

Arora NK, Khare E, Maheshwari DK (2010) Plant growth promoting rhizobacteria: constraints in bioformulation, commercialization, and future strategies. In: Maheshwari DK (ed) Plant growth and health promoting bacteria. Springer, Berlin, pp 97–116

Arthurs S, Dara SK (2019) Microbial biopesticides for invertebrate pests and their markets in the United States. J Invertebr Pathol 165:13–21

Atsumi S, Miyamoto K, Yamamoto K, Narukawa J, Kawai S, Sezutsu H, Kobayashi I, Uchino K, Tamura T, Mita K, Kadono-Okuda K (2012) Single amino acid mutation in an ATP-binding cassette transporter gene causes resistance to *Bt* toxin Cry1Ab in the silkworm, *Bombyx mori*. Proc Natl Acad Sci U S A 109(25):E1591–E1598

Balasubramanian P, Jayakumar R, Shambharkar P, Unnamalai N, Pandian SK, Kumaraswami NS, Ilangovan R, Sekar V (2002) Cloning and characterization of the crystal protein-encoding gene of *Bacillus thuringiensis* subsp. *yunnanensis*. Appl Environ Microbiol 68:408–411

Ballardo C, Abraham J, Barrena R, Artola A, Gea T, Sánchez A (2016) Valorization of soy waste through SSF for the production of compost enriched with *Bacillus thuringiensis* with biopesticide properties. J Environ Manag 169:126–131

Ballardo C, Barrena R, Artola A, Sánchez A (2017) A novel strategy for producing compost with enhanced biopesticide properties through solid-state fermentation of biowaste and inoculation with *Bacillus thuringiensis*. Waste Manag 70:53–58

Ballardo C, del Carmen Vargas-García M, Sánchez A, Barrena R, Artola A (2020) Adding value to home compost: Biopesticide properties through *Bacillus thuringiensis* inoculation. Waste Manag 106:32–43

Balog A, Hartel T, Loxdale HD, Wilson K (2017) Differences in the progress of the biopesticide revolution between the EU and other major crop-growing regions. Pest Manag Sci 73 (11):2203–2208

Barbosa LCB, Farias DL, Silva IDMG, Melo FL, Ribeiro BM, Aguiar RWDS (2015) Draft genome sequence of *Bacillus thuringiensis* 147, a Brazilian strain with high insecticidal activity. Mem Inst Oswaldo Cruz 110(6):822–823

Barkhade UP, Thakare AS (2010) Protease mediated resistance mechanism to Cry1C and Vip3A in *Spodoptera litura*. Egypt Acad J Biol Sci A Entomol 3(2):43–50

Bates SL, Zhao JZ, Roush RT, Shelton AM (2005) Insect resistance management in GM crops: past, present and future. Nat Biotechnol 23(1):57–62

Batra CP, Mittal PK, Adak T (2000) Control of *Aedes aegypti* breeding in desert coolers and tires by use of *Bacillus thuringiensis* var. *israelensis* formulation. J Am Mosq Control Assoc 16 (4):321–323

Baum JA, Johnson TB, Carlton BC (1999) *Bacillus thuringiensis* - Natural and recombinant bioinsecticide products. Methods Biotechnol 5:189–209

Baum JA, Chu CR, Rupar M, Brown GR, Donovan WP, Huesing JE, Ilagan O, Malvar TM, Pleau M, Walters M, Vaughn T (2004) Binary toxins from *Bacillus thuringiensis* active against the western corn rootworm, *Diabrotica virgifera virgifera* LeConte. Appl Environ Microbiol 70 (8):4889–4898

Baxter SW, Zhao JZ, Gahan LJ, Shelton AM, Tabashnik BE, Heckel DJ (2005) Novel genetic basis of field-evolved resistance to *Bt* toxins in *Plutella xylostella*. Insect Mol Biol 14:327–334. https://doi.org/10.1111/j.1365-2583.2005.00563.x

Beard CE, Ranasinghe C, Akhurst RJ (2001) Screening for novel *cry* genes by hybridization. Lett Appl Microbiol 33:241–245. https://doi.org/10.1046/j.1472-765X.2001.00982.x

Ben-Dov E (2014) *Bacillus thuringiensis* subsp. *israelensis* and its dipteran-specific toxins. Toxins 6(4):1222–1243

Berliner E (1915) Ueber die schlaffsucht der *Ephestia kuhniella* und *Bac. thuringiensis* n. sp. Z Angew Entomol 2:21–56

Bernardi O, Bernardi D, Ribeiro RS, Okuma DM, Salmeron E, Fatoretto J, Medeiros FC, Burd T, Omoto C (2015) Frequency of resistance to Vip3Aa20 toxin from *Bacillus thuringiensis* in *Spodoptera frugiperda* (Lepidoptera: Noctuidae) populations in Brazil. Crop Prot 76:7–14

Bernardi O, Bernardi D, Horikoshi RJ, Okuma DM, Miraldo LL, Fatoretto J, Medeiros FC, Burd T, Omoto C (2016) Selection and characterization of resistance to the Vip3Aa20 protein from *Bacillus thuringiensis* in *Spodoptera frugiperda*. Pest Manag Sci 72(9):1794–1802

Betz FS, Hammond BG, Fuchs RL (2000) Safety and advantages of *Bacillus thuringiensis*-protected plants to control insect pests. Regul Toxicol Pharmacol 32(2):156–173

Biosafety Clearing-House (2014) Gene and DNA Sequence, Cry1A.105. https://bch.cbd.int/database/record.shtml?documentid-43771. Accessed 16 Feb 2017

Bird LJ, Akhurst RJ (2004) Relative fitness of Cry1A-resistant and-susceptible *Helicoverpa armigera* (Lepidoptera: Noctuidae) on conventional and transgenic cotton. J Econ Entomol 97 (5):1699–1709

Brar SK, Verma M, Tyagi RD, Valéro JR, Surampalli RY, Banerji SK (2004) Development of sludge based stable aqueous *Bacillus thuringiensis* formulations. Water Sci Technol 50 (9):229–236

Brar SK, Verma M, Tyagi RD, Valéro JR (2006) Recent advances in downstream processing and formulations of *Bacillus thuringiensis* based biopesticides. Process Biochem 41(2):323–342

Bravo A, Soberon M (2008) How to cope with insect resisitance to *Bt* toxins? Trends Biotechnol 26 (10):573–579

Bravo A, Sánchez J, Kouskoura T, Crickmore N (2002) N-terminal activation is an essential early step in the mechanism of action of the *Bacillus thuringiensis* Cry1Ac insecticidal toxin. J Biol Chem 277(27):23985–23987

Bravo A, Gómez I, Conde J, Munoz-Garay C, Sánchez J, Miranda R, Zhuang M, Gill S, Soberón M (2004) Oligomerization triggers binding of a *Bacillus thuringiensis* Cry1Ab pore-forming toxin to aminopeptidase N receptor leading to insertion into membrane microdomains. Biochim Biophys Acta 1667(1):38–46

Bravo A, Gill SS, Soberón M (2007) Mode of action of *Bacillus thuringiensis* Cry and Cyt toxins and their potential for insect control. Toxicon 49(4):423–435

Bravo A, Likitvivatanavong S, Gill SS, Soberón M (2011) *Bacillus thuringiensis*: a story of a successful bioinsecticide. Insect Biochem Mol Biol 41(7):423–431

Brown KL, Whiteley HR (1992) Molecular characterization of two novel crystal protein genes from *Bacillus thuringiensis* subsp. *thompsoni*. J Bacteriol 174(2):549–557

Brownbridge M, Buitenhuis R (2019) Integration of microbial biopesticides in greenhouse floriculture: The Canadian experience. J Invertebr Pathol 165:4–12

Burkness EC, Dively G, Patton T, Morey AC, Hutchison WD (2010) Novel Vip3A *Bacillus thuringiensis* (*Bt*) maize approaches high-dose efficacy against *Helicoverpa zea* (Lepidoptera: Noctuidae) under field conditions: implications for resistance management. GM Crops 1 (5):337–343

Caccia S, Hernández-Rodríguez CS, Mahon RJ, Downes S, James W, Bautsoens N, Van Rie J, Ferre J (2010) Binding site alteration is responsible for field-isolated resistance to *Bacillus thuringiensis* Cry2A insecticidal proteins in two *Helicoverpa* species. PLoS One 5(4):e9975

Carrière Y, Fabrick JA, Tabashnik BE (2016) Can pyramids and seed mixtures delay resistance to Bt crops? Trends Biotechnol 34(4):291–302

Carroll J, Ellar DJ (1993) An analysis of *Bacillus thuringiensis* δ-endotoxin action on insect-midgut-membrane permeability using a light-scattering assay. Eur J Biochem 214(3):771–778

CERA (2012) GM Crop Database. Center for Environmental Risk Assessment (CERA), ILSI Research Foundation, Washington, DC. http://cera-gmc.org/index.php?action=gm_crop_database

Chakrabarty S, Jin M, Wu C, Chakraborty P, Xiao Y (2020) *Bacillus thuringiensis* vegetative insecticidal protein family Vip3A and mode of action against pest Lepidoptera. Pest Manag Sci 76(5):1612–1617

Chakroun M, Banyuls N, Bel Y, Escriche B, Ferré J (2016a) Bacterial vegetative insecticidal proteins (Vip) from entomopathogenic bacteria. Microbiol Mol Biol Rev 80(2):329–350

Chakroun M, Banyuls N, Walsh T, Downes S, James B, Ferré J (2016b) Characterization of the resistance to Vip3Aa in *Helicoverpa armigera* from Australia and the role of midgut processing and receptor binding. Sci Rep 6(1):1–11

Chandrasena DI, Signorini AM, Abratti G, Storer NP, Olaciregui ML, Alves AP, Pilcher CD (2018) Characterization of field-evolved resistance to *Bacillus thuringiensis*-derived Cry1Fδ-endotoxin in *Spodoptera frugiperda* populations from Argentina. Pest Manag Sci 74(3):746–754

Choudhary B, Gaur K (2010) *Bt* cotton in India: a country profile, ISAAA Series of Biotech Crop Profiles. ISAAA, Ithaca, NY

Choudhary B, Gaur K (2015) Biotech Cotton in India, 2002 to 2014, ISAAA Series of Biotech Crop Profiles. ISAAA Biotech Information Centre, New York

Corbin DR, Romano CP (2006) Monsanto Technology LLC. Plants transformed to express Cry2A δ-endotoxins. US Patent 7,064,249

Crickmore N, Zeigler DR, Feitelson J, Schnepf E, Van Rie J, Lereclus D, Baum J, Dean DH (1998) Revision of the nomenclature for the *Bacillus thuringiensis* pesticidal crystal proteins. Microbiol Mol Biol Rev 62(3):807–813

Crickmore N, Berry C, Panneerselvam S, Mishra R, Connor TR, Bonning BC (2020) A structure-based nomenclature for *Bacillus thuringiensis* and other bacteria-derived pesticidal proteins. J Invertebr Pathol:107438

Dahi HF (2013) Assessment of the effects of transgenic *Bt* cotton that contains Cry1Ac and Cry2Ab toxins on the abundance of non-target organisms community. J Nat Sci 11(2):117–122

Damalas CA, Koutroubas SD (2018) Current status and recent developments in biopesticide use. Agriculture 8(1):13

De Barjac HD (1981) Identification of H-serotypes of *Bacillus thuringiensis*. In: Burges HD (ed) Microbial control of pests and plant diseases: 1970–1980.d. Academic Press, London, pp 35–43

De Barjac H, Frachon E (1990) Classification of *Bacillus thuringiensis* strains. Entomophaga 35 (2):233–240

de Maagd RA, Bravo A, Berry C, Crickmore N, Schnepf HE (2003) Structure, diversity, and evolution of protein toxins from spore-forming entomopathogenic bacteria. Ann Rev Gen 37:409–433

deVendomois JS, Roullier F, Dominique C, Seralini G (2009) A comparison of the effects of three GM corn varieties on mammalian health. Int J Biol Sci 5(7):706–726

Devi PV, Ravinder T, Jaidev C (2005) Cost-effective production of *Bacillus thuringiensis* by solid-state fermentation. J Invertebr Pathol 88(2):163–168

Dhingra H, Chaudhary K (2011) Production of *Bacillus thuringiensis* S6 biomass using different cheap nitrogen sources. Acta Agric Serbica 16(31):3–8

Dhurua S, Gujar GT (2011) Field-evolved resistance to *Bt* toxin Cry1Ac in the pink bollworm, *Pectinophora gossypiella* (Saunders)(Lepidoptera: Gelechiidae), from India. Pest Manag Sci 67 (8):898–903

DiFonzo C, Krupke C, Shields E, Tilmon K, Tooker J (2016) An open letter to the seed industry regarding the efficacy of Cry1F against western bean cutworm. Cornell Field Crop News

DiFonzo, C., Porter, P. and Tilmon, K. 2018. The Handy *Bt* trait table; 2018. https://lubbock.tamu.edu/fles/2018/01/BtTraitTableJan2018.pdf

Dively GP, Venugopal PD, Finkenbinder C (2016) Field-evolved resistance in corn earworm to Cry proteins expressed by transgenic sweet corn. PLoS One 11(12):e0169115

Donovan WP, Dankocsik CC, Gilbert MP, Gawron-Burke MC, Groat RG, Carlton BC (1988a) Amino acid sequence and entomocidal activity of the P2 crystal protein. An insect toxin from *Bacillus thuringiensis* var *kurstaki*. J Biol Chem 263(1):561–567

Donovan WP, Dankocsik C, Gilbert MP (1988b) Molecular characterization of a gene encoding a 72-kilodalton mosquito-toxic crystal protein from *Bacillus thuringiensis* subsp. *israelensis*. J Bacteriol 170(10):4732–4738

Donovan WP, Donovan JC, Slaney AC (2000) *Bacillus thuringiensis cryET33* and *cryET34* compositions and uses therefor. Monsanto Co, U.S. Patent 6,063,756

Donovan WP, Engleman JT, Donovan JC, Baum JA, Bunkers GJ, Chi DJ, Clinton WP, English L, Heck GR, Ilagan OM, Krasomil-Osterfeld KC (2006) Discovery and characterization of Sip1A: A novel secreted protein from *Bacillus thuringiensis* with activity against coleopteran larvae. Appl Microbiol Biotechnol 72(4):713–719

Downes S, Mahon R (2012) Successes and challenges of managing resistance in *Helicoverpa armigera* to *Bt* cotton in Australia. GM Crops Food 3(3):228–234

Downes S, Walsh T, Tay WT (2016) *Bt* resistance in Australian insect pest species. Curr Opin Insect Sci 15:78–83

DPPQS (2017) Directorate of Plant Protection Quarantine and Storage, Ministry of Agriculture and Farmers Welfare, Government of India. http://ppqs.gov.in/divisions/integrated-pest-managment/bio-control-labs

Dulmage HT (1970) Insecticidal activity of HD-1, a new isolate of *Bacillus thuringiensis* var. *alesti*. J Invertebr Pathol 15(2):232–239

Eichenseer H, Strohbehn R, Burks JC (2008) Frequency and severity of western bean cutworm (Lepidoptera: Noctuidae) ear damage in transgenic corn hybrids expressing different *Bacillus thuringiensis* Cry toxins. J Econ Entomol 101(2):555–563

Ejiofor AO (1991) Production of *Bacillus thuringiensis* serotype H-14 as bioinsecticide using a mixture of spent brewer's yeast and waste cassava starch as the fermentation medium. Disc Innov 3(2):85–88

Ekino K, Okumura S, Ishikawa T, Kitada S, Saitoh H, Akao T, Oka T, Nomura Y, Ohba M, Shin T, Mizuki E (2014) Cloning and characterization of a unique cytotoxic protein parasporin-5 produced by *Bacillus thuringiensis* A1100 strain. Toxins 6(6):1882–1895

El-Bendary MA, Moharam ME, Mohamed SS, Hamed SR (2016) Pilot-scale production of mosquitocidal toxins by *Bacillus thuringiensis* and *Lysinibacillus sphaericus* under solid-state fermentation. Biocontrol Sci Tech 26(7):980–994

El-Bendary MA, Elsoud MMA, Hamed SR, Mohamed SS (2017) Optimization of mosquitocidal toxins production by *Bacillus thuringiensis* under solid state fermentation using Taguchi orthogonal array. Acta Biol Szegediensis 61(2):135–140

Ellis RT, Stockhoff BA, Stamp L, Schnepf HE, Schwab GE, Knuth M, Russell J, Cardineau GA, Narva KE (2002) Novel *Bacillus thuringiensis* binary insecticidal crystal proteins active on western corn rootworm, *Diabrotica virgifera virgifera* LeConte. Appl Environ Microbiol 68 (3):1137–1145

Entwistle P, Bailey MC, Cory JC, Higgs SC (1993) *Bacillus thurigiensis* an environmental biopesticidas theory and practice. John Wiley & Sons, Chichester

Estada U, Ferre´ J (1994) Binding of insecticidal crystal proteins of *Bacillus thuringiensis* to the midgut brush border of the cabbage looper, *Trichoplusia ni* (Hu¨bner) (Lepidoptera: Noctuidae), and selection for resistance to one of the crystal proteins. Appl Environ Microbiol 60 (10):3840–3846

Estruch JJ, Warren GW, Mullins MA, Nye GJ, Craig JA, Koziel MG (1996) Vip3A, a novel *Bacillus thuringiensis* vegetative insecticidal protein with a wide spectrum of activities against lepidopteran insects. Proc Natl Acad Sci U S A 93(11):5389–5394

Fabrick J, Oppert C, Lorenzen MD, Morris K, Oppert B, Jurat-Fuentes JL (2009) A novel *Tenebrio molitor* cadherin is a functional receptor for *Bacillus thuringiensis* Cry3Aa toxin. J Biol Chem 284:18401–18410. https://doi.org/10.1074/jbc.M109.001651

Fabrick JA, Ponnuraj J, Singh A, Tanwar RK, Unnithan GC, Yelich AJ, Li X, Carrière Y, Tabashnik BE (2014) Alternative splicing and highly variable cadherin transcripts associated with field-evolved resistance of pink bollworm to *Bt* cotton in India. PLoS One 9(5):e97900

Fabrick JA, Mathew LG, LeRoy DM, Hull JJ, Unnithan GC, Yelich AJ, Carrière Y, Li X, Tabashnik BE (2020) Reduced cadherin expression associated with resistance to *Bt* toxin Cry1Ac in pink bollworm. Pest Manag Sci 76(1):67–74

Fand BB, Nagrare VS, Gawande SP, Nagrale DT, Naikwadi BV, Deshmukh V, Gokte-Narkhedkar N, Waghmare VN (2019) Widespread infestation of pink bollworm, *Pectinophora gossypiella* (Saunders) (Lepidoptera: Gelechiidae) on *Bt* cotton in Central India: a new threat and concerns for cotton production. Phytoparasitica 47(3):313–325

Farias JR, Andow DA, Horikoshi RJ, Sorgatto RJ, Santos ACD, Omoto C (2016) Dominance of Cry1F resistance in *Spodoptera frugiperda* (Lepidoptera: Noctuidae) on TC1507 Bt maize in Brazil. Pest Manag Sci 72(5):974–979

FICCI (Federation of Indian Chambers of Commerce and Industry (2015) Study on substandard, spurious/counterfeit pesticides in India, 66 p. https://croplife.org/wpcontent/uploads/2015/10/Study-on-sub-standard-spurious-counterfeit-pesticides-in- India.pdf

Finer J, Dhillon T (2007) Transgenic plant production. In: Stewart CN (ed) Plant biotechnology and genetics: principles, techniques, and applications. John Wiley & Sons, Inc., Hoboken, NJ, pp 245–274

Fischhoff DA, Bowdish KS, Perlak FJ, Marrone PG, McCormick SM, Niedermeyer JG, Dean DA, Kusano-Kretzmer K, Mayer EJ, Rochester DE (1987) Insect tolerant transgenic tomato plants. Biotechnology 5(8):807

Flagel LE, Swarup S, Chen M, Bauer C, Wanjugi H, Carroll M, Hill P, Tuscan M, Bansal R, Flannagan R, Clark TL (2015) Genetic markers for western corn rootworm resistance to Bt toxin. G3 5(3):399–405

Foncerrada L, Sick AJ, Payne JM (1992) Novel coleopteran-active *Bacillus thuringiensis* isolate and a novel gene encoding a coleopteran-active toxin. Eur Pat Appl 0498537A2:1–5

Frank DL, Zukoff A, Barry J, Higdon ML, Hibbard BE (2013) Development of resistance to eCry3.1 Ab-expressing transgenic maize in a laboratory-selected population of western corn rootworm (Coleoptera: Chrysomelidae). J Econ Entomol 106(6):2506–2513

Frutos R, Rang C, Royer M (1999) Managing insect resistance to plants producing *Bacillus thuringiensis* toxins. Crit Rev Biotechnol 19(3):227–276

Gahan LJ, Gould F, Heckel DG (2001) Identification of a gene associated with *Bt* resistance in *Heliothis virescens*. Science 293:857–860

Gahan LJ, Pauchet Y, Vogel H, Heckel DG (2010a) An ABC transporter mutation is correlated with insect resistance to *Bacillus thuringiensis* Cry1Ac toxin. PLoS Genet 6:1–11

Gahan LJ, Pauchet Y, Vogel H, Heckel DG (2010b) An ABC transporter mutation is correlated with insect resistance to *Bacillus thuringiensis* Cry1Ac toxin. PLoS Genet 6(12):e1001248

Ganesh KN, Reyaz AL, Balakrishnan N (2018) Molecular characterization of an indigenous lepidopteran toxic *Bacillus thuringiensis* strain T532. J Biol Control 32(4):246–251

Gassmann AJ, Petzold-Maxwell JL, Keweshan RS, Dunbar MW (2011) Field-evolved resistance to *Bt* maize by western corn rootworm. PLoS One 6(7):e22629

Gassmann AJ, Petzold-Maxwell JL, Clifton EH, Dunbar MW, Hoffmann AM, Ingber DA, Keweshan RS (2014) Field-evolved resistance by western corn rootworm to multiple *Bacillus thuringiensis* toxins in transgenic maize. Proc Natl Acad Sci 111(14):5141–5146

Gassmann AJ, Shrestha RB, Jakka SR, Dunbar MW, Clifton EH, Paolino AR, Ingber DA, French BW, Masloski KE, Dounda JW, St. Clair CR (2016) Evidence of resistance to Cry34/35Ab1 corn by western corn rootworm (Coleoptera: Chrysomelidae): root injury in the field and larval survival in plant-based bioassays. J Econ Entomol 109(4):1872–1880

Gassmann AJ, Shrestha RB, Kropf AL, St Clair CR, Brenizer BD (2020) Field-evolved resistance by western corn rootworm to Cry34/35Ab1 and other *Bacillus thuringiensis* traits in transgenic maize. Pest Manag Sci 76(1):268–276

Georghiou GP (1972) The evolution of resistance to pesticides. Annu Rev Ecol Syst 3(1):133–168

Georghiou GP, Wirth MC (1997) Influence of exposure to single versus multiple toxins of *Bacillus thuringiensis* subsp. *israelensis* on development of resistance in the mosquito *Culex quinquefasciatus* (Diptera: Culicidae). Appl Environ Microbiol 63:1095–1101

Glare TR, O'Callaghan M (2019) Microbial biopesticides for control of invertebrates: Progress from New Zealand. J Invertebr Pathol 165:82–88

Goldberg LJ, Margalit J (1977) A bacterial spore demonstrating rapid larvicidal activity against *Anopheles sergentii, Uranotaenia unguiculata, Culex univitattus, Aedes aegypti and Culex pipiens*. Mosq News 37(3):355–358

Gómez I, Sánchez J, Miranda R, Bravo A, Soberón M (2002) Cadherin-like receptor binding facilitates proteolytic cleavage of helix α-1 in domain I and oligomer pre-pore formation of *Bacillus thuringiensis* Cry1Ab toxin. FEBS Lett 513:242–246

Gómez I, Pardo-López L, Munoz-Garay C, Fernandez LE, Pérez C, Sánchez J, Soberón M, Bravo A (2007) Role of receptor interaction in the mode of action of insecticidal Cry and Cyt toxins produced by *Bacillus thuringiensis*. Peptides 28(1):169–173

Gomis-Cebolla J, Wang Y, Quan Y, He K, Walsh T, James B, Downes S, Kain W, Wang P, Leonard K, Morgan T (2018) Analysis of cross-resistance to Vip3 proteins in eight insect colonies, from four insect species, selected for resistance to *Bacillus thuringiensis* insecticidal proteins. J Invertebr Pathol 155:64–70

González J Jr, Dulmage HT, Carlton BC (1981) Correlation between specific plasmids and δ-endotoxin production in *Bacillus thuringiensis*. Plasmid 5(3):351–365

Gopalakrishnan C, Gangavisalakshy PN (2005) Field efficacy of commercial formulations of *Bacillus thuringiensis* var. *kurstaki* against *Papilio demoleus* L. on citrus. Entomon 30(1):93

Goudar G, Alagawadi AR, Krishnaraj PU, Goud KB (2012) Characterization of *Bacillus thuringiensis* isolates of Western Ghats and their insecticidal activity against diamond back moth (*Plutella xylostella* L.). Karnataka J Agric Sci 25(2)

Gouffon CV, Van Vliet A, Van Rie J, Jansens S, Jurat-Fuentes JL (2011) Binding sites for *Bacillus thuringiensis* Cry2Ae toxin on heliothine brush border membrane vesicles are not shared with Cry1A, Cry1F, or Vip3A toxin. Appl Environ Microbiol 77(10):3182–3188

Gould F, Martinez-Ramirez A, Anderson A, Ferré J, Silva FJ, Moar, WJ (1992) Broad-spectrum resistance to *Bacillus thuringiensis* toxins in *Heliothis virescens*. Proc Natl Acad Sci U S A 89:7986–7990

Grimi DA, Ocampo F, Martinelli S, Head GP (2015) Detection and characterization of *Diatraea saccharalis* resistant to Cry1A.105 protein in a population of Northeast San Luis province in Argentina. Congreso Argentino de Entomología; Posadas, Misiones, Argentina

Grimi DA, Parody B, Ramos ML, Machado M, Ocampo F, Willse A, Martinelli S, Head G (2018) Field-evolved resistance to *Bt* maize in sugarcane borer (*Diatraea saccharalis*) in Argentina. Pest Manag Sci 74(4):905–913

Guan P, Ai P, Dai X, Zhang J, Xu L, Zhu J, Li Q, Deng Q, Li S, Wang S, Liu H (2012) Complete genome sequence of *Bacillus thuringiensis* serovar *sichuansis* strain MC28. Rice Research Institute, Sichuan Agricultural University, Sichuan, China

Guan P, Dai X, Zhu J, Li Q, Li S, Wang S, Li P, Zheng A (2014) *Bacillus thuringiensis* subsp. *sichuansis* strain MC28 produces a novel crystal protein with activity against *Culex quinquefasciatus* larvae. World J Microbiol Biotechnol 30(4):1417–1421

Gulzar A, Pickett B, Sayyed AH, Wright DJ (2012) Effect of temperature on the fitness of a Vip3A resistant population of *Heliothis virescens* (Lepidoptera: Noctuidae). J Econ Entomol 105:964–970

Guo Z, Kang S, Zhu X, Xia J, Wu Q, Wang S, Xie W, Zhang Y (2015) Down-regulation of a novel ABC transporter gene (Pxwhite) is associated with Cry1Ac resistance in the diamondback moth, *Plutella xylostella* (L.). Insect Biochem Mol Biol 59:30–40

Guo Z, Sun D, Kang S, Zhou J, Gong L, Qin J, Guo L, Zhu L, Bai Y, Luo L, Zhang Y (2019) CRISPR/Cas9-mediated knockout of both the PxABCC2 and PxABCC3 genes confers high-level resistance to *Bacillus thuringiensis* Cry1Ac toxin in the diamondback moth, *Plutella xylostella* (L.). Insect Biochem Mol Biol 107:31–38

Guo Y, Carballar-Lejarazú R, Sheng L, Fang Y, Wang S, Liang G, Hu X, Wang R, Zhang F, Wu S (2020) Identification and characterization of aminopeptidase-N as a binding protein for Cry3Aa in the midgut of *Monochamus alternatus* (Coleoptera: Cerambycidae). J Econ Entomol 113 (5):2259–2268

Gupta S, Dikshit AK (2010) Biopesticides: an ecofriendly approach for pest control. J Biopest 3:186–188

Gupta M, Kaur S, Debbarma A (2020) Frequency distribution of lepidopteran-specific secretory genes in indigenous *Bacillus thuringiensis* isolates recovered from diverse soil habitats of India. J Entomol Zool Stud 8(4):674–681

Haitao L, Rongmei L, Jiguo G, Jie Z (2019) *Bacillus thuringiensis* insecticidal gene *cry79Aa1*, expression protein and application thereof. Publication Number 109929015

Han S, Craig JA, Putnam CD, Carozzi NB, Tainer JA (1999) Evolution and mechanism from structures of an ADP-ribosylating toxin and NAD complex. Nat Struct Biol 6(10):932–936

Han H, Li G, Wang Z, Zhang J, He K (2009) Cross-resistance of Cry1 Ac-selected Asian corn borer to other *Bt* toxins. Acta Phytophylacica Sinica 36(4):329–334

Hansen G, Wright MS (1999) Recent advances in the transformation of plants. Trends Plant Sci 4 (6):226–231

Hartl DL (1988) A primer of population genetics. Sinauer Associates, Inc., Sunderland, MA, 221 pp. ISBN: 0878933042

Hasanain AM (2017) Development of a cheap media for *Bacillus thuringiensis* growth. Int Journal of Biotechnol Bioeng 3(6):216–223

Hatting JL, Moore SD, Malan AP (2019) Microbial control of phytophagous invertebrate pests in South Africa: current status and future prospects. J Invertebr Pathol 165:54–66

Head GP, Carroll MW, Evans SP, Rule DM, Willse AR, Clark TL, Storer NP, Flannagan RD, Samuel LW, Meinke LJ (2017) Evaluation of SmartStax and SmartStax PRO maize against western corn rootworm and northern corn rootworm: efficacy and resistance management. Pest Manag Sci 73(9):1883–1899

Heckel DG (1994) The complex genetic basis of resistance to *Bacillus thuringiensis* toxin in insects. Biocontrol Sci Tech 4(4):405–417

Heckel DG (2012) Learning the ABCs of Bt: ABC transporters and insect resistance to *Bacillus thuringiensis* provide clues to a crucial step in toxin mode of action. Pestic Biochem Physiol 104 (2):103–110

Hernández-Rodríguez CS, Van Vliet A, Bautsoens N, Van Rie J, Ferré J (2008) Specific binding of *Bacillus thuringiensis* Cry2A insecticidal proteins to a common site in the midgut of *Helicoverpa* species. Appl Environ Microbiol 74(24):7654–7659

Herrnstadt C, Gilroy TE, Sobieski DA, Bennett BD, Gaertner FH (1987) Nucleotide sequence and deduced amino acid sequence of a coleopteran-active delta-endotoxin gene from *Bacillus thuringiensis* subsp. *san diego*. Gene 57(1):37–46

Higuchi M, Haginoya K, Yamazaki T, Miyamoto K, Katagiri T, Tomimoto K, Shitomi Y, Hayakawa T, Sato R, Hori H (2007) Binding of *Bacillus thuringiensis* Cry1A toxins to brush border membrane vesicles of midgut from Cry1Ac susceptible and resistant *Plutella xylostella*. Comp Biochem Physiol B Biochem Mol Biol 147:716–724

Hilbeck A, Moar W, Pusztai-Garet M, Bigler F (1998) Toxicity of *Bacillus thuringiensis* Cry1Ab toxin to *Chrysoperla carnea*. Environ Entomol 27:1255–1263

Hilbeck A, Baumgartner M, Fried PM, Bigler F (1999) Effects of transgenic *Bt* corn fed prey on immature development of *Chrysoperla carnea*. Environ Entomol 27:480–487

Höfte H, Whiteley HR (1989) Insecticidal crystal proteins of *Bacillus thuringiensis*. Microbiol Mol Biol Rev 53(2):242–255

Höfte, H., Seurinck, J., Van Houtven, A. and Vaeck, M. 1987. Nucleotide sequence of a gene encoding an insecticidal protein of *Bacillus thuringiensis* var. *tenebrionis* toxic against Coleoptera. Nucl Acids Res 15(\): 7183

Hollensteiner J, Poehlein A, Spröer C, Bunk B, Sheppard AE, Rosentstiel P, Schulenburg H, Liesegang H (2017) Complete genome sequence of the nematicidal *Bacillus thuringiensis* MYBT18246. Stand Genomic Sci 12(1):48

Hsu PD, Lander ES, Zhang F (2014) Development and applications of CRISPR-Cas9 for genome engineering. Cell 157(6):1262–1278

Hua G, Zhang R, Bayyareddy K, Adang MJ (2009) *Anopheles gambiae* alkaline phosphatase is a functional receptor of *Bacillus thuringiensis* Cry11Ba toxin. Biochemistry 48(41):9785–9793

Huang F, Higgins RA, Buschman LL (1997) Baseline susceptibility and changes in susceptibility to *Bacillus thuringiensis* subsp. *kurstaki* under selection pressure in European corn borer (Lepidoptera: Pyralidae). J Econ Entomol 90(5):1137–1143

Huang DF, Zhang J, Song FP, Lang ZH (2007) Microbial control and biotechnology research on *Bacillus thuringiensis* in China. J Invertebr Pathol 95(3):175–180

Huang F, Ghimire MN, Leonard BR, Daves C, Levy R, Baldwin J (2012) Extended monitoring of resistance to *Bacillus thuringiensis* Cry1Ab maize in *Diatraea saccharalis* (Lepidoptera: Crambidae). GM Crops Food 3(3):245–254

Huang F, Qureshi JA, Meagher RL Jr, Reisig DD, Head GP, Andow DA, Ni X, Kerns D, Buntin GD, Niu Y, Yang F (2014) Cry1F resistance in Fall armyworm *Spodoptera frugiperda*: single gene versus pyramided *Bt* maize. PLoS One 9(11):e112958

Huang J, Xu Y, Zuo Y, Yang Y, Tabashnik BE, Wu Y (2020) Evaluation of five candidate receptors for three Bt toxins in the beet armyworm using CRISPR-mediated gene knockouts. Insect Biochem Mol Biol 121:103361

Husz B (1930) Field Experiments on the application of *Bacillus thuringiensis* against the corn borer. Internat Corn Borer Invest Sci Rep 1927-28(3):91–98

Ishiwata S (1901) On a type of severe Xacherie (sotto disease). Dainihon Sanshi Kaiho 114:1–5

Ishiwata S (1905) Concerning 'Sotto-Kin' a bacillus of a disease of the silkworm. Rept Assoc Seric 160:1–8

Ito T, Sahara K, Bando H, Asano S (2002) Cloning and expression of novel crystal protein genes *cry39A* and *39orf2* from *Bacillus thuringiensis* subsp. *aizawai* Bun1-14 encoding mosquitocidal proteins. J Insect Biotechnol Sericol 71(3):123–128

Ito A, Sasaguri Y, Kitada S, Kusaka Y, Kuwano K, Masutomi K, Mizuki E, Akao T, Ohba M (2004) A *Bacillus thuringiensis* crystal protein with selective cytocidal action to human cells. J Biol Chem 279(20):21282–21286

Ito T, Sahara K, Bando H, Asano S (2005) Cloning and expression of novel mosquitocidal crystal protein genes *cry39A, cry40A*, and *cry40B* from *B.thuringiensis* subsp. *aizawai* Bun1-14. In: Binh ND, Akhurst RJ, Dean DH (eds) Biotechnology of *Bacillus thuringiensis*, vol 5. Proceedings of the 5th Pacific Rim Conference on the Biotechnology of *Bacillus thuringiensis* and its Environmental Impact, pp 183–186

Ito T, Ikeya T, Sahara K, Bando H, Asano SI (2006) Cloning and expression of two crystal protein genes, *cry30Ba1* and *cry44Aa1*, obtained from a highly mosquitocidal strain, *Bacillus thuringiensis* subsp. *entomocidus* INA288. Appl Environ Microbiol 72(8):5673–5676

Jakka SR, Shrestha RB, Gassmann AJ (2016) Broad-spectrum resistance to *Bacillus thuringiensis* toxins by western corn rootworm (*Diabrotica virgifera virgifera*). Sci Rep 6(1):1–9

Jamil S, Shahzad R, Rahman SU, Iqbal MZ, Yaseen M, Ahmad S, Fatima R (2020) The level of Cry1Ac endotoxin and its efficacy against *Helicoverpa armigera* in Bt cotton at large scale in Pakistan. GM Crops Food 12(1):1–17

Jiang W, Bikard D, Cox D, Zhang F, Marraffini LA (2013) CRISPR-assisted editing of bacterial genomes. Nat Biotechnol 31(3):233

Jiménez AI, Reyes EZ, Cancino-Rodezno A, Bedoya-Pérez LP, Caballero-Flores GG, Muriel-Millan LF, Likitvivatanavong S, Gill SS, Bravo A, Soberón M (2012) *Aedes aegypti* alkaline phosphatase ALP1 is a functional receptor of *Bacillus thuringiensis* Cry4Ba and Cry11Aa toxins. Insect Biochem Mol Biol 42(9):683–689

Jimenez-Juarez N, Munoz-Garay C, Gomez I, Gill S, Soberon M, Bravo A (2008) The pre-pore from *Bacillus thuringiensis* Cry1Ab toxin is necessary to induce insect death in *Manduca sexta*. Peptides 29(2):318–323

Jin L, Zhang H, Lu Y, Yang Y, Wu K, Tabashnik BE, Wu Y (2015) Large-scale test of the natural refuge strategy for delaying insect resistance to transgenic Bt crops. Nat Biotechnol 33 (2):169–174

Jinek M, Chylinski K, Fonfara I, Hauer M, Doudna JA, Charpentier E (2012) A programmable dual-RNA–guided DNA endonuclease in adaptive bacterial immunity. Science 337 (6096):816–821

Jing X, Yuan Y, Wu Y, Wu D, Gong P, Gao M (2019) Crystal structure of *Bacillus thuringiensis* Cry7Ca1 toxin active against *Locusta migratoria manilensis*. Protein Sci 28:609–619. https://doi.org/10.1002/pro.3561

Jose D, Shekarappa, Patil RK (2008) Evaluation of biorational pesticides for the management of stem borer, *Chilo partellus* Swinhoe in sweet sorghum. Karnataka J Agric Sci 21(2):293–294

Juárez-Pérez V, Porcar M, Orduz S, Delécluse A (2003) Cry29A and Cry30A: Two novel δ-endotoxins isolated from *Bacillus thuringiensis* serovar *medellin*. Syst Appl Microbiol 26 (4):502–504

Karimi J, Dara SK, Arthurs S (2019) Microbial insecticides in Iran: History, current status, challenges and perspective. J Invertebr Pathol 165:67–73

Kaur G, Guo J, Brown S, Head GP, Price PA, Paula-Moraes S, Ni X, Dimase M, Huang F (2019) Field-evolved resistance of *Helicoverpa zea* (Boddie) to transgenic maize expressing pyramided Cry1A.105/Cry2Ab2 proteins in Northeast Louisiana, the United States. J Invertebr Pathol 163:11–20

Kaviyapriya M, Lone RA, Balakrishnan N, Rajesh S, Ramalakshmi A (2019) Cloning and characterization of insecticidal *cry/vip* genes from an indigenous *Bacillus thuringiensis* isolate T29 and evaluation of its toxicity to maize fall armyworm *Spodoptera frugiperda*. J Entomol Zool Stud 7(3):1314–1321

Kawalek MD (1998) Cloning and characterization of mosquitocidal genes from *Bacillus thuringiensis* subsp. *jegathesan*. Ph.D. Dissertation. University of California, Riverside

Kereša S, Grdiša M, Barić M, Barčić J, Marchetti S (2008) Transgenic plants expressing insect resistance genes. Sjemenarstvo 25(2):139–153

Kesavan R, Easwaramoorthy S, Santhalakshmi G (2003) Evaluation of different formulations of *Bacillus thuringiensis* against sugarcane early shoot borer *Chilo infuscatellus*. Sugar Tech 5 (1–2):51–55

Khajuria C, Buschman LL, Chen LL, Siegfried MS, Zhu KY (2011) Identification of a novel aminopeptidase P-like gene (*OnAPP*) possibly involved in *Bt* toxicity and resistance in a major corn pest (*Ostrinia nubilalis*). PLoS One 6(8):e23983. https://doi.org/10.1371/journal.pone.0023983

Kim JS, Je YH (2012) Milling effect on the control efficacy of spray-dried *Bacillus thuringiensis* technical powder against diamondback moths. Pest Manag Sci 68(3):321–323

Kim YH, Issa MS, Cooper AM, Zhu KY (2015) RNA interference: applications and advances in insect toxicology and insect pest management. Pestic Biochem Physiol 120:109–117

Kirouac M, Vachon V, Noël JF, Girard F, Schwartz JL, Laprade R (2002) Amino acid and divalent ion permeability of the pores formed by the *Bacillus thuringiensis* toxins Cry1Aa and Cry1Ac in insect midgut brush border membrane vesicles. Biochim Biophys Acta Biomembr. https://doi.org/10.1016/S0005-2736(02)00342-5

Kirsch K, Schmutterer H (1988) Low efficacy of a *Bacillus thuringiensis* (Berl.) formulation in controlling the diamondback moth, *Plutella xylostella* (L.), in the Philippines. J Appl Entomol 105:249–255

Knowles BH (1994) Mechanism of action of *Bacillus thuringiensis* insecticidal δ-endotoxins. Adv Insect Physiol 24:275–308

Knowles BH, Ellar DJ (1987) Colloid-osmotic lysis is a general feature of the mechanism of action of *Bacillus thuringiensis* δ-endotoxins with different insect specificity. BBA Gen Subj 924 (3):509–518. https://doi.org/10.1016/0304-4165(87)90167-X

Kongsuwan K, Gough J, Kemp D, McDevitt A, Akhurst R (2005) Characterization of a new *Bacillus thuringiensis* endotoxin, Cry47Aa, from strains that are toxic to the Australian sheep blowfly, *Lucilia cuprina*. FEMS Microbiol Lett 252(1):127–136

Koni PA, Ellar DJ (1993) Cloning and characterization of a novel *Bacillus thuringiensis* cytolytic delta-endotoxin. J Mol Biol 229(2):319–327

Kranthi KR (2015) Pink bollworm strikes *Bt*-cotton. Cotton Stat News 37:1–5

Kranthi KR, Kranthi S, Ali S, Banerjee SK (2000) Resistance to CrylAc δ-endotoxin of *Bacillus thuringiensis* in a laboratory selected strain of *Helicoverpa armigera* (Hubner). Curr Sci 78:1001–1004

Kranthi S, Satija U, Pusadkar P, Rishi K, Shastri CS, Ansari S, Santosh HB, Monga D, Kranthi KR (2017) Non-*Bt* seeds provided by seed companies in India-are they suitable as refuge for Bt-cotton? Curr Sci 112(10):1992–1993

Krieg A, Huger A, Langenbruch G, Schnetter W (1983) *Bacillus thuringiensis* var. *tenebrionis*, a new pathotype effective against larvae of Coleoptera. *Zeitschrift für angewandte Entomologie*. https://doi.org/10.1111/j.1439-0418.1983.tb03704.x

Kumar S, Kaur J (2017) Efficacy of *Beauveria bassiana* and *Bacillus thuringiensis* as ecosafe alternatives to chemical insecticides against sunflower capitulum borer, *Helicoverpa armigera* (Hübner). J Entomol Zool Stud 5:185–188

Kumar S, Singh A (2015) Biopesticides: present status and the future prospects. J Fertil Pestic 6 (2):100–129

Kumar GS, Bhaskarb LV, Satishc Y, Rehamand SJ (2016) Evaluation of liquid formulations of *Bt* against gram pod borer, *Helicoverpa armigera* (Hubner) and spotted pod borer, *Maruca vitrata* (Geyer) in pigeonpea. J Appl Biol Biotechnol 4(1):39–42

Kumar KK, Sridhar J, Murali Baskaran RK, Senthilnathan S, Kaushal P, Dara SK, Arthurs S (2019) Microbial biopesticides for insect pest management in India: Current status and future prospects. J Invertebr Pathol 165:74–81

Kumar JS, Jayaraj J, Shanthi M, Theradimani M, Venkatasamy B, Irulandi S, Prabhu S (2020) Potential of standard strains of *Bacillus thuringiensis* against the tomato pinworm, *Tuta absoluta* (Meyrick) (Lepidoptera: Gelechiidae). Egypt J Biol Pest Control 30(1):1–7

Lachhab K, Tyagi RD, Valéro JR (2001) Production of *Bacillus thuringiensis* biopesticides using wastewater sludge as a raw material: effect of inoculum and sludge solids concentration. Process Biochem 37:197–208

Lambert B, Höfte H, Annys K, Jansens S, Soetaert P, Peferoen M (1992) Novel *Bacillus thuringiensis* insecticidal crystal protein with a silent activity against coleopteran larvae. Appl Environ Microbiol 58(8):2536–2542

Lee HK, Gill SS (1997) Molecular cloning and characterization of a novel mosquitocidal protein gene from *Bacillus thuringiensis* subsp. *fukuokaensis*. Appl Environ Microbiol 63 (12):4664–4670

Lee MK, Rajamohan F, Gould F, Dean DH (1995) Resistance to *Bacillus thuringiensis* CryIA d-endotoxins in a laboratory-selected *Heliothis virescens* strain is related to receptor alteration. Appl Environ Microbiol 61:3836–3842

Lee MK, Miles P, Chen JS (2006) Brush border membrane binding properties of *Bacillus thuringiensis* Vip3A toxin to *Heliothis virescens* and *Helicoverpa zea* midguts. Biochem Biophys Res Commun 339:1043–1047

Lee H, Park SY, Zhang ZJ (2013) An overview of genetic transformation of soybean. In: A comprehensive survey of international soybean research-genetics, physiology, agronomy and nitrogen relationships. IntechOpen, London. https://doi.org/10.5772/51076

Lee SB, Aimanova KG, Gill SS (2014) Alkaline phosphatases and aminopeptidases are altered in a Cry11Aa resistant strain of *Aedes aegypti*. Insect Biochem Mol Biol 54:112–121

Lefko SA, Nowatzki TM, Thompson SD, Binning RR, Pascual MA, Peters ML, Simbro EJ, Stanley BH (2008) Characterizing laboratory colonies of western corn rootworm (Coleoptera: Chrysomelidae) selected for survival on maize containing event DAS-59122-7. J Appl Entomol 132:189–204

Leonard RB, Graves JB, Sparks TC, Pavloff AM (1988) Variation in resistance of field populations of tobacco budworm and bollworm (Lepidoptera: Noctuidae) to selected insecticides. J Econ Entomol 81(6):1521–1528

Lereclus D, Mahillon J, Menou G, Lecadet MM (1986) Identification of Tn 4430, a transposon of *Bacillus thuringiensis* functional in *Escherichia coli*. MGG Mol Gen Genet 204(1):52–57. https://doi.org/10.1007/BF00330186

Li Q, Zou T, Ai P, Pan L, Fu C, Li P, Zheng A (2015) Complete genome sequence of *Bacillus thuringiensis* HS18-1. J Biotechnol 214:61–62

Li Y, Shu C, Shan YM, Geng LL, Song FP, Zhang J (2017) Complete genome sequence of *Bacillus thuringiensis Bt*185, a potential soil insect biocontrol agent. J Integr Agric 16(3):749–751

Li G, Huang J, Ji T, Tian C, Zhao X, Feng H (2020) Baseline susceptibility and resistance allele frequency in *Ostrinia furnacalis* related to Cry1 toxins in the Huanghuaihai summer corn region of China. Pest Manag Sci 76(12):4311–4317

Liesner L, Antilla L, Whitlow M, Staten RT (2011) Arizona pink bollworm eradication program 2010 update. In: 2011 Beltwide Cotton Conferences. National Cotton Council of America, Memphis, TN, pp 1143–1149

Liesner LR, Fairchild D, Solis J (2018) Arizona pink bollworm eradication program 2017 update. In: 2018 Beltwide Cotton Conferences. National Cotton Council of America, Memphis, TN, pp 718–720

Liesner LR, Fairchild D, Brengle J (2019) Arizona pink bollworm eradication program 2018 update. In: 2019 Beltwide Cotton Conferences. National Cotton Council of America, Memphis, TN, pp 371–373

Likitvivatanavong S, Chen J, Evans AM, Bravo A, Soberon M, Gill SS (2011) Multiple receptors as targets of Cry toxins in mosquitoes. J Agric Food Chem 59(7):2829–2838. https://doi.org/10.1021/jf1036189

Liu Y, Tabashnik BE (1997) Inheritance of resistance to the *Bacillus thuringiensis* toxin Cry1C in the diamondback moth. Appl Environ Microbiol 63(6):2218–2223

Liu YB, Tabashnik BE, Dennehy TJ, Patin AL, Sims MA, Meyer SK, Carrière Y (2001a) Effects of *Bt* cotton and Cry1Ac toxin on survival and development of pink bollworm (Lepidoptera: Gelechiidae). J Econ Entomol 94:1237–1242

Liu YB, Tabashnik BE, Meyer SK, Carrière Y, Bartlett AC (2001b) Genetics of pink bollworm resistance to *Bacillus thuringiensis* toxin Cry1Ac. J Econ Entomol 94:248–252

Liu Z, Fu S, Ma X, Baxter SW, Vasseur L, Xiong L, Huang Y, Yang G, You S, You M (2020) Resistance to *Bacillus thuringiensis* Cry1Ac toxin requires mutations in two *Plutella xylostella* ATP-binding cassette transporter paralogs. PLoS Pathog 16(8):e1008697

Losey JE, Rayor LS, Carter ME (1999) Transgenic pollen harms monarch larvae. Nature 399:214

Lu YH (2010) Mirid bug outbreaks in multiple crops correlated with wide scale adoption of *Bt* cotton in China. Science 328:1151–1154

Ludwick DC, Meihls LN, Ostlie KR, Potter BD, French L, Hibbard BE (2017) Minnesota field population of western corn rootworm (Coleoptera: Chrysomelidae) shows incomplete resistance to Cry34Ab1/Cry35Ab1 and Cry3Bb1. J Appl Entomol 141(1–2):28–40

Ma W, Zhang T (2019) Next-generation transgenic cotton: pyramiding RNAi with *Bt* counters insect resistance. In: Transgenic cotton. Humana Press, New York, NY, pp 245–256

Mahon RJ, Downes SJ, James B (2012) Vip3A resistance alleles exist at high levels in Australian targets before release of cotton expressing this toxin. PLoS One 7(6):e39192

Makart L, Commans F, Gillis A, Mahillon J (2017) Horizontal transfer of chromosomal markers mediated by the large conjugative plasmid pXO16 from *Bacillus thuringiensis* serovar *israelensis*. Plasmid 91:76–81

Marrone PG (2014) The market and potential for biopesticides. In: Biopesticides: state of the art and future opportunities. American Chemical Society, Washington, DC, pp 245–258

Martin PA, Travers RS (1989) Worldwide abundance and distribution of *Bacillus thuringiensis* isolates. Appl Environ Microbiol 55(10):2437–2442

Martins SE, Monnerat RG, Queiroz PR, Dumas VF, Braz SV, Aguiar RWD, Gomes ACMM, Sánchez J, Bravo A, Ribeiro BM (2010) Midgut GPI-anchored proteins with alkaline phosphatase activity from the cotton boll weevil (*Anthonomus grandis*) are putative receptors for the Cry1B protein of *Bacillus thuringiensis*. Insect Biochem Mol Biol 40(2):138–145

Mascarin GM, Lopes RB, Delalibera Í Jr, Fernandes ÉKK, Luz C, Faria M (2019) Current status and perspectives of fungal entomopathogens used for microbial control of arthropod pests in Brazil. J Invertebr Pathol 165:46–53

Mattes O (1927) Parasitare Krankheiten der Mehlmottenlarven und Versuche uber ihre Verwendbarkeit als biologisches Bekiampfungsmittel. Sitzber Ges Beforder Ges Naturw Marburg 62:381–417

McClintock JT, Schaffer CR, Sjoblad RD (1995) A comparative review of the mammalian toxicity of *Bacillus thuringiensis*based pesticides. Pestic Sci 45(2):95–105

McGaughey WH (1985) Insect resistance to the biological insecticide *Bacillus thuringiensis*. Science 229:193–195

McGaughey WH, Beeman RW (1988) Resistance to *Bacillus thuringiensis* in colonies of Indianmeal moth and almond moth (Lepidoptera:Pyralidae). J Econ Entomol 81:28–33

Meihls LN, Higdon ML, Siegfried BD, Miller NJ, Sappington TW, Ellersieck MR, Spencer TA, Hibbard BE (2008) Increased survival of western corn rootworm on transgenic corn within three generations of on-plant greenhouse selection. Proc Natl Acad Sci U S A 105(49):19177–19182

Meihls LN, Higdon ML, Ellersieck M, Hibbard BE (2011) Selection for resistance to mCry3A-expressing transgenic corn in western corn rootworm. J Econ Entomol 104(3):1045–1054

Meihls LN, Higdon ML, Ellersieck MR, Tabashnik BE, Hibbard BE (2012) Greenhouse-selected resistance to Cry3Bb1-producing corn in three western corn rootworm populations. PLoS One 7 (12):e51055

Mendelshon M, Kough J, Vaituzis Z, Mathews K (2003) Are Bt crops safe? Nat Biotechnol 21 (9):1003–1009

Metalnikov S, Chorine V (1929) On the infection of the gypsy moth and certain other insects with Bacterium thuringiensis. International Corn Borer Investigations. Sci Rep 2:60–61

Mishra J, Tewari S, Singh S, Arora NK (2015) Biopesticides: where we stand? In: Arora NK (ed) Plant microbes symbiosis: applied facets. Springer, New Delhi, pp 37–75

Mishra J, Dutta V, Arora NK (2020) Biopesticides in India: Technology and sustainability linkages. 3 Biotech 10:1–12. https://doi.org/10.1007/s13205-020-02192-7

Mittal PK (2003) Biolarvicides in vector control: challenges and prospects. J Vector Borne Dis 40 (1/2):20

Mizuki E, Park YS, Saitoh H, Yamashita S, Akao T, Higuchi K, Ohba M (2000) Parasporin, a human leukemic cell-recognizing parasporal protein of *Bacillus thuringiensis*. Clin Diagn Lab Immunol 7(4):625–634

Moar WJ, Pusztai-Carey M, van Faassen H, Bosch D, Frutos R, Rang C, Luo K, Adang MJ (1995) Development of *Bacillus thuringiensis* CryIC resistance by *Spodoptera exigua* (Hu¨bner) (Lepidoptera: Noctuidae). Appl Environ Microbiol 61:2086–2092

Mohan KS (2017) An area-wide approach to pink bollworm management on *Bt* cotton in India – a dire necessity with community participation. Curr Sci 112(10):2016–2020

Mohan KS (2018) Refuge-in-bag for Bt cotton. Curr Sci 114(4):726

Mohan KS, Ravi KC, Suresh PJ, Sumerford D, Head GP (2015) Field resistance to the *Bacillus thuringiensis* protein Cry1Ac expressed in Bollgard® hybrid cotton in pink bollworm, *Pectinophora gossypiella* (Saunders), populations in India. Pest Manag Sci 72(4):738–746. https://doi.org/10.1002/ps.4047

Monika T, Govind G (2017) Formulation for optimizing *Bacillus thuringiensis* production. J Biol Control 31(1):38–49

Morin S, Biggs RW, Sisterson MS, Shriver L, Ellers-Kirk C, Higginson D, Holley D, Gahan LJ, Heckel DG, Carriere Y, Dennehy TJ (2003) Three cadherin alleles associated with resistance to *Bacillus thuringiensis* in pink bollworm. Proc Natl Acad Sci U S A 100(9):5004–5009

Morris ON, Kanagaratnam P, Converse V (1997) Suitability of 30 Agricultural products and by-Products as nutrient sources for laboratory production of *Bacillus thuringiensis* subsp. *aizawai* (HD133). J Invertebr Pathol 70(2):113–120

Morse RJ, Yamamoto T, Stroud RM (2001) Structure of Cry2Aa suggests an unexpected receptor binding epitope. Structure 9(5):409–417

Mu¨ller-Cohn J, Chaufaux J, Buisson C, Gilois N, Sanchis V, Lereclus D (1996) *Spodoptera littoralis* (Lepidoptera: Noctuidae) resistance to CryIC and cross-resistance to other *Bacillus thuringiensis* crystal toxins. J Econ Entomol 89:791–797

Nagamatsu Y, Okamura S, Saitou H, Akao T, Mizuki E (2010) Three Cry toxins in two types from *Bacillus thuringiensis* strain M019 preferentially kill human hepatocyte cancer and uterus cervix cancer cells. Biosci Biotechnol Biochem 74(3):494–498

Naik VC, Kumbhare S, Kranthi S, Satija U, Kranthi KR (2018) Field evolved resistance of pink bollworm, *Pectinophora gossypiella* (Saunders) (Lepidoptera: Gelechiidae), to transgenic *Bacillus thuringiensis* (Bt) cotton expressing crystal 1Ac (Cry1Ac) and Cry2Ab in India. Pest Manag Sci 74(11):2544–2554

Naik VCB, Pusadkar PP, Waghmare ST, Raghavendra KP, Kranthi S, Kumbhare S, Nagrare VS, Kumar R, Prabhulinga T, Gokte-Narkhedkar N, Waghmare VN (2020) Evidence for population expansion of cotton pink bollworm *Pectinophora gossypiella* (Saunders)(Lepidoptera: Gelechiidae) in india. Sci Rep 10(1):1–11

Narva KE, Payne JM, Schwab GE, Hickle LA, Galasan T, Sick AJ (1991) Novel *Bacillus thuringiensis* microbes active against nematodes, and genes encoding novel nematode-active toxins cloned from *Bacillus thuringiensis* isolates. European Patent Office, EP0462721

Narva KE, Schwab GE, Payne JM (1995) Gene encoding nematode-active toxin PS63B cloned from *Bacillus thuringiensis* isolate. Mycogen Corp., U.S. Patent 5,439,881

Narva KE, Siegfried BD, Storer NP (2013) Transgenic approaches to western corn rootworm control. In: Yellow Biotechnology II. Springer, Berlin, pp 135–162

National Research Council (1986) Pesticide resistance: Strategies and tactics for management. National Academies Press, Washington, DC

Nethravathi CJ, Hugar PS, Krishnaraj PU, Vastrad AS (2010) Bioefficacy of native *Bacillus thuringiensis* isolates against cabbage leaf webber, *Crocidolomia binotalis* Z. Karnataka J Agric Sci 23(1)

Noguera PA, Ibarra JE (2010) Detection of new *cry* genes of *Bacillus thuringiensis* by use of a novel PCR primer system. Appl Environ Microbiol 76(18):6150–6155

O'Callaghan M, Glare TR, Burgess EPJ, Malone LA (2004) Effects of plants genetically modifies for insect resistance on non-targert organisms. Annu Rev Entomol 50(1):271–292

Ocelotl J, Sánchez J, Gómez I, Tabashnik BE, Bravo A, Soberón M (2017) ABCC2 is associated with *Bacillus thuringiensis* Cry1Ac toxin oligomerization and membrane insertion in diamond-back moth. Sci Rep 7(1):1–9

Okumura S, Saitoh H, Ishikawa T, Wasano N, Yamashita S, Kusumoto KI, Akao T, Mizuki E, Ohba M, Inouye K (2005) Identification of a novel cytotoxic protein, Cry45Aa, from *Bacillus thuringiensis* A1470 and its selective cytotoxic activity against various mammalian cell lines. J Agric Food Chem 53(16):6313–6318

Omoto C, Bernardi O, Salmeron E, Sorgatto RJ, Dourado PM, Crivellari A, Carvalho RA, Willse A, Martinelli S, Head GP (2016) Field-evolved resistance to Cry1Ab maize by *Spodoptera frugiperda* in Brazil. Pest Manag Sci 72(9):1727–1736

Osman GEH, Already R, Assaeedi ASA, Organji SR, El-Ghareeb D, Abulreesh HH, Althubiani AS (2015) Bioinsecticide *Bacillus thuringiensis* a comprehensive review. Egypt J Biol Pest Control 25(1):271–288

Ostrem JS, Pan Z, Flexner JL, Owens E, Binning R, Higgins LS (2016) Monitoring susceptibility of western bean cutworm (Lepidoptera: Noctuidae) field populations to *Bacillus thuringiensis* Cry1F protein. J Econ Entomol 109(2):847–853

Oswald KJ, Wade French B, Nielson C, Bagley M (2011) Selection for Cry3Bb1 resistance in a genetically diverse population of nondiapausing western corn rootworm (Coleoptera: Chrysomelidae). J Econ Entomol 104(3):1038–1044

Pacheco S, Gómez I, Arenas I, Saab-Rincon G, Rodríguez-Almazán C, Gill SS, Bravo A, Soberón M (2009) Domain II loop 3 of *Bacillus thuringiensis* Cry1Ab toxin is involved in a "Ping Pong" binding mechanism with *Manduca sexta* aminopeptidase-N and cadherin receptors. J Biol Chem. https://doi.org/10.1074/jbc.M109.024968

Padua LE, Ohba M, Aizawa K (1984) Isolation of a *Bacillus thuringiensis* strain (serotype 8a: 8b) highly and selectively toxic against mosquito larvae. J Invertebr Pathol 44(1):12–17

Palma L, Munoz D, Berry C, Murillo J, Cabellero P (2014) *Bacillus thuringiensis* toxins: an overview of their biocidal activity. Toxins 6:3296–3325

Pardo-López L, Gómez I, Muñoz-Garay C, Jiménez-Juarez N, Soberón M, Bravo A (2006) Structural and functional analysis of the pre-pore and membrane-inserted pore of Cry1Ab toxin. J Invertebr Pathol 92(3):172–177. https://doi.org/10.1016/j.jip.2006.02.008

Pardo-Lopez L, Soberon M, Bravo A (2013) *Bacillus thuringiensis* insecticidal three-domain Cry toxins: mode of action, insect resistance and consequences for crop protection. FEMS Microbiol Rev 37(1):3–22

Patin AL, Dennehy TJ, Sims MA, Tabashnik BE, Liu YB, Antilla L, Gouge D, Henneberry TJ, Staten R (1999) Status of pink bollworm susceptibility to *Bt* in Arizona. In: *Proceedings of Beltwide Cotton Conferences*. Vol. 2. National Cotton Council of America, Memphis, TN, pp 991–996

Payne J, Sick AJ (1993) *Bacillus thuringiensis* isolate active against lepidopteran pests, and genes encoding novel lepidopteran-active toxins. U.S. Patent 5,188,960

Payne JM, Narva KE, Fu J (1996) *Bacillus thuringiensis* genes encoding nematode-active toxins. U.S. Patent 5,589.382

Peng DH, Pang CY, Wu H, Huang Q, Zheng JS, Sun M (2015) The expression and crystallization of Cry65Aa require two C-termini, revealing a novel evolutionary strategy of *Bacillus thuringiensis* Cry proteins. Sci Rep 5:8291

Pereira EJ, Lang BA, Storer NP, Siegfried BD (2008) Selection for Cry1F resistance in the European corn borer and cross-resistance to other Cry toxins. Entomol Exp Appl 126 (2):115–121

Perlak FJ, Fuchs RL, Dean DA, McPherson SL, Fischhoff DA (1991) Modification of the coding sequence enhances plant expression of insect control protein genes. Proc Natl Acad Sci U S A 88(8):3324–3328

Petolino JF, Arnold NL (2009) Whiskers-mediated maize transformation. In: Transgenic maize. Humana Press, Totowa, NJ, pp 59–67

Pickett B (2009) Studies on resistance to vegetative (Vip3A) and crystal (Cry1A) insecticidal toxins of *Bacillus thuringiensis* in *Heliothis virescens* (Fabricius). Ph.D. thesis, Imperial College London, London. http://europepmc.org/abstract/eth/510744. Accessed 12 Nov 2015

Pickett BR, Gulzar A, Ferré J, Wright DJ (2017) *Bacillus thuringiensis* Vip3Aa toxin resistance in *Heliothis virescens* (Lepidoptera: Noctuidae). Appl Environ Microbiol 83(9):e03506–e03516. https://doi.org/10.1128/AEM.03506-16

Pigott CR, Ellar DJ (2007) Role of receptors in *Bacillus thuringiensis* crystal toxin activity. Microbiol Mol Biol Rev 71(2):255–281

Poopathi S, Kumar KA (2003) Novel fermentation media for production of *Bacillus thuringiensis* subsp. *israelensis*. J Econ Entomol 96(4):1039–1044

Poopathi S, Anupkumar K, Arunachalam N, Sekar V, Tyagi BK (2003) A small scale mosquito control field trial with the biopesticides *Bacillus sphaericus* and *Bacillus thuringiensis* serovar israelensis produced from a new culture medium. Biocontrol Sci Tech 13(8):743–748

Prabakaran G, Balaraman K (2006) Development of a cost-effective medium for the large scale production of *Bacillus thuringiensis* var *israelensis*. Biol Control 36(3):288–292

Priest FG (1992) Biological control of mosquitoes and other biting flies by *Bacillus sphaericus* and *Bacillus thuringiensis*. J Appl Bacteriol 72(5):357–369

Rajagopal R, Arora N, Sivakumar S, Rao NG, Nimbalkar SA, Bhatnagar RK (2009) Resistance of *Helicoverpa armigera* to Cry1Ac toxin from *Bacillus thuringiensis* is due to improper processing of the protoxin. Biochem J 419(2):309–316

Ramalakshmi A, Udayasuriyan V (2010) Diversity of *Bacillus thuringiensis* isolated from Western Ghats of Tamil Nadu state, India. Curr Microbiol 61(1):13–18

Ramanujam B, Rangeshwaran R, Sivakmar G, Mohan M, Yandigeri MS (2014) Management of insect pests by microorganisms. Proc Indian Natl SciAcad 80(2):455–471

Ramasamy A, Suresh M, Mallesh MSH (2020) Toxicity evaluation of aphidicidal crystalliferous toxins from *Bacillus thuringiensis* strains: a molecular study. Ann Microbiol 70(1):1–14

Rana S, Mohankumar S (2016) Silencing of Chitin Synthase A gene in *Spodoptera litura* (Fabricius) using RNAi approach. Madras Agric J 103:237–241
Rana S, Rajurkar AB, Kumar KK, Mohankumar S (2020) Comparative analysis of chitin synthaseA dsRNA mediated RNA interference for management of crop pests of different families of Lepidoptera. Front Plant Sci 11:427
Rangeshwaran R, Gorky A, Viswakethu V, Karkera A, Sivakumar G, Mohan M (2014) Cry gene and plasmid profiling of *Bacillus thuringiensis* isolated from Indian soils. J Biol Control 28:185–191
Rao AR, Rao PK, Jyotsna KP (2015) Efficacy of certain natural insecticides against citrus leaf miner, *Phyllocnistis citrella* Stainton as prophylactic and curative measures on sathgudi sweet orange. Pest Manag Horticult Ecosyst 21(1):11–15
Raymond B, Johnston PR, Nielsen-LeRoux C, Lereclus D, Crickmore N (2010) *Bacillus thuringiensis*: an impotent pathogen? Trends Microbiol 18(5):189–194
Reyaz AL, Arulselvi PI (2016) Cloning, characterization and expression of a novel haplotype *cry2A*-type gene from *Bacillus thuringiensis* strain SWK1, native to Himalayan valley Kashmir. J Invertebr Pathol 136:1–6
Reyaz AL, Gunapriya L, Arulselvi PI (2017) Molecular characterization of indigenous *Bacillus thuringiensis* strains isolated from Kashmir valley. 3 Biotech 7(2):143
Reyaz AL, Balakrishnan N, Udayasuriyan V (2018) A new observation on feeding behaviour of pink bollworm and its application in screening *Bt*-resistant population. 3 Biotech 8(5):237
Reyaz A, Balakrishnan N, Udayasuriyan V (2019) Genome sequencing of *Bacillus thuringiensis* isolate T414 toxic to pink bollworm (*Pectinophora gossypiella* Saunders) and its insecticidal genes. Microb Pathog 134:103553
Reyaz AL, Balakrishnan N, Udayasuriyan V (2020) A novel *Bacillus thuringiensis* isolate toxic to cotton pink bollworm (*Pectinophora gossypiella* Saunders). Microb Pathog 150:104671
Rose R, Dively GP, Pettis J (2007) Effects of *Bt* corn pollen on honeybees: emphasis on protocol development. Apidologie 38:1–11
Rosellini D (2012) Selectable markers and reporter genes: a well furnished toolbox for plant science and genetic engineering. Crit Rev Plant Sci 31(5):401–453
Rosso ML, Delécluse A (1997) Contribution of the 65-kilodalton protein encoded by the cloned gene *cry19A* to the mosquitocidal activity of *Bacillus thuringiensis* subsp. *jegathesan*. Appl Environ Microbiol 63(11):4449–4455
Rupar MJ, Donovan WP, Chu CR, Pease E, Tan Y, Slaney AC, Malvar TM, Baum JA (2003) Coleopteran-toxic polypeptide compositions and insect-resistant transgenic plants. Monsanto Technology LLC. U.S. Patent 6,555,655.
Salama HS, Foda MS, Dulmage HT, El-Sharaby A (1983) Novel fermentation media for production of δ-endotoxins from *Bacillus thuringiensis*. J Invertebr Pathol 41(1):8–19
Salazar-Magallon JA, Hernandez-Velazquez VM, Alvear-Garcia A, Arenas-Sosa I, Pena-Chora G (2015) Evaluation of industrial by-products for the production of *Bacillus thuringiensis* strain GP139 and the pathogenicity when applied to *Bemisia tabaci* nymphs. Bull Insectol 68 (1):103–109
San Kim H, Saitoh H, Yamashita S, Akao T, Park YS, Maeda M, Tanaka R, Mizuki E, Ohba M (2003) Cloning and characterization of two novel crystal protein genes from a *Bacillus thuringiensis* serovar *dakota* strain. Curr Microbiol 46(1):0033–0038
Sanchis V, Chaufaux J, Lereclus D (1996) Amélioration biotechnologique de *Bacillus thuringiensis*: les enjeux et les risques. Annales de l'Institut Pasteur/Actualités 7(4):271–284
Schnepf HE, Whiteley HR (1981) Cloning and expression of the *Bacillus thuringiensis* crystal protein gene in *Escherichia coli*. Proc Natl Acad Sci U S A 78(5):2893–2897
Schnepf HE, Wong HC, Whiteley HR (1985) The amino acid sequence of a crystal protein from *Bacillus thuringiensis* deduced from the DNA base sequence. J Biol Chem 260(10):6264–6272
Schnepf E, Crickmore N, Van Rie J, Lereclus D, Baum J, Feitelson J, Zeigler D, Dean D (1998) *Bacillus thuringiensis* and its pesticidal crystal proteins. Microbiol Mol Biol Rev 62(3):775–806
Sebesta K, Farkas J, Horska K, Vankova J (1981) Thuringiensin, the beta-exotoxin of *Bacillus thuringiensis*. In: Burges HD (ed) Microbial control of pests and plant diseases 1970–1980. Academic Press, London, pp 249–282

Shabbir MZ, Quan Y, Wang Z, Bravo A, Soberón M, He K (2018) Characterization of the Cry1Ah resistance in Asian corn borer and its cross-resistance to other *Bacillus thuringiensis* toxins. Sci Rep 8(1):1–9. https://doi.org/10.1038/s41598-017-18586-2

Shelton AM, Tang JD, Roush RT, Earle ED (1998) Can we manage resistance to Bt-engineered plants? Results of greenhouse and field tests. Proceedings of the Sixth Australian Applied Entomological Research Conference pp. 258–266

Shelton AM, Tang JD, Roush RT, Metz TD, Earle ED (2000) Field tests on managing resistance to *Bt*-engineered plants. Nat Biotechnol 18(3):339–342

Shishir A, Roy A, Islam N, Rahman A, Khan SN, Hoq M (2014) Abundance and diversity of *Bacillus thuringiensis* in Bangladesh and their *cry* genes profile. Front Environ Sci 2:20

Shojaaddini M, Moharramipour S, Khodabandeh M, Talebi A (2010) Development of a cost effective medium for production of *Bacillus thuringiensis* bioinsecticide using food barley. J Plant Protect Res 50(1):9–14

Shrestha RB, Gassmann AJ (2019) Field and laboratory studies of resistance to *Bt* corn by western corn rootworm (Coleoptera: Chrysomelidae). J Econ Entomol 112(5):2324–2334

Shrestha RB, Dunbar MW, French BW, Gassmann AJ (2018) Effects of field history on resistance to *Bt* maize by western corn rootworm, *Diabrotica virgiferavirgifera* LeConte (Coleoptera: Chrysomelidae). PLoS One 13(7):e0200156

Singh KI, Debbarma A, Singh HR (2015) Field efficacy of certain microbial insecticides against *Plutella xylostella* Linnaeus and *Pieris brassicae* Linnaeus under cabbage-crop-ecosystem of Manipur. J Biol Control 29(4):194–202

Siqueira HA, Moellenbeck D, Spencer T, Siegfried BD (2004) Cross-resistance of Cry1Ab-selected *Ostrinia nubilalis* (Lepidoptera: Crambidae) to *Bacillus thuringiensis* δ-endotoxins. J Econ Entomol 97(3):1049–1057

Sivakumar S, Rajagopal R, Venkatesh GR, Srivastava A, Bhatnagar RK (2007) Knockdown of aminopeptidase-N from *Helicoverpa armigera* larvae and in transfected Sf21 cells by RNA interference reveals its functional interaction with *Bacillus thuringiensis* insecticidal protein Cry1Ac. J Biol Chem 282(10):7312–7319

Slaney AC, Robbins HL, English L (1992) Mode of action of *Bacillus thuringiensis* toxin CryIIIA: an analysis of toxicity in *Leptinotarsa decemlineata* (Say) and *Diabrotica undecimpunctata howardi* Barber. Insect Biochem Mol Biol 22(1):9–18

Smith RH, Hood EE (1995) *Agrobacterium tumefaciens* transformation of monocotyledons. Crop Sci 35(2):301–309

Smith JL, Lepping MD, Rule DM, Farhan Y, Schaafsma AW (2017) Evidence for field-evolved resistance of *Striacosta albicosta* (Lepidoptera: Noctuidae) to Cry1F *Bacillus thuringiensis* protein and transgenic corn hybrids in Ontario, Canada. J Econ Entomol 110(5):2217–2228

Smulevitch SV, Osterman AL, Shevelev AB, Kaluger SV, Karasin AI, Kadyrov RM, Zagnitko OP, Chestukhina GG, Stepanov VM (1991) Nucleotide sequence of a novel δ-endotoxin gene *cryIg* of *Bacillus thuringiensis* ssp. *galleriae*. FEBS Lett 293(1–2):25–28

Soberón M, Pardo-López L, López I, Gómez I, Tabashnik E, Bravo A (2007) Engineering modified Bt toxins to counter insect resistance. Science. https://doi.org/10.1126/science.1146453

Song X, Kain W, Cassidy D, Wang P (2015) Resistance to *Bacillus thuringiensis* toxin Cry2Ab in *Trichoplusia ni* is conferred by a novel genetic mechanism. Appl Environ Microbiol 81 (15):5184–5195

Soonsanga S, Luxananil P, Promdonkoy B (2020) Modulation of Cas9 level for efficient CRISPR-Cas9 mediated chromosomal and plasmid gene deletion in *Bacillus thuringiensis*. Biotechnol Lett 42(4):625–632

Sparks TC (1981) Development of insecticide resistance in *Heliothis zea* and *Heliothis virescens* in North America. Bull ESA 27(3):186–192

Sparks TC, Graves JB, Leonard BR (1993) Insecticide resistance and the tobacco budworm: past, present and future. Rev Pestic Toxicol 2:149–183

Stone TB, Sims SR, Marrone PG (1989) Selection of tobacco budworm for resistance to a genetically engineered *Pseudomonas fluorescens* containing the 8-endotoxin of *Bacillus thuringiensis* subsp. *kurstaki*. J Invertebr Pathol 53:228–234

Storer NP, Van Duyn JW, Kennedy GG (2001) Life history traits of *Helicoverpa zea* (Lepidoptera: Noctuidae) on non-*Bt* and *Bt* transgenic corn hybrids in eastern North Carolina. J Econ Entomol 94(5):1268–1279

Storer NP, Babcock JM, Schlenz M, Meade T, Thompson GD, Bing JW, Huckaba RM (2010) Discovery and characterization of field resistance to *Bt* maize: *Spodoptera frugiperda* (Lepidoptera: Noctuidae) in Puerto Rico. J Econ Entomol 103(4):1031–1038

Storer NP, Kubiszak ME, King JE, Thompson GD, Santos AC (2012) Status of resistance to *Bt* maize in *Spodoptera frugiperda*: lessons from Puerto Rico. J Invertebr Pathol 110(3):294–300

Subbanna ARNS, Chandrashekara C, Stanley J, Mishra KK, Mishra PK, Pattanayak A (2019) Bio-efficacy of chitinolytic *Bacillus thuringiensis* isolates native to northwestern Indian Himalayas and their synergistic toxicity with selected insecticides. Pestic Biochem Physiol 158:166–174

Subbaratnam GV, Radhika P (2005) Global view of insecticide resistance management in cotton - A review. Pest Manage Econ Zool 13:163–183

Sun Y, Zhao Q, Xia L, Ding X, Hu Q, Federici BA, Park HW (2013) Identification and characterization of three previously undescribed crystal proteins from *Bacillus thuringiensis* subsp. *jegathesan*. Appl Environ Microbiol 79(11):3364–3370

Sun Y, Yang P, Jin H, Liu H, Zhou H, Qiu L, Lin Y, Ma W (2020) Knockdown of the aminopeptidase N genes decreases susceptibility of *Chilo suppressalis* larvae to Cry1Ab/Cry1Ac and Cry1Ca. Pestic Biochem Physiol 162:36–42

Suyanandana P, Potacharoen W, Arunpairojana V, Boonsong P, Vattanakul J (1996) Solid state cultivation of *Bacillus thuringiensis* for insect larvicide in agriculture and public health. Khon Kaen Agric J 24:190–196

Syed T, Askari M, Meng Z, Li Y, Abid MA, Wei Y, Guo S, Liang C, Zhang R (2020) Current insights on vegetative insecticidal proteins (Vip) as next generation pest killers. Toxins 12 (8):522

Tabashnik BE (1994) Evolution of resistance to *Bacillus thuringiensis*. Annu Rev Entomol 39 (1):47–79

Tabashnik BE, Carrière Y (2017) Surge in insect resistance to transgenic crops and prospects for sustainability. Nat Biotechnol 35(10):926–935

Tabashnik BE, Carrière Y (2019) Global patterns of resistance to *Bt* crops highlighting pink bollworm in the United States, China, and India. J Econ Entomol 112(6):2513–2523

Tabashnik BE, Gould F (2012) Delaying corn rootworm resistance to *Bt* corn. J Econ Entomol 105 (3):767–776

Tabashnik BE, Patin AL, Dennehy TJ, Liu YB, Carrière Y, Sims MA, Antilla L (2000) Frequency of resistance to *Bacillus thuringiensis* in field populations of pink bollworm. Proc Natl Acad Sci U S A 97(24):12980–12984

Tabashnik BE, Van Rensburg JBJ, Carrière Y (2009) Field-evolved insect resistance to *Bt* crops: definition, theory, and data. J Econ Entomol 102(6):2011–2025

Tabashnik BE, Sisterson MS, Ellsworth PC, Dennehy TJ, Antilla L, Liesner L, Whitlow M, Staten RT, Fabrick JA, Unnithan GC, Yelich AJ (2010) Suppressing resistance to *Bt* cotton with sterile insect releases. Nat Biotechnol 28(12):1304

Tabashnik BE, Brévault T, Carrière Y (2013) Insect resistance to Bt crops: lessons from the first billion acres. Nat Biotechnol 31(6):510–521

Tabashnik BE, Liesner LR, Ellsworth PC, Unnithan GC, Fabrick JA, Naranjo SE, Li X, Dennehy TJ, Antilla L, Staten RT, Carrière Y (2020) Transgenic cotton and sterile insect releases synergize eradication of pink bollworm a century after it invaded the United States. Proc Natl Acad Sci U S A 118(1)

Tan Y, Donovan WP (2000) Deletion of *aprA* and *nprA* genes for alkaline protease A and neutral protease A from *Bacillus thuringiensis*: effect on insecticidal crystal proteins. J Biotechnol 84 (1):67–72

Tan F, Zhu J, Tang J, Tang X, Wang S, Zheng A, Li P (2009) Cloning and characterization of two novel crystal protein genes, *cry54Aa1* and *cry30Fa1*, from *Bacillus thuringiensis* strain BtMC28. Curr Microbiol 58(6):654–659

Tirado-Montiel MDL, Tyagi RD, Valero JR (2001) Wastewater treatment sludge as a raw material for the production of *Bacillus thuringiensis* based biopesticides. Water Res (Oxford) 35 (16):3807–3816

Unalmis S, Ayvaz A, Yilmaz S, Azizoglu U (2015) Molecular screening and bioactivity of native *Bacillus thuringiensis* isolates. J Agric Sci Technol 17:1197–1207

Vachon V, Laprade R, Schwartz J-L (2012) Current models of the mode of action of *Bacillus thuringiensis* insecticidal crystal proteins: a critical review. J Invertebr Pathol 111(1):1–12

Vaeck M, Reynaerts A, Höfte H, Jansens S, De Beuckeleer M, Dean C, Zabeau M, Van Montagu M, Leemans J (1987) Transgenic plants protected from insect attack. Nature 328 (6125):33

Van Frankenhuyzen K (2009) Insecticidal activity of *Bacillus thuringiensis* crystal proteins. J Invertebr Pathol 101(1):1–16

Van Rensburg JBJ (2007) First report of field resistance by the stem borer, *Busseola fusca* (Fuller) to *Bt*-transgenic maize. S Afr J Plant Soil 24(3):147–151

Vassallo CN, Figueroa Bunge F, Signorini AM, Valverde-Garcia P, Rule D, Babcock J (2019) Monitoring the evolution of resistance in *Spodoptera frugiperda* (Lepidoptera: Noctuidae) to the Cry1F protein in Argentina. J Econ Entomol 112(4):1838–1844

Venkatesan T, Pattar S (2017) Intellectual property guidelines and commercialization of biocontrol agents/biopesticides technologies. In: Varshney R, Gupta A, Patil J (eds) Training manual on identification, mass production and utilization of parasitoids, predators and entomopathogens for sustainable insect pest management. ICAR-NBAIR, Bengaluru, pp 115–126

Vidyarthi AS, Tyagi RD, Valero JR, Surampalli RY (2002) Studies on the production of B. thuringiensis based biopesticides using wastewater sludge as a raw material. Water Res 36(19):4850–4860

Vilas-Bôas GT, Peruca APS, Arantes OMN (2007) Biology and taxonomy of *Bacillus cereus, Bacillus anthracis* and *Bacillus thuringiensis*. Can J Microbiol 53(6):673–687

Vimala Devi PS, Vineela V (2015) Suspension concentrate formulation of *Bacillus thuringiensis* var. *kurstaki* for effective management of *Helicoverpa armigera* on sunflower (*Helianthus annuus*). Biocontrol Sci Tech 25(3):329–336

Vouk V (1930) The fight against the corn borer in Yugoslavia Internat Corn Borer Invest Sci Rep 1927-28 3

Waalwijk C, Dullemans AM, van Workum ME, Visser B (1985) Molecular cloning and the nucleotide sequence of the M r 28 000 crystal protein gene of *Bacillus thuringiensis* subsp. *israelensis*. Nucleic Acids Res 13(22):8207–8217

Wabiko H, Raymond KC, Bulla JR, L.A. (1986) *Bacillus thuringiensis* entomocidal protoxin gene sequence and gene product analysis. DNA 5(4):305–314

Walsh TK, Downes SJ, Gascoyne J, James W, Parker T, Armstrong J and Mahon, R.J. 2014. Dual Cry2Ab and Vip3A resistant strains of *Helicoverpa armigera* and *Helicoverpa punctigera* (Lepidoptera: Noctuidae); testing linkage between loci and monitoring of allele frequencies. J Econ Entomol, 107:1610–1617

Walters FS, Stacy CM, Lee MK, Palekar N, Chen JS (2008) An engineered chymotrypsin/cathepsin G site in domain I renders *Bacillus thuringiensis* Cry3A active against western corn rootworm larvae. Appl Environ Microbiol 74(2):367–374

Wan P, Huang Y, Wu H, Huang M, Cong S, Tabashnik BE, Wu K (2012) Increased frequency of pink bollworm resistance to *Bt* toxin Cry1Ac in China. PLoS One 7(1):e29975

Wan P, Xu D, Cong S, Jiang Y, Huang Y, Wang J, Wu H, Wang L, Wu K, Carrière Y, Mathias A (2017) Hybridizing transgenic *Bt* cotton with non-*Bt* cotton counters resistance in pink bollworm. Proc Natl Acad Sci U S A 114(21):5413–5418

Wang Z, Wang Y, Cui H, Xia Y, Altosar J (2002) Toxicological evaluation of transgenic rice flour from *Bacillus thuringiensis*. J Sci Food Agric 82(7):738–744

Wang P, Zhao JZ, Rodrigo-Simón A, Kain W, Janmaat AF, Shelton AM, Ferré J, Myers J (2007) Mechanism of resistance to *Bacillus thuringiensis* toxin Cry1Ac in a greenhouse population of the cabbage looper, *Trichoplusia ni*. Appl Environ Microbiol 73(4):1199–1207

Wang Y, Wang Y, Wang Z, Bravo A, Soberón M, He K (2016) Genetic basis of Cry1F-resistance in a laboratory selected Asian corn borer strain and its cross-resistance to other *Bacillus thuringiensis* toxins. PLoS One 11(8):e0161189

Wang J, Wang H, Liu S, Liu L, Tay WT, Walsh TK, Yang Y, Wu Y (2017) CRISPR/Cas9 mediated genome editing of *Helicoverpa armigera* with mutations of an ABC transporter gene *HaABCA2* confers resistance to *Bacillus thuringiensis* Cry2A toxins. Insect Biochem Mol Biol 87:147–153

Wang L, Ma Y, Wan P, Liu K, Xiao Y, Wang J, Cong S, Xu D, Wu K, Fabrick JA, Li X (2018a) Resistance to *Bacillus thuringiensis* linked with a cadherin transmembrane mutation affecting cellular trafficking in pink bollworm from China. Insect Biochem Mol Biol 94:28–35

Wang Y, Liu Y, Zhang J, Crickmore N, Song F, Gao J, Shu C (2018b) Cry78Aa, a novel *Bacillus thuringiensis* insecticidal protein with activity against *Laodelphax striatellus* and *Nilaparvata lugens*. J Invertebr Pathol 158:1–5

Wang L, Ma Y, Guo X, Wan P, Liu K, Cong S, Wang J, Xu D, Xiao Y, Li X, Tabashnik BE (2019) Pink bollworm resistance to *Bt* toxin Cry1Ac associated with an insertion in cadherin exon 20. Toxins 11(4):186

Wang J, Wang J, He J, Wu J, Sun J, Wang R, Zhang W (2020a) Vip3Aa from *Bacillus thuringiensis* subsp. *kurstaki* HD1 is toxic to *Aedes aegypti* (Diptera: Culicidae). J Invertebr Pathol 171:107342

Wang L, Ma Y, Wei W, Wan P, Liu K, Xu M, Cong S, Wang J, Xu D, Xiao Y, Li X (2020b) Cadherin repeat 5 mutation associated with *Bt* resistance in a field-derived strain of pink bollworm. Sci Rep 10(1):1–10

Wang J, Zuo YY, Li LL, Wang H, Liu SY, Yang YH, Wu YD (2020c) Knockout of three aminopeptidase N genes does not affect susceptibility of *Helicoverpa armigera* larvae to *Bacillus thuringiensis* Cry1A and Cry2A toxins. Insect Sci 27(3):440–448

Wang X, Xu Y, Huang J, Jin W, Yang Y, Wu Y (2020d) CRISPR-Mediated knockout of the *ABCC2* gene in *Ostrinia furnacalis* confers high-level resistance to the *Bacillus thuringiensis* Cry1Fa Toxin. Toxins 12(4):246

Ward ES, Ellar DJ (1987) Nucleotide sequence of a *Bacillus thuringiensis* var. *israelensis* gene encoding a 130 kDa delta-endotoxin. Nucl Acids Res 15(17):7195

Welch KL, Unnithan GC, Degain BA, Wei J, Zhang J, Li X, Tabashnik BE, Carrière Y (2015) Cross-resistance to toxins used in pyramided *Bt* crops and resistance to *Bt* sprays in *Helicoverpa zea*. J Invertebr Pathol 132:149–156

Whalon ME, Miller D, Hollingworth L, Grafius RM, E. J. and Miller, J. R.. (1993) Selection of a Colorado potato beetle (Coleoptera; Chrysomelidae) strain resistant to *Bacillus thuringiensis*. J Econ Entomol 86:226–233

Widner WR, Whiteley HR (1989) Two highly related insecticidal crystal proteins of *Bacillus thuringiensis* subsp. *kurstaki* possess different host range specificities. J Bacteriol 171 (2):965–974

Wilcks A, Jayaswal N, Lereclus D, Andrup L (1998) Characterization of plasmid pAW63, a second self-transmissible plasmid in *Bacillus thuringiensis* subsp. *kurstaki* HD73. Microbiology 144 (Pt 5):1263–1270. https://doi.org/10.1099/00221287-144-5-1263

Wojciechowska JA, Lewitin E, Revina LP, Zalunin IA, Chestukhina GG (1999) Two novel delta-endotoxin gene families *cry26* and *cry28* from *Bacillus thuringiensis* ssp. *finitimus*. FEBS Lett 453(1–2):46–48

Wolfenbarger DA (1981) Development of resistance in *Heliothis* spp. in the Americas, Australia, Africa, and Asia. Bull ESA 27(3):181–185

Wu Y (2014) Detection and mechanisms of resistance evolved in insects to Cry toxins from *Bacillus thuringiensis*. Adv Insect Physiol 47:297–342

Wu KM, Guo YY (2005) The evolution of cotton pest management practices in China. Annu Rev Entomol 50:31–52

Xiao Y, Zhang T, Liu C, Heckel DG, Li X, Tabashnik BE, Wu K (2014) Mis-splicing of the *ABCC2* gene linked with *Bt* toxin resistance in *Helicoverpaarmigera*. Sci Rep 4:6184

Xu X, Yu L, Wu Y (2005) Disruption of a cadherin gene associated with resistance to Cry1Ac δ-endotoxin of *Bacillus thuringiensis* in *Helicoverpa armigera*. Appl Environ Microbiol 71 (2):948–954

Xu L, Wang Z, Zhang J, He K, Ferry N, Gatehouse AMR (2010) Cross-resistance of Cry1Ab-selected Asian corn borer to other Cry toxins. J Appl Entomol 134(5):429–438

Xu C, Wang BC, Yu Z, Sun M (2014) Structural insights into *Bacillus thuringiensis* Cry, Cyt and parasporin toxins. Toxins 6(9):2732–2770

Yamashita S, Katayama H, Saitoh H, Akao T, Park YS, Mizuki E, Ohba M, Ito A (2005) Typical three-domain Cry proteins of *Bacillus thuringiensis* strain A1462 exhibit cytocidal activity on limited human cancer cells. J Biochem 138(6):663–672

Yang Y, Chen H, Wu Y, Yang Y, Wu S (2007) Mutated cadherin alleles from a field population of *Helicoverpa armigera* confer resistance to *Bacillus thuringiensis* toxin Cry1Ac. Appl Environ Microbiol 73(21):6939–6944

Yang F, Kerns DL, Head GP, Leonard BR, Levy R, Niu Y, Huang F (2014) A challenge for the seed mixture refuge strategy in *Bt* maize: impact of cross-pollination on an ear-feeding pest, corn earworm. PLoS One 9(11):e112962

Yang F, Kerns DL, Leonard BR, Oyediran I, Burd T, Niu Y, Huang F (2015) Performance of Agrisure® Viptera™ 3111 corn against *Helicoverpa zea* (Lepidoptera: Noctuidae) in seed mixed plantings. Crop Prot 69:77–82

Yang F, González JCS, Little N, Reisig D, Payne G, Dos Santos RF, Jurat-Fuentes JL, Kurtz R, Kerns DL (2020a) First documentation of major Vip3Aa resistance alleles in field populations of *Helicoverpa zea* (Boddie)(Lepidoptera: Noctuidae) in Texas, USA. Sci Rep 10(1):1–8

Yang F, Head GP, Price PA, Santiago González JC, Kerns DL (2020b) Inheritance of *Bacillus thuringiensis* Cry2Ab2 protein resistance in *Helicoverpa zea* (Lepidoptera: Noctuidae). Pest Manag Sci 76(11):3676–3684

Ye W, Zhu L, Liu Y, Crickmore N, Peng D, Ruan L, Sun M (2012) Mining new crystal protein genes from *Bacillus thuringiensis* on the basis of mixed plasmid-enriched genome sequencing and a computational pipeline. Appl Environ Microbiol 78(14):4795–4801

Yezza A, Tyagi RD, Valero JR, Surampalli RY, Smith J (2004) Scale-up of biopesticide production processes using wastewater sludge as a raw material. J Ind Microbiol Biotechnol 31 (12):545–552

Yezza A, Tyagi RD, Valéro JR, Surampalli RY (2005) Production of *Bacillus thuringiensis*-based biopesticides in batch and fed batch cultures using wastewater sludge as a raw material. J Chem Technol Biotechnol 80(5):502–510

Yokoyama T, Tanaka M, Hasegawa M (2004) Novel *cry* gene from *Paenibacillus lentimorbus* strain Semadara inhibits ingestion and promotes insecticidal activity in *Anomala cuprea* larvae. J Invertebr Pathol 85(1):25–32

Zghal RZ, Kharrat M, Rebai A, Khedher SB, Jallouli W, Elleuch J, Ginibre C, Chandre F, Tounsi S (2018) Optimization of bio-insecticide production by Tunisian *Bacillus thuringiensis israelensis* and its application in the field. Biol Control 124:46–52

Zhang X, Candas M, Griko NB, Rose-Young L, Bulla LA (2005) Cytotoxicity of *Bacillus thuringiensis* Cry1Ab toxin depends on specific binding of the toxin to the cadherin receptor BT-R1 expressed in insect cells. Cell Death Differ 12(11):1407–1416

Zhang X, Candas M, Griko NB, Taussig R, Bulla A (2006) A mechanism of cell death involving an adenylyl cyclase/PKA signaling pathway is induced by the Cry1Ab toxin of *Bacillus thuringiensis*. Proc Natl Acad Sci. https://doi.org/10.1073/pnas.0604017103

Zhang S, Cheng H, Gao Y, Wang G, Liang G, Wu K (2009) Mutation of an aminopeptidase N gene is associated with *Helicoverpa armigera* resistance to *Bacillus thuringiensis* Cry1Ac toxin. Insect Biochem Mol Biol 39(7):421–429

Zhang J, Liu X, Zhang J, Li D, Sun Y, Guo Y, Ma E, Zhu KY (2010) Silencing of two alternative splicing-derived mRNA variants of chitin synthase 1 gene by RNAi is lethal to the oriental migratory locust, *Locusta migratoria manilensis* (Meyen). Insect Biochem Mol Biol 40(11):824–833

Zhang H, Yin W, Zhao J, Jin L, Yang Y, Wu S, Tabashnik BE, Wu Y (2011) Early warning of cotton bollworm resistance associated with intensive planting of *Bt* cotton in China. PLoS One 6 (8):e22874

Zhang HN, Wu SW, Yang YH, Tabashnik BE, Wu YD (2012) Non-Recessive *Bt* toxin 5 resistance conferred by an intracellular cadherin mutation in field-selected populations of 6 cotton bollworm. PLoS One 7:e53418

Zhang W, Qiu L, Gong A, Cao Y, Wang B (2013) Solid-state fermentation of kitchen waste for production of *Bacillus thuringiensis*-based bio-pesticide. Bioresources 8(1):1124–1135

Zhang J, Jin M, Yang Y, Liu L, Yang Y, Gómez I, Bravo A, Soberón M, Xiao Y, Liu K (2020) The Cadherin protein is not involved in susceptibility to *Bacillus thuringiensis* Cry1Ab or Cry1Fa toxins in *Spodoptera frugiperda*. Toxins 12(6):375

Zhao JZ, Cao J, Li Y, Collins HL, Roush RT, Earle ED, Shelton AM (2003) Transgenic plants expressing two *Bacillus thuringiensis* toxins delay insect resistance evolution. Nat Biotechnol 21(12):1493–1497

Zhao JZ, Cao J, Collins HL, Bates SL, Roush RT, Earle ED, Shelton AM (2005) Concurrent use of transgenic plants expressing a single and two *Bacillus thuringiensis* genes speeds insect adaptation to pyramided plants. Proc Natl Acad Sci 102(24):8426–8430

Zhao J, Jin L, Yang Y, Wu Y (2010) Diverse cadherin mutations conferring resistance to *Bacillus thuringiensis* toxin Cry1Ac in *Helicoverpa armigera*. Insect Biochem Mol Biol 40(2):113–118

Zhou H, Hu W, Huang Q, Abouzaid M, Jin H, Sun Y, Qiu L, Zhang W, Lin Y, Ma W (2020a) Knockdown of cadherin genes decreases susceptibility of *Chilo suppressalis* larvae to *Bacillus thuringiensis* produced crystal toxins. Insect Mol Biol 29(3):301–308

Zhou J, Guo Z, Kang S, Qin J, Gong L, Sun D, Guo L, Zhu L, Bai Y, Zhang Z, Zhou X (2020b) Reduced expression of the P-glycoprotein gene *PxABCB1* is linked to resistance to *Bacillus thuringiensis* Cry1Ac toxin in *Plutella xylostella* (L.). Pest Manag Sci 76(2):712–720

Zhou Y, Wu Z, Zhang J, Wan Y, Jin W, Li Y, Fang X (2020c) Cry80Aa1, a novel *Bacillus thuringiensis* toxin with mosquitocidal activity to *Culex pipiens pallens*. J Invertebr Pathol 107386

Zhu KY, Palli SR (2020) Mechanisms, applications, and challenges of insect RNA interference. Annu Rev Entomol 65:293–311

Zhu J, Zheng AP, Tan FR, Wang SQ, Deng QM, Li SC, Wang LX, Li P (2010) Characterisation and expression of a novel holotype crystal protein gene, *cry56Aa1*, from *Bacillus thuringiensis* strain Ywc2-8. Biotechnol Lett 32(2):283–288

Zhu X, Lei Y, Yang Y, Baxter SW, Li J, Wu Q, Wang S, Xie W, Guo Z, Fu W, Zhang Y (2015) Construction and characterisation of near-isogenic *Plutella xylostella* (Lepidoptera: Plutellidae) strains resistant to Cry1Ac toxin. Pest Manag Sci 71(2):225–233

Zhu J, Zhu Z, Zhang B, Liu X, Liu Y, Zheng A, Wang S, Li S, Deng Q, Liu H, Liang Y (2020) Characterization of the novel insecticidal crystal protein Cyt3Aa1 from *Bacillus thuringiensis* TD516. Adv Microbiol 10(7):349

Zhuang L, Zhou S, Wang Y, Chang M (2011) Mosquito biolarvicide production by sequential fermentation with dual strains of *Bacillus thuringiensis* subsp. *israelensis* and *Bacillus sphaericus* using sewage sludge. Bioresour Technol 102(2):1574–1580

Zou H, Ding S, Zhang W, Yao J, Jiang L, Liang J (2016) Study on influence factors in *Bacillus thuringiensis* production by semi-solid state fermentation using food waste. Procedia Environ Sci 31:127–135

Zouari N, Ali SBS, Jaoua S (2002) Production of delta-endotoxins by *Bacillus thuringiensis* strains exhibiting various insecticidal activities towards Lepidoptera and Diptera in gruel and fish meal media. Enzym Microb Technol 31(4):411–418

Zukoff SN, Ostlie KR, Potter B, Meihls LN, Zukoff AL, French L, Ellersieck MR, Wade French B, Hibbard BE (2016) Multiple assays indicate varying levels of cross resistance in Cry3Bb1-selected field populations of the western corn rootworm to mCry3A, eCry3.1Ab, and Cry34/35Ab1. J Econ Entomol 109(3):1387–1398

The Endophytes

4

Abdoolnabi Bagheri, Majeed Askari Seyahooei, and
Yaghoub Fathipour

Contents

A. Bagheri · M. Askari Seyahooei
Plant Protection Research Department, Hormozgan Agricultural and Natural Resources Research and Education Center, Agricultural Research Education and Extension Organization (AREEO), Bandar Abbas, Iran

Y. Fathipour (✉)
Department of Entomology, Faculty of Agriculture, Tarbiat Modares University, Tehran, Iran
e-mail: fathi@modares.ac.ir

Omkar (ed.), *Microbial Approaches for Insect Pest Management*,
https://doi.org/10.1007/978-981-16-3595-3_4

Abstract

Chemical control may result in perishing of the microbiota population, a useful group of microorganisms with an important role in plant growth promotion and insect pest control. In the last two decades, endophytes have received a special attention. These microorganisms include bacteria, fungi, or actinomycetes, which dwell within robust plant tissues by having a symbiotic association. All or most plants possess endophytes, and in most cases endophytes are seed transmitted and begin to promote growth and plant health as soon as the seeds germinate. As the particular mechanisms by which endophytic microbes perform various functions in plants differ in different microbes and plants and that the endophytic population is greatly affected by climatic conditions and location, therefore, to encourage the widespread use of fungal entomopathogen-based biopesticides, there is a need for products with activity against multiple pests in addition to improved delivery methods and increased persistence. Endophytic microbes are often functional in the way, which may carry nutrients from the soil into plants, modulate plant development, increase plant tolerance, suppress virulence of pathogens, increase pests and disease resistance in plants, and suppress the development of competitor plant species. Control of insect pests by biocontrol agents, such as entomopathogenic microorganisms or those that inhibit/antagonize microorganisms pathogenic to plants, is a safe way helping us to reduce or eliminate the use of chemical products in agricultural systems. Various methods and experimental protocols have been tested to artificially inoculate endophytic entomopathogenic fungi (EPF) into crop plants, including spraying leaves with conidial suspension, soaking seeds in conidial suspension, injecting fungi inoculum into stems, dipping of seedling roots in conidial suspension, and soil drenching with conidial suspension. To reach a good result by using endophytes, it is needed to make sure that a symbiotic relationship is established between the plant and the microorganism. Failure to colonize the plants might be due to innate characteristics of the fungal isolate or host plant genetics, leaf surface chemistry, and competition with other endophytes naturally present within plants. Another issue that may cause an endophyte not to be appropriately colonized in a plant is the time of endophyte inoculation. Some endophytes may stop the growth of nonadapted host plants and eventually cause their death and play a role as a weed control agent. In addition, endophytes can slow down the rate of disease transmission either by decreasing vector population or by reducing the transmission and replication of viral diseases. In total, endophytes, as a new area in nonchemical approaches, have been considered as a novel ecofriendly pest management alternative which can be used practically in the pest management programs in the near future.

Keywords

Endophytes · Rhizophagy · *Beauveria* · *Metarhizium* · Entomopathogens

Learning Objectives

1. Excessive use of chemical pesticides and their adverse effects have compelled us to think to reduce chemical consumption and the use of other measures of pest management.
2. Endophytes are an important group of widespread and diverse plant symbionts that reside inside the plant tissues without any harm or diseases in the host plants.
3. Endophytes benefit their host plants in different manners, including increasing growth, increasing resistance to pests and diseases, and other abiotic stresses.
4. The use of endophytes offers a great potential for increasing the resistance of plants to herbicides.

4.1 Introduction

Chemical pesticides used in agriculture excessively for a long period of time (Grigoletti Junior et al. 2000), pose human health risks, and cause strong environmental imbalances in many cases by destructing natural enemies of crop pests in various agricultural ecosystems. In addition, chemical control may result in perishing of the microbiota population, which are either symbionts or beneficial to plants. More important residues of these chemicals frequently reported in the environment, cause pest resistance, and brought up serious concerns for the environmental scientists (Ethur et al. 2007) and the safety of foods. The biopesticides have been shown to be a viable alternatives for synthetic pesticides and a key element of environment-friendly pest management (Glare et al. 2012). In other words, decreasing efficacy of the pesticides as well as associated risks of pesticide residues on the edible part of plants highlighted the necessity of more effective and safer alternative control measures. The interest in the use of microorganisms in agriculture has been increased significantly in the last recent years, because both in plant growth promotion and insect biocontrol, among other applications, they are potential substitutes of chemical products, thus leading to environmental conservation (Peixoto Neto et al. 2002; Souza 2001). Among biopesticides, endophytes have received a special attention. Endophytes comprise a diverse polyphyletic group of microorganisms, which exhibit more than one type of life history in distinct life stages (Arnold and Lewis 2005). Endophytic microorganisms refer to the group of microorganisms, which grow in the intercellular spaces of higher plants, are recognized as one of the most promising groups of microorganisms in the terms of diversity and pharmaceutical potential (Wagenaar and Clardy 2001). These microorganisms include bacteria, fungi, or actinomycetes, which dwell within robust plant tissues by having a symbiotic association. Endophytic population is greatly affected by climatic conditions and the location where the host plant grows. They were isolated from scale primordia, meristem and resin ducts (Pirttilä et al. 2000; Pirttilä et al. 2003), leaf segments with midrib and roots (Hata et al. 2002), and from stem, bark, leaf blade, petiole (Hata and Sone 2008), and buds (Pirttilä et al. 2008). All or most plants possess endophytes, and in most cases, endophytes are seed transmitted and begin to promote growth and plant health as soon as the seeds germinate. The particular mechanisms by which endophytic microbes perform various functions in plants differ in different microbes and plants. The widespread capacity of many microbes to produce plant signal molecules (such as nitric oxide) growth regulators (such as auxins and ethylene) could be another reflection of the coevolutionary association of microbes and plants. Microbial endophytes and soil microbes could be employed to improve plant health and enhance productivity directly in commercial crop plants. Benefits could also be realized when endophytes reduce pathogens, insect damage, and competition with weedy plants (Jaber and Ownley 2018). Endophytes usually occur in above-ground plant tissues but also occasionally in roots (for example, dark septate endophytic fungi have been isolated from various plants) and are different from mycorrhizae by lacking external hyphae (Mandyam and Jumpponen 2005; Tedersoo et al. 2009). Endophytic microorganisms are potentially useful to agriculture and industry, particularly to the food and pharmaceutical sectors. Many selected endophyte species exhibit the potential to be used in agrochemical industries, besides serving as genetic vectors (Souza et al. 2004). These microorganisms are able to produce toxins, antibiotics, and other pharmaceuticals, besides performing other functions important to the host, such as providing higher resistance to stress conditions, changing physiological properties, and producing phytohormones (Azevedo et al. 2000).

A better understanding of the ecology of fungal entomopathogens would stimulate the development and uptake of more commercially available biopesticides based on these fungi in mainstream agriculture (Vega et al. 2009; Glare et al. 2012; Lacey et al. 2015). Furthermore, to encourage the widespread use of fungal entomopathogen-based biopesticides, there is a need for products with activity against multiple pests in addition to improved delivery methods and increased persistence (Glare et al. 2012).

4.2 What Is an Endophyte?

Endophytes are microbes (mostly bacteria and fungi) present asymptomatically within the tissues of healthy plants. Endophytes may colonize plants by means of horizontal transmission, when leaves accumulate numerous infections shortly after emergence by means of epiphytic germination of fungal propagules, followed by cuticular penetration or entry through stomata, or vertically when endophytes grow systematically throughout roots, stems, and leaves, and infect the seed progeny of an infected plant (Zabalgogeazcoa 2008). All or most plants possess endophytes, and in most cases, endophytes are seed transmitted and begin to promote growth and plant health as soon as seeds germinate (Johnston-Monje and Raizada 2011; Verma et al. 2018). Mendes and Azevedo (2007) extended the concept of endophytes and classified them into two main types: Type 1, which does not produce structures external to their hosts, and Type II, which produces external structures, such as mycorrhizal fungi and bacteria, that produce nitrogen fixation nodules. Endophytic fungi have been reported in different climatic conditions ranging from temperate to tropical and in different plant groups, i.e., grasses, agricultural crops, tropical trees, and also in soils (Muvea 2015). However, it has been demonstrated that the tropical highland region harbors more endophytes diversity as compared to moist transitional and dry transitional agro-ecological zones (Akello 2012). The loss of endophytic microbes from crop plants during domestication and long-term cultivation could be remedied by the transfer of endophytes from wild relatives of crops to crop species. Increasing atmospheric carbon dioxide levels could reduce the efficiency of the rhizophagy cycle due to the repression of reactive oxygen used to extract nutrients from microbes in roots (White et al. 2019b). In overall, the total amount of endophytic fungi represents something about 65% of all estimated 1.5 million fungal species (Hawksworth 2001). This indicates that endophytic fungi consider the majority groups of fungal diversity.

Agro-ecologically, there are differences in the distribution and abundance of fungal endophytes. In agricultural systems, many abiotic and biotic factors are modified by management techniques, which strongly impact fungal communities. For instance, studies have shown that some practices, such as tillage, monocropping, and fertilization, negatively influence the abundance and diversity of fungi (Helgason et al. 1998; Verbruggen et al. 2012).

One of the beneficial roles of endophytes is the production of secondary metabolites. Alkaloids, steroids, terpenoids, isocoumarins, quinones, flavonoids, phenylpropanoids, lignans, peptides, phenolics, aliphatics, and volatile organic compounds, etc. are the range of metabolites produced by the endophytes (Kusari et al. 2013). It is hypothesized that the secondary metabolite synthesis genes are transferred to the endophytes, and hence the endophytes have the pathway genes for the synthesis of secondary metabolites. Previous reports have suggested that endophytes develop genetic systems to allow them the transfer of information within the group of endophytes and between the endophytes and their host plants (Borges et al. 2009). Also, long-term coexistence with their hosts resulted in a coevolutionary process through which these microorganisms have acquired interesting capabilities, such as powerful transformation. For instance, some endophytes can synthesize biologically active substances similar to the secondary metabolites produced by their hosts (Wang and Dai 2011). The host secondary metabolites produced by endophytes may compete with other invading pathogens (Shweta et al. 2013) and also to provide plant defenses against pathogens. Therefore, the endophytes may increase fitness benefits by providing higher fitness to the hosts (Moussa et al. 2016). The entophyte *Bacillus* sp. exhibits directly or indirectly the suppression of a broad spectrum of phytopathogens through the production of secondary metabolites, namely, difficidin, polyketides, and bacillaene (Nakkeeran et al. 2019).

Endophytic microbes are often functional in the way, which may carry nutrients from the soil into plants, modulate plant development, increase plant tolerance, suppress virulence of pathogens, increase disease resistance in plants, and suppress the development of competitor plant species (Ikram et al. 2018; Li et al. 2018a, b; Compant et al. 2010; Kandel et al. 2017). Further, the interaction between endophytes and plants involves the production of several secondary metabolites or bioactive substances of industrial interest for the development of pathogen control systems or with diverse biological activities (Nisa et al. 2015; Venugopalan and Srivastava 2015). The diverse group of endophytes is also useful in nutrient-poor environments and when plants are under stress due to drought or pathogen attacks (Rodrigues et al. 2000; Saikkonen et al. 2004). Irrespective of plant host or endophyte genera, symbiosis has resulted in increased plant biomass production and reduction of disease in plants (Rodrigues et al. 2000). Up to now reported natural products from endophytic microbes comprise antibiotics, antipathogens, immunosuppressants, anticancer compounds, antioxidant agents, and other biologically active substances. In spite of a focused interest in synthetic products, bioactive natural products maintain an enormous impact on current medicine. Around 60% of the new drugs registered during 1981–2002 by the FDA as anticancer, antimigraine, and antihypertensive agents are either natural products or based on natural products (Newman et al. 2003).

Endophytic fungi in recent evidence suggest that they can play symbiotic roles in nature, such as antagonists of plant pests and diseases, increased drought tolerance, and plant growth. Several studies indicate that endophytes reduce the attack of insects and pathogenic fungi against the host plants (Landum et al. 2016; Jaber and Ownley 2018). Of course, some fungi might be pathogenic on the main host species, but symptomless endophytes on other hosts. This differential behavior may result from differences in fungal gene expression in response to the plant, or from the differences in the ability of the plant to respond to the fungus (Sieber 2007). Webber (1981) described *Phomopsis oblonga*, an endophytic fungus, protected elm trees against the beetle, *Physocnemum brevillineum,* which is a vector of the Elm Dutch disease caused by the pathogenic fungus *Ceratocystis ulmi*. According to Gai et al. (2011), the bacterial communities associated with vector insects and plants differ in abundance through the annual season. Endophytic bacteria could influence disease development by reducing the insect transmission efficiency due to competition with pathogens in host plants and also in insect foreguts. The use of endophytic fungal entomopathogens as seed treatments introduced at an early stage of plant development overcomes several inherent problems usually encountered when using fungal entomopathogens as contact biocontrol agents. These include exposure to detrimental environmental conditions (e.g., damaging UV radiation, reduced humidity, and excessive rainfall), compatibility with other control measures, and the challenge of synchronizing the biocontrol agents with the target pests. In addition to their promising dual biocontrol potential against insect pests and plant diseases, fungal entomopathogens as endophytes may also offer protection against cryptic pests (e.g., insect borers) that would otherwise be difficult to control by topical application (Jaronski 2010). They can also provide additional benefits, such as accelerating seedling emergence and improved plant growth (Sasan and Bidochka 2012; Lopez and Sword 2015; Jaber and Enkerli 2016; Jaber and Enkerli 2017).

Interspecies variation of endophytic fungi at different parts of the host plant is largely attributed to variation in physiological conditions and texture difference of host tissue (Aly et al. 2010). Some studies have revealed that a majority of endophytic fungi are not host specific; instead, they have a wide range of hosts and that the abundance and dominant species on each host plant may be different. The colonization rate and interspecies diversity may vary with different parts of host plants and also correlate with the age of the hosts and seasons (Sun et al. 2008). The endophytes have been shown to exhibit organ and tissue specificity due to their adaption to the altered physiological environment in different plant tissues. Variations in the endophytic profile can be caused by different parameters, such as seasonal changes, stresses on the host plant, and plant organs (Mocali et al. 2003). It is noteworthy that intraspecies diversity of endophytic fungi within the same part of the host plant is also abundant. The distribution and diversity of endophytes are done by a culture-based study based on the colonization frequency. A culture-based study of ten sea grass species revealed that *Aspergillus terreus* was the most dominant species in rhizomes of the sea grasses (Venkatachalam et al. 2015).

The bacterial endophytes are present in the roots of most plants in higher numbers compared with above-ground tissues (Rosenblueth and Martínez-Romero 2006). Many seeds carry a diversity of endophytic bacteria (Quadt-Hallmann et al. 1997) and plants that propagate vegetatively (such as, potatoes or sugarcane) transmit endophytes to the next generation. Bacterial endophytes do not inhabit living vegetal cells but colonize intercellular spaces and xylem vessels (Ryan et al. 2008). Endophytic bacteria can establish a mutualistic association with their hosts (Quadt-

Hallmann et al. 1997), and increase crop yields, degrade contaminants and produce novel substances or fixed nitrogen (Rosenblueth and Martínez-Romero 2006). Endophytes usually have a systemic movement.

Endophytic bacteria can promote plant growth through nitrogen fixation (e.g., Sevilla et al. 2001), production of phytohormones, by enhancing nutrient availability (Sturz et al. 2000; Verma et al. 2001; Lee et al. 2004; Pirttilä et al. 2003) or by biocontrol of phytopathogens in the root zone (through the production of antifungal or antibacterial agents, siderophore production, nutrient competition and induction of systematic acquired host resistance or immunity) or in the vascular system (Quadt-Hallmann et al. 1997).

4.2.1 Colonization Cycle

Colonization of plant tissues by fungal endophytes involves several steps, including host recognition, spore germination, penetration of the plant surface, and tissue colonization (Petrini 1991). It is likely that it exhausts the host plant resources leaving none available for the plant pathogen when it attempts to colonize. Change in the order of plant colonization, for example, when the pathogen colonizes the plant before the endophyte can shift the endophyte pathogen interaction from disease suppression to disease facilitation (see Adame-Álvarez et al. 2014). In other words, once inside the plant, an endophyte occupies a niche with relatively low competition from other microorganisms, provided the endophyte gets there first (Haggag 2010). Initial endophytic colonization also induces plants to produce lignin and other cell wall deposits as a mechanical defense response and this too might consequently prevent or limit infection by disease-causing plant pathogens (see Schulz and Boyle 2005).

Bacterial endophytes are capable of colonizing different seed parts including the embryo. These endophytes likely mobilize and grow in the developing seedlings during germination and early seedling growth. As seedlings emerge and plant growth begins, interactions between the roots and the soil microbiome commence. Plant exudates fuel microbial activities in the rhizosphere, which facilitate the attachment and entry of bacteria into the plant roots. Eventually, certain endophytes initiate colonization of tissues beyond the roots, such as the stems and leaves, and ultimately throughout the plant endosphere. Some bacterial endophytes also colonize flowers and seeds and most likely get transferred vertically from the maternal endophyte community into the offspring. It has been shown that endophytes could colonize corresponding seeds after the flowers were inoculated. Moreover, endophytes passed on to seeds, resumed endophytic activity after the seeds were planted (Kandel et al. 2017).

4.2.2 Mechanisms Underlying Endophyte Resistance

Endophytes can affect plant disease in several ways, including direct suppression of plant pathogens, induction of systemic plant resistance, and promotion of plant growth. Regarding induction of systemic plant resistance, it is believed that the plant colonization by inoculated fungi can at first be recognized by the plant as potential invaders leading to the triggering of immune responses with the synthesis of specific regulatory elements, such as transcription factors involved in resistance against herbivores (Canassa et al. 2019). The induction of proteins related to plant defense or stress response in *Phoenix dactylifera* leaves colonized by *Beauveria bassiana* has also been reported (Gómez-Vidal et al. 2009).

The promotion of plant growth is another mechanism by which endophytes confer protection against pests and diseases and other biotic and abiotic stresses. For instance, disease damages may increase under abiotic stresses. Endophytes can promote the plant growth under stress conditions and mitigate the negative effects of destructive agents. In this context, in a study that tested the effect of inoculation of three fugal endophytes (*Aspergillus niger* MG890603, *Paecilomyces formosus* MG904988, and *Alternaria alternata* MG907039) isolated from mastic trees (*Pistacia mutica*) on the promotion of plant growth in sweet pepper under salinity conditions a significant improvement was observed in the presence of endophytes. This study revealed a decreasing trend in the levels of measured traits including, root and stem length, fresh and dry weight of root, shoot weight, chlorophyll a, chlorophyll b, and total chlorophyll while the amount of catalase enzyme activity, peroxidase, and proline increased by increasing salinity. Inoculation of the above-mentioned endophytes reduced the negative impact of salinity and the highest performance was achieved when all endophytes were inoculated (Kavehnia et al. 2018a, b). In another study, we inoculated cucumber seedlings from Emperor cultivar with three bacterial endophytes (*Bacillus subtilis*, *Rhizobium pusense,* and *Agrobacterium tumefaciens*) isolated from wild almond, *Amygdalus scoparia*, to study its salinity resistance in association with bacterial endophytes. Results exhibited a considerable and significant decrease in the negative impact of salinity was inoculated with the bacterial endophytes. Main traits including fresh weight, dry weight, chlorophyll a, total chlorophyll, catalase, and peroxidase were found between endophyte-associated seedlings (E+) and endophyte-free seedlings (E-) ones (Peikari 2018). We also investigated the symbiotic association of *Penicillium chrysogenum* (fungal endophyte) and *Exiguobacterium aurantiacum* (bacterial endophyte) on some growth and physiological attributes of tomato plant (*Solanum lycopersicum* L.) cultivar 8320 SEMINIS and showed that both endophytes significantly increased fresh and dry weight, stem height and diameter, number of leaves, chlorophyll content (SPAD) and fluorescence chlorophyll, chlorophyll a, chlorophyll b, carotenoids, and relative leaf content. In addition, they had a significant synergistic effect on the mentioned traits in tomato plants in simultaneous application compared to when they were inoculated separately (Aghaei Dargiri et al. 2021a).

Fig. 4.1 Endophytes are able to provide their host plants under drought and salinity stress conditions and cause the root system to develop

In a study on the effects of inoculating mandarin endophytic fungi (*Penicillium citrinum*, *Aureobasidium pullulans,* and *Dothideomycetes* sp.), individually and in combination with each other on the reactive oxygen species (ROS) scavenging and antioxidant functions in *Citrus reticulata* under drought stress (four irrigations interval including 2, 4, 6, and 8 days), we found that drought stress significantly reduced growth, chlorophyll, carotenoid content, and chlorophyll fluorescence (Fv/Fm) of the plants lacking endophytes. Combined applications of three fungal endophytes (*P. citrinum* + *A. pullulans* + *Dothideomycetes* sp.) significantly improved the above-mentioned parameters under drought stress. H_2O_2, O_2-, and lipid peroxidation levels were significantly reduced in the plants inoculated with fungal endophytes. Drought stress significantly increased the activities of ascorbate peroxidase (APX), superoxide dismutase (SOD), and glutathione reductase (GR) and levels of ascorbate (ASA) and glutathione (GSH), and decreased activities of catalase (CAT), monodehydroascorbate reductase (MDHAR), and dehydroascorbate reductase (DHAR). Fungal endophytes inoculated drought-stressed seedlings enhanced the above-mentioned indicators as compared to the drought-stressed plants without fungal endophytes, as well as in the ratios of reduced ascorbate/dehydroascorbic acid (ASA/DHA), and reduced glutathione/oxidized glutathione (GSH/GSSG). Overall, fungal endophyte inoculation improved drought tolerance and reduced the accumulation of ROS by increasing their scavenging via improving the redox state of ascorbate and glutathione, and promotion of antioxidant enzyme activity (Sadeghi et al. 2020) (Fig. 4.1).

Antibiosis, by the production of secondary metabolites, has been shown to confer protection against disease-causing plant pathogens and insect pests (Ownley et al.

2010). Fungal entomopathogens are a rich source of secondary metabolites with antimicrobial, insecticidal, and cytotoxic activities (Gibson et al. 2014). For instance, *Beauveria bassiana* produces numerous secondary metabolites, including beauvericin, bassianin, beauverolides, bassianolides, oosporein, bassianolone (Ownley et al. 2010). Among these metabolites, beauvericin in particular has broad and significant multiple bioactivities and can also be produced by several entomopathogenic fungal genera, such as *Paecilomyces*, *Isaria,* and *Fusarium* (Jaber and Ownley 2018). Beauvericin is produced during broth culture by *B. bassiana* strain 11–98 (Jaber and Ownley 2018), an endophytic strain found to suppress damping-off caused by *Rhizoctonia solani* and *Pythium myriotylum* in tomato and cotton (Jaber and Ownley 2018). It was reported by Shrivastava et al. (2015) that tomato plants endophytically colonized by *B. bassiana* showed higher levels of monoterpenes and sesquiterpenes compared to control plants and larvae of *Spodoptera exigua* (Hübner) (Lepidoptera: Noctuidae) feeding on fungal colonized plants had lower weight than those feeding on control plants, suggesting that the observed difference in the levels of terpenoids may be related to a defense response of fungus-inoculated plants.

Specifically, *B. bassiana* is able to produce a range of secondary metabolites, such as beauvericin (Grove and Pople 1980; Wang and Xu 2012), bassianolides, bassiacridin, bassianin, beauverolides, bassianolone, and others (Canassa et al. 2019). Such metabolites extracted in vitro from the mycelia of an endophytic isolate of *B. bassiana* (isolated from *Orthorhinus cylindrirostris* Fabricius (Coleoptera: Curculionidae) caused mortality and reduced reproduction of *Aphis gossypii* Glover (Hemiptera: Aphididae) (Canassa et al. 2019).

4.3 Root Endophytes and the Role of Host Plants on Diversity and Density of Endophytes

Roots of terrestrial plants are associated with mycorrhizal and nonmycorrhizal root-endophytic fungi. Mycorrhizae are distinguished from endophytic fungi by lacking external hyphae or mantels (Saikkonen et al. 1998). The nonmycorrhiza has been suggested to impact plant growth and development (bioregulation), plant nutrition (biofertilization), and plant tolerance and resistance to abiotic and biotic stresses (bioprotection) (Blumenstein 2015). Dark septate endophytes are a group of root endophytes (Blumenstein 2015), which contain mycorrhiza-forming and nonmycorrhizal root colonizers and occur worldwide (Blumenstein 2015). In other words, arbuscular mycorrhizal fungi (AMF) and dark septate endophytes (DSE) are two fungal groups that can directly influence plant success in a given environment. While AMF are obligate symbionts of living plant roots, DSE are facultative fungal symbionts that can live on organic debris and in biological soil crusts in addition to plant roots (de Mesquita et al. 2018). One of the best-studied members is the species *Piriformospora indica* (Franken 2012). Its plant growth-promoting effects have been revealed for various hosts, and its application to plant production has been proposed (Varma et al. 1999, 1999). For instance, barley plants colonized by *P. indica* were more resistant to pathogens and more tolerant to salt stress and showed higher yield (Waller et al. 2005). It has been suggested that *P. indica* may protect a wide variety of plants against fungal pathogens: root pathogens might be directly inhibited by antagonistic activities of the endophyte, which is able to produce ROS (reactive oxygen species) and synthesize antioxidants (Waller et al. 2005). It has been demonstrated that *P. indica* root colonization systemically induces resistance, which may provide protection against pathogens in the above-ground plant parts (Waller et al. 2005).

Although, variation in endophyte assemblages in above-ground tissues varied with host growth habit (stems are the richest tissue in woody plants and roots were the richest tissue in graminoids) (Harrison and Griffin 2020), it is generally believed that through secretion of exudates, plants alter the numbers and diversity of microbes on root surfaces and in the rhizosphere (Broeckling et al. 2008). Plants are known to increase the secretion of exudates in nutrient-limiting soils, likely leading to increased microbial activity around roots and increased "microbial mining" for nutrients (Bowsher et al. 2016). Root exudates attract bacteria in particular that will grow in a biofilm in the root exudates (White et al. 2019b). In this sense, root exudates act as signal molecules that attract a diverse community of microbes to the exudate zone and biofilm around the root tip meristem (White et al. 2019b). Through the continued secretion of root exudates, plants are cultivating microbes, and when nutrients are scarce, plants increase the cultivation of microbes by producing more exudates (Bowsher et al. 2016). It has been reported that plants can release significant amounts of photosynthates or exudates from their roots, which influence microbial communities in the rhizosphere. Root exudates, including organic acids, amino acids, and proteins, may be involved in recruiting bacterial endophytes from the rhizosphere. Root exudates are likely to contain substrates that initiate early communication between host plants and bacterial endophytes and consequently steer the colonization process. For example, evidence of the involvement of oxalate in the recruitment of the beneficial bacterial strain *Burkholderia phytofirmans* (PsJN) by host plants has been reported (Kandel et al. 2017).

4.4 Artificially Endophytic Entomopathogenic Fungi Application

Various methods and experimental protocols have been tested to artificially inoculate endophytic entomopathogenic fungi (EPF) into crop plants, including spraying leaves with conidial suspension, soaking seeds in conidial suspension, injecting fungi inoculum into stems, dipping of seedling roots in conidial suspension, and soil drenching with conidial suspension (Mantzoukas and Eliopoulos 2020).

The foliar application is the most common method with many promising results. However, certain drawbacks have been reported and must be taken into serious consideration. The foliar endophytes can reduce the insect population by producing alkaloids that are toxic (Tan and Zou 2001; Strobel 2003). The major concern of this method is the extremely localized colonization that is often limited to the foliar parts of the plant, with EPF being absent from stems and roots (Parsa et al. 2013; Yan et al. 2015). Apart from that poor efficiency of hyphal penetration into leaf tissues has been reported (Posada et al. 2007; Muvea et al. 2014) possibly due to the low density of stomata (natural entries for fungal infection), leaf surface structure, and specific cuticular components.

Soaking seeds in conidial suspension before propagation is another inoculation method that has been successfully applied to many major crops. However, there have been reports where inoculation through the seed resulted in some or no colonization of the stem or leaf (Tefera and Vidal 2009; Qayyum et al. 2015). This has been attributed to the negative effect of soil microorganisms that may act antagonistically toward the EPF. Stem injection has also been evaluated as an inoculation method of EPF on various plants (Bing and Lewis 1991; Posada et al. 2007). When *B. bassiana* was inoculated with this method, the highest postinoculation recovery was yielded in coffee seedlings, compared with foliar spraying or soil drenching (Posada et al. 2007), and efficient *Helicoverpa armigera* (Hübner) (Lepidoptera: Noctuidae) control was provided for tomato plants (Qayyum et al. 2015). Dipping roots in conidial suspension has proven an effective inoculation method, although results were often contradictory when compared with other methods, such as foliar application or seed treatment (Muvea et al. 2014; Russo et al. 2015). The success of this method has been reported to be greatly dependent on the plant species. The soil drenching technique includes the watering of seedlings with conidial suspension. Similar to root dipping, the low colonization rate that is often recorded with this method has been linked to the interaction between EPF and other competing soil microorganisms. The use of sterile growth media instead of nonsterile soil significantly has enhanced the success of this method (Tefera and Vidal 2009).

The combination of endophytic EPF with other biocontrol agents, such as predators and parasitoids, has been proposed as a promising approach to increase the efficiency of this ecofriendly approach (Akutse et al. 2013; Jaber and Araj 2018; González-Mas et al. 2019a, 2019b). For instance, in a study carried out to explore the effectiveness of the combined use of *B. bassiana*, *Metarhizium brunneum*, and the aphid endoparasitoid *Aphidius colemani* Viereck (Hymenoptera: Braconidae) against the green peach aphid *Myzus persicae* Sulzer (Homoptera: Aphididae) in sweet pepper (Tan and Zou 2001). Similar research has documented successful combinations of endophytic EPF species with entomophagous insects against leaf miners in beans (Akutse et al. 2013). Regarding the use of less-known endophytes against plant pathogens, it has been demonstrated that endophytes derived from closely related plants were more efficient because the endophytes and pathogens of a plant species are closely related evolutionarily (Petrini et al. 1993).

Muvea (2015) reported fewer thrips on onions inoculated with *Clonostachys rosea* ICIPE 707, *Trichoderma asperellum* M2RT4, *Trichoderma atroviride* ICIPE 710, *Trichoderma harzianum* 709, *Hypocrea lixii* F3ST1, and *Fusarium* sp. ICIPE 712 isolates compared with those inoculated with *Fusarium* sp. ICIPE 717 and the control. By studying the behavior of thrips on endophytically colonized onion plants, they found that female *Thrips tabaci* Lindeman (Thysanoptera: Thripidae) preferred endophyte-free (E^-) over endophyte-inoculated (E^+) plants (in Y-tube olfactometer assay, thrips showed about 3.6-fold preference for E^- plants). In addition, the number of feeding punctures and eggs was more on E^- than on E^+ plants, and oviposition was reduced sixfold on E^+ plants within a 72 h experimental period. The endophyte inoculation affected the behavior of thrips larvae and in individual larval choice experiments, significantly more first-instar and second preferred to feed on leaf sections of E^- compared with the E^+ plants. Also, in a settlement preference assay with groups of second instars, larvae preferred leaf sections from E^- over E^+ plants with increased time.

The negative effect of E+ plants on ovipositional preference of other pests, such as *Liriomyza huidobrensis* (Blanchard) (Diptera: Agromyzidae) on *Vicia faba* L. and *Phaseolus vulgaris* L. has been reported by Akutse et al. (2013). Gurulingappa et al. (2010) showed that feeding of *A. gossypii* on *B. bassiana* Balsamo Vuillemin or *Lecanicillium lecanii* Zimmerman colonized cotton leaves slowed down the reproductive rates and reduced fecundity and longevity of the insects. In the field conditions, inoculation of maize plants with endophytic *B. bassiana* affected larval development and reduced damages caused by stem borers *Ostrinia nubilalis* (Hübner) (Lepidoptera: Pyralidae) and *Sesamia calamistis* Hampson (Lepidoptera: Noctuidae) (Bing and Lewis 1991; Cherry et al. 2004).

4.5 Can Endophytes Always Colonize Host Plant?

Although there are many good instances regarding successful colonization of both fungal and bacterial endophytes in different plant species (Shrivastava et al. 2015; Russo et al. 2015; Silva et al. 2020), sometimes plants may not be colonized by endophytes. To accurately assess the ability of a root endophyte to colonize plant roots, ensuring the endophytic association is crucial (Barelli et al. 2018). Failure to colonize the plants might be due to innate characteristics of the fungal isolate (Posada et al. 2007) or host plant genetics (Arnold and Lewis 2005), leading to potentially unique outcomes for each plant genome–endophyte genome interaction. Leaf surface chemistry (Griffin 2007; Posada et al. 2007) and competition with other endophytes naturally present within plants (Jaber and Ownley 2018) could also lead to differential colonization rates of plants by fungal isolates (Akutse et al. 2013; Vidal and Jaber 2015; Mutune et al. 2016). Another issue that may cause an endophyte not to be appropriately colonized in a plant is the time of endophyte inoculation. In addition, endophytes may have a different spatiotemporal distribution, i.e., diversity and frequency of endophytes may be changed based on the different parts of a plant (leaves vs petiole, and twigs) or in different time periods (Thongsandee et al. 2012). For instance, while the occurrence of *Phyllosticta* sp. in both leaves and petioles of *Gingko biloba* L. was first detected in August and peaked in October (none in the month of May), *Phomopsis* sp. was isolated from twigs throughout the growing season. These results show that the distribution of these two endophytic fungi has a different spatiotemporal distribution (Thongsandee et al. 2012). Chareprasert et al. (2006) showed that endophyte richness can vary from one plant species to another and within the same species, it may have a season-based fluctuation. They found that matured leaves of teak (*Tectona grandis* L.) and rain tree (*Samanea saman* Merr.) had higher numbers of genera and endophyte species, with higher colonization frequency, than young leaves while their occurrence in leaves increased during the rainy season. Guo et al. (2008) stated the endophytic composition of plants may vary according to environmental conditions, geographic location, and seasons. Endophyte colonization not only is influenced by environmental factors, such as temperature and relative humidity, but also by soil microorganisms (Bing and Lewis 1991). Other factors, e.g., age and species of plants, growth medium, conidial density and species of the EPF, and method of inoculation may have a vital role in the inoculation of EPF into plant tissues (Mantzoukas and Eliopoulos 2020). Li et al. (2018, 2018) evaluated the distribution of fungal endophytes in roots of *Stipa krylovii* across six vegetation types in the grassland of northern China and found that environmental parameters had more contribution in variation of the communities than the vegetation type or geographical distance.

4.6 Instances of Endophyte Inoculation

Fungal entomopathogens play an important role in reducing herbivory following their colonization of plants as endophytes. There are many instances in which endophytes have been successfully applied against pests and diseases. Lewis and Cossentine (1986) reported the season-long suppression of the European corn borer *O. nubilalis* in maize *Zea mays* L. (Poaceae), measured as reduced tunneling by the insect, to the establishment of *B. bassiana* as an endophyte following application of an aqueous suspension of the fungus to the plants. Subsequent work by Lewis and colleagues using the same model system indicated successful re-isolation of *B. bassiana* from internal plant tissues after application of the fungus using different inoculation methods and examined the in planta growth and movement of the fungus (Arnold and Lewis 2005). In other instances, plant colonization by *B. bassiana* has been reported to reduce damage caused by the lepidopteran cob and stem-borers *O. nubilalis* and *Sesamia calamistis* in maize (Bing and Lewis 1991; Cherry et al. 2004); the tomato fruitworm *Helicoverpa zea* (Boddie) (Lepidoptera: Noctuidae) in tomato (Powell et al., 2007); the banana weevil, *Cosmopolites sordidus* Germar (Coleoptera: Curculionidae) in banana (Akello et al. 2008); the poppy stem gall wasp, *Iraella luteipes* Thompson (Hymenoptera: Cynipidae) in opium poppy (Quesada-Moraga et al., 2009); and the stem weevil *Apion corchori* Marshall (Coleoptera: Curculionidae) in white jute (Biswas et al. 2013). Although the above examples were mostly related to the use of *B. bassiana* against pests, there are many successful examples regarding the negative effects of *L. lecanii* and *Aspergillus parasiticus* against the cotton aphid, *A. gossypii* and the Australian plague locust, *Chortoicetes terminifera* Walker (Orthoptera: Acrididae) in cotton and wheat (Gurulingappa et al. 2010); *Hypocrea lixii*, *Gibberella moniliformis*, *Fusarium oxysporum,* and *Trichoderma asperellum* against the pea aphid, *Acyrthosiphon pisum* Harris (Hemiptera: Aphididae); the black bean aphid, *Aphis fabae* Scopoli

(Hemiptera: Aphididae) and the pea leaf miner, *L. huidobrensis* in broad bean (Akello and Sikora 2012; Akutse et al. 2013); *Clonostachys rosea, H. lixii, T. harzianum, Trichoderma asperellum, Trichoderma atroviride* and *Fusarium* sp. against the onion thrips, *T. tabaci* in onion (Muvea et al. 2014); *Purpureocillium lilacinum* against *A. gossypii* and the cotton bollworm, *H. zea* in cotton (Castillo-Lopez et al., 2014; Lopez and Sword 2015); *Metarhizium robertsii* and *Isaria fumosorosea* against the Mediterranean corn stalk borer, *Sesamia nonagrioides* Lefebre (Lepidoptera: Noctuidae) in sweet sorghum (Mantzoukas et al. 2015).

As endophytes can promote the growth of plants by the production of phytohormones, siderophore, ACC deaminase, hydrolytic enzymes, in addition to a direct impact on pests and diseases, they can also reduce the damage of pests and diseases indirectly. In rice, plant growth-promoting endophytic fungi, *Aspergillus fumigatus* TS1 and *Fusarium proliferatum* BRL1, have been shown to produce gibberellins and regulate plant endogenous hormones on colonization in rice (Bilal et al. 2018). In this context, Ji et al. (2014) have reported 576 bacterial endophytic isolates from leaves, shoots, and roots of 10 rice cultivars and found that 12 isolates, when treated to rice seedlings, improved plant growth and increased height, dry weight, and antagonistic effects against fungal pathogens. Lubna et al. (2018) evaluated the endophytic fungus *Aspergillus niger* isolated from *Cannabis sativa* to improve the growth of rice. The fungi were observed to show growth-promoting traits, such as the presence of siderophores, phosphate solubilization, and the production of indole acetic acid (IAA) and gibberellins and were further found to promote the growth of rice. The mechanism of growth promotion was attributed to the production of different forms of gibberellins and IAA by the endophyte. The presence of GA pathway genes (P50–1, P450–3, P450–4, ggs2, and des) was also confirmed using semiquantitative RT-PCR (Lubna et al. 2018).

4.7 The Roles of Endophytes in Plants

4.7.1 Phytostimulation

Plants require 16 essential elements, like C, H, N, O, and P and 11 more. These essential elements are available to plants for their growth and development in chemical form, which they obtain from the atmosphere, soil, water, and organic matter. Endophytes can play an important role in the uptake of these nutrients (Malinowski et al. 2000). Sadeghi et al. (2019) reported that inoculation of mandarin by endophytic fungi increased the level of Ca, K, and P in the plants. Endophytic bacteria produce a wide range of phytohormones, such as auxins, cytokinins, and gibberellic acids. *Burkholderia vietnamiensis* is a diazotrophic endophytic bacterium that has been isolated from wild cottonwood (*Populus trichocarpa*) that can produce indole acetic acid (IAA), and promotes the growth of plants. Xin et al. (2009) showed that *B. vietnamiensis* inoculated (*Populus trichocarpa*) grew on nitrogen free media, gained more dry weight and more nitrogen content compared with noninoculated plants, showing its ability to promote the growth of plants (Xin et al. 2009). A new strain of fungus *Cladosporium sphaerospermum* isolated from roots of *Glycine max* (L) Merr. (with high amounts of bioactive GA3, GA4, and GA7), induced maximum plant growth in both rice and soybean varieties (Hamayun et al. 2009). It has been shown that the endophyte inoculation can increase the level of auxin and gibberellin in mandarin (Sadeghi et al. 2019). The highest and lowest amounts of auxin were reported when mandarin plants were inoculated by *Aureobasidium pullulans* and *Dothideomycetes* sp., respectively. Similarly, the highest and lowest amount of gibberellins was found in *P. citrinum* and the lowest in *A. pullulans*.

4.7.2 Endophytes Modulate Plant Development

Controlled experiments have revealed that seedlings of grasses cleaned of most of their endophytic microbes lose the root gravitropic response (i.e., roots do not grow downward), and seedlings frequently are diminished in size with reduced or no root hair formation (Verma et al. 2017; Verma et al. 2018), showing modulation of seedling development by endophytes is likely the result of the evolution of plants in continuous symbiosis with microbes that colonize plant tissues and thus reliably participate in the development process. The widespread capacity of many microbes to produce plant signal molecules (such as nitric oxide), growth regulators (such as auxins and ethylene) could be another reflection of the co-evolutionary association of microbes and plants. It has been found that root hairs elongate until all microbes have been ejected from hairs (White et al. 2018). Root hair elongation may be triggered by nitric oxide or ethylene production by the intracellular microbes protoplasts that cluster in the tip of the elongating hair, but this has not been proven (White et al. 2018). Endophytic microbes in plants have also been shown to enhance root growth and increase root branching, further leading to increased plant growth (Compant et al. 2010; Kandel et al. 2017; Irizarry and White 2018). It has been shown that the inoculated mandarin with fungal endophytes (*P. citrinum*, *A. pullunts*, *Dothideomycetes* sp.) had the well-developed roots (Sadeghi et al. 2019). The effects of endophytes on root growth are generally attributed to the production of growth regulators by microbes; however, enhanced nutrient acquisition from microbes may equally contribute to enhanced plant growth (White et al. 2019b; Aghaei Dargiri et al. 2021a, 2021b, 2021c, 2021d, 2021e; Kavehnia et al. 2018a, 2018b; Peikari 2018; Baghazadeh Daryaii et al. 2021).

Another type of nutritional endophytic symbiosis involves microbes that inhabit both endophytic tissues and extend out into the soil. Dark septate endophytes and mycorrhizal fungi establish this kind of symbiosis with many families of plants. Hyphae of these fungi grow endophytically in roots, and the mycelia extending into soil acquire nutrients and mobilize them back to plants (White et al. 2018).

In the rhizophagy cycle, plants cultivate microbes that function as carriers of nutrients and support plant growth. In an experiment, it was found that grass plants obtained approximately 30% of nitrogen from rhizophagy (Hill et al. 2013; Paungfoo-Lonhienne et al. 2013; White et al. 2015).

The endophytes promote the growth of the host plant by the production of phytohormones, siderophore, ACC deaminase, hydrolytic enzymes, etc. Plant growth-promoting endophytic fungi *Aspergillus fumigatus* TS1 and *F. proliferatum* BRL1 have been shown to produce gibberellins and regulate plant endogenous hormones on colonization in rice (Bilal et al. 2018). Sadeghi et al. (2019) showed that three endophytic fungi, *P. citrinum*, *A. pullunts*, *Dothideomycete*s sp., increased the level of superoxide dismutase (SOD), ascorbate peroxidase (APX), and glutathione reductase (GR), ascorbate (ASA), and glutathione (GSH) in mandarin. Ji et al. (2014) have reported 576 bacterial endophytic isolates from leaves, shoots, and roots of 10 rice cultivars and found that 12 isolates, when treated to rice seedlings, improved plant growth and increased height, dry weight, and antagonistic effects against fungal pathogens. Lubna et al. (2018) evaluated the endophytic fungus *Aspergillus niger* isolated from *Cannabis sativa* to improve the growth of rice. The fungi were observed to show growth-promoting traits, such as the presence of siderophores, phosphate solubilization, and the production of indole acetic acid (IAA) and gibberellins and were further found to promote the growth of rice. The mechanism of growth promotion was attributed to the production of different forms of gibberellins and IAA by the endophyte. Presence/absence of the GA pathway genes (P50–1, P450–3, P450–4, ggs2, and des) was investigated and confirmed using semiquantitative RT-PCR (Gupta 2016).

4.7.3 Plants Use Microbes to Mine for Soil Metals

Plant root exudates are known to enhance mobility of metals and nutrients by (i) acidification due to proton (H+) release or by forming organic/amino acid-metal/mineral complexes; (ii) intracellular binding compounds (e.g., phytochelatins, organic acids, and amino acids); (iii) electron transfer by enzymes in the rhizosphere (e.g., redox reactions); and (iv) indirectly stimulating rhizosphere microbial activity (e.g., survival, growth, propagation, and functioning), therefore enhancing phytoremediation efficiency (Ma et al. 2016). In natural ecosystems, most nutrients, such as N, P, and S, are bound in organic molecules and are therefore minimally bioavailable for plants. To access these nutrients, plants are dependent on the growth of soil microbes, such as bacteria and fungi, which possess the metabolic machinery to depolymerize and mineralize organic forms of N, P, and S. The contents of these microbial cells are subsequently released, either through turnover and cell lysis, or via protozoic predation. This liberates inorganic N, P, and S forms into the soil, including ionic species, such as ammonium, nitrate, phosphate, and sulfate that are the preferred nutrient forms for plants. In natural settings, these microbial nutrient transformations are key drivers of plant growth and can sometimes be the rate-limiting step in ecosystem productivity (Jacoby et al. 2017).

In fact, root exudates provide microbes with an energy source, and in return, microbes stimulate exudation from plant roots. In the coevolutionary process, plants and their associated microbes coexist or compete for survival in the changing environment, and their relationships, either beneficial or detrimental are of significant importance for both partners (Ma et al. 2016).

Warner and Lolkema (2002) showed that plant roots secrete organic acids, including acetic acid, citric acid, and malic acid, that have a high affinity for metals including iron, zinc, copper, and magnesium (Warner and Lolkema 2002). For instance, many microbes (e.g., *Bacillus* spp.) possess high-affinity transporters that enable them to detect and absorb these organic acid–metal complexes (Warner and Lolkema 2002). In fact, microbes benefit nutritionally by absorbing the organic acid–metal complexes, in which they acquire carbon nutrients in the organic acids and mineral nutrients simultaneously. The entry of the microbes into the root cells permits plants to extract the metals from the microbes. Harvesting of metals from the soil microbes via the rhizophagy cycle likely gives plants the critical soil nutrients needed for sustenance and growth (Warner and Lolkema 2002).

4.7.4 Rhizophagy Microbes Take Nutrients from Other Soil Microbes

Rhizophagy microbes, such as *Bacillus* spp., have the capacity to extract nutrients from other soil microbes by causing nutrient leakage from their cells. This enables them to access nutrients contained in the soil microbial community and carry those nutrients back to the plant. Rhizophagy microbes take nutrients from other microbes using "hemolysins" (biosurfactants) that form pores in microbe membranes, causing them to leak nutrients (Braun and Focareta 1991). *Bacillus* spp. frequently possesses hemolysins that are lipopeptides, which act as biosurfactants that increase membrane porosity and induce nutrient leakage from affected cells, typically fungi (White Jr et al. 2014).

Recent research on the effects of rising atmospheric CO_2 levels on the nutrient content of major food crops shows an inverse relationship between CO_2 level and the efficiency of nutrient extraction from soils (Myers et al. 2019). C-3 photosynthesis pathway plants are particularly affected by high CO_2 levels, having reduced content in nitrogen and minerals including magnesium, zinc, and iron (Myers et al. 2019). This effect of CO_2 in reducing nutrient acquisition by plants may be explained by the suppressive effect of CO_2 on NADPH oxidase involved in the rhizophagy cycle (White et al. 2018). Reactive oxygen (primarily superoxide) in the rhizophagy cycle functions to extract nutrients from microbes that enter root cells (White et al. 2018). Carbon dioxide suppresses the formation of superoxide needed to extract nutrients from microbes (Kogan et al. 1997). Increasing the level of CO_2 by 50% in air around seedlings of wheat, tomato, and tall fescue seedlings (with C-3photosynthesis pathway) substantially reduced the amount of reactive oxygen (superoxide) secreted by root cells onto microbes, resulting in fewer nutrients being extracted from intracellular microbes. It is believed that increased CO_2 in greenhouses, especially older greenhouses, as a result of fuel could be one of the reasons for the declining efficiency of endophytes and the reduction of element uptake.

4.7.5 Mechanisms for Endophyte-Mediated Diseases Suppression

One of the different ways in which endophytes improve plant health is suppressing pathogen growth and fitness. This includes several mechanisms, e.g., direct antagonism by competition with pathogens for space and nutrients through the production of antimicrobial metabolites and through induction of systemic resistance or increasing resistance in plants against pathogens via upregulation of host defense genes. There are increasing numbers of studies that suggest that endophytes (fungi and bacteria) provide a defense to host plants against pathogens and other pests from seed germination to the end of the host plant life. Bacterial endophytes of genus *Pseudomonas* including *Pseudomonas aeruginosa* and *Pseudomonas fluorescens*, produce a variety of antifungal compounds including phenazine-1-carboxylic acid, 2, 4-diacetylphloroglucinol, pyrrolnitrin, pyoluteorin, and volatiles, like hydrogen cyanide compounds, that significantly inhibit the growth of fungal pathogens.

Species of the genus *Bacillus* are among the important disease control agents because they synthesize a variety of biologically active molecules that are potential inhibitors of phytopathogens. A variety of lipopeptides that they produce induce leakage in fungal hyphal membranes that greatly reduce their virulence as pathogens of plants. Many of the antifungal compounds produced by endophytes target membranes of fungi, inducing nutrient leakage, resulting in reduced virulence of the fungi. Endophytic symbionts also may improve plant resistance and protect plants against a broad spectrum of pathogens, particularly through induced systemic defense (ISR) by upregulating salicylic acid (SA) and jasmonate (JA) pathways and ethylene or pathogenesis-related (PR) proteins.

Regarding forest tree pathogens, it has been revealed that endophytes may function as competitors or antagonists (Arnold et al. 2003; Mejía et al. 2008) that occupy the same tissues. One of the best-known examples of a commercial biological control of a forest pathogen by an antagonistic fungus is the use of *Phlebiopsis gigantea* (Fr.) Jül in control of butt and stem rot disease of conifer trees, caused by the soil-borne pathogens *Heterobasidion parviporum* Niemelä & Korhonen and *H. annosum* sensu stricto (s.s.) (Fr.) Bref. (Korhonen 1978). Another example of biological control of a tree disease, not related to forestry but with importance to the fruit industry, involves antagonism of the fire blight pathogen *Erwinia amylovora* by a closely related bacterium, *Erwinia herbicola* (Vanneste et al. 1992).

Several isolated bacterial endophytes have shown efficiency in antagonism or inducing resistance against the oomycete pathogen in crops, such as grapevine, lettuce, sunflower, and maize. Endophytic *Bacillus asahii* isolated from cucumber has shown 42.1% control efficacy against the downy mildew disease in cucumber in field trials (Sun et al. 2008). Puopolo et al. (2014) have reported that *Lysobacter capsici* AZ78 produced a 2,5-diketopiperazine, which showed antagonistic activity against *Plasmopara viticola* and *Phytophthora infestans*. Waqas et al. (2014) have reported the increased production of abscisic acid and jasmonic acid on colonization with *Paecilomyces formosus* to combat heat stress and improve plant growth in rice. The endophytes display disease suppression either by competing with the pathogens for colonization or by directly antagonizing the pathogen by producing antimicrobial compounds or by inducing systemic resistance in the host by the production of defense-related enzymes. Moussa et al. (2016) isolated endophytic *Phoma* sp. from finger millet roots, which showed antifungal activity against *Fusarium graminearum*.

4.7.6 Endophytes Alter Oxidative Stress Tolerance in Plants

Environmental stresses trigger plant cells to form reactive oxygen species (ROS; including superoxide, hydroperoxyl radicals, hydrogen peroxide, and hydroxyl radicals). The release of ROS within plant tissues and cells can cause oxidative damage to plant proteins, nucleic acids, and membranes. Some endophytes induce stress tolerance to both biotic and abiotic stresses. At the early stages of endophytic colonization, plant defense responses are activated to produce ROS. A q-PCR analysis showed that bacteria at the early stages of colonization caused upregulated transcript levels of ROS-degrading genes, including superoxide dismutase and glutathione reductase. The upregulation of host ROS-degrading genes may further reduce oxidative damage to plants by pathogens that induce or produce ROS. For example, tall fescue (*Festuca arundinacea*) grass tissues infected by the endophytic fungus *Epichloë coenophiala* have higher concentrations of osmoprotective mannitol and other antioxidant fungal carbohydrates involved in the protection of plants under oxidative stress. Endophytic fungus, *Piriformospora indica*, has been shown to induce abiotic stress tolerance in many plants. *Piriformospora indica* infected Chinese cabbage (*Brassica rapa*) treated with polyethylene glycol to mimic drought stress, exhibited upregulation of antioxidant enzymes peroxidases, catalases, and superoxide dismutases in leaves within 24 h. The expression of drought-protective genes DREB2A, CBL1, RD29A, and ANAC072 was upregulated in leaves of endophyte-containing plants. By meta-genome analysis of rice endophytes, it was found that the presence of numerous genes encoding enzymes involved in protection from excessive ROS including glutathione synthases and also glutathione-S-transferases. The important role of endophytes in reducing the oxidative stress generated in plants in metal contaminated soils has been well known. The infection of soybean by endophytic *Paecilomyces formosus* significantly reduced lipid peroxidation, and increased formation of peroxidase, polyphenol oxidase, catalase, and superoxide dismutase in Ni contaminated substrates.

4.8 Pest Suppression

The control of insect pests and diseases by means of biological processes, such as the use of entomopathogenic microorganisms or those that inhibit/antagonize microorganisms pathogenic to plants, is an alternative that may help to reduce or eliminate the use of chemical products in agricultural systems (Azevedo et al. 2000). The nature of modern agriculture is basically against ecosystem equilibrium by using chemical fertilizers, insecticides, herbicides, and antibiotics in large scales. In other words, although products, such as insecticides and fungicides, can help us to control pests and phytopathogenic microorganisms, they can play a double-edged sword and cause eliminating important species of insects that control other pests and microorganisms that are performing a crucial role in the environment, inhibiting the growth and the multiplication of other microorganisms. One group of microorganisms that is affected by these anthropogenic modifications is the endophytes (Lacava and Azevedo 2014).

Reduced plant damage is achieved by endophytic EPF through many mechanisms including the retardation of the developmental rate of the pest, inhibition of insect food consumption rate, reduction of larval survival, and decreased reproduction rate. For instance, it has been revealed that some endophytes produce and fill plants with compounds that reduce herbivory by insects and other herbivores. Species of fungal endophytes in the genus *Epichloë* (Clavicipitaceae) intercellularly inhabit aerial parts of plants (i.e., leaves, culms, and seeds) and produce a variety of alkaloids that deter feeding by herbivores. These endophytes have found application in increasing pest tolerance in commercial forage and turf grasses. Similarly, in plants commonly referred to as "locoweeds" in the family Fabaceae, endophytic fungi of the genus *Undifilum* (Pleosporaceae) produce the toxic alkaloid swainsonine, a powerful anti-herbivore compound and toxin (White et al. 2019b). The toxin production by endophytes is well correlated with their ability to repel insects. Therefore, several toxins produced by endophytic fungi confer protection to host against different herbivores. The production of toxins leaves the plant unpalatable to various pests, like aphids, beetles, and grasshoppers (Joshi et al. 2018).

There have been several attempts to explain the reduction in the consumption by insects when feeding on inoculated plants. It has been proposed that the production of secondary metabolites, the production of superoxides, changes in the phytosterol profile of plants, or the induction of an indirect systemic response could be responsible for this change of behavior in insects (Lopez and Sword 2015). In this sense, the investigation performed by Shrivastava et al. (2015) demonstrated that plants inoculated with *B. bassiana* showed higher levels of terpenoids, which are considered secondary metabolites with antiherbivore properties (Fig. 4.2).

4.8.1 Mechanisms for Endophyte-Mediated Pest Management

Studying the endophytic relationship of fungi with higher plant indicates the capacity of endophytic fungus to repel insects, induce weight loss, growth and development reduction, and even to increase pest death rate, which is in correlation with the toxin production. In several cases, it was shown that the mode of action of certain fungi was based on the capability to render the plant unpalatable to several types of pests, like aphids,grasshoppers, and beetles. In fact, several toxins are produced by endophytic fungi and these substances confer host plant protection against different herbivores (Azevedo et al. 2000).

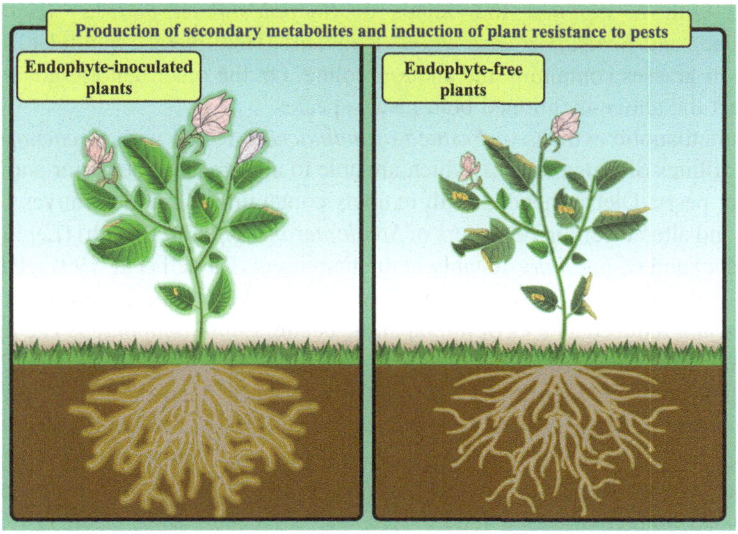

Fig. 4.2 Plants inoculated with endophytes have a higher level of resistance than endophyte-free plants. This issue can be related to the production of secondary metabolites that can prolong the growth period of pests and reduce the reproductive potential of adults

Examples of toxin production by endophytic fungi, notably those colonizing grasses are considerably abundant in the literature. For instance, Miller (1986) showed that the protection of Canadian fir against the spruce budworms resulted from the production of toxic secondary metabolites by endophytic fungi. Prestidge and Gallagher (1988) established a relationship between the presence of the fungus *Acremonium lolii* in *Lolium perenne* and the growth, survival, and feeding behavior of *Listronotus bonariensis* (Kuschel) (Coleoptera: Curculionidae) larvae. In this case, the reduction in insect attacks toward infected plants was due to a strong toxin, lolitrem B, in which it is also toxic to mammals. This toxin once added to insect diets can reduce insect growth and survival. Its assimilation occurs by ingestion but not by absorption through the insect integument. In certain cases, the production of toxin by the endophytic fungi was a plausible explanation for interactions resulting in natural insect control. Therefore, Clark et al. (1989) showed that among 900 samples of fungal isolates derived from *Abies balsamea* L. and red spruce *Picea rubens* Sarg., five of them produce toxic substances and three of them produce powerful toxins once extracted and given to insects, cause death and decrease development rate of *Choristoneura fumiferana* (Clemens) (Lepidoptera: Tortricidae). Siegel et al. (1990) verified the presence of the alkaloids N-formyl, N-acetilloline, peramine, lolitrem B and ergovaline during plant attack by aphids. Several grasses infected with *Acremonium* spp. and *Epichloë typhina* have been analyzed. These fungi generally produce alkaloids, peramine, and ergovaline. Peramine, lolitrem B, and ergovaline have been found in *Lolium* and *Festuca* infected with *Acremonium coenophialum* and *A. lolii* and in *Festuca longifolia* infected with *E. typhina*. Individuals of *Rhopalosiphum padi*(L.) (Hemiptera: Aphididae) and *Schizaphis graminum* (Rondani) (Hemiptera: Aphididae) did not survive in grasses containing the alkaloid loline. On the other hand, ergovaline did not affect the above-mentioned both insect species.

The methanolic extracts of *Festuca arundinacea* infected with *A. coenophialum* contain lolines of fungal origin, which are able to alter feeding behavior and weight of insect pests. Diets amended with extracts containing loline derivatives reduced weight and altered certain behaviors of *Spodoptera frugiperda* (Smith) (Lepidoptera: Noctuidae) and *O. nubilalis*, notably in the first insect (Riedell et al. 1991). Patterson et al. (1991) observed in *Lolium* and *Festuca* that the production of alkaloids by *Acremonium* reduced attacks of the Japanese beetle, *Popillia japonica*. The majority of works related to toxin production were performed in grasses. However, Calhoun et al. (1992), for the first time identified toxic products synthesized by endophytic fungi in woody plants and that were able to modify growth and death rates in larvae of the spruce budworm *C. fumiferana* fed on balsam fir. The endophytes in this case were identified as *Phyllosticta* and *Hormonema dematioides* and the toxic compounds were mainly heptelidic acid and rugulosin. Bills et al. (1992) also detected the existence of tremorgenic toxins in tropical woody plants infected with an endophytic fungus from the genus *Phomopsis*.

4.9 Control of Weeds by Endophytes

Some endophytes stop the growth of nonadapted host plants and eventually cause their death. Therefore, the symbiotic relationships between a host plant and its endophytic microbes are unique, which can become a liability for the plant (White et al. 2019a). Endobiome interference occurs where entry of nonadapted microbial endophytes into plant cells and tissues results in repressed plant growth and disruption of functions of the endophyte-host symbiosis (White et al. 2019a). In a case of endobiome interference, a fungal endophyte (*Aureobasidium pullulans*) was isolated from roots of a weedy yet native species *Froelichia gracilis* (Hook.) Moq. (Amaranthaceae). When *A. pullulans* was introduced by seedling inoculation into the cells and tissues of seedling roots of the exotic plant species, *Amaranthus hypochondriacus*, resulted in growth repression of seedlings (White et al. 2019a). Furthermore, when the bacterial endophyte, *Micrococcus luteus*, originally isolated from tomato seeds and seedlings, was transferred to seedlings, where they entered into seedling root cells of multiple plant species (including *Phragmites australis* s (Cav.) Trin. Ex. Steudel), *Poa annua* L., *Fallopia japonica* (Houtt.) Ronse Decr., *Rumex crispus* L., and *Taraxacum officinale* L.), it reduced the density of native endophytic bacteria and then decreased seedling growth (Kowalski et al. 2015). Endobiome interference could be a common phenomenon in natural plant communities and could be a way that a plant reduces the growth of competitor plants. Similarly, if it is used as a management treatment, endobiome interference may have the potential to reduce the invasive character of invasive and weedy plant species (Kowalski et al. 2015).

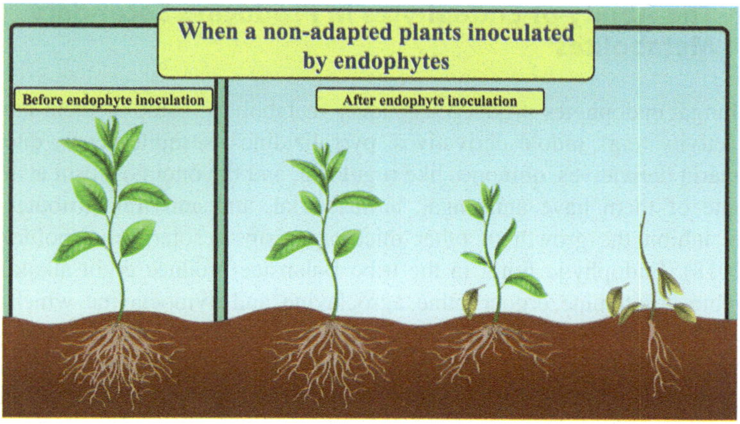

Fig. 4.3 Endophytes, when introduced to nonhost plants, may lead to the weakening and death of the inoculated plant (this approach can be widely used against weeds)

Since some endophytes have shown to produce compounds that are phytotoxic to nonhost species, this phenomenon can be used practically to control undesirable weed species. Endophyte-based weed control may have unique advantages over the application of pathogens (application of propagation materials, such as fungal spores or bacterial suspensions), such as the improved ability of candidate microbes to persist in field conditions through having a more consistent ecological niche within their host plant, or the provision of other benefits to their host, such as nutrient acquisition or disease resistance (Harding and Raizada 2015). This phenomenon is known as endobiome interference. Endobiome interference occurs where entry of nonadapted microbial endophytes into plant cells and tissues results in repressed plant growth and disruption of functions of the endophyte-host symbiosis. In a case of endobiome interference, when a fungal endophyte (*Aureobasidium pullulans*) isolated from roots of a weedy yet native species *Froelichia gracilis* (Amaranthaceae) was introduced by seedling inoculation into the cells and tissues of seedling roots of the exotic plant species *Amaranthus hypochondriacus*, it caused the growth repression of the seedlings. In another example, when the bacterial endophyte, *Micrococcus luteus*, originally isolated from tomato seeds and seedlings, was transferred to seedlings of multiple plant species (including *Phragmites australis*, *Poa annua*, *Fallopia japonica*, *Rumex crispus*, and *Taraxacum officinale*), it reduced native endophytic bacteria and seedling growth. Endobium interference is probably a common phenomenon in natural plant communities where plants reduce the growth of competitor plants. Similarly, if it is used as a management therapy, endobium interference may have the potential to reduce the invasive character of invasive plant and weed species (White et al. 2019a) (Fig. 4.3).

4.10 The Ability of Endophytes in Producing Secondary Metabolites

Many fungal endophytes produce secondary metabolites, some of them have anti-insect activity (e.g., indole derivatives, pyrrolizidines, sesquiterpenes, diterpenes, isocoumarin derivatives, quinones like rugulosin, and flavonoids) (Joshi et al. 2018) and some of them have antifungal, antibacterial, and antiviral attributes which strongly inhibit the growth of other microorganisms (esoteric metabolite) (Joshi et al. 2018). Endophytic fungi in the tribe Balansiae produce ergot alkaloids viz. ergonovine, ergotamine, ergocryptine, agroclavine, and elymoclavine, which cause a reduction in larval weight and leaf area consumption of *S. frugiperda* at concentrations of 77–100 mg liter^{-1} (Kumar et al. 2008). Peramine, a pyrrolopyrazine alkaloid with insecticidal activity against argentine stem weevil, *L. bonariensis*, has been isolated and characterized from several endophytic fungi present in the stem and leaf of tall fescue, ryegrass (*Festuca arundinacea* Schreb.) and other grasses (Kumar et al. 2008). However, different endophyte species, depending on the plant hosting it, produce different types of alkaloids and the endophytes sporulation abilities affect their impact on the levels of insect herbivory and damage caused to plants (Clement et al. 2005; Tintjer and Rudgers 2006). For example, *Epichlöe festucae* produces ergovaline and lolines in *Festuca gigantea* (L.) Vill., while in *Festuca glauca* Vill., it produces ergovaline and peramine (Siegel and Bush 1996). The lolines are mainly active against insects and do not affect other nontarget organisms, e.g., mammals (Dahlman et al. 1991).

Three sesquiterpene lactones, heptelidic acid (HA) and their derivatives, HA chlorohydrins and hydro-HA isolated from *Abies balsamea*, were shown to be toxic to *C. fumiferana* larvae in concentrations ranging from 5 to 15μM. Several toxic isocoumarins and related metabolites were isolated and characterized from *Conoplea elegantula* endophytic on *Picea mariana*. Recently, these compounds were also isolated from *Mycosphaerella* spp. endophytic on *Picea rubens* (Joshi et al. 2018). Senthilkumar et al. (2014) isolated different types of phytochemicals (ethyl ester, phthalic acid, octyl 2-pentyl ester, and dodecanoic acid) from *Phomopsis* sp. isolated from *Tectona grandis* (teak), which show insecticidal activity. Similarly, Bensaci et al. (2015) reported that *Cladosporium oxysporum* also showed insecticidal activity against *A. fabae*. The topical application of the extracts of *Emericella nidulans*, *Aspergillus oryzae*, *Aspergillus tamarii*, and *Aspergillus versicolor* on *Spodoptera litura* F. (Lepidoptera: Noctuidae) larvae showed insecticidal activity (Abraham et al. 2015). Li et al. (2012) reported that *Aspergillus fumigatus* isolated from the bark of *Melia azedarach* produced 39 secondary metabolites. Nine of them steered antifeedant activity against armyworm (*Mythimna separata* (Walker) (Lepidoptera: Noctuidae)) larvae. Among these nine, fumitremorgin B (50.0%) and verruculogen (55.0%) exhibited the best activity. Generally, the secondary metabolites act on activating the glutamate-gated chloride channel of insects especially that control locomotion, feeding, and mediating sensory inputs into behavior. Nodulisporic was the first compound isolated from an endophyte, *Nodulisporium* sp., from the plant *Bontia daphnoides* L. The endophytic fungi *Claviceps purpurea* whose secretions contain ergotoxine and related alkaloids that stimulate smooth muscles also shows significant insecticidal activity against *A. gossypii* (Prakash and Srinivasan 2020).

4.11 How Do Endophytes Help their Host Plants Grow?

Research with controlled experiments has revealed that seedlings of grasses cleaned of most of their endophytic microbes lose the root gravitropic response (i.e., roots do not grow downward), and seedlings are frequently diminished in size with reduced or no root hair formation. Accordingly, the reinoculation of axenic or near axenic seedlings with microbes that internally colonize seedlings, resulted in the reacquisition of the gravitropic response of roots and increased plant stature and root hair development. Several experiments suggested that root hairs elongate until all microbes have been ejected from hairs. Another type of nutritional endophytic symbiosis involves microbes that inhabit both endophytic tissues and extend out into soil. Dark septate endophytes and mycorrhizal fungi establish this kind of symbiosis with many families of plants. Hyphae of these fungi grow endophytically in roots, and the mycelia extending into soil acquire nutrients and mobilize it back to plants (White et al. 2019b).

Iron is a necessary cofactor for many enzymatic reactions and is an essential nutrient for virtually all organisms. In aerobic conditions, iron exists predominantly in its ferric state (Fe^{3+}) and reacts to form highly insoluble hydroxides and oxyhydroxides that are largely unavailable to plants and microorganisms. To acquire sufficient iron, siderophores produced by bacteria can bind Fe^{3+} with a high affinity to solubilize this metal for its efficient uptake. Bacterial siderophores are low-molecular-weight compounds with high Fe^{3+} chelating affinities responsible for the solubilization and transport of this element into bacterial cells. In a state of iron limitation, the siderophore-producing microorganisms are also able to bind and transport the iron siderophore complex by the expression of specific proteins. Siderophores can induce resistance mechanisms in the plant through plant-growth promotion (Lacava and Azevedo 2014). According to Verma et al. (2011), three endophytic actinobacteria strains isolated from the root tissues of *Azadirachta indica* plants were selected through tests for their potential as biocontrol and plant-growth-promoting agents. It was also observed that the seed treated with the spore suspension of three selected endophytic strains of *Streptomyces* significantly promoted plant growth.

4.12 Entomopathogenic Fungi

An increasing number of recent studies show that entomopathogenic fungi, often considered only as insect pathogens, play other roles in nature including endophytism, plant disease antagonism, plant growth promotion, and rhizosphere colonization. Such additional roles, recently discovered to be played by entomopathogenic fungi, provide opportunities for multiple uses of these fungi in integrated pest management (IPM) strategies (Jaber and Ownley 2018). Of particular interest is the ability displayed by various genera of entomopathogenic fungi to colonize a wide variety of plant species in different families, both naturally and artificially following inoculation, and confer protection against not only insect pests but also plant pathogens. Several emerging roles played by fungal entomopathogens provide the promising potential for their indirect, multifaceted, and cost-effective use in sustainable agriculture, e.g., as biofertilizers, vertically transmitted fungal endophytes, and dual microbial control agents in plant diseases and arthropod pests (Jaber and Ownley 2018).

Regarding the potential of entomopathogenic fungi to affect insects as well as their endophytic role, these two potentials can be well used against pests to reduce their damage. For instance, endophytic colonization by *B. bassiana* and *M. brunneum*, following the foliar application of conidia caused additional mortality in the larvae of the beet armyworm, *Spodoptera littoralis* (Boisduval) (Lepidoptera: Noctuidae) (Resquín-Romero et al. 2016) and nymphs of the sweet potato whitefly *Bemisia tabaci* (Gennadius) (Hemiptera: Aleyrodidae) (Garrido-Jurado et al. 2017). Therefore, the combined action of foliar sprays and endophytic colonization could improve the overall efficacy of commercially available mycopesticides based on these fungi. It could also, more importantly, help overcome some of the constraints associated with the topical application of mycopesticides, such as inoculum or dosage selection, adverse effects of abiotic factors, and potential effects on nontarget organisms (Lacey et al. 2015). Mycosis of insect cadavers recovered from plant tissues colonized with fungal entomopathogens has only been reported in a very small number of studies (Jaber and Ownley 2018). In some cases, the negative effects of endophytic fungal entomopathogens on herbivorous insects have been attributed to induce systemic plant resistance (Jaber and Ownley 2018). While systemic plant resistance has not yet been described against herbivorous insects, it has already been described for resistance to fungal and bacterial plant diseases induced by fungal entomopathogens, such as *B. bassiana* and *Lecanicillium* species. It is possible that similar systemic resistance is elicited by these fungi against insect herbivores as well (Jaber and Ownley 2018). Notably, feeding deterrence or antibiosis due to fungal metabolites secreted in plants has been widely suggested as the mode of action in several studies investigating endophytic entomopathogenic fungi–herbivorous insect interactions (Jaber and Ownley 2018). This suggestion has been supported by the absence of fungal sporulation (mycosis) on insects that have died when feeding on endophytically colonized plants by fungal entomopathogens. Despite this, only a few of these studies have identified and quantified the fungal secondary metabolites produced in plant tissues colonized by entomopathogenic fungi. For example, in plant production of destruxins (DTXs) was measured in cowpea plants endophytically colonized by *M. robertsii* ARSEF 2575 12 days after fungal inoculation (Golo et al. 2014). Similarly, destruxin A was quantified in melon (Garrido-Jurado et al. 2017) and potato (Ríos-Moreno et al. 2016) leaves inoculated by several strains of *M. brunneum* 72 and 96 h postinoculation, respectively. The latter study has, however, reported that the amount of destruxin A produced by *M. brunneum* within plant tissues was very small compared to the degree of plant colonization by the fungus, indicating that destruxin A production by the fungus in plant might only be ephemeral (Ríos-Moreno et al. 2016).

4.12.1 Activity as Biocontrol Agents against Plant Diseases

Some endophytic fungal entomopathogens, particularly *B. bassiana* and *Lecanicillium* spp., may have antagonistic activity against plant pathogens in addition to their well-known biocontrol activity against insect pests. This issue shows that these entomopathogens have a promising potential to be developed as biopesticides for multiple purposes in IPM strategies (Jaber and Ownley 2018). Pretreatment of cotton seedlings with *B. bassiana* strain resulted in reduced severity of bacterial blight caused by *Xanthomonas axonopodis* pv. *malvacearum* (Xam) (Griffin et al. 2006; Ownley et al. 2008). More recently, several strains of *B. bassiana* were found to significantly reduce the incidence and severity of the *Zucchini yellow mosaic virus* (ZYMV; genus Potyvirus, family Potyviridae) in squash (Jaber and Salem 2014) and downy mildew caused by *Plasmopara viticola* (Berk. and Curt.) Berl. &de Toni. (Oomycota: Peronosporaceae) in grapevines (Jaber 2015) following foliar inoculation of plants with conidial suspensions of the tested strains. Only a handful of studies have shown the pathogenicity of *Lecanicillium* spp. against plant pathogens including *Pythium ultimum* Trow (Oomycota: Pythiales) (Benhamou and Brodeur, 2001) and powdery mildew *Sphaerotheca fuliginea* (Schlecht.: Fr.) Pollacci (Leotiomycetes: Erysiphaceae) (Hirano et al. 2008) that have been attributed to limited endophytic colonization of cucumber roots by *Lecanicillium* spp. DAOM 198499 and *Lecanicillium muscarium* B-2, respectively. Most recently, an endophytic isolate of *Phialemonium inflatum* (formerly *Paecilomyces inflatus*) has been reported to suppress penetration, galling, and reproduction of the root-knot nematode, *Meloidogyne incognita* (Kofoid & White) Chitwood (Tylenchida: Meloidogynidae) in cotton plants following seed treatment with the conidial suspension of the fungus overnight (Zhou et al. 2018).

4.12.2 *Metarhizium* and *Beauveria*

The earliest studies with entomopathogenic fungi occurred in the early 1800s and concentrated on developing ways for managing diseases threatening the silkworm industry in France. Agostino Bassi (1773–1856) demonstrated that *B. bassiana* was the infectious agent of the muscardine disease of silkworms (Vega et al. 2009). The stimulus for the idea of using fungal insect pathogens to manage pest insects came largely from the ensuing silkworm-disease studies, after finding that the fungus also infected other insects (Audoin 1837). Subsequently, Pasteur (1874) and LeConte (1873) suggested that fungi could be used against insects. In Russia, Elie Metchnikoff (1845–1916) conducted studies on an insect disease of wheat cockchafers that he called green muscardine, and identified the infecting agent as

Entomophthora anisopliae (=*Metarhizium anisopliae*). This fungus was mass-produced by Krassilstschik (1888) and used in the field against the sugar beet weevil. Most reports on the effects of endophytes on insect herbivores have concentrated on turf and agronomic grasses infected with endophytic clavicipitalean fungi (Ascomycota: Hypocreales: Clavicipitaceae), which systemically infect mostly grasses in the Poaceae, Juncaceae, and Cyperaceae. For example, *Neotyphodium*-infected perennial ryegrass (*Lolium perenne* L.) and tall fescue (*Festuca arundinacea* Schreb.) have been shown to have negative effects on over 40 insect species in six orders (Vega et al. 2009).

Entomopathogenic fungi are commonly found in a diverse array of habitats and are known to infect many different taxa of arthropods. They have also been found as rhizosphere colonizers in the surrounding environment of the host plant. In addition, recent evidence suggests that certain EPF species have the potential to engage in fungus–plant interactions, as fungal endophytes or plant disease antagonists, without causing any immediate negative effect or even promoting growth of host plants (Vega et al. 2008). The entomopathogenic fungi are considered as important bio-control agents (BCAs). They are traditionally applied in an inundative approach, but recent studies have shown that EPF play diverse roles in nature including as endophytes (Vega et al. 2009). In another definition, entomopathogenic fungi have been defined as a unique and highly specialized group of microbial agents that possess several desirable traits favoring their development as biopesticides (Lacey et al. 2015). Although there are almost 700 species in about 100 genera of fungal entomopathogens (Humber 2007), the majority of the commercially produced fungi are only based on a few species of *Beauveria*, *Metarhizium*, *Isaria*, and *Lecanicillium*, *Acremonium*, *Cladosporium*, *Clonostachys*, *Cordyceps,* and *Paecilomyces* (Lacey et al. 2015) that are tested as biocontrol agents (Vega et al. 2008). These fungi have been found in *Citrus* spp., *Glycine max*, *Theobroma cacao*, *Saccharum*, *Vitis labrusca*, *Coffea arabica*, and *Z. mays* (Lacava and Azevedo 2014). Although many entomopathogenic fungal endophytes might not be very abundant in most plant species, some taxa like *B. bassiana* (Bals.-Criv.) Vuill. (Ascomycota: Hypocreales) have a wide range of plant hosts and worldwide distribution, which can live as a plant endophyte and usually does not cause visible damage to the host (Allegrucci et al. 2017). It has been naturally isolated from several plant species, and artificially introduced into many others, such as tomato (*Solanum lycopersicum* L.), banana (*Musa paradisiaca* L.) coffee (*Coffea arabic*a L.), sorghum (*Sorghum bicolor* Kuntze), pine (*Pinus radiate* D. Don), tobacco (*Nicotiana tabacum* L.), corn (*Zea mays* L.), wheat (*Triticum aestivum* L.), and soybeans (*Glycine max* L.) by using various techniques (Allegrucci et al. 2017).

Colonization of plant tissues by *B. bassiana* has proved to provide a good protection against insect damage and inhibition of insect establishment and development (Vega et al. 2008). In addition, protection against phytopathogens has been documented by applying *B. bassiana*. In Argentina, *B. bassiana* has been registered as entomopathogenic in Lepidoptera, Hemiptera, Orthoptera, Diptera, and Coleoptera (Allegrucci et al. 2017). Allegrucci et al. (2017) reported for the first time the ability of *B. bassiana* LPSC 1067 to colonize endophytically tomato plants and its potential to infect *Tuta absoluta* (Meyrick) (Lepidoptera: Gelechiidae) by direct contact with conidia. Direct contact by leaf spraying showed a higher mortality rate and a lower MST value than indirect contact. Other isolates of *B. bassiana* have been established previously as endophytes in various plants using different inoculation methods. Some good instances are available regarding potatoes by foliar spraying (Wagner and Lewis 2000), tomatoes by stem injections (Bing and Lewis 1991), or coating seeds with *B. bassiana* conidial suspensions and in opium poppies after spraying leaves or coating seeds with *B. bassiana* conidial suspensions (Quesada-Moraga et al. 2006). However, some studies indicated that leaves are poor routes of entry for this fungus in some plants, such as coffee (Posada et al. 2007).

Allegrucci et al. (2017) indicated that *B. bassiana* was effectively established as an endophyte in tomato plants when inoculated either by leaf spraying, root dipping, or seed immersion, and was reisolated from leaves 7, 14, and 28 days after its inoculation. They found that leaf spraying was the most effective inoculation technique and the highest percentage of colonization was recorded 7 days after inoculation. They demonstrated that the selected isolate of *B. bassiana* was able to colonize tomato plant tissues and was reisolated from new leaves, which confirmed the establishment of the fungus in the plant tissues and its potential to move throughout them. However, the endophytic colonization, estimated by the percentage of recovery of *B. bassiana* after inoculation, decreased over time. Similarly, inoculation of bean seeds, *P. vulgaris*, by *B. bassiana* significantly reduced the growth and reproduction of the spider mite *Tetranychus urticae* Koch (Acari: Tetranychidae) (Dash et al. 2018). They reported the negative impacts on population growth and reproduction of *T. urticae* when they were kept on bean plants (*P. vulgaris*) grown from seeds inoculated by three isolates of *B. bassiana* (B12, B13, B16), and isolates of *I. fumosorosea* (isolate17) and *L. lecanii* (isolate L1), compared to noninoculated control plants. A significant reduction in larval development, adult longevity, and female fecundity of spider mites when reared on *B. bassiana* treated plants; in addition, increased bean plant heights and biomass were reported (Dash et al. 2018). Similarly, reduced insect herbivore population growth on fungal inoculated plants compared to control plants has also been reported by Gathage et al. (2016) who found lower infestation levels of *Liriomyza* leaf miners [(*L. huidobrensis*, *Liriomyza sativae* Blanchard and *Liriomyza trifolii* (Burgess) Diptera: Agromyzidae)] (in *P. vulgaris* plants endophytically colonized with *B. bassiana* isolate G1LU3 compared to control; besides lower numbers of pupae were also observed. Qayyum et al. (2015) reported a high mortality of *H. armigera* when fed tomato plants colonized by *B. bassiana* isolate WG-40. Similarly, *B. bassiana* isolates ITCC 5408 and ITCC 6063 as endophytes reduced the stem weevil, *Apion corchori* Marshall (Coleoptera: Curculionidae) in white jute, *Corchorus capsularis* L. (Biswas et al. 2013). A reduction of the population growth rate of *Chortoicetes terminifera* (Walker) (Orthoptera: Acrididae) nymphs was reported by Gurulingappa et al. (2010) when they fed wheat leaves colonized by a *B. bassiana* strain. Furthermore, *B. bassiana* isolate G41 reduced larval survivorship of banana weevil, *Cosmopolites sordidus* Chevrolat (Coleoptera: Curculionidae) in banana (Akello et al. 2008). Endophytic colonization by *B. bassiana* isolate 0007 significantly reduced damage caused by *S. calamistis* (Cherry et al. 2004); and *B. bassiana* isolate ARSEF3113 by *O. nubilalis* (Bing and Lewis 1991), both in maize.

Bamisile et al. (2019) tested the pathogenicity of two fungal strains of *B. bassiana* and one strain of *I. fumosorosea* against adults of Asian citrus psyllid (*Diaphorina citri* Kuwayama (Hemiptera: Psyllidae)) and found to induce 50% reduction in the survival rate of *D. citri* adults within 5 days of exposure. They found that one of the *B. bassiana* strains (BB Fafu-13) was sustained up to 12 weeks in the colonized seedlings, whereas the other *B. bassiana* strain (BB Fafu-16) was only recovered up to 8 weeks postinoculation. However, *I. fumosorosea* (IF Fafu-1) failed to colonize the plant. Both strains of *B. bassiana* induced significant improvement in plant height and flush production in endophytically colonized seedlings and caused 10–15% *D. citri* adult mortality within 7 days of exposure (No mycosis was detected on any of the dead psyllids). Female *D. citri* fed on *B. bassiana* inoculated plants laid fewer eggs compared with those fed on endophyte-free seedlings. In addition, a reduction in adult emergence was recorded on *B. bassiana* treated plants (Bamisile et al. 2019). They also found the systemic colonization of the various citrus plant parts including leaves, stems, and roots by *B. bassiana* BB Fafu-13 strain through its reisolation from plant parts other than the treated ones. It was observed that all of the nymphs that emerged from the eggs laid by female *D. citri* fed on BB Fafu-13 challenged seedlings failed to emerge as adults after 17 days, as against the control seedlings, where a significantly higher mean number of adult emergences was recorded. Over time, they recorded an increase in mean number of nymphs and adults at the third generation across all treatments as a result of a decline in the titer of *B. bassiana*.

Sánchez-Rodríguez et al. (2018) showed that the endophytic *B. bassiana* caused the mortality rate of cotton leafworm (*S. littoralis*) larvae to increase up to 57% in treated cotton leaves. Similar outcome was found in the study of Resquín-Romero et al. (2016), where 25% to 46.7% mortality of *S. littoralis* larvae was reported following treatment of alfalfa, melon, and tomato plants with *B. bassiana*. In addition, Rondot and Reineke (2018) found a significant reduction in infestation rate and growth of vine mealybug (*Planococcus ficus* Signoret (Hemiptera: Pseudococcidae)) following treatment of grapevine (*Vitis vinifera* L.) with endophytic *B. bassiana*.

Russo et al. (2019) showed that *Rachiplusia nu* (lepidoptera: Noctuidae) larvae consumed less of corn leaves endophytically colonized by *B. bassiana*. They declared that *B. bassiana* can provide multiple benefits to *Z. mays* L. and can play an important role in future integrated pest management programs. Likewise *Beauveria*, the genus *Metarhizium* is considered as an important entomopathogen and endophyte, which is able to colonize a wide variety of plants and can cause increased plant growth and protect plants against pests. In other words, the genus *Metarhizium* (Hypocreales: Clavicipitaceae) is a main entomopathogen and endophytic symbiont of plants; i.e., besides causing mortality of economically important arthropod pests, these fungus species are also able to colonize a wide variety of plant species, causing increased plant growth (Canassa et al. (2019) and protection of plants against pests and phytopathogens (Jaber and Ownley 2018). There are good reports regarding the successful inoculation of *Metarhizium anisopliae* (Metchnikoff) Sorokin and *Metarhizium robertsii* J.F. Bisch., Rehner & Humber with fungal establishment in different plant species (Sasan and Bidochka 2012; Batta 2013; Bamisile et al. 2018). In a study on the effects of aqueous and granular formulations of *B. bassiana* in controlling the damage of *O. nubilalis* feeding on *Z. mays*, it was found that a small percentage of insects had mycosis (Bing and Lewis 1991), and it was proposed, as no conidia were found inside the host plant, that the mode of action involves fungal metabolites, which cause insect feeding deterrence or antibiosis (Cherry et al. 2004).

Besides causing negative effects on arthropod pests, both *B. bassiana* and *Metarhizium* spp. have shown to improve plant growth (Canassa et al. (2019), leading to higher yields (Jaber and Araj 2018). In this context, it has been revealed that *Metarhizium* spp. are able to transfer nitrogen from infected insects in the soil to plants via mycelium-root connections in a tritrophic association between host insect, fungus, and plant in the rhizosphere (Behie and Bidochka 2014), resulting in an increase in the overall plant productivity. The *M. robertsii* established as an endophyte in stems and leaves of sorghum, *Sorghum bicolor* L. (Moench) (Poaceae), could reduce infestation levels by the larvae of *Sesamia nonagrioides* (Lefebre) (Lepidoptera: Noctuidae) compared to the control and suppressed tunneling by 87% (Mantzoukas et al. 2015). Furthermore, *Beauveria* and *Metarhizium* frequently occupy different niches in plants in such a way that *Metarhizium* spp. being restricted almost exclusively to the root system while *B. bassiana* establishes as an endophyte within all plant tissues (Behie et al. 2015), indicating a potential for complementary localization in crops and effects against pests.

In comparison with *B. bassiana*, there are fewer reports on the plant inoculations with *Metarhizium* spp. to cause negative effects against the herbivorous insect. For instance, Jaber and Araj (2018) reported that the inoculation of *M. brunneum* strain BIPESCO5 in sweet pepper (*Capsicum annuum* L.) by plant root drench resulted in fewer aphids, *M. persicae*, including prolonged development time and reduced reproduction compared to aphid feeding on control plants. The inoculations of *M. anisopliae* isolate ICIPE 20 in bean (*P. vulgaris*) by seed soaking reduced the damage of bean stem maggot, *Ophiomyia phaseoli* Tryon (Diptera: Agromyzidae) (Mutune et al. 2016). Similarly, Leckie et al. (2008) reported that larvae of *H. zea* had delayed development, lower weight, and higher mortality when fed on diets containing mycelia of a *B. bassiana* isolate compared to control larvae, and beauvericin was detected in the broth cultures added into the diet. Likewise, *B. bassiana, Metarhizium* spp. can produce secondary metabolites, particularly destruxins (Roberts 1981).

Golo et al. (2014) detected destruxins in roots, stems, and leaves of cowpea plants (*Vigna unguiculate*(L.) Walp.) inoculated with *M. robertsii* ARSEF 2575 at 12 days after seed inoculation. Ríos-Moreno et al. (2016) and Resquín-Romero et al. (2016) detected destruxin A in potato and tomato leaves, when endophytically colonized by a *M. brunneum* isolate. Similarly, Garrido-Jurado et al. (2017) detected destruxin A in melon leaves endophytically colonized by a *M. brunneum* isolate, and also in

B. tabaci nymphs fed on the melon leaves. Nonentomopathogenic fungi are also reported to have negative impacts on *T. urticae* based on defensive inductions in the plant. Furthermore, Jaber and Araj (2018) also confirmed growth promotion by *M. brunneum* (commercial strain BIPESCO5) in sweet pepper plants besides the negative effects on the development and fecundity of the aphid *M. persicae*. Consistent increase in plant growth during infestation with two successive *M. persicae* generations reported by Jaber and Araj (2018), indicating the ability of these fungi to promote growth under experimentally imposed biotic stress. The positive effects of *M. robertsii* and *B. bassiana* on bean growth have also been reported by Canassa et al. (2019), when the seeds were treated by these two fungi.

4.12.3 Coapplication of Entomopathogens or Individual Using (with Emphasis on *M. Robertsii* and *B. Bassiana*)

Coapplication of multiple (two or more) endophytes may not lead to further control of a pest compared to when the plant is inoculated with an individual endophyte. For example, Canassa et al. (2019) found no significant differences in reducing the population of *T. urticae* and improving plant growth when bean seeds were inoculated with one endophyte (*M. robertsii* or *B. bassiana*) or both. Gurulingappa et al. (2010) reported that feeding by *A. gossypii* on cotton leaves colonized by either *B. bassiana* or *L. lecanii* Zimmermann slowed aphid reproduction and consumption of wheat leaves.

Regarding other endophytes, Martinuz et al. (2012) showed that the endophytes *Rhizobiumetli* and *F. oxysporum* individually induced systemic resistance against *A. gossypii*, but inoculation by both microbes did not have a significant additive biocontrol effect compared with the individual treatments. Similarly, colonization of strawberries by two individual mycorrhizal species of *Glomus* spp. reduced the growth and survival of larvae of *Otiorhynchus sulcatus* F. (Coleoptera: Curculionidae); however, the combination of the two species did not lead to an additional reduction (Gange 2001). However, apart from the antiherbivory effects of endophytes, there are many instances in which the combined use of endophytes has led to better plant growth and enhanced nutrient uptakes (Sadeghi et al. 2020; Aghaei Dargiri et al. 2021a, 2021b).

4.12.4 The Interactions between Endophytic Fungal Entomopathogens, Insect Pests, and Natural Enemies

Pathogens and arthropod natural enemies may contribute to the suppression of insect pest populations either as individual species or as species complexes. However, because natural enemies of insects have evolved and function in a multitrophic context, it is important to assess interactions within complexes of natural enemies if they are to be exploited effectively in pest management. Natural enemies can interact either synergistically/additively or antagonistically (Roy and Pell 2000). The interactions among endophytic fungal entomopathogens, arthropod pests, and their natural enemies have been explored for different natural enemies. For instance, Bixby-Brosi and Potter (2012) showed that plant secondary chemicals can alter herbivore suitability for parasitoids by weakening or stunting the host, delaying its development, or when larval parasitoids encounter ingested phytotoxins in the body of their host. They tested different parasitoids that exploited the same host species feeding on the same plant, including the encyrtid wasp, *Copidosoma bakeri* (Howard) (Hymenoptera: Encyrtidae), and the tachinid fly *Linnaemya comta* (Diptera: Tachinidae), known as a slow-developing polyembryonic egg-larval parasitoid and a fast-developing solitary species, respectively. These parasitoids both parasitize *Agrotis ipsilon* (Hufnagel) (Lepidoptera: Noctuidae), a generalist noctuid feeding on perennial ryegrass containing an alkaloid-producing fungal endophyte. They found that the outcome of endophyte-mediated tritrophic interactions was different for different parasitoid species. Indeed, fewer parasitized cutworms yielded *C. bakeri* broods, and those host mummies were smaller, formed more slowly, and contained fewer adults when the hosts fed on endophytic as opposed to endophyte-free grass. In contrast, *L. comta* fitness parameters were similar regardless of the type of grass upon which their host fed, confirming the issue that *C. bakeri*, because of its more prolonged developmental association with the host, would suffer greater fitness costs when *A. ipsilon* feeds on perennial ryegrass containing an alkaloid-producing fungal endophyte.

In another study, Akutse et al. (2014) investigated the effects of the fungal endophytes *B. bassiana* (isolates ICIPE 279, G1LU3, S4SU1) and *Hypocrea lixii* (isolate F3ST1) on the life-history of *Phaedrotoma scabriventris* Nixon (Hymenoptera: Braconidae) and *Diglyphus isaea* (Walker) (Hymenoptera: Eulophidae), parasitoids of the pea leaf miner *L. huidobrensis* on endophytically inoculated faba bean, *V. faba* and found no significant difference between the control and the endophyte-inoculated plants in terms of parasitism rates of *P. scabriventris* and *D. isaea* and adult' survival times. Jaber and Araj (2018) studied the effects of the fungal entomopathogens, *B. bassiana* and *M. brunneum,* on parasitism of the aphid endoparasitoid *Aphidius colemani* Vier. (Hymenoptera, Braconidae, Aphidiinae) parasitizing the green peach aphid *M. persicae* on sweet pepper *Capsicum annum* L. Their findings indicated that the percentage mummification and adult emergence of *A. colemani* progenies parasitizing second-generation aphid reared on inoculated or control plants were not affected by plant colonization with *B. bassiana* and *M. brunneum.* In addition, no differences were observed in development time, percentage female, and adult longevity of *A. colemani* progeny among inoculated and control plants. Furthermore, it was exhibited by Schausberger et al. (2012) that mycorrhizal inoculated plants infested with *T. urticae* were more attractive than nonmycorrhizal plants to the spider mite predator, *Phytoseiulus persimilis*. It was suggested that this effect was mediated by the increased production of ß-ocimene and ß-caryophyllene, indicating that the predatory mites learned to recognize the plant response (Patiño-Ruiz and Schausberger 2014) and show greater oviposition rates on these plants resulting in enhanced *T. urticae* suppression (Hoffmann et al. 2011). Canassa et al. (2019) showed no significant differences in the predation rate of the predatory mite, *Phytoseiulus persimilis* Athias-Henriot (Acari: Phytoseiidae) on *T. urticae* fed on endophyte-treated or untreated plants even though the predators were most likely to feed on spider mites from fungal treated plants during the first half of the trial, and on spider mites from control plants during the remainder of the trial.

Scorsetti et al. (2017) evaluated the compatibility and interaction between *Eriopis connexa* (Germar) (Coleoptera: Coccinellidae) and the entomopathogenic fungus *B. bassiana*. When *B. bassiana* was directly sprayed on *E. connexa*, significant differences were found between different life stages in terms of mortality (the highest and the lowest mortality was observed in the first instar and pupal stage, respectively). In addition, significant differences in the development time and fecundity were found between insects fed on *B. bassiana*-infected prey and those were fed with uninfected prey. Accordingly, *Eriopis connexa* had longer development time and lower fecundity on prey treated with *B. bassiana*.

When anthropogenic fungi are used directly against pests, the presence of insect natural enemies may have an impact on local transmission of a fungal pathogen. The presence of a foraging adult coccinellid, for example, resulted in a substantial increase in the local transmission of the aphid pathogen, *Erynia neoaphidis,* within a population of pea aphids, *A. pisum* on individual bean plants in the laboratory (Roy and Pell 2000). When anthropopathogens are used as endophytes, it is expected to induce a wide range of changes in the composition of plant nutrients and/or defensive compounds. These changes could influence interactions between the plant and higher trophic levels. González-Mas et al. (2019a, 2019b) evaluated the predation/parasitism efficacy of larvae of the lacewing, *Chrysoperla carnea* (Stephens) (Neuroptera: Chrysopidae), and the braconid parasitoid, *Aphidius colemani* Viereck (Hymenoptera: Braconidae: Aphidiinae), when offered aphids that had been challenged by the entomopathogenic fungus *B. bassiana*. Aphids were either inoculated directly with a fungal suspension (lacewing bioassay only) or had been feeding on melon plants endophytically colonized by *B. bassiana*. They indicated that *B. bassiana* application did not significantly influence the number of aphid prey consumed by lacewings, or the time took them to consume each aphid. In a choice bioassay, *C. carnea* larvae preferred to feed on aphids reared on *B. bassiana*-colonized plants compared with control plants. In another choice assay, the number of aphids parasitized by *A. colemani* and their sex ratio was not influenced by whether the aphids had been feeding on *B. bassiana*-colonized plants or not. Their findings supported the hypothesis that endophytic entomopathogenic fungi can be used in combination with natural enemies, such as predators and parasitoids, in integrated pest management programs (Fig. 4.4).

4.12.5 Niche Partitioning and Long Lasting Protection by Endophyte Application Type

Several factors can cause niche differentiation between endophytes, which may attenuate competition and thus allow for a high fungal diversity on the same host

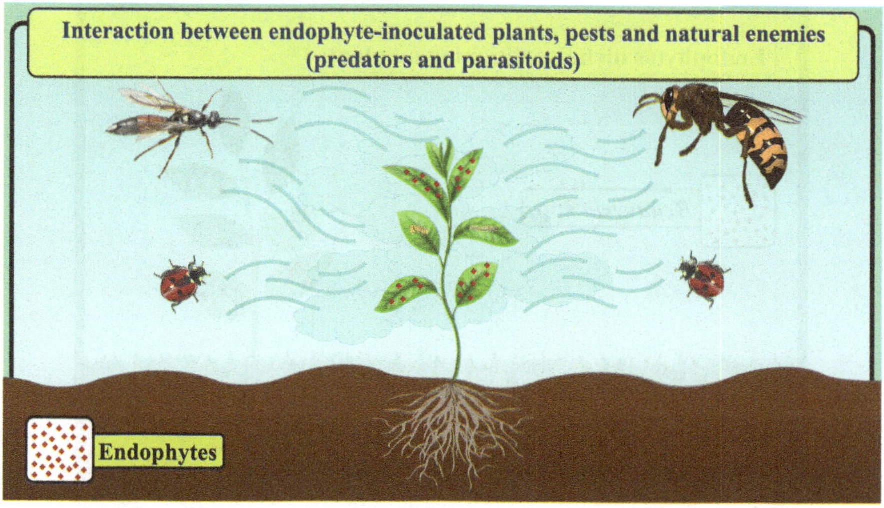

Fig. 4.4 Inoculated plants with endophytes may have a positive or negative interaction with the third trophic members. Therefore in an integrated pest management program, the type of relationship between endophyte-inoculated plants and natural enemies should be determined before any decisions

species. One factor is space, which is with respect to endophytes hierarchically structured from continent to region, to habitat, to host individual, to host organ, and further down to the level of host cells. Two of these levels, i.e., the habitat type and the host organ, were studied by Ernst et al. (2011), regarding *Microdochium bolleyi* and *Microdochium phragmitis*, preferentially colonize the same organ, i.e., roots. They showed while *M. bolleyi* occurs more frequently on roots at dry sites, *M. phragmitis* occurs more frequently on roots at flooded sites.

Numerous studies show that there is a specialization in niche partitioning for two genera *Metarhizium* and *Beauveria*. For instance, Canassa et al. (2019) indicated although *M. robertsii* ESALQ 1622 and *B. bassiana* ESALQ 3375, both were able to colonize the bean plants, *M. robertsii* only being recovered in the roots and from soil, whereas *B. bassiana* recovered from soil and from the three different parts of *P. vulgaris* (in individual or in combined inoculation). Similar spatial segregation patterns of the fungal genera were reported by Behie et al. (2015) under laboratory and field conditions, where *M. robertsii* was restricted to the roots of haricot bean plants (*P. vulgaris*) whereas *B. bassiana* was found throughout the plant, indicating specific variation in the endophytes ability to colonize different plant tissues. Similarly, Akello and Sikora (2012) showed, unlike an isolate of *M. anisopliae* that colonized only the roots, a *B. bassiana* isolate colonized different parts of the plant *V. faba*. Other studies have also shown that *B. bassiana* can establish an endophyte throughout the entire plant (Fig. 4.5). However, although most of the reports confirm the above-mentioned instances, there are some cases in which this

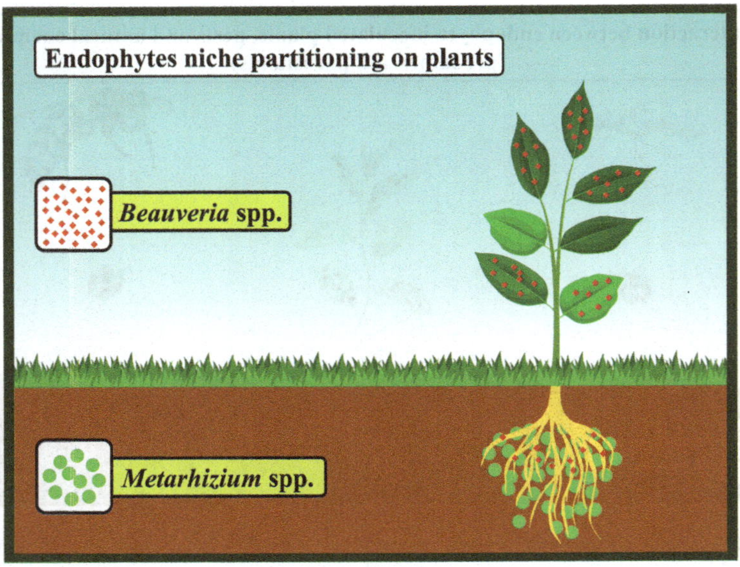

Fig. 4.5 Parts of plants are colonized by *B. bassiana* and *M. anisopliae*. While *Beauveria bassiana* is present in all parts of the plant, *Metarhizium anisopliae* only colonizes the root. This partitioning may be different for other species of the two genera

issue has not been addressed. For example, Greenfield et al. (2016) found *M. anisopliae* and *B. bassiana* both colonize the roots of cassava plants.

Our attitude toward entomopathogenic fungi (in the form of endophytes or fungi that kills pests in a contact manner) as well as our attitude toward their application are of immense importance. For instance, the application of entomopathogens as endophytes rather than their conventional foliar application enables them to colonize plants systemically, thereby offering continuous protection and enhanced persistence (Bing and Lewis 1991; Akello et al. 2008) and considerably low inoculum is required (Athman 2006). Backman and Sikora (2008) outlined that integrated pest management on seeds reduces costs and environmental impact, while allowing the biological agent to build up momentum for biological control. Posada et al. (2007) found that direct injection of *B. bassiana* conidial suspensions had the highest postinoculation recovery in coffee seedlings than foliar sprays and soil drenches. Muvea (2015) showed that there were differences in the level of colonization of different plant parts by fungal isolates. For instance, root sections had higher colonization compared with stems and leaves. Guo et al. (2008) expressed that the difference in the colonization could be due to tissue specificity exhibited by endophytic fungi and their adaptation to particular physiological conditions of the plants.

4.13 Endophyte Can Reduce the Transmission of Viral Diseases

Endophytes not only slow down the rate of disease transmission by decreasing the vector population but also can play an important role in reducing the transmission and replication of viral diseases. For instance, inoculation of meadow ryegrass (*Lolium pretense* (Huds.) = *Festuca pratensis*) with *Neotyphodium* sp. endophytes reduced the population of aphids and protect the plant from barley yellow dwarf virus (BYDV) infections (Lehtonen et al. 2006). Muvea (2015) conducted a series of experiments in the greenhouse where a colony of viruliferous *T. tabaci* was studied in terms of feeding and transmission of *iris yellow spot virus* (IYSV) on E- and E+ onion plants. Transmission of the disease was studied both through infection of the whole plant (thrips fed on leaves) and through leaf discs. In the first case, the disease transmission was evaluated after 2 weeks post thrips exposure and in the latter case they tested the transmission of IYSV using leaf disc assays and individual thrips. Control healthy plants (without endophyte and virus) were tested simultaneously for baseline titers. They indicated that endophytes could colonize onion plants successfully and reduce the rate of virus infection. Similarly, a reduction in aphid population and barley yellow dwarf virus (BYDV) transmission rate on *Neotyphodium uncinatum* inoculated Meadow ryegrass compared to control plants was reported by Lehtonen et al. (2006). Jaber and Salem (2014) evaluated mechanical transmission of *Zucchini Yellow Mosaic Virus* on cucurbit plants inoculated with *B. bassiana* isolates and indicated a reduced disease incidence and severity on endophyte inoculated plants. Rúa et al. (2013) showed that the reduction in the negative effects of viral diseases on plants is a function of the relationship between the virus, endophytes, and the genetic pattern of the host plant. Endophyte infection in addition to potentially providing protection against virus infection by decreasing vector abundance may also mitigate viral effects on host below-ground allocation and thereby enhance host tolerance to viral infection.

Maize lethal necrosis (MLN), a disease caused by the coinfection of maize plants with Maize Chlorotic Mottle Virus (MCMV) (Tombusviridae: Machlomovirus) and Sugarcane Mosaic Virus (SCMV) (Potyviridae: Potyvirus) was successfully managed by endophyte inoculation. Kiarie et al. (2020) examined the potential of 10 fungal isolates to colonize maize plants and induce resistance against MCMV and SCMV. They found that isolates of *Trichoderma harzianum*, *Trichoderma atroviride*, and *Hypocrea lixii* could colonize well different plant sections. Although all plants singly or dually inoculated with SCMV and MCMV tested positive for the viruses by reverse transcription-polymerase chain reaction (RT-PCR), maize plants inoculated by *T. harzianum* and *Metarhizium anisopliae* resulted in up to 1.4 and 2.7-fold reduced SCMV severity and titer levels, respectively, over the controls. However, the inoculated plants were not able to mitigate the severity of MCMV.

4.14 Seeds and Endophytes

Plant seeds carry embryonic plants and nutrients for the early stages of seedling growth; in some plants seeds also carry small communities of symbiotic microbes (primarily bacteria and fungi) that are needed for defense from pathogens, modulation of plant development, and nutrient acquisition in seedlings. Seedlings usually become more susceptible to biotic (diseases and pests) and abiotic stresses (oxidative stresses, drought, and heavy metals) in the absence of their microbes (White et al. 2019b). Seed-vectored microbes can suppress diseases (Verma et al. 2018) in different manners: (1) by direct colonization of potentially pathogenic soilborne fungi and suppression of their growth and virulence, (2) colonization of seedlings resulting in upregulation of defense-related genes that makes plants more resistant to disease, and (3) excluding pathogenic microbes by monopolizing space and/or production of antibiotics or toxins.

Bacteria and fungi associated with seed tissue also influence the development of seedlings and increase the resistance of plants to pests and diseases. Although this process is not well understood, microbes colonize the seedlings and increase gravitropic response, root elongation rate, root branching, and root hair elongation. On the other hand, the rate of colonization of different plant tissues in seed inoculation method is higher than other methods. Muvea (2015) showed that seed inoculation resulted in 1.47 times higher mean percentage postinoculation recovery of all the endophytes tested as compared to seedling inoculation. In addition, biotic and abiotic stresses affect plants negatively by increasing internally generated reactive oxygen species (ROS), which leads to increased internal oxidative damage in plants to membranes, proteins, and nucleic acids and eventually to cell death (Hamilton et al. 2012). Seed-vectored microbes colonize seedlings and elicit a reactive oxygen defense response in plants that causes seedlings to upregulate stress resistance and antioxidant genes, resulting in seedlings that are more tolerant to oxidative stresses than seedlings without the microbes (Irizarry and White 2018).

Multiple operations on seeds cause seeds to lose a significant portion of their endophytes. For instance, acid treatment of cotton seeds to remove fibers removes natural seed-vectored microbes and makes cotton seedlings more vulnerable to stress and disease. Acquisition of microbes from seeds of uncultivated plants in the cotton family greatly decreased stress intensity and improved disease resistance in cotton seedlings. The high levels of diseases and pests that plague cotton could be the result of the loss of symbiotic microbes from cotton seeds. The negative effects of removing endophytes in corn seeds on the viability of seedlings have also been well studied. Maize has been intensively cultivated and modified to the extent that external seed structures that vector microbes like hulls once present in ancestral teosinte have been lost. Modern hybrid maize varieties require higher inputs of nitrogen and pesticides to produce crops than older flint-type Indian maize or tropical maize and this may be the result of loss of symbiotic endophytes from hybrid maize varieties. Some grass seeds, such as Bermuda grass (*Cynodon dactylon* (L.)) are routinely cleaned of the microbial-rich seed husk and covered with fungicides, leaving grass seedlings without their natural endophytes. The effects of long-term use of inorganic fertilizers, fungicides, or other agrochemicals on endophytic microbes in crop plants have not been well studied and it is possible that long-term agrochemical use has caused a loss of symbiotic endophytic microbes from many crop species (White et al. 2019b).

Inoculation of seeds by appropriate endophytes is one of the promising methods to produce morphologically and physiologically high-quality plants. By studying tomato seeds (cultivar 8320) inoculated by *Penicillium chrysogenum* and *Aurantiacum exiguobacterium*, it was revealed that these endophytes significantly increased fresh and dry weight, stem height and diameter, leaf number, chlorophyll content (SPAD), and chlorophyll fluorescence of the tomato plants. Also it was found that these two endophytes had a synergistic interaction (Aghaei Dargiri et al. 2021a, 2021b). In another study, endophyte-inoculated tomato seeds sown under salinity condition had higher levels of key osmolytes, total soluble carbohydrates, and free proline compared to endophyte-free plants under salinity stress (Aghaei Dargiri et al. 2021c). In overall, some plants lack certain species of endophytes or the endophytes are so low that they may not be able to establish a symbiotic relationship with their host. To remedy losses of essential endophytic microbes and reduce reliance on agrochemicals in crop cultivation, it could be necessary to obtain endophytic microbes from wild relatives of crops and reintroduce them into crops perhaps as seed treatments (White et al. 2019b).

4.15 How Do Endophytes Confer Resistance to their Host Plants?

Past efforts to improve plant tolerance to abiotic stress through breeding and genetic engineering have had limited success owing to the genetic complexity of stress responses. Progress is now anticipated through plant endophyte interaction by which endophytes impart tolerance to plants in a habitat-specific manner. This feature, referred to as "habitat-adapted symbiosis," has now been widely recognized in the laboratory. The endophytes from these habitats confer habitat-specific stress tolerance to plants. This habitat-specific phenomenon provides an intergenomic epigenetic mechanism for plant adaptation and survival in high-stress habitats. Oxidative stress protection by increased production of antioxidants produced either by the microbes or by hosts in response to microbes, ethylene reduction by the production of ACC deaminase, ammonia or ammonium detoxification and consequent oxidative stress avoidance. Stress tolerance through endophyte-mediated osmotic adjustment and microbe-enhanced abiotic stress tolerance are some mechanisms underlying by which endophytes enable host plants to mitigate the adverse effects of stresses (Dey et al. 2019).

Regarding disease resistance, endophytes confer the disease resistance to the hosts by different manners including direct antagonism by competition with pathogens for space and nutrients, through production of antimicrobial metabolites and through induction of systemic resistance or increasing resistance in plants against pathogens via upregulation of host defense genes (Irizarry and White

2017; Hardoim et al. 2015). Valuable information is available regarding the endophytes (fungi and bacteria), providing defense to host plants against pathogens and other pests beginning at seed germination and lasting the life of the plant (White et al. 2019b). Bacterial endophytes of genus *Pseudomonas*, including *P. aeruginosa* and *P. fluorescens*, produce a variety of antifungal compounds, including phenazine-1-carboxylic acid, 2,4-diacetylphloroglucinol, pyrrolnitrin, pyoluteorin, and volatiles, like hydrogen cyanide compounds that significantly inhibit the growth of fungal pathogens (White et al. 2019b). Species of the genus *Bacillus* are important disease control agents because they synthesize a variety of biologically active molecules that are potential inhibitors of phytopathogens (Ongena and Jacques 2008). A variety of lipopeptides that are produced by *Bacillus* spp. induce leakage in fungal hyphal membranes that greatly reduce the virulence of pathogens (Ongena and Jacques 2008). This may result in a "quorum-quenching" effect where pathogenic fungi remain avirulent rather than causing disease. Many of the antifungal compounds produced by endophytes target membranes of fungi, inducing nutrient leakage, resulting in reduced virulence of the fungi (White et al. 2019b). Endophytic symbionts also may improve plant resistance and protect plants against a broad spectrum of pathogens, particularly through induced systemic defense (ISR) by upregulating salicylic acid (SA) and jasmonate (JA) pathways and ethylene or pathogenesis-related protein (Bastias et al. 2017).

4.16 Endophytic Bacteria

Although fungi are the most frequently isolated endophytes, endophytic bacteria can colonize an ecological niche similar to what are colonized by plant pathogens but do not cause damage to their hosts. Endophytic bacteria have the ability to promote growth and inhibit plant disease, and as they are in intimate contact with the plant they are an attractive choice as biological control agents (Haggag 2010). For example, Sturz et al. (1999) found that 61 of 192 endophytic bacterial isolates from potato stem tissues were effective biocontrol agents against *Clavibacter michiganensis* subsp. *sepedonicus*.

A number of endophytic actinobacteria were previously isolated by culture dependent methods, with the major genera being *Streptomyces*, *Microbispora*, *Micromonospora*, and Nocardioides (Coombs and Franco 2003). A number of these isolates were capable of suppressing fungal pathogens of wheat in vitro and in planta, including *Rhizoctonia solani*, *Pythium* spp., and *Gaeumannomyces graminis* var. *tritici*, indicating their potential use as biocontrol agents (Coombs et al. 2004). Yang et al. (2013) isolated endophytic *Paenibacillus xylanilyticus*, *Paenibacillus polymyxa,* and *Bacillus subtilis* from seedling, squaring, and boll-setting stages of cotton. The combined application of three endophytic bacteria was found to control the effects of *Verticillium dahlia,* which causes *Verticillium* wilt of cotton. *Bacillus subtilis* strain isolated from wheat was found to exhibit high antifungal activity against *Gaeumannomyces graminis* var. *tritici,* which causes take-all disease in wheat. Field experiments showed that endophyte-inoculated plants were found to give 55.3% protection against take-all disease compared to uninoculated controls (Liu et al. 2009). A total of 60 different endophytic bacteria were isolated from *Cymbopogon citratus* L., *Azadirachta indica* (L.) Adelb., *Phyllanthus emblica* L., *Boerhavia diffusa* L., *Boerhavia repens* L., *Pisum sativum* L., *Sorghum bicolor* (L.), and *Parthenium hysterophorus* L. Among the endophytes, *Pseudomonas fluorescens* and *Bacillus* sp. showed the highest protection (68% and 63%, respectively) against downy mildew disease in pearl millet (Chandrashekhara et al. 2007).

4.16.1 Obligatory or Facultative Endophytic Bacteria

Endophytic bacteria can develop their entire cycle in a host plant, depending on it for development and reproduction, in this case called obligatory endophytes, or develop part of their cycle outside a host plant, called facultative endophytes. The division of the term into facultative and obligatory endophyte was proposed to distinguish, respectively, strains capable of colonizing both the surface and the inside of roots and able to survive in soil, from the ones that do not survive in the soil, but colonize inside and shoots of plant tissues without causing pathogenicity symptoms (Baldani et al. 1997).

Despite systematically colonizing plants, endophytic bacteria have a preference to colonize certain tissues. Kuklinsky-Sobral et al. (2004) observed in soybean that the density and diversity of endophytic bacteria vary according to tissue, plant development stage, seasonal changes, and host genotype, where the observed bacterial density was higher in roots and lower in leaves. Spatiotemporal change in density of bacterial endophytes has been indicated by Aghaei Dargiri et al. (2021f) and Baghazadeh Daryaii et al. (2021).

4.16.2 The Role of Rhizobacteria in the Control of Pest Insects

In addition to supplying plants, rhizobacteria can also play an important role in pest control. Bong and Sikorowski (1991) found an alteration in larval growth and reduction in the emergence of adults of *H. zea* as a result of *Pseudomonas maltophilia* infection. Later, Thuler et al. (2006) verified that the isolates EN4 of *Kluyvera ascorbata* and EN5 of *Alcaligenes piechaudii*, little reported in the literature on insects, reduced the viability of *Plutella xylostella* (L.) (Lepidoptera: Plutellidae) in about 80 to 50%, respectively, indicating a broad field of research to explore the potentials of endophytic bacteria. Praça (2012) studied the insecticidal activity of four strains of *Bacillus thuringiensis* Berliner including S1450, S1905, S2122, and S2124 and found that the *B. thuringiensis* strains S1905 and S2122 caused100% of mortality in caterpillars in the third instar of *P. xylostella* in the evaluation performed 48 h after caterpillars exposed to selective bioassays, whereas S2124 caused 58.33% of mortality after the same period and 98.33% of mortality after 96 h. The capacity for biocontrol and plant growth promotion of these endophytic microorganisms can be achieved from various mechanisms, e.g., their biological nitrogen fixation (Huergo et al. 2008), phosphate solubilization (Rodriguez et al. 2004), production of growth hormones like auxins, gibberellins, and cytokinins (Donate-Correa et al., 2004) and synthesis of siderophores (Vessey 2003).

The potential to use Plant Growth-Promoting Bacteria (PGPB) in the control of pest insects has been attributed to stimuli generated in the plant itself, through the action in different metabolic routes including salicylic acid, jasmonic acid, and ethylene. These compounds act as elicitors to induce defense and/or resistance, which are kept inactive in their absence. This process, called resistance induction, causes the plant to produce or increase the production of proteinase-inhibitor compounds (pathogens produce extracellular proteinases and in response to their action plants synthesize inhibitors, like serine, cysteine, and aspartate), glycoalkaloids, polyphenols, etc. (de Oliveira Araújo 2015).

4.16.3 The Efficacy of Different Bacterial Strains against Pests

Macedo et al. (2012) in study on selecting and characterizing native strains of *B. thuringiensis* toxic to *Diatraea saccharalis* (F.) (Lepidoptera: Crambidae) observed that the strains causing more than 75% of mortality after dilution of 50 times were S602, S1264, and S1301 (S602 and S1264 were the most toxic strains, showing statistically similar LC50 values). Melatti et al. (2010) by testing the selected strains of *B. thuringiensis* for the control of cotton aphid (*A. gossypii*) found that the strains S29, S40, S616, S1576 (*Bacillus aizawai*) and S1168 (*Bacillus kurstaki*) were the most toxic to *A. gossypii*, causing mortality higher than 50%. In addition, S29 and S1168 were the most effective in the selective bioassay, causing mortalities of 76 and 73% against *A. gossypii*, respectively.

Different strains of an endophyte may be different in terms of pathogenicity on pests, thus to achieve a proper pest management, it is necessary to find and use high efficient strains. Polanczyk et al. (2003) studied 58 subspecies of *B. thuringiensis* against *S. frugiperda* and found that only *B. thuringiensis* var. *morrisoni* caused 80% of mortality in caterpillars. Berlitz et al. (2003) tested 24 isolates of *B. thuringiensis* for the control of *S. frigiperda* and obtained the best mortality rates (between 31.6 and 100%) with only five isolates. Campanini et al. (2012) studied the pathogenicity of isolates of *B. thuringiensis* against *S. frigiperda* and *Sphenophorus levis* Vaurie (Coleoptera: Curculionidae) and found that the isolates IB17.3 and IB8.2 were highly efficient in the control of caterpillars of *S. frugiperda*, and that the isolate IB26.2 had the lowest efficacy in the control of larvae of *S. levis* (all of them with average mortality rates higher than 75%). In another study, it was revealed that only two natural isolates of *B. thuringiensis* identified by Azambuja and Fiuza (2003) caused 37 and 50% of mortality against velvet bean caterpillar, *Anticarsia gemmatalis* Hübner (Lepidoptera: Noctuidae) (Azambuja and Fiuza 2003). Praça et al. (2004) expressed that among the 300 tested strains of *B. thuringiensis*, only coapplication of S234 and S997 led to the control of larvae of *S. frugiperda*,

A. gemmatalis, *Anthonomus grandis* Boheman (Coleoptera: Curculionidae), *Aedes aegypti*(L.) (Diptera: Culicidae) and *Culex quinquefasciatus* Say (Diptera: Culicidae).

4.17 Recombinant Endophyte

Transgenically modifying endophyte genomes can be a useful strategy for genetic manipulation of host plants. Genes introduced into endophytic microbes confer new characteristics, which may be useful in biocontrol of plant pathogens, pest control, growth promotion of host plants, and/or production of medicines for humans or animals (Zhao et al. 2010; White et al. 2019b). Compared to transgenic plant, endophyte has several advantages, including (a) when inoculating endophyte into plant, the plant genome is not changed, similar to what happens in transgenesis (White et al. 2019b); (b) Recombinant endophytes require a shorter time to be developed because it is easier to modify a microorganism than a plant; (c) The modified endophytic microorganism is not transmitted to progeny seed of inoculated plants and stays restricted to the inner parts of the plant, which is safer because of no dissemination to next generation or other plants. Promising instances on the use of transgenic endosymbiotes are available. For example, the endophytic bacterium, *Clavibacter xyli* subsp. *cynodontis*, which colonizes the xylem of several plant species, was transgenically modified to express the *B. thuringiensis* gene encoding endotoxin for the control of insects. In another example, an endophytic *Burkholderia pyrrocinia* JK-SH007 was transformed with the Bt endotoxin gene to express the insecticidal protein against the second stage of *Bombyx mori* L. (Lepidoptera: Bombycidae) instar silkworms. Further, an endophytic *Pseudomonas putida* (WCS358r) was modified with an antifungal gene and introduced into wheat with a resultant reduction in fungal populations in soil including pathogenic *Fusarium* spp. (White et al. 2019b). Zhang et al. (2011) cloned the Pinellia ternate agglutinin (PTA) gene into SJ-10 (SJ-10 known as *Enterobacter cloacae* by morphological, physiological, biochemical, and 16 s rDNA characteristics, was isolated from rice seedlings) for expression. The positive transformant, selected by antibiotic resistance, was evaluated using PCR, SDS-PAGE, and Western blot assay. After inoculation, rSJ-10 could colonize rice plants so that they expressed PTA, and then the rice was shown to have insecticidal activity against the white-backed planthopper (WBPH; *Sogota furcifera* (Horváth) (Hemiptera: Delphacidae). Their results showed that rSJ-10 could significantly decrease the survival and fecundity of WBPH fed on rice seedlings. They showed that at day 19, the fecundity of WBPH inoculated with rSJ-10, or with wild-type SJ-10 decreased by 86.1%, and 25.6%, respectively. In addition, at day 22, numbers of WBPH on rice in the control were 19.4 times greater than on rice inoculated with rSJ-10 and at day 26, the rice seedlings all died in the control group, but the seedlings inoculated with rSJ-10 grew well (White et al. 2019b); (d) Genetically engineered endophytes have wide applicability and could be used for many plant species; (e) Endophyte can multiply highly inside the plant, often resulting in up to 10^6–10^8 CFU g^{-1} of inoculated plant, its antiherbivory activity will be as highly active as transgenic plant (Azevedo et al. 2000). In overall, as an alternative to transgenic plants, genetically engineered endophytic microbes might provide a pathway for plants to benefit from foreign genes to cope with diseases or pests and diminish the negative effects of environmental pollutants (Barac et al. 2004).

4.18 Possibility of Using Endophytes and Endophytic Products

Endophytic microbes have shown that they can be integrated with each other or with other endophytic products (Akutse et al. 2014; Gathage et al. 2016) and may thus be used as a complementary tool in IPM programs. Endophytic and/or plant growth-promoting bacteria can be used in combination with other microorganisms in the control of pest insects in agriculture. In this context, Broderick et al. (2000) identified an increase of 35% in the mortality of the lepidopterous *Lymantria dispar* (L.) (Lepidoptera: Erebidae) when using *B. thuringiensis* and zwittermycin A of *Bacillus cereus*, which is responsible for the synergetic effect of the microorganisms. Wraight and Ramos (2005) also showed synergism of 35.2, 33.8, and 21.1% when commercial products based on *B. thuringiensis* and on the fungus *B. bassiana* were simultaneously used against *Leptinotarsa decemlineata* (Say) (Coleoptera: Chrysomelidae). They revealed that the interaction may have resulted from the intoxication caused by entomopathogen, inhibiting insect feeding, thus causing stress, and physiological effects, which facilitated fungus penetration in the insect. Similar effects were also observed by Ma et al. (2008), when the Cry 1Ac protein of *B. thuringiensis* was used with *B. bassiana*. They reported the deleterious effects in the mortality of larvae of *Ostrinia furnacalis* (Guenée) (Lepidoptera: Crambidae), besides the decrease in the formation of pupae and emergence of the adult insects.

4.19 Does the Use of Endophytes Provide a Permanent Immunity?

Fungal entomopathogens may not fully colonize all plant tissues or persist for long periods of time due to multipartite interactions with other bacterial and fungal inhabitants within the host plants (see Schulz et al. 2015). A deeper knowledge about the extent and persistence of entomopathogenic fungi inside plants is required and constitutes the basis for determining the degree of plant protection as well as other benefits conferred by these fungi as endophytes. Previous studies have reported the endophytic colonization of plants by *B. bassiana* to continue for as long as 3 months in jute (Biswas et al. 2013), 8 months in coffee (Posada et al. 2007), and 9 months in radiata pine (Brownbridge et al. 2012). Under field conditions, *B. bassiana*, *M. robertsii*, and *I. fumosorosea* were reisolated 30 days after foliar spray of sweet sorghum (Mantzoukas et al. 2015). However, the extent and persistence of endophytic fungal colonization within plants can be improved by repeated application of the microbial agent through foliar spray or soil drench. Same as natural enemies, serial subculturing on artificial media reduces viability and virulence, which may therefore alter functionality of fungal isolates including their endophytic capacity to colonize plants. Storing agar plugs of fungal isolates in tubes of sterile distilled water can maintain viability over long periods of time (Richter et al. 2016). Alternatively, passage through insect hosts (Nahar et al. 2008) or reisolation of endophytes from inoculated plants (Busby et al. 2016) may serve to refresh the ecological function of fungal isolates as endophytes. Bamisile et al. (2019) showed that *B. bassiana* was retained for up to 3 months in colonized citrus plants. This result is similar to the study of Brownbridge et al. (2012) and Posada et al. (2007), where *B. bassiana* was reported to be retained in radiata pine and coffee up to nine and 8 months, respectively. Biswas et al. (2012) also reported sustenance of *B. bassiana* up to 8 months in jute plants. A report of *B. bassiana* and *M. brunneum* colonizing *V. faba* for 1 month is also available (Jaber and Enkerli 2016).

4.20 Specialization in Endophytes and Hosts

Screening strains or isolates in order to select the ones most adapted to endophytism in a specific host plant or even cultivar may enhance the colonization rate of plant tissues. Selecting superior endophytic strains with high virulence against one or more pests could subsequently facilitate the development of these strains for the wider management of multiple plant pests. Furthermore, the differential expression of fungal genotype-plant genotype interactions under different environmental conditions should be taken into account. For example, even though experimental studies that investigate endophytic fungal entomopathogens using sterile planting substrates have merit and thus a strong "internal validity," they will probably have a limited "external validity" because sterile soils do not exist in nature (Parsa et al. 2018).

In addition to innate characteristics of the fungal isolate (Posada et al. 2007) and host plant genetics (Arnold and Lewis 2005), different issues may affect the colonization rates of plants such as leaf surface chemistry (Posada et al. 2007) and competition with other endophytes naturally occurring within plants (Posada et al. 2007; Schulz et al. 2015; Jaber and Enkerli 2016). Of course, even sometimes poor endophytic establishment does not pose a problem for its performance. For instance, Akutse et al. (2013) also reported that despite poor colonization of different parts of *P. vulgaris*, two isolates of *B. bassiana* had negative effects on the number of pupae and emergence of *L. huidobrensis*. Isolates of *M. anisopliae* that could not be confirmed to colonize bean plants endophytically still resulted in reduced feeding, oviposition, pupation, and the emergence of the bean stem maggot *Ophiomyia phaseoli* Tryon (Diptera: Agromyzidae) (Mutune et al. 2016).

The specificity between certain endophytic fungi and host plants led Leuchtmann (1993) to suggest a further study on the occurrence of physiological races in endophytic fungi that could be important to the development of new biological controls. Surveys have been carried out aiming to the discovery of new toxins useful to insect control. Two new active toxins against the Spruce budworm *C. fumiferana* were found in an unidentified endophytic fungus infecting the wintergreen *Gaultheria procumben* L. (Azevedo et al. 2000). Alkaloids from *Neotyphodium lolii* and *Lolium perenne* L. are capable of altering insect behavior. Several of these alkaloids were added to the diet of adult individuals of the *Heteronychus arator* F. (Coleoptera: Scarabaeidae). Peramine, lolitrem B, lysergol-type alkaloids, festuclavine, and lysergic acid showed no effects on the insect. Ergonovine showed moderate effects whereas ergotamine, ergovaline from the ergot-type alkaloid family seem to be responsible for the plant resistance (Azevedo et al. 2000). Miles et al. (1998) showed that endophytic isolates of *Neotyphodium* sp. produce N-formilonine and a paxiline analogous in the host *Echinopogum ovatus*. These compounds show insecticidal activity against *L. bonariensis* and other insects (Azevedo et al. 2000).

4.21 How May Environmental Factors Change the Diversity and Frequency of Endophytes?

Basic knowledge of the variation in endophyte species and strains and their frequencies over a geographic range of environmental conditions may provide insights into the long-term nature of the interactions of endophytes and their hosts. In addition, the genetics of host plants may vary over the range of a species and may interact with variation in endophyte species or strain to affect the persistence of the plant–endophyte symbiota. Indeed, host and endophyte genotypic combinations, especially in maternally transmitted endophytes, may have coevolved with each other to increase fitness, and thus may be adapted to local environmental conditions (Shymanovich and Faeth 2019).

Endophytic population varies from plant to plant and from species to species. Within the same species, it not only varies from region to region but also differs with change in climatic conditions of the same region (Aghaei Dargiri et al. 2021f; Baghazadeh Daryaii et al. 2021). Furthermore, the frequency of an endophyte may vary over months. For instance, it has been revealed that *Epichloë alsodes* frequency can change with July Max temperature, July precipitation, and soil nitrogen and phosphorous (Shymanovich and Faeth 2019). Chareprasert et al. (2006) studied temporal changes in the relative frequency of total endophytic fungi and found matured leaves of teak (*Tectona grandis* L.) and rain tree (*Samanea saman* Merr.) had a greater number of genera and species, with higher colonization frequency than those in the young leaves and their occurrence in leaves increased during the rainy season. In addition to variability and frequency, environmental conditions may even alter the amount of metabolites (toxin) produced by endophytes. For instance, Breen (1992) verified that changes in temperature and seasons modify toxin levels in the plant. This is the case of the peramine toxin produced in *Lolium perenne L.* by *A. lolii*. The antixenosis toward the aphid *S. graminum* is dependent on peramine and the endophyte concentration so, a natural consequence is that the effect on the insect-pest will also be modified (Azevedo et al. 2000).

Sadeghi et al. (2019) in a study on the spatiotemporal distribution of endophytic fungi associated with leaf, stem, trunk, and root of mandarin (*Citrus reticulata* cv. Siyahoo) in four geographically isolated mandarin growing areas of Hormozgan province of Iran including Siyahoo, Ahmadi, Sikhoran, and Roudan, found 702 fungal isolates from leaf, stem, trunk, and root of healthy mandarin trees. These isolates were divided into 26 distinct morphotypes based on morphological characteristics. Accordingly, 10 different fungal orders from 5 fungal classes were identified, i.e., Saccharomycetes (Saccharomycetales), Eurotiomycetes (Eurotiales), Dothideomycetes (Capnodiales, Pleosporales, Dothideales), and Sordariomycetes (Diaporthales, Hypocreales, Microascales, Togniniales), all from Ascomycota represented 97.2% and Ustilaginomycetes (Ustilaginales) from Basidiomycota represented 2.8% of the isolates. The *Aureobasidium pullulans, Penicillium citrinum*, and *Dothideomycetes* sp. were the most frequent isolates. The trunk and leaf showed the highest and lowest total colonization frequency and species richness of endophytic fungi, respectively, in all sampling periods. The results showed that the colonization frequency of endophytes in Hormozgan province was higher in autumn than that in spring, winter, and summer. The trunk showed the maximum diversity of endophytes over all seasons. The Shannon–Wiener (H′) and Simpson indices had a significant correlation with sampling sites and tissue types and the maximum value of Shannon and Simpson indices ($H' = 3.05$ and $1 - D = 0.94$) was found in the specimens collected from Siyahoo. They found that the three factors (season, location, and tissue type) all together could determine the fungal endophyte composition of *C. reticulate* (Sadeghi et al. 2019).

4.22 Disadvantages of Endophyte Application

Currently, no one knows how and to what extent plant tissues are colonized by endophytic entomopathogenic fungi (EEPFs), and whether the colonization per se or changes in plant metabolism mediated by these fungi, contribute to the reduced herbivore damage. For use as a biocontrol agent, the efficacy of the product should be guaranteed and following an application, pest abundance reductions need to be consistent and at a comparable level to chemical insecticides. An additional major hurdle is the potential of some of the entomopathogenic fungi to produce a wide array of compounds, such as mycotoxins with biological activity against other organisms including humans (White et al. 2019b). Some of the bioactive compounds produced by endophytes, while beneficial to the plant, can be detrimental to livestock and birds (Pirelli et al. 2016). In livestocks, fescue toxicosis is associated with ergot alkaloids produced by both *E. coenophiala* (in tall fescue) and *E. festucae* var. *lolii* (in perennial ryegrass). These toxins are vasoconstrictors; they constrict blood vessels and reduce circulation to body extremities. Symptoms of fescue toxicosis include low average daily gains, reduced fertility, rough hair coats, and a preference for shade or muddy areas, as animals are less able to regulate body temperature (Pirelli et al. 2016). Of course, considering that the ancestors of some breeding plants have endophytes that these endophytes have been destroyed due to plant breeding, it may be possible to reach the types of endophytes, their use does not have a problem for other nontarget organisms. Therefore, the lack of an overall awareness of the general presence of communities of endophytic microbes in tissues of plants has been a hindrance in advancing the exploration of applications of endophytes in crops (White et al. 2019b). Using endophytes in practical forest protection creates challenges. In nature, it is likely that endophyte communities rather than just a single endophyte, may contribute to resistant phenotypes. The dynamics of microbial communities over time and space adds to the challenge. If IPM measures to control tree diseases included endophytes, the question remains on how the endophyte community could be engineered in forests (Blumenstein 2015).

It is as yet unclear as to whether endophytes introduced as BCAs on plants may be effective in reducing disease and pests, but another important aspect is to understand if they have adverse effects on the natural microbial community of the host when the plant is under environmental stress. The introduction of endophytes that have not coevolved with the host plant may result in the loss of beneficial organisms and so negatively impact the host plant. Furthermore, it is important to consider whether the gains provided by the endophyte outweigh the costs associated with it. For example, gall wasps are a problematic species to trees. However, *Apiognomonia errabunda*, the dominant endophyte in beech leaves, has been found to cause abscission of galls by forming necrotic tissue around the affected area; but this may in time prove to be more harmful to the host than the gall. Furthermore, there is also a risk that some endophytes may not be as useful as hoped in integrated pest management systems, as they may affect the efficacy of other BCAs employed to combat pest species (Rabiey et al. 2019). Bultman et al. (1997) found that although endophytes proved effective against plant herbivores, they had repercussions higher up the trophic chain, significantly affecting the performance of parasitoids by reducing pupal mass, which would reduce the parasitoids' success as a BCA (Rabiey et al. 2019).

4.23 Conclusions

Endophytes, along with other nonchemical approaches, are a suitable pest management option that can be used practically in the pest management programs. Comprehensive studies on endophyte identification in closely related plant species will contribute to the success of endophytes in pest control systems. In addition, although the available information focuses more on the efficacy of different endophytic varieties in pest control, it should not be neglected that the success of this approach also depends on the variety of plants and pests. It means that a variety of endophytes currently inefficient against a pest may be effective against another population of that pest.

Endophytes could help cultivate crops with less fertilizers, fungicides, insecticides, or herbicides. Supplementing microbial diversity through microbe amendments to soils and plants that function to bring nutrients to plants (e.g., through the rhizophagy cycle), while simultaneously suppressing virulence in pathogens, deterring insect feeding, and reducing growth of competitor weeds can result in less environmental contamination and agricultural practices that are more parsimonious with natural processes. It was also found that endophytes have the ability to integrate with natural enemies although before any integration, their successful combination should be investigated and guaranteed. In overall, according to the available information, it can be inferred that the use of entomopathogenes in the form of endophytes has many advantages than the method of their direct application against pests, because they can obtain the ability to promote plant growth and suppress the growth of plant diseases.

4.24 Points to Remember

1. Different endophytes may have different niches, therefore their ecological niche should be considered when selecting endophytes for specific purposes.
2. Choosing the right time and method is very effective in inoculating endophytes.
3. Different endophytes may have synergistic, antagonistic or neutral effects, then the endophytes should be examined before inoculation.
4. The antiherbivore effects of *Beauveria* and *Metarhizium* in the form of endophytes are much greater than when they kill herbivores in the form of physical contact.
5. *Beauveria* and *Metarhizium* in addition to insecticidal attributes, they can also be effective in reducing damage caused by plant pathogens.
6. Before using endophytes in integrated pest management programs, it is necessary to ensure their compatibility with other control methods such as biocontrol agents.
7. Inoculation of endophytes does not provide permanent immunity to host plants and should be repeated regularly.
8. The use of some endophytes should be done with caution due to the production of some compounds that may be harmful to other nontarget organisms.

Acknowledgement The authors are grateful to Mousa Abdollahipour, PhD student of Entomology at Tarbiat Modares University and Fatemeh Koohpayma for their great assistance.

References

Abraham S, Basukriadi A, Pawiroharsono S, Sjamsuridzal W (2015) Insecticidal activity of ethyl acetate extracts from culture filtrates of mangrove fungal endophytes. Mycobiology 43 (2):137–149

Adame-Álvarez RM, Mendiola-Soto J, Heil M (2014) Order of arrival shifts endophyte–pathogen interactions in bean from resistance induction to disease facilitation. FEMS Microbiol Lett 355 (2):100–107

Aghaei Dargiri S, Samsampour D, Seyahooei A, Bagheri A (2021a) Evaluation of the effect of symbiosis of fungal *Penicillium chrysogenum* and bacterial *Exiguobacteium aurantiacum* endophytes on improvement of the morpho-physiological characteristics of tomato seedlings. Iran Soc Plant Physiol. Under review

Aghaei Dargiri S, Samsampour D, Seyahooei A & Bagheri A (2021b) The effect of bacterial endophyte (*Exiguobacterium aurantiacum*) isolated from *Salsola imbricata* on growth characteristics of tomato seedlings under salinity stress. J Ferdowsi Hortic Sci Mashhad. Under review

Aghaei Dargiri S, Samsampour D, Seyahooei A, Bagheri A (2021c) The role of the fungal endophyte *Penicillium Chrysogenum* in the reduction of salinity stress in tomato plants. Sari Univ Agric Sci Nat Res J Crop Breeding. Under review

Aghaei Dargiri S, Samsampour D, Seyahooei A, Bagheri A (2021d) Reduction of salinity effect on tomato growth and yield by inoculation fungal endophyte *Curvularia hawaiiensis* isolated from halophyte plant *Halocnemum strobilaceum* of Qeshm Island. J Trends Plant Sci. Under review

Aghaei Dargiri S, Samsampour D, Seyahooei A, Bagheri A (2021e) The effect of endophyte fungal symbiosis on tomato and their role in salinity stress tolerance. J Stud Mycol. Under review

Aghaei Dargiri S, Samsampour D, Seyahooei A, Bagheri A (2021f) Investigation of spatiotemporal diversity of fungal endophytes of halophyte plants in Hormozgan province. Journal Under review

Akello J (2012) Biodiversity of fungal endophytes associated with maize, sorghum and Napier grass and the influence of biopriming on resistance to leaf mining, stem boring and sap sucking insect pests. ZEF, Bonn

Akello J, Sikora R (2012) Systemic acropedal influence of endophyte seed treatment on *Acyrthosiphon pisum* and *Aphis fabae* offspring development and reproductive fitness. Biol Control 61(3):215–221

Akello J, Dubois T, Coyne D, Kyamanywa S (2008) Effect of endophytic *Beauveria bassiana* on populations of the banana weevil, Cosmopolites sordidus, and their damage in tissue-cultured banana plants. Entomol Exp Appl 129(2):157–165

Akutse KS, Maniania NK, Fiaboe KKM, Van den Berg J, Ekesi S (2013) Endophytic colonization of *Vicia faba* and *Phaseolus vulgaris* (Fabaceae) by fungal pathogens and their effects on the life-history parameters of *Liriomyza huidobrensis* (Diptera: Agromyzidae). Fungal Ecol 6 (4):293–301

Akutse KS, Fiaboe KK, Van den Berg J, Ekesi S, Maniania NK (2014) Effects of endophyte colonization of *Vicia faba* (Fabaceae) plants on the life–history of leafminer parasitoids *Phaedrotoma scabriventris* (Hymenoptera: Braconidae) and *Diglyphus isaea* (Hymenoptera: Eulophidae). PLoS One 9(10):e109965

Allegrucci N, Velazquez MS, Russo ML, Pérez ME, Scorsetti AC (2017) Endophytic colonisation of tomato by the entomopathogenic fungus *Beauveria bassiana*: the use of different inoculation techniques and their effects on the tomato leafminer *Tuta absoluta*. Gelechiidae, Lepidoptera

Aly AH, Debbab A, Kjer J, Proksch P (2010) Fungal endophytes from higher plants: a prolific source of phytochemicals and other bioactive natural products. Fungal Divers 41(1):1–16

Arnold AE, Lewis LC (2005) Ecology and evolution of fungal endophytes, and their roles against insects. Insect-Fungal Associations: Ecology and Evolution. Oxford University Press, New York, pp 74–96

Arnold AE, Mejía LC, Kyllo D, Rojas EI, Maynard Z, Robbins N, Herre EA (2003) Fungal endophytes limit pathogen damage in a tropical tree. Proc Natl Acad Sci U S A 100 (26):15649–15654

Athman SY (2006) Host-endophyte-pest interactions of endophytic *fusarium oxysporum* antagonistic to *Radopholus similis* in banana (*Musa* spp.). Doctoral dissertation, University of Pretoria

Azambuja AO, Fiuza LM (2003) Isolation of *Bacillus thuringiensise* patogenici-dade to *Anticarsia gemmatalis* (Lepidoptera, Noctuidae). Shows of Scientific Initiation of Unisinos. June. São Leopoldo. RS

Azevedo JL, Maccheroni W Jr, Pereira JO, de Araújo WL (2000) Endophytic microorganisms: a review on insect control and recent advances on tropical plants. Electron J Biotechnol 3 (1):15–16

Backman PA, Sikora RA (2008) Endophytes: an emerging tool for biological control. Biol Control 46(1):1–3

Baghazadeh Daryaii L, Samsampour D, Bagheri A, Sohrabipour J (2021) Endophytic bacteria associated with some seaweed species of the Persian Gulf and Oman Sea (South of Iran) and their utilization in Mexican lime seedlings growth. J Appl Phycol. Under review

Baldani J, Caruso L, Baldani VL, Goi SR, Döbereiner J (1997) Recent advances in BNF with non-legume plants. Soil Biol Biochem 29(5–6):911–922

Bamisile BS, Dash CK, Akutse KS, Keppanan R, Afolabi OG, Hussain M, Qasim M, Wang L (2018) Prospects of endophytic fungal entomopathogens as biocontrol and plant growth promoting agents: an insight on how artificial inoculation methods affect endophytic colonization of host plants. Microbiol Res 217:34–50

Bamisile BS, Dash CK, Akutse KS, Qasim M, Ramos Aguila LC, Wang F, Keppanan R, Wang L (2019) Endophytic *Beauveria bassiana* in foliar-treated *Citruslimon* plants acting as a growth suppressor to three successive generations of *Diaphorina citri* Kuwayama (Hemiptera: Liviidae). Insects 10(6):176

Barac T, Taghavi S, Borremans B, Provoost A, Oeyen L, Colpaert JV, Vangronsveld J, van der Lelie D (2004) Engineered endophytic bacteria improve phytoremediation of water-soluble, volatile, organic pollutants. Nat Biotechnol 22(5):583–588

Barelli L, Moreira CC, Bidochka MJ (2018) Initial stages of endophytic colonization by *Metarhizium* involves rhizoplane colonization. Microbiology 164(12):1531–1540

Bastias DA, Martínez-Ghersa MA, Ballaré CL, Gundel PE (2017) *Epichloë* fungal endophytes and plant defenses: not just alkaloids. Trends Plant Sci 22(11):939–948

Batta YA (2013) Efficacy of endophytic and applied *Metarhizium anisopliae* (Metch.) Sorokin (Ascomycota: Hypocreales) against larvae of *Plutella xylostella* L.(Yponomeutidae: Lepidoptera) infesting *Brassica napus* plants. Crop Prot 44:128–134

Behie SW, Bidochka MJ (2014) Ubiquity of insect-derived nitrogen transfer to plants by endophytic insect-pathogenic fungi: an additional branch of the soil nitrogen cycle. Appl Environ Microbiol 80(5):1553–1560

Behie SW, Jones SJ, Bidochka MJ (2015) Plant tissue localization of the endophytic insect pathogenic fungi *Metarhizium* and *Beauveria*. Fungal Ecol 13:112–119

Benhamou N, Brodeur J (2001) Pre-inoculation of Ri T-DNA transformed cucumber roots with the mycoparasite, Verticillium lecanii, induces host defense reactions against Pythium ultimum infection. Physiol Mol Plant Pathol 58(3):133–146

Bensaci OA, Daoud H, Lombarkia N, Rouabah K (2015) Formulation of the endophytic fungus Cladosporium oxysporum Berk. and MA Curtis, isolated from Euphorbia bupleuroides subsp. luteola, as a new biocontrol tool against the black bean aphid (Aphis fabae Scop.). J Plant Prot Res 55(1):80–87

Berlitz DL, Azambuja DL, Azambuja AO, Antonio AC, Oliveira JV, Fiuza LM (2003) New insulated and cry proteins from *Bacillus thuringiensis* applied in the control of caterpillars of *Spodoptera frugiperda*. III Brazilian Congress of Irrigated Rice and XXV meeting of the Irrigated Rice Culture, Camboriú. Analls. Camboriú

Bilal L, Asaf S, Hamayun M, Gul H, Iqbal A, Ullah I, Lee IJ, Hussain A (2018) Plant growth promoting endophytic fungi *Asprgillus fumigatus* TS1 and *Fusarium proliferatum* BRL1 produce gibberellins and regulates plant endogenous hormones. Symbiosis 76(2):117–127

Bills GF, Giacobbe RA, Lee SH, Peláez F, Tkacz JS (1992) Tremorgenic mycotoxins, paspalitrem A and C, from a tropical Phomopsis. Mycol Res 96(11):977–983

Bing LA, Lewis LC (1991) Suppression of *Ostrinia nubilalis* (Hübner) (Lepidoptera: Pyralidae) by endophytic *Beauveria bassiana* (Balsamo) Vuillemin. Environ Entomol 20(4):1207–1211

Biswas C, Dey P, Satpathy S, Satya P, Mahapatra BS (2013) Endophytic colonization of white jute (*Corchorus capsularis*) plants by different *Beauveria bassiana* strains for managing stem weevil (*Apion corchori*). Phytoparasitica 41(1):17–21

Bixby-Brosi AJ, Potter DA (2012) Endophyte-mediated tritrophic interactions between a grass-feeding caterpillar and two parasitoid species with different life histories. Arthropod Plant Interact 6(1):27–34

Blumenstein K (2015) Endophytic fungi in elms. 2015(106)

Bong CFJ, Sikorowski PP (1991) Effects of cytoplasmic polyhedrosis virus and bacterial contamination on growth and development of the corn earworm, *Helicoverpa zea* (Lepidoptera: Noctuidae). J Invertebr Pathol 57(3):406–412

Borges WDS, Borges KB, Bonato PS, Said S, Pupo MT (2009) Endophytic fungi: natural products, enzymes and biotransformation reactions. Curr Org Chem 13(12):1137–1163

Bowsher AW, Ali R, Harding SA, Tsai CJ, Donovan LA (2016) Evolutionary divergences in root exudate composition among ecologically-contrasting *Helianthus* species. PLoS One 11(1): e0148280

Braun V, Focareta T (1991) Pore-forming bacterial protein hemolysins (cytolysins). Crit Rev Microbiol 18(2):115–158

Breen JP (1992) Temperature and seasonal effects on expression of *Acremonium* endophyte-enhanced resistance to *Schizaphis graminum* (Homoptera: Aphididae). Environ Entomol 21 (1):68–74

Broderick NA, Goodman RM, Raffa KF, Handelsman JO (2000) Synergy between zwittermicin A and *Bacillus thuringiensis* subsp. *kurstaki* against gypsy moth (Lepidoptera: Lymantriidae). Environ Entomol 29(1):101–107

Broeckling CD, Broz AK, Bergelson J, Manter DK, Vivanco JM (2008) Root exudates regulate soil fungal community composition and diversity. Appl Environ Microbiol 74(3):738–744

Brownbridge M, Reay SD, Nelson TL, Glare TR (2012) Persistence of *Beauveria bassiana* (Ascomycota: Hypocreales) as an endophyte following inoculation of radiata pine seed and seedlings. Biol Control 61(3):194–200

Bultman TL, Borowicz KL, Schneble RM, Coudron TA, Bush LP (1997) Effect of a fungal endophyte on the growth and survival of two *Euplectrus parasitoids*. Oikos:170–176

Busby PE, Ridout M, Newcombe G (2016) Fungal endophytes: modifiers of plant disease. Plant Mol Biol 90(6):645–655

Calhoun LA, Findlay JA, Miller JD, Whitney NJ (1992) Metabolites toxic to spruce budworm from balsam fir needle endophytes. Mycol Res 96(4):281–286

Campanini EB, Davolos CC, Alves EDC, Lemos MVF (2012) Characterization of new strains of *Bacillus thuringiensis* for the control of important insect pests in agriculture. Bragantia 71 (3):362–369

Canassa F, Tall S, Moral RA, de Lara IA, Delalibera I Jr, Meyling NV (2019) Effects of bean seed treatment by the entomopathogenic fungi *Metarhizium robertsii* and *Beauveria bassiana* on plant growth, spider mite populations and behavior of predatory mites. Biol Control 132:199–208

Castillo Lopez D, Zhu-Salzman K, Ek-Ramos MJ, Sword GA (2014) The entomopathogenic fungal endophytes Purpureocillium lilacinum (formerly Paecilomyces lilacinus) and Beauveria bassiana negatively affect cotton aphid reproduction under both greenhouse and field conditions. PLoS One 9(8):103891

Chandrashekhara SN, Deepak SA, Amruthesh KN, Shetty PN, Shetty SH (2007) Endophytic bacteria from different plant origin enhance growth and induce downy mildew resistance in pearl millet. Asian J Plant Pathol 1(1):1–11

Chareprasert S, Piapukiew J, Thienhirun S, Whalley AJ, Sihanonth P (2006) Endophytic fungi of teak leaves *Tectona grandis* L. and rain tree leaves *Samanea saman* Merr. World J Microbiol Biotechnol 22(5):481–486

Cherry AJ, Banito A, Djegui D, Lomer C (2004) Suppression of the stem-borer *Sesamia calamistis* (Lepidoptera; Noctuidae) in maize following seed dressing, topical application and stem injection with African isolates of *Beauveria bassiana*. Int J Pest Manag 50(1):67–73

Clark CL, Miller JD, Whitney NJ (1989) Toxicity of conifer needle endophytes to spruce budworm. Mycol Res 93(4):508–512

Clement SL, Elberson LR, Bosque-Pérez NA, Schotzko DJ (2005) Detrimental and neutral effects of wild barley–Neotyphodium fungal endophyte associations on insect survival. Entomol Exp Appl 114(2):119–125

Compant S, Clément C, Sessitsch A (2010) Plant growth-promoting bacteria in the rhizo-and endosphere of plants: their role, colonization, mechanisms involved and prospects for utilization. Soil Biol Biochem 42(5):669–678

Coombs JT, Franco CM (2003) Isolation and identification of actinobacteria from surface-sterilized wheat roots. Appl Environ Microbiol 69(9):5603–5608

Coombs JT, Michelsen PP, Franco CM (2004) Evaluation of endophytic actinobacteria as antagonists of *Gaeumannomyces graminis* var. *tritici* in wheat. Biol Control 29(3):359–366

Dahlman DL, Eichenseer H, Siegel MR (1991) Chemical perspectives on endophytegrass interactions and their implications for insect herbivory. In: Barbosa P, Krischik VA, Jones CG (eds) Microbial mediation of plant-herbivore interactions. John Wiley and Sons, New York, pp 227–252

Dash CK, Bamisile BS, Keppanan R, Qasim M, Lin Y, Islam SU, Hussain M, Wang L (2018) Endophytic entomopathogenic fungi enhance the growth of *Phaseolus vulgaris* L. (Fabaceae) and negatively affect the development and reproduction of *Tetranychus urticae* Koch (Acari: Tetranychidae). Microb Pathog 125:385–392

de Mesquita CPB, Sartwell SA, Ordemann EV, Porazinska DL, Farrer EC, King AJ, Spasojevic MJ, Smith JG, Suding KN, Schmidt SK (2018) Patterns of root colonization by arbuscular mycorrhizal fungi and dark septate endophytes across a mostly-unvegetated, high-elevation landscape. Fungal Ecol 36:63–74

de Oliveira Araújo É (2015) Rhizobacteria in the control of pest insects in agriculture. African J Plant Sci 9(9):368–373

Dey P, Datta D, Saha D, Parida S, Panda D (2019) Plant-endophyte interaction and its application to abiotic stress management of crop plants. Int J Curr Microbiol App Sci 8(7):2708–2716

Ernst M, Neubert K, Mendgen KW, Wirsel SG (2011) Niche differentiation of two sympatric species of *Microdochium* colonizing the roots of common reed. BMC Microbiol 11(1):1–13

Ethur LZ, Blume E, Muniz MFB, Flores MGV (2007) Selection of fungal antagonists to *Fusarium solani* and *Fusarium oxysporum* commercial substrate for seedlings. Sci Rural 37(6):1794–1797

Franken P (2012) The plant strengthening root endophyte *Piriformospora indica*: potential application and the biology behind. Appl Microbiol Biotechnol 96(6):1455–1464

Gai CS, Dini-Andreote F, Andreote FD, Lopes JRS, Araújo WL, Miller TA, Azevedo JL, Lacava PT (2011) Endophytic bacteria associated to sharpshooters (Hemiptera: Cicadellidae), insect vectors of *Xylella fastidiosa* subsp. *pauca*. J Pathol Microbiol 2(109):2

Gange AC (2001) Species-specific responses of a root-and shoot-feeding insect to arbuscular mycorrhizal colonization of its host plant. New Phytol 150(3):611–618

Garrido-Jurado I, Resquín-Romero G, Amarilla SP, Ríos-Moreno A, Carrasco L, Quesada-Moraga E (2017) Transient endophytic colonization of melon plants by entomopathogenic fungi after foliar application for the control of *Bemisia tabaci* Gennadius (Hemiptera: Aleyrodidae). J Pest Sci 90(1):319–330

Gathage JW, Lagat ZO, Fiaboe KKM, Akutse KS, Ekesi S, Maniania NK (2016) Prospects of fungal endophytes in the control of *Liriomyza* leafminer flies in common bean *Phaseolus vulgaris* under field conditions. BioControl 61(6):741–753

Gibson DM, Donzelli BG, Krasnoff SB, Keyhani NO (2014) Discovering the secondary metabolite potential encoded within entomopathogenic fungi. Nat Prod Rep 31(10):1287–1305

Glare T, Caradus J, Gelernter W, Jackson T, Keyhani N, Köhl J, Marrone P, Morin L, Stewart A (2012) Have biopesticides come of age? Trends Biotechnol 30(5):250–258

Golo PS, Gardner DR, Grilley MM, Takemoto JY, Krasnoff SB, Pires MS, Fernandes EK, Bittencourt VR, Roberts DW (2014) Production of destruxins from *Metarhizium* spp. fungi in artificial medium and in endophytically colonized cowpea plants. PLoS One 9(8):e104946

Gómez-Vidal S, Salinas J, Tena M, Lopez-Llorca LV (2009) Proteomic analysis of date palm (*Phoenix dactylifera* L.) responses to endophytic colonization by entomopathogenic fungi. Electrophoresis 30(17):2996–3005

González-Mas N, Cuenca-Medina M, Gutiérrez-Sánchez F, Quesada-Moraga E (2019a) Bottom-up effects of endophytic *Beauveria bassiana* on multitrophic interactions between the cotton aphid, *Aphis gossypii*, and its natural enemies in melon. J Pest Sci 92:1271–1281

González-Mas N, Sánchez-Ortiz A, Valverde-García P, Quesada-Moraga E (2019b) Effects of endophytic entomopathogenic ascomycetes on the life-history traits of *Aphis gossypii* Glover and its interactions with melon plants. Insects 10(6):165

Greenfield M, Gómez-Jiménez MI, Ortiz V, Vega FE, Kramer M, Parsa S (2016) *Beauveria bassiana* and *Metarhizium anisopliae* endophytically colonize cassava roots following soil drench inoculation. Biol Control 95:40–48

Griffin MR (2007) Beauveria bassiana, a cotton endophyte with biocontrol activity against seedling disease. PhD dissertation, University of Tennessee

Griffin MR, Ownley BH, Klingeman WE, Pereira RM (2006) Evidence of induced systemic resistance with *Beauveria bassiana* against *Xanthomonas* in cotton. Phytopathology 96(6)

Grigoletti Junior A, dos Santos ÁF, Auer CG (2000) Perspectives on the use of biological control against forest diseases. Floresta 30(1/2):155–165

Grove JF, Pople M (1980) The insecticidal activity of beauvericin and the enniatin complex. Mycopathologia 70(2):103–105

Guo LD, Huang GR, Wang Y (2008) Seasonal and tissue age influences on endophytic fungi of *Pinus tabulaeformis* (Pinaceae) in the Dongling Mountains, Beijing. J Integr Plant Biol 50 (8):997–1003

Gupta VK (2016) New and future developments in microbial biotechnology and bioengineering: *aspergillus* system properties and applications. Elsevier, Amsterdam

Gurulingappa P, Sword GA, Murdoch G, McGee PA (2010) Colonization of crop plants by fungal entomopathogens and their effects on two insect pests when in planta. Biol Control 55(1):34–41

Haggag WM (2010) Role of entophytic microorganisms in biocontrol of plant diseases. Life Sci J 7 (2):57–62

Hamayun M, Khan SA, Ahmad N, Tang DS, Kang SM, Na CI, Sohn EY, Hwang YH, Shin DH, Lee BH, Kim JG (2009) *Cladosporium sphaerospermum* as a new plant growth-promoting endophyte from the roots of *Glycine max* (L.) Merr. World J Microbiol Biotechnol 25(4):627–632

Hamilton CE, Gundel PE, Helander M, Saikkonen K (2012) Endophytic mediation of reactive oxygen species and antioxidant activity in plants: a review. Fungal Divers 54(1):1–10

Harding DP, Raizada MN (2015) Controlling weeds with fungi, bacteria and viruses: a review. Front Plant Sci 6:659

Hardoim PR, Van Overbeek LS, Berg G, Pirttilä AM, Compant S, Campisano A, Döring M, Sessitsch A (2015) The hidden world within plants: ecological and evolutionary considerations for defining functioning of microbial endophytes. Microbiol Mol Biol Rev 79(3):293–320

Harrison JG, Griffin EA (2020) The diversity and distribution of endophytes across biomes, plant phylogeny and host tissues: how far have we come and where do we go from here? Environ Microbiol

Hata K, Sone K (2008) Isolation of endophytes from leaves of *Neolitsea sericea* in broadleaf and conifer stands. Mycoscience 49(4):229–232

Hata K, Atari R, Sone K (2002) Isolation of endophytic fungi from leaves of *Pasania edulis* and their within-leaf distributions. Mycoscience 43(5):0369–0373

Hawksworth DL (2001) The magnitude of fungal diversity: the 1.5 million species estimate revisited. Mycol Res 105(12):1422–1432

Helgason T, Daniell TJ, Husband R, Fitter AH, Young JPW (1998) Ploughing up the wood-wide web? Nature 394(6692):431–431

Hill PW, Marsden KA, Jones DL (2013) How significant to plant N nutrition is the direct consumption of soil microbes by roots? New Phytol 199(4):948–955

Hirano E, Koike M, Aiuchi D, Tani M (2008) Pre-inoculation of cucumber roots with *Verticillium lecanii* (*Lecanicillium muscarium*) induces resistance to powdery mildew. Res Bull Obihiro Univ 29:82–94

Hoffmann D, Vierheilig H, Schausberger P (2011) Arbuscular mycorrhiza enhances preference of ovipositing predatory mites for direct prey-related cues. Physiol Entomol 36(1):90–95

Huergo LF, Monteiro RA, Bonatto AC, Rigo LU, Steffens MBR, Cruz LM, Chubatsu LS, Souza EM, Pedrosa FO (2008) Regulation of nitrogen fixation in *Azospirillum brasilense*. CASSÁN, FD; GARCIA DE SALAMONE,I. *Azospirillum sp.*: cell physiology, plant interactions and agronomic research in Argentina. Asociación Argentina de Microbiologia, Buenos Aires, pp 17–35

Humber RA (2007) Reecent phylogenetically based reclassifications of fungal pathogens of invertebrates. ARSEF, New York, p 5. http://www.sipweb.org/fungi/humber.pdf

Ikram M, Ali N, Jan G, Jan FG, Rahman IU, Iqbal A, Hamayun M (2018) IAA producing fungal endophyte *Penicillium roqueforti* Thom., enhances stress tolerance and nutrients uptake in wheat plants grown on heavy metal contaminated soils. PLoS One 13(11):e0208150

Irizarry I, White JF (2017) Application of bacteria from non-cultivated plants to promote growth, alter root architecture and alleviate salt stress of cotton. J Appl Microbiol 122(4):1110–1120

Irizarry I, White JF (2018) *Bacillus amyloliquefaciens* alters gene expression, ROS production and lignin synthesis in cotton seedling roots. J Appl Microbiol 124(6):1589–1603

Jaber LR (2015) Grapevine leaf tissue colonization by the fungal entomopathogen *Beauveria bassiana* and its effect against downy mildew. BioControl 60(1):103–112

Jaber LR, Araj SE (2018) Interactions among endophytic fungal entomopathogens (Ascomycota: Hypocreales), the green peach aphid *Myzuspersicae* Sulzer (Homoptera: Aphididae), and the aphid endoparasitoid *Aphidiuscolemani* Viereck (hymenoptera: Braconidae). Biol Control 116:53–61

Jaber LR, Enkerli J (2016) Effect of seed treatment duration on growth and colonization of *Vicia faba* by endophytic *Beauveria bassiana* and *Metarhizium brunneum*. Biol Control 103:187–195

Jaber LR, Enkerli J (2017) Fungal entomopathogens as endophytes: can they promote plant growth? Biocontrol Sci Tech 27(1):28–41

Jaber LR, Ownley BH (2018) Can we use entomopathogenic fungi as endophytes for dual biological control of insect pests and plant pathogens? Biol Control 116:36–45

Jaber LR, Salem NM (2014) Endophytic colonisation of squash by the fungal entomopathogen *Beauveria bassiana* (Ascomycota: Hypocreales) for managing zucchini yellow mosaic virus in cucurbits. Biocontrol Sci Tech 24(10):1096–1109

Jacoby R, Peukert M, Succurro A, Koprivova A, Kopriva S (2017) The role of soil microorganisms in plant mineral nutrition—current knowledge and future directions. Front Plant Sci 8:1617

Jaronski ST (2010) Ecological factors in the inundative use of fungal entomopathogens. BioControl 55(1):159–185

Ji SH, Gururani MA, Chun SC (2014) Isolation and characterization of plant growth promoting endophytic diazotrophic bacteria from Korean rice cultivars. Microbiol Res 169(1):83–98

Johnston-Monje D, Raizada MN (2011) Conservation and diversity of seed associated endophytes in *Zea* across boundaries of evolution. Ethnography ecology. PLoS One 6(6):e20396

Joshi S, Sahgal M, Sahu S, Prakash A (2018) Fungal endophytes and their secondary metabolites: role in sustainable agriculture. In: Fungi and their role in sustainable development: current perspectives. Springer, Singapore, pp 121–146

Kandel SL, Joubert PM, Doty SL (2017) Bacterial endophyte colonization and distribution within plants. Microorganisms 5(4):77

Kavehnia A, Samsampour D, Askari Seyahooei M (2018a) Isolation, morphological and molecular identification of some important fungal endophytes of *Pistacia mutica* in Hormozgan province and their effects on inducing salinity stress tolerance in pepper seedling var. California wonder310. MSc thesis. Hormozgan University

Kavehnia A, Samsampour D, Askari Seyahooei M (2018b) Isolation and identification of morphological and molecular characteristics of some important endophyte bacteria of *Amygdalus scoparia* in Hormozgan province and their effect on increasing inducing salinity tolerance in cucumber var. Emperor MSc thesis Hormozgan University

Kiarie S, Nyasani JO, Gohole LS, Maniania NK, Subramanian S (2020) Impact of fungal endophyte colonization of maize (Zea mays L.) on induced resistance to thrips-and aphid-transmitted viruses. Plants 9(4):416

Kogan A, Grachev SV, Eliseeva SV, Bolevich S (1997) Carbon dioxide—a universal inhibitor of the generation of active oxygen forms by cells (deciphering one enigma of evolution). Izvestiia Akademii nauk. Seriia Biologicheskaia (2):204–217

Korhonen K (1978) Intersterility groups of *Heterobasidion annosum* (Vol. 94, no. 6). Communicationes Instituti Forestalis Fenniae 94:1–25

Kowalski KP, Bacon C, Bickford W, Braun H, Clay K, Leduc-Lapierre M, Lillard E, McCormick MK, Nelson E, Torres M, White J (2015) Advancing the science of microbial symbiosis to support invasive species management: a case study on *Phragmites* in the Great Lakes. Front Microbiol 6:95

Krassilstschik J (1888) La production industrielle des parasites végétaux pour la destruction des insectes nuisibles. Bull Biol Fr Belg 19:461–472

Kuklinsky-Sobral J, Araújo WL, Mendes R, Geraldi IO, Pizzirani-Kleiner AA, Azevedo JL (2004) Isolation and characterization of soybean-associated bacteria and their potential for plant growth promotion. Environ Microbiol 6(12):1244–1251

Kumar S, Kaushik N, Edrada-Ebel R, Ebel R, Proksch P (2008) Endophytic fungi for pest and disease management. In: Integrated management of diseases caused by fungi, phytoplasma and bacteria. Springer, Dordrecht, pp 365–387

Kusari S, Pandey SP, Spiteller M (2013) Untapped mutualistic paradigms linking host plant and endophytic fungal production of similar bioactive secondary metabolites. Phytochemistry 91:81–87

Lacava PT, Azevedo JL (2014) Biological control of insect-pest and diseases by endophytes. In: Advances in Endophytic research. Springer, New Delhi, pp 231–256

Lacey LA, Grzywacz D, Shapiro-Ilan DI, Frutos R, Brownbridge M, Goettel MS (2015) Insect pathogens as biological control agents: back to the future. J Invertebr Pathol 132:1–41

Landum MC, do Rosário Félix M, Alho J, Garcia R, Cabrita MJ, Rei F, Varanda CM (2016) Antagonistic activity of fungi of *Olea europaea* L. against *Colletotrichum acutatum*. Microbiol Res 183:100–108

Leckie BM, Ownley BH, Pereira RM, Klingeman WE, Jones CJ, Gwinn KD (2008) Mycelia and spent fermentation broth of *Beauveria bassiana* incorporated into synthetic diets affect mortality, growth and development of larval *Helicoverpa zea* (Lepidoptera: Noctuidae). Biocontrol Sci Tech 18(7):697–710

Leconte JL (1873) Hints for the promotion of economic entomology in the United States. Am Nat 7 (12):710–722

Lee S, Flores-Encarnacion M, Contreras-Zentella M, Garcia-Flores L, Escamilla JE, Kennedy C (2004) Indole-3-acetic acid biosynthesis is deficient in *Gluconacetobacter diazotrophicus* strains with mutations in cytochrome c biogenesis genes. J Bacteriol 186(16):5384–5391

Lehtonen PT, Helander M, Siddiqui SA, Lehto K, Saikkonen K (2006) Endophytic fungus decreases plant virus infections in meadow ryegrass (*Lolium pratense*). Biol Lett 2(4):620–623

Leuchtmann A (1993) Systematics, distribution, and host specificity of grass endophytes. Nat Toxins 1(3):150–162

Lewis LC, Cossentine JE (1986) Season long intraplant epizootics of entomopathogens, Beauveria bassiana and Nosema pyrausta, in a corn agroecosystem [European cornborer (Ostrinia nubilalis)]. Entomophaga 31(4):363–369

Li XJ, Zhang Q, Zhang AL, Gao JM (2012) Metabolites from *Aspergillus fumigatus*, an endophytic fungus associated with *Melia azedarach*, and their antifungal, antifeedant, and toxic activities. J Agric Food Chem 60(13):3424–3431

Li H, Guan Y, Dong Y, Zhao L, Rong S, Chen W, Lv M, Xu H, Gao X, Chen R, Li L (2018a) Isolation and evaluation of endophytic *Bacillus tequilensis* GYLH001 with potential application for biological control of *Magnaporthe oryzae*. PLoS One 13(10):e0203505

Li X, Wang J, Zhang S, Wang H, Li X, Li X, Zhang H (2018b) Distribution of fungal endophytes in roots of *Stipa krylovii* across six vegetation types in grassland of northern China. Fungal Ecol 31:47–53

Liu L, Liu S, Chen X, Guo L, Che Y (2009) Pestalofones A–E, bioactive cyclohexanone derivatives from the plant endophytic fungus *Pestalotiopsis fici*. Bioorg Med Chem 17(2):606–613

Lopez DC, Sword GA (2015) The endophytic fungal entomopathogens *Beauveria bassiana* and *Purpureocillium lilacinum* enhance the growth of cultivated cotton (*Gossypium hirsutum*) and negatively affect survival of the cotton bollworm (*Helicoverpa zea*). Biol Control 89:53–60

Lubna AS, Hamayun M, Gul H, Lee IJ, Hussain A (2018) *Aspergillus niger* CSR3 regulates plant endogenous hormones and secondary metabolites by producing gibberellins and indoleacetic acid. J Plant Interact 13(1):100–111

Ma XM, Liu XX, Ning X, Zhang B, Han F, Guan XM, Tan YF, Zhang QW (2008) Effects of *Bacillus thuringiensis* toxin Cry1Ac and *Beauveria bassiana* on Asiatic corn borer (Lepidoptera: Crambidae). J Invertebr Pathol 99(2):123–128

Ma Y, Oliveira RS, Freitas H, Zhang C (2016) Biochemical and molecular mechanisms of plant-microbe-metal interactions: relevance for phytoremediation. Front Plant Sci 7:918

Macedo CLD, Martins ÉS, Macedo LLPD, Santos ACD, Praça LB, Góis LABD, Monnerat RG (2012) Seleção e caracterização de estirpes de Bacillus thuringiensis eficientes contra a Diatraea saccharalis (Lepidoptera: Crambidae). Pesqui Agropecu Bras 47(12):1759–1765

Malinowski DP, Alloush GA, Belesky DP (2000) Leaf endophyte *Neotyphodium coenophialum* modifies mineral uptake in tall fescue. Plant Soil 227(1–2):115–126

Mandyam K, Jumpponen A (2005) Seeking the elusive function of the root-colonising dark septate endophytic fungi. Stud Mycol 53:173–189

Mantzoukas S, Eliopoulos PA (2020) Endophytic entomopathogenic fungi: A valuable biological control tool against plant pests. Appl Sci 10(1):360

Mantzoukas S, Chondrogiannis C, Grammatikopoulos G (2015) Effects of three endophytic entomopathogens on sweet sorghum and on the larvae of the stalk borer *Sesamia nonagrioides*. Entomol Exp Appl 154(1):78–87

Martinuz A, Schouten A, Menjivar RD, Sikora RA (2012) Effectiveness of systemic resistance toward *Aphis gossypii* (Hom., Aphididae) as induced by combined applications of the endophytes *fusarium oxysporum* Fo162 and *rhizobium etli* G12. Biol Control 62(3):206–212

Mejía LC, Rojas EI, Maynard Z, Van Bael S, Arnold AE, Hebbar P, Samuels GJ, Robbins N, Herre EA (2008) Endophytic fungi as biocontrol agents of *Theobroma cacao* pathogens. Biol Control 46(1):4–14

Melatti VM, Praça LB, Martins ES, Sjii E, Berry C, de Pontes RGMS (2010) Selection of Bacillus thuringiensis strains toxic against cotton aphid, Aphis gossypii glover (Hemiptera: Aphididae). Embrapa Recursos Genéticos e Biotecnologia-Artigo em periódico indexado (ALICE)

Mendes R, Azevedo JL (2007) Valor biotecnológico de fungos endofíticos isolados de plantas de interesse econômico. Micologia: avanços no conhecimento. Sociedade Brasileira de Micologia, Recife, Brasil, pp 129–140

Miles CO, di Menna ME, Jacobs SW, Garthwaite I, Lane GA, Prestidge RA, Marshall SL, Wilkinson HH, Schardl CL, Ball OJP, Latch GC (1998) Endophytic fungi in indigenous Australasian grasses associated with toxicity to livestock. Appl Environ Microbiol 64 (2):601–606

Miller JD (1986) Toxic metabolites of epiphytic and endophytic fungi of conifer needles. In: Fokkema NJ, Van Den Heuvel J (eds) Microbiology of the phyllosphere

Mocali S, Bertelli E, Di Cello F, Mengoni A, Sfalanga A, Viliani F, Caciotti A, Tegli S, Surico G, Fani R (2003) Fluctuation of bacteria isolated from elm tissues during different seasons and from different plant organs. Res Microbiol 154(2):105–114

Moussa M, Ebrahim W, El-Neketi M, Mándi A, Kurtán T, Hartmann R, Lin W, Liu Z, Proksch P (2016) Tetrahydroanthraquinone derivatives from the mangrove-derived endophytic fungus *Stemphylium globuliferum*. Tetrahedron Lett 57(36):4074–4078

Mutune B, Ekesi S, Niassy S, Matiru V, Bii C, Maniania NK (2016) Fungal endophytes as promising tools for the management of bean stem maggot *Ophiomyia phaseoli* on beans *Phaseolus vulgaris*. J Pest Sci 89(4):993–1001
Muvea AM (2015) The effect of fungal endophytes on thrips and tospovirus epidemiology (doctoral dissertation). Gottfried Wilhelm Leibniz Universität Hannover, Hannover
Muvea AM, Meyhöfer R, Subramanian S, Poehling HM, Ekesi S, Maniania NK (2014) Colonization of onions by endophytic fungi and their impacts on the biology of *Thrips tabaci*. PLoS One 9(9):e108242
Myers SS, Zanobetti A, Kloog I, Huybers P, Leakey AD, Bloom AJ, Carlisle E, Dietterich LH, Fitzgerald G, Hasegawa T, Holbrook NM (2019) Increasing CO2 threatens human nutrition (vol 510, pg 139, 2014). Nature 574(7778):E14–E14
Nahar PB, Kulkarni SA, Kulye MS, Chavan SB, Kulkarni G, Rajendran A, Yadav PD, Shouche Y, Deshpande MV (2008) Effect of repeated in vitro sub-culturing on the virulence of *Metarhizium anisopliae* against *Helicoverpa armigera* (Lepidoptera: Noctuidae). Biocontrol Sci Tech 18 (4):337–355
Nakkeeran S, Vinodkumar S, Renukadevi P, Rajamanickam S, Jogaiah S (2019) Bioactive molecules from *Bacillus spp.*: an effective tool for plant stress management. In: Bioactive molecules in plant defense. Springer, Cham, pp 1–23
Newman DJ, Cragg GM, Snader KM (2003) Natural products as sources of new drugs over the period 1981–2002. J Nat Prod 66(7):1022–1037
Nisa H, Kamili AN, Nawchoo IA, Shafi S, Shameem N, Bandh SA (2015) Fungal endophytes as prolific source of phytochemicals and other bioactive natural products: a review. Microb Pathog 82:50–59
Ongena M, Jacques P (2008) *Bacillus lipopeptides*: versatile weapons for plant disease biocontrol. Trends Microbiol 16(3):115–125
Ownley BH, Griffin MR, Klingeman WE, Gwinn KD, Moulton JK, Pereira RM (2008) *Beauveria bassiana*: endophytic colonization and plant disease control. J Invertebr Pathol 98(3):267–270
Ownley BH, Gwinn KD, Vega FE (2010) Endophytic fungal entomopathogens with activity against plant pathogens: ecology and evolution. BioControl 55(1):113–128
Parsa S, Ortiz V, Vega FE (2013) Establishing fungal entomopathogens as endophytes: towards endophytic biological control. J Visual Exp 74:e50360
Parsa S, Ortiz V, Gómez-Jiménez MI, Kramer M, Vega FE (2018) Root environment is a key determinant of fungal entomopathogen endophytism following seed treatment in the common bean, *Phaseolus vulgaris*. Biol Control 116:74–81
Pasteur L (1874) Observations (au sujet des conclusions de M. Dumas) relatives au phylloxera. C R Hebd Seances Acad Sci 79:1233–1234
Patiño-Ruiz JD, Schausberger P (2014) Spider mites adaptively learn recognizing mycorrhiza-induced changes in host plant volatiles. Exp Appl Acarol 64(4):455–463
Patterson CG, Potter DA, Fannin FF (1991) Feeding deterrency of alkaloids from endophyte-infected grasses to Japanese beetle grubs. Entomol Exp Appl 61(3):285–289
Paungfoo-Lonhienne C, Schmidt S, Webb RI, Lonhienne TG (2013) Rhizophagy—a new dimension of plant–microbe interactions. Mol Microb Ecol Rhizosphere 1:1199–1207
Peikari A, Samsampour D, Askari Seyahooei M (2018) Isolation and identification of morphological and molecular characteristics of some important endophyte bacteria of Amygdalus scoparia in Hormozgan province and their effect on increasing inducing salinity tolerance in cucumber var. Emperor. MSc thesis. Hormozgan University
Peixoto Neto PAS, Azevedo JL, Araujo WL (2002) Endophytic microorganisms. Biotechnol Sci Dev 29:62–77
Petrini O (1991) Fungal endophytes of tree leaves. In: Microbial ecology of leaves. Springer, New York, NY, pp 179–197
Petrini O, Sieber TN, Toti L, Viret O (1993) Ecology, metabolite production, and substrate utilization in endophytic fungi. Nat Toxins 1(3):185–196

Pirelli GJ, Anderson NP, Craig AM, Young CA (2016) Endophyte toxins in grass and other feed sources

Pirttilä AM, Laukkanen H, Pospiech H, Myllylä R, Hohtola A (2000) Detection of intracellular bacteria in the buds of *Scotch pine* (*Pinus sylvestris* L.) by in situ hybridization. Appl Environ Microbiol 66(7):3073–3077

Pirttilä AM, Pospiech H, Laukkanen H, Myllylä R, Hohtola A (2003) Two endophytic fungi in different tissues of Scots pine buds (*Pinus sylvestris* L.). Microb Ecol 45(1):53–62

Pirttilä AM, Podolich O, Koskimäki JJ, Hohtola E, Hohtola A (2008) Role of origin and endophyte infection in browning of bud-derived tissue cultures of Scots pine (*Pinus sylvestris* L.). Plant Cell Tissue Organ Cult 95(1):47–55

Polanczyk RA, da Silva RFP, Fiuza LM (2003) Screening of *Bacillus thuringiensis* isolates pathogenic to *Spodoptera frugiperda* (JE Smith) (Lepidoptera: Noctuidae). Arq Inst Biol 70:69–72

Posada F, Aime MC, Peterson SW, Rehner SA, Vega FE (2007) Inoculation of coffee plants with the fungal entomopathogen *Beauveria bassiana* (Ascomycota: Hypocreales). Mycol Res 111 (6):748–757

Powell WA, Klingeman WE, Ownley BH, Gwinn KD, Dee M, Flanagan PC (2007) Endophytic Beauveria bassiana in tomatoes yields mycosis in tomato fruitworm larvae. HortScience 42 (4):933–933

Praça LB (2012) Interactions between *Bacillus thuringiensis* strains and hybrids of cabbage for the control of *Plutella xylostella* and plant growth promotion. Brasília: UNB, 2012 (Doctoral dissertation, Thesis (doctoral)-Faculty of Agronomy and Veterinary Medicine, University of Brasília, Brasília. 141p)

Praça LB, Batista AC, Martins ES, Siqueira CB, Dias DGS, Gomes ACMM, Falcão R, Monnerat RG (2004) Effective strains of *Bacillus thuringiensis* against insects of orders Lepidoptera, Coleoptera and Diptera. Pesq Agrop Braz 39(1):11–16

Prakash GB, Srinivasan T (2020) Endophytic secondary metabolites for biological control: a latest perspective. In: Plant microbiome paradigm. Springer, Cham, pp 293–307

Prestidge RA, Gallagher RT (1988) Endophyte fungus confers resistance to ryegrass: argentine stem weevil larval studies. Ecol Entomol 13(4):429–435

Puopolo G, Cimmino A, Palmieri MC, Giovannini O, Evidente A, Pertot I (2014) Lysobacter capsici AZ78 produces cyclo (l-Pro-l-Tyr), a 2, 5-diketopiperazine with toxic activity against sporangia of Phytophthora infestans and Plasmopara viticola. J Appl Microbiol 117(4):1168–1180

Qayyum MA, Wakil W, Arif MJ, Sahi ST, Dunlap CA (2015) Infection of *Helicoverpa armigera* by endophytic *Beauveria bassiana* colonizing tomato plants. Biol Control 90:200–207

Quadt-Hallmann A, Hallmann J, Kloepper JW (1997) Bacterial endophytes in cotton: location and interaction with other plant-associated bacteria. Can J Microbiol 43(3):254–259

Quesada-Moraga E, Landa BB, Muñoz-Ledesma J, Jiménez-Diáz RM, Santiago-Alvarez C (2006) Endophytic colonisation of *opium poppy*, *Papaver somniferum*, by an entomopathogenic *Beauveria bassiana* strain. Mycopathologia 161(5):323–329

Quesada-Moraga E, Munoz-Ledesma FJ, Santiago-Alvarez C (2009) Systemic protection of Papaver somniferum L. against Iraella luteipes (Hymenoptera: Cynipidae) by an endophytic strain of Beauveria bassiana (Ascomycota: Hypocreales). Environ Entomol 38(3):723–730

Rabiey M, Hailey LE, Roy SR, Grenz K, Al-Zadjali MA, Barrett GA, Jackson RW (2019) Endophytes vs tree pathogens and pests: can they be used as biological control agents to improve tree health? Eur J Plant Pathol:1–19

Resquín-Romero G, Garrido-Jurado I, Delso C, Ríos-Moreno A, Quesada-Moraga E (2016) Transient endophytic colonizations of plants improve the outcome of foliar applications of mycoinsecticides against chewing insects. J Invertebr Pathol 136:23–31

Richter DL, Dixon TG, Smith JK (2016) Revival of saprotrophic and mycorrhizal basidiomycete cultures after 30 years in cold storage in sterile water. Can J Microbiol 62(11):932–937

Riedell WE, Kieckhefer RE, Petroski RJ, Powell RG (1991) Naturally-occurring and synthetic loline alkaloid derivatives: insect feeding behavior modification and toxicity. J Entomol Sci 26 (1):122–129

Ríos-Moreno A, Garrido-Jurado I, Resquín-Romero G, Arroyo-Manzanares N, Arce L, Quesada-Moraga E (2016) Destruxin A production by *Metarhizium brunneum* strains during transient endophytic colonisation of *Solanum tuberosum*. Biocontrol Sci Tech 26(11):1574–1585

Roberts DW (1981) Toxins of entomopathogenic fungi. In: Burges HD (ed) Microbial control of pests and plant diseases 1970–1980, pp 441–464

Rodrigues KF, Hesse M, Werner C (2000) Antimicrobial activities of secondary metabolites produced by endophytic fungi from *Spondias mombin*. J Basic Microbiol 40(4):261–267

Rodriguez H, Gonzalez T, Goire I, Bashan Y (2004) Gluconic acid production and phosphate solubilization by the plant growth-promoting bacterium *Azospirillum* spp. Naturwissenschaften 91(11):552–555

Rondot Y, Reineke A (2018) Endophytic *Beauveria bassiana* in grapevine *Vitis vinifera* (L.) reduces infestation with piercing-sucking insects. Biol Control 116:82–89

Rosenblueth M, Martínez-Romero E (2006) Bacterial endophytes and their interactions with hosts. Mol Plant-Microbe Interact 19:827–837

Roy HE, Pell JK (2000) Interactions between entomopathogenic fungi and other natural enemies: implications for biological control. Biocontrol Sci Tech 10(6):737–752

Rúa MA, McCulley RL, Mitchell CE (2013) Fungal endophyte infection and host genetic background jointly modulate host response to an aphid-transmitted viral pathogen. J Ecol 101 (4):1007–1018

Russo ML, Pelizza SA, Cabello MN, Stenglein SA, Scorsetti AC (2015) Endophytic colonisation of tobacco, corn, wheat and soybeans by the fungal entomopathogen *Beauveria bassiana* (Ascomycota, Hypocreales). Biocontrol Sci Tech 25(4):475–480

Russo ML, Scorsetti AC, Vianna MF, Cabello M, Ferreri N, Pelizza S (2019) Endophytic effects of *Beauveria bassiana* on corn (*Zea mays*) and its herbivore, *Rachiplusia nu* (lepidoptera: Noctuidae). Insects 10(4):110

Ryan RP, Germaine K, Franks A, Ryan DJ, Dowling DN (2008) Bacterial endophytes: recent developments and applications. FEMS Microbiol Lett 278(1):1–9

Sadeghi F, Samsampour D, Seyahooei MA, Bagheri A, Soltani J (2019) Diversity and spatiotemporal distribution of fungal endophytes associated with *Citrus reticulata* cv. Siyahoo Curr Microbiol 76(3):279–289

Sadeghi F, Samsampour D, Seyahooei MA, Bagheri A, Soltani J (2020) Fungal endophytes alleviate drought-induced oxidative stress in mandarin (*Citrus reticulata* L.): toward regulating the ascorbate–glutathione cycle. Sci Hortic 261:108991

Saikkonen K, Faeth SH, Helander M, Sullivan TJ (1998) Fungal endophytes: a continuum of interactions with host plants. Annu Rev Ecol Syst 29(1):319–343

Saikkonen K, Wäli P, Helander M, Faeth SH (2004) Evolution of endophyte–plant symbioses. Trends Plant Sci 9(6):275–280

Sánchez-Rodríguez AR, Raya-Díaz S, Zamarreño ÁM, García-Mina JM, del Campillo MC, Quesada-Moraga E (2018) An endophytic *Beauveria bassiana* strain increases spike production in bread and durum wheat plants and effectively controls cotton leafworm (*Spodoptera littoralis*) larvae. Biol Control 116:90–102

Sasan RK, Bidochka MJ (2012) The insect-pathogenic fungus *Metarhizium robertsii* (Clavicipitaceae) is also an endophyte that stimulates plant root development. Am J Bot 99 (1):101–107

Schausberger P, Peneder S, Jürschik S, Hoffmann D (2012) Mycorrhiza changes plant volatiles to attract spider mite enemies. Funct Ecol 26(2):441–449

Schulz B, Boyle C (2005) The endophytic continuum. Mycol Res 109(6):661–686

Schulz B, Haas S, Junker C, Andrée N, Schobert M (2015) Fungal endophytes are involved in multiple balanced antagonisms. Curr Sci:39–45

Scorsetti AC, Pelizza S, Vianna F, Fogel MN, Schneider MI (2017) Interactions between the entomopathogenic fungus *Beauveria bassiana* and the Neotropical predator *Eriopis connexa* (Coleoptera: Coccinellidae): implications in biological control of pest. J Plant Prot Res

Senthilkumar N, Murugesan S, Babu DS, Rajeshkannan C (2014) GC-MS analysis of the extract of endophytic fungus, *Phomopsis* sp. isolated from tropical tree species of India, *Tectona grandis* L. Int J Innov Sci Eng Technol 3:10176–10179

Sevilla M, Burris RH, Gunapala N, Kennedy C (2001) Comparison of benefit to sugarcane plant growth and 15N2 incorporation following inoculation of sterile plants with *Acetobacter diazotrophicus* wild-type and nif mutant strains. Mol Plant-Microbe Interact 14(3):358–366

Shrivastava G, Ownley BH, Augé RM, Toler H, Dee M, Vu A, Köllner TG, Chen F (2015) Colonization by arbuscular mycorrhizal and endophytic fungi enhanced terpene production in tomato plants and their defense against a herbivorous insect. Symbiosis 65(2):65–74

Shymanovich T, Faeth SH (2019) Environmental factors affect the distribution of two Epichloë fungal endophyte species inhabiting a common host grove bluegrass (Poa alsodes). Ecol Evol 9 (11):6624–6642

Shweta S, Bindu JH, Raghu J, Suma HK, Manjunatha BL, Kumara PM, Ravikanth G, Nataraja KN, Ganeshaiah KN, Shaanker RU (2013) Isolation of endophytic bacteria producing the anti-cancer alkaloid camptothecine from *Miquelia dentata* Bedd.(Icacinaceae). Phytomedicine 20 (10):913–917

Sieber TN (2007) Endophytic fungi in forest trees: are they mutualists? Fungal Biol Rev 21 (2–3):75–89

Siegel MR, Bush LP (1996) Defensive chemicals in grass-fungal endophyte associations. In: Phytochemical diversity and redundancy in ecological interactions. Springer, Boston, MA, pp 81–119

Siegel MR, Latch GCM, Bush LP, Fannin FF, Rowan DD, Tapper BA, Bacon CW, Johnson MC (1990) Fungal endophyte-infected grasses: alkaloid accumulation and aphid response. J Chem Ecol 16(12):3301–3315

Silva ACL, Silva GA, Abib PHN, Carolino AT, Samuels RI (2020) Endophytic colonization of tomato plants by the entomopathogenic fungus *Beauveria bassiana* for controlling the south American tomato pinworm, *Tuta absoluta*. CABI Agric Biosci 1(1):1–9

Souza ML (2001) Use of endophytic microorganisms in agriculture. Biotechnol Sci Dev 21:28–31

Souza AQL, Astolfi Filho S, Belem Pinheiro ML, Sarquis MIM, Pereira JO (2004) Antimicrobial activity of endophytic fungi isolated from toxic plants in the Amazon: *Palicourea longiflora* (aubl.) rich and *Strychnos cogens* bentham. Acta Amazon 34(2):185–195

Strobel GA (2003) Endophytes as sources of bioactive products. Microbes Infect 5(6):535–544

Sturz A, Christie B, Matheson B, Arsenault W, Buchanan N (1999) Endophytic bacterial communities in the periderm of potato tubers and their potential to improve resistance to soil-borne plant pathogens. Plant Pathol 48(3):360–369

Sturz AV, Christie BR, Nowak J (2000) Bacterial endophytes: potential role in developing sustainable systems of crop production. Crit Rev Plant Sci 19(1):1–30

Sun L, Qiu F, Zhang X, Dai X, Dong X, Song W (2008) Endophytic bacterial diversity in rice (*Oryza sativa* L.) roots estimated by 16S rDNA sequence analysis. Microb Ecol 55(3):415–424

Tan RX, Zou WX (2001) Endophytes: a rich source of functional metabolites. Nat Prod Rep 18 (4):448–459

Tedersoo L, Pärtel K, Jairus T, Gates G, Põldmaa K, Tamm H (2009) Ascomycetes associated with ectomycorrhizas: molecular diversity and ecology with particular reference to the Helotiales. Environ Microbiol 11(12):3166–3178

Tefera T, Vidal S (2009) Effect of inoculation method and plant growth medium on endophytic colonization of sorghum by the entomopathogenic fungus *Beauveria bassiana*. BioControl 54 (5):663–669

Thongsandee W, Matsuda Y, Ito S (2012) Temporal variations in endophytic fungal assemblages of *Ginkgo biloba* L. J For Res 17(2):213–218

Thuler RT, Barros R, Mariano R, Vendramim JD (2006) Effect of plant growth-promoting rhizobacteria (PGPR) on the development of *Plutella xylostella* (L.) (Lepidopetera: Plutellidae) in cabbage. Científica (Jaboticabal) 34(2):217–222

Tintjer T, Rudgers JA (2006) Grass–herbivore interactions altered by strains of a native endophyte. New Phytol 170(3):513–521

Vanneste JL, Yu J, Beer SV (1992) Role of antibiotic production by *Erwinia herbicola* Eh252 in biological control of *Erwinia amylovora*. J Bacteriol 174(9):2785–2796

Varma A, Verma S, Sahay N, Bütehorn B, Franken P (1999) *Piriformospora indica*, a cultivable plant-growth-promoting root endophyte. Appl Environ Microbiol 65(6):2741–2744

Vega FE, Posada F, Aime MC, Pava-Ripoll M, Infante F, Rehner SA (2008) Entomopathogenic fungal endophytes. Biol Control 46(1):72–82

Vega FE, Goettel MS, Blackwell M, Chandler D, Jackson MA, Keller S, Koike M, Maniania NK, Monzon A, Onley BH, Pell JK, Rangel DEN, Roy HE (2009) Fungal entomopathogens: new-insights on their ecology. Fungal Ecol 2:149–159

Venkatachalam A, Govinda Rajulu MB, Thirunavukkarasu N, Suryanarayanan TS (2015) Endophytic fungi of marine algae and seagrasses: a novel source of chitin modifying enzymes. Mycosphere 6(3):345–355

Venugopalan A, Srivastava S (2015) Endophytes as in vitro production platforms of high value plant secondary metabolites. Biotechnol Adv 33(6):873–887

Verbruggen E, Van Der Heijden MG, Weedon JT, Kowalchuk GA, Röling WF (2012) Community assembly, species richness and nestedness of arbuscular mycorrhizal fungi in agricultural soils. Mol Ecol 21(10):2341–2353

Verma SC, Ladha JK, Tripathi AK (2001) Evaluation of plant growth promoting and colonization ability of endophytic diazotrophs from deep water rice. J Biotechnol 91(2–3):127–141

Verma VC, Singh SK, Prakash S (2011) Bio-control and plant growth promotion potential of siderophore producing endophytic *Streptomyces* from *Azadirachta indica* A. Juss. J Basic Microbiol 51(5):550–556

Verma SK, Kingsley K, Irizarry I, Bergen M, Kharwar RN, White JF Jr (2017) Seed-vectored endophytic bacteria modulate development of rice seedlings. J Appl Microbiol 122 (6):1680–1691

Verma SK, Kingsley K, Bergen M, English C, Elmore M, Kharwar RN, White JF (2018) Bacterial endophytes from rice cut grass (*Leersia oryzoides* L.) increase growth, promote root gravitropic response, stimulate root hair formation, and protect rice seedlings from disease. Plant Soil 422 (1–2):223–238

Vessey JK (2003) Plant growth promoting rhizobacteria as biofertilizers. Plant Soil 255(2):571–586

Vidal S, Jaber LR (2015) Entomopathogenic fungi as endophytes: plant–endophyte–herbivore interactions and prospects for use in biological control. Curr Sci:46–54

Wagenaar MM, Clardy J (2001) Dicerandrols, new antibiotic and cytotoxic dimers produced by the fungus Phomopsis l ongicolla isolated from an endangered mint. J Nat Prod 64(8):1006–1009

Wagner BL, Lewis LC (2000) Colonization of corn, *Zea mays*, by the entomopathogenic fun gus *Beauveria bassiana*. Appl Environ Microbiol 66(8):3468–3473

Waller F, Achatz B, Baltruschat H, Fodor J, Becker K, Fischer M, Heier T, Hückelhoven R, Neumann C, Von Wettstein D, Franken P (2005) The endophytic fungus *Piriformospora indica* reprograms barley to salt-stress tolerance, disease resistance, and higher yield. Proc Natl Acad Sci 102(38):13386–13391

Wang Y, Dai CC (2011) Endophytes: a potential resource for biosynthesis, biotransformation, and biodegradation. Ann Microbiol 61(2):207–215

Wang Q, Xu L (2012) Beauvericin, a bioactive compound produced by fungi: a short review. Molecules 17(3):2367–2377

Waqas M, Khan AL, Lee IJ (2014) Bioactive chemical constituents produced by endophytes and effects on rice plant growth. J Plant Interact 9(1):478–487

Warner JB, Lolkema JS (2002) Growth of *Bacillus subtilis* on citrate and isocitrate is supported by the Mg2+−citrate transporter CitM. Microbiology 148(11):3405–3412

Webber J (1981) A natural biological control of Dutch elm disease. Nature 292(5822):449–451
White JF Jr, Torres MS, Sullivan RF, Jabbour RE, Chen Q, Tadych M, Irizarry I, Bergen MS, Havkin-Frenkel D, Belanger FC (2014) Occurrence of *Bacillus amyloliquefaciens* as a systemic endophyte of vanilla orchids. Microsc Res Tech 77(11):874–885
White JF, Chen Q, Torres MS, Mattera R, Irizarry I, Tadych M, Bergen M (2015) Collaboration between grass seedlings and rhizobacteria to scavenge organic nitrogen in soils. Ann Botany Plants 7
White JF, Kingsley KL, Verma SK, Kowalski KP (2018) Rhizophagy cycle: an oxidative process in plants for nutrient extraction from symbiotic microbes. Microorganisms 6(3):95
White JF, Kingsley KL, Butterworth S, Brindisi L, Gatei JW, Elmore MT, Verma SK, Yao X, Kowalski KP (2019a) Seed-vectored microbes: their roles in improving seedling fitness and competitor plant suppression. In: Seed endophytes. Springer, Cham, pp 3–20
White JF, Kingsley KL, Zhang Q, Verma R, Obi N, Dvinskikh S, Elmore MT, Verma SK, Gond SK, Kowalski KP (2019b) Endophytic microbes and their potential applications in crop management. Pest Manag Sci 75(10):2558–2565
Wraight SP, Ramos ME (2005) Synergistic interaction between *Beauveria bassiana*-and *Bacillus thuringiensis* tenebrionis-based biopesticides applied against field populations of Colorado potato beetle larvae. J Invertebr Pathol 90(3):139–150
Xin G, Zhang G, Kang JW, Staley JT, Doty SL (2009) A diazotrophic, indole-3-acetic acid-producing endophyte from wild cottonwood. Biol Fertil Soils 45(6):669–674
Yan JF, Broughton SJ, Yang SL, Gange AC (2015) Do endophytic fungi grow through their hosts systemically? Fungal Ecol 13:53–59
Yang P, Sun ZX, Liu SY, Lu HX, Zhou Y, Sun M (2013) Combining antagonistic endophytic bacteria in different growth stages of cotton for control of *Verticillium* wilt. Crop Prot 47:17–23
Zabalgogeazcoa I (2008) Fungal endophytes and their interaction with plant pathogens
Zhang X, Li J, Qi G, Wen K, Lu J, Zhao X (2011) Insecticidal effect of recombinant endophytic bacterium containing *Pinellia ternata* agglutinin against white backed planthopper, *Sogatella furcifera*. Crop Prot 30(11):1478–1484
Zhao X, Qi G, Zhang X, Lan N, Ma X (2010) Controlling sap-sucking insect pests with recombinant endophytes expressing plant lectin. Nature Precedings, p1
Zhou W, Wheeler TA, Starr JL, Valencia CU, Sword GA (2018) A fungal endophyte defensive symbiosis affects plant-nematode interactions in cotton. Plant Soil 422(1–2):251–266

The Symbionts

5

Marzieh Kashkouli, Mohammad Mehrabadi, and
Yaghoub Fathipour

Contents

Abstract

All insects, including insect pests, possess symbiotic bacteria inside their body, particularly those feeding on restricted diets, such as plant sap, vertebrate blood, or woody material. These symbiotic associations span a spectrum of types that differ with respect to the effect of the symbiont on the host, such as providing essential nutrients, defending from natural enemy, increasing host resistance against unfavorable environmental conditions, and detoxifying insecticides.

M. Kashkouli · M. Mehrabadi (✉) · Y. Fathipour (✉)
Department of Entomology, Faculty of Agriculture, Tarbiat Modares University, Tehran, Iran
e-mail: m.mehrabadi@modares.ac.ir; fathi@modares.ac.ir

Omkar (ed.), *Microbial Approaches for Insect Pest Management*,
https://doi.org/10.1007/978-981-16-3595-3_5

Insect symbionts offer an opportunity to deal with the anticipated elevated demand for novel pest management strategies through manipulation of the symbionts or the host–symbiont associations. Manipulation of these microbial partners can reduce the pest status of and vectorial capacity of insects. Targeting essential symbionts (providing essential nutritional elements) required by the insect, as a control strategy, results in insect mortality or suppression of insect growth or fecundity. Using heterologous microorganisms (transferring microorganisms from one species to other) and genetic manipulation of microbial symbionts (paratransgenesis) are also of interest to control insect vectors of human diseases. Here, we discuss different symbiont-based strategies that can be exploited for insect pest and vector control.

Keywords

Heritable symbionts · Bacterial symbionts · Fungal symbionts · Insect-symbiont interactions · Mutualism · Sex-ratio distorters

Learning Objectives

1. Most insects, including pests, are associated with different symbiotic microorganisms. These associations may lie anywhere on a continuum between parasitism and mutualism.
2. The symbionts play a prominent role in insect ecology, such as aiding in the digestion of food or providing nutrients, influencing insect–plant interactions, host population, heat tolerance, and pesticide detoxification, as well as protection from natural enemies.
3. Insect symbionts offer an opportunity to deal with the anticipated elevated demand for novel pest management strategies through manipulation of the symbionts or the host–symbiont associations.
4. Main approaches for novel pest management strategies include: (i) utilizing heterologous microorganisms, (ii) paratransgenesis, (iii) insect incompatibility technique (IIT), and (iv) the disruption of microbial symbionts required by insect pests.
5. This chapter outlines various types of interactions between insects and microorganisms, which provides novel insight into the design of pest management strategies. Furthermore, detailed studies and recent advances in the control of pests and vectors founded on manipulation of microbial partners are fully described.

5.1 Introduction

Insects can be serious economic pests in agriculture through their activities as herbivores of major crop plants or as vectors of plant pathogens. In addition, insects' function as vectors for human pathogens, causing severe diseases. To control insect pest populations, novel approaches, which are effective, environment friendly, and socially acceptable are required. Insect symbionts offer an opportunity to deal with the anticipated elevated demand for novel pest management strategies through manipulation of the symbionts or the host–symbiont associations.

Approaches, such as high throughput sequencing, functional and comparative omics, and gene editing technologies have provided insights into the microbial functions in insect–symbiont interactions. Studies on these functions have yielded valuable information about how mutualistic symbionts may confer additional properties to their hosts, enabling them to survive in novel ecological conditions, having novel reactions toward natural enemies and pesticides, expanding host plant range, and affecting other trophic levels. Among these, most of the best-described associations are based on nutritional and defensive services provided by the symbionts. Deeply investigated cases involve *Buchnera* as nutrition providing symbionts in aphids and *Wolbachia* as protecting symbionts in flies. The knowledge of insect–microbe interaction provides novel insight into the design of the pest management strategies for the management of insect-related problems.

This chapter primarily outlines diverse functions of the symbionts on insect host ecology. Then, we discuss the detailed studies in the host protection against natural enemies mediated by the symbionts. In the last part of this chapter, recent advances and detailed studies in the application of the symbionts for novel pest management are discussed. These approaches include: (i) utilizing heterologous microorganisms, (ii) paratransgenesis, (iii) insect incompatibility technique (IIT), and (iv) the disruption of microbial symbionts required by insect pests. Here, we reviewed the well-known examples and the progress of using these strategies for the management of vector-borne diseases, in the context of the demand for novel methods of insect pest control that are both durable and environmentally benign. These approaches could improve and/or be a substitute for the traditional methods of insect pest management.

5.2 Insect Symbionts, Different Types, and Roles

5.2.1 Definition of Symbiosis and Different Types

Symbiosis describes the persistent and intimate association between members of different species (Haine 2008). The word "symbiosis"was first defined in 1879 by the German botanist and mycologist Heinrich Anton de Bary in his monograph "Die Erscheinung der Symbiose" (De Bary 1879). De Bary's definition of symbiosis was: "a phenomenon in which dissimilar organisms live together" (Oulhen et al. 2016). The modern concept of symbiosis was defined in 1952 by the pioneer of symbiosis research Paul Buchner in the latest edition of his book "Endosymbiose der Tiere mit pflanzlichen Mikroorganismen" (Buchner 1952). In his seminal book, Buchner described hundreds of different symbioses of insects with microorganisms with an emphasis on anatomy (Feldhaar 2011). Buchner's definition of endosymbiosis was: "By endosymbiosis we understand a regulated, harmonious cohabitation of two nonrelated partners, in which one of them lives in the body of the other usually more highly organized being, and in which the mutual adaptation has reached such a high degree of intimacy, that the supposition is justified, it could be a useful arrangement for the host" (Koch 1960).

Symbiotic relationships range from mutually beneficial (mutualism) to neutral (commensalism) or parasitic (parasitism) associations that differ with respect to the effect of the symbiont on the host (Baumann 2005). Commensal has no discernible impact on host health or fitness, the disease-causing pathogen is thought to benefit at the expense of its host and beneficial mutualist is known to have a mutual benefit with the host (Richards and Brooks 1958). The potential durable interactions between hosts and symbionts may lie anywhere on a continuum between parasitism and mutualism (Haine 2008).

5.2.2 Diversity of Symbiotic Associations in Insects

Insects live together with many different symbionts, both inside and outside their bodies, in a variety of ways (Baumann 2005; Su et al. 2013). Some microbial symbionts live outside of the host's body on the food source, assisting in its breakdown to simple substrates suitable for consumption or enriching the diet with assimilated nitrogen (Sudakaran et al. 2017). Such interactions are common in the gardens or galleries of fungus-farming insects. The present discussion is focused on the interactions inside insect bodies referred to as "symbiont" afterward. The symbiont includes gut microbes (Engel and Moran 2013), extracellular symbiosis, and intracellular symbiosis (also called endosymbionts). The deeper inside the microbe resides, the more intimate the interaction with the host insect (Ishikawa 2006), so the endosymbiosis is the most intimate association between two different organisms, and it is generally reasoned that the association is maintained through the host's generations (Su et al. 2013).

A diverse range of symbioses occur across insect taxa based on biology and evolutionary history (Sudakaran et al. 2017). First, many obligate (i.e., primary or P-) symbionts are essential for host survival and reproduction (Haine 2008) by providing essential nutrients to their insect hosts that are rare or absent in the host's diet (Engel and Moran 2013). These symbionts are often housed in a specialized organ, called a bacteriome, and tend to be purely vertically transmitted (Moran et al. 2008). The best-known P-symbionts are *Buchneraaphidicola* in aphids, *Portieraaleyrodidarum* in whiteflies, *Carsonellaruddii* in psyllids, and *Tremblayaprinceps* in mealybugs (Su et al. 2013).

The second category comprises of facultative (i.e., secondary or S-) symbionts that are generally not essential for their host but may have prominent effects on important traits of the host, such as reproductive capacity, defense, thermal tolerance, or nutrition of their hosts (Moran et al. 2008). Some subdivisions for S-symbionts based on their function are facultative mutualist, facultative reproductive manipulator, and facultative of unknown effect (Moran et al. 2008; Wernegreen 2012; Su et al. 2013). These symbionts are erratically distributed within various tissues that can be located intra- or extracellularly (Sudakaran et al. 2017). Most of the S-symbionts are predominantly vertically transmitted; however, horizontal transmission occasionally occurs between individuals and between species (Haine 2008; Ferrari and Vavre 2011).

The third group encompasses extracellular microbes that infect insect guts and are localized in the gut lumen or specialized posterior midgut structures, called crypts or caeca (Fukatsu and Hosokawa 2002; Sudakaran et al. 2017). These symbionts are often orally and vertically transmitted to newborn nymphs by post-hatching transmission mechanisms, such as egg smearing, symbiont-containing capsules, coprophagy, or acquisition from the environment or conspecifics by social transmission (Buchner 1965a; Fukatsu and Hosokawa 2002; Kashkouli et al. 2020a; b). Members of this third category can contribute to their host's fitness through nutrient provisioning, breaking down of plant polymers, nitrogen recycling, or detoxification of plant defenses (Sudakaran et al. 2017).

5.2.3 The Importance of the Symbionts in Insect Ecology

The symbionts play a prominent role in insect ecology by aiding in the digestion of food or providing nutrients, influencing insect–plant interactions, host population, heat tolerance, and pesticide detoxification, as well as protecting from natural enemies (Feldhaar 2011). In the following, we briefly summarize the knowledge of these functional roles of symbionts in insects (Fig. 5.1).

Providing Essential Nutrition and Food Digestion

Many insects that rely on a single food source, like blood, plant sap, and wood, harbor beneficial microbes, which can digest the food or supplement nutrients (Akman et al. 2002; Engel and Moran 2013). Many blood-feeding arthropods harbor obligate symbionts providing B vitamins, such as *Wigglesworthiaglossinidia* in tsetse flies (Akman et al. 2002), *Rhodococcusrhodnii* in *Rhodniusprolixus* (Eichler and Schaub 2002), and *Wolbachia* in Cimicidae (Sudakaran et al. 2017).

Virtually all plant sap-feeding insects have symbionts providing essential nutrition (Ferrari and Vavre 2011) because plant sap is nutritionally poor in vitamins and amino acids and rich in carbohydrates (Sasaki et al. 1996). The nutritional role of symbionts has been investigated extensively for aphids. Most aphid species, with the exception of aphids from the tribe Cerataphidini, possess intracellular bacteria of the genus *Buchnera* (Brownlie and Johnson 2009). Annotation of the complete genome sequence of several *B. aphidicola* isolates revealed that these bacteria have the genetic capacity for the biosyntheses of several amino acids essential for the host (Shigenobu et al. 2000; Douglas 2009; Smith and Moran 2019). In *Carsonellaruddii*

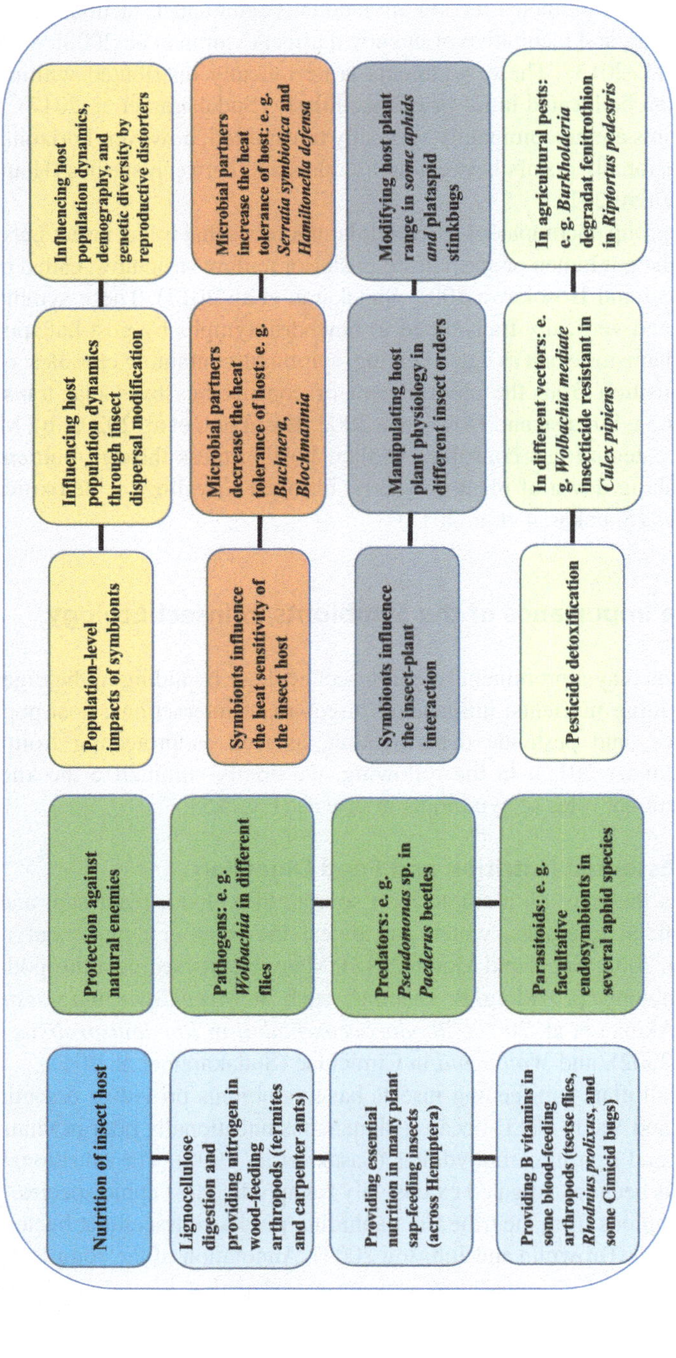

Fig. 5.1 The importance of the symbionts in insect ecology

of psyllids and *Sulciamuelleri* of auchenorrhynchan sap feeders, similar gene content and order has been maintained among strains, especially for essential amino acid biosynthesis (McCutcheon and Moran 2012; Hansen and Moran 2014). Furthermore, plataspid bug *Megacoptapunctatissima* and pentatomid stinkbugs *Acrosternum arabicum* and *Halyomorphahalys* harbor crypt-associated symbionts "*Candidatus* Ishikawaellacapsulata," "Ca. Pantoeapersica," and "Ca. Pantoeacarbekii," respectively (Kashkouli et al. 2020a; b; Nikoh et al. 2011; Kenyon et al. 2015). These symbionts feature reduced genomes that encode the biosynthesis of essential amino acids and vitamins (Kashkouli et al. 2020a; b; Nikoh et al. 2011; Kenyon et al. 2015; Sudakaran et al. 2017).

Among wood-feeding arthropods, termites provide clear cases of nutritional roles of gut microorganisms (Zucchi et al. 2012; Engel and Moran 2013). These symbionts, including bacteria, archaea, and protists, are housed in an enlarged portion of the hindgut called the paunch (Ishikawa 2006) and contribute to lignocellulose digestion producing high levels of acetate, and also provide their hosts with nitrogen (Engel and Moran 2013). In addition, the carpenter ants are related to the symbionts in the genus *Blochmannia*, which provide their hosts with amino acids (Ferrari and Vavre 2011).

Influence on Insect–Plant Interaction

There is growing evidence of the importance of insect-associated microbes in insect–plant interactions. Mutualists may affect host plant range and enable insects to manipulate plant physiology, so it was suggested to be "hidden players" in insect–plant interactions (Frago et al. 2012). Worldwide surveys of *Acyrthosiphonpisum* populations have revealed an association between infection with *Regiellainsecticola* and the use of clover as a host plant (Oliver et al. 2010). Injection of *R. insecticola* from a clover-adapted pea aphid to vetch aphid *Megouracrassicauda* allowed the latter that normally could not feed on clover to use this host plant (Frago et al. 2012). As in aphids, symbiont exchange between the two plataspidstinkbugs, a pest species, *Megacoptapunctatissima,* and a closely related non-pest species, M. cribraria, resulted in the complete reversal of the bugs' performances: the pest species suffered low egg hatch rate, whereas the non-pest species restored normal egg hatch rate and showed good performance, indicating that the symbiont strain, not the host genotype, governs host plant specialization (Su et al. 2013; Sudakaran et al. 2017).

Insect symbionts may actively manipulate plant physiology and antiherbivore defenses to their insect hosts' advantage (Frago et al. 2012). Perhaps the best-known example is ambrosia beetles, which, in partnership with their fungal symbionts, can overcome bark defenses (Miller et al. 2019). In corn root borer, the regulation of defense-related genes in the plant was influenced by the symbiont *Wolbachia* (Dicke et al. 2020). Similarly, the symbiont "Ca. Liberibacterpsyllaurous" in tomato psyllid, *Bactericercacockerelli,* could suppress jasmonic acid (JA) and salicylic acid (SA) defensive signaling pathways of tomato plants (Casteel et al. 2012). The wheat aphid *Sitobionmiscanthi* harbors the symbiont *Hamiltonella,* which reduces the activity levels of the defense-related enzymes polyphenol oxidase (PPO) and peroxidase (POD) in the wheat plants (Li et al. 2019). Also, *Serratiasymbiotica* infection in *Acyrthosiphonpisum* may facilitate aphid adaptation to host plants (Wang et al. 2020). On the other hand, facultative symbionts in *Bemisiatabaci* and *Leptinotarsadecemlineata* manipulate host plant physiology through salivary effectors that attenuate induced defenses to the benefit of their hosts (Mala et al. 2017). The apple leaf-mining moth, *Phyllonorycterblancardella* relies on *Wolbachia* to cope with nutritional constraints in senescent leaves; *Wolbachia*-infected larvae are able to induce so-called "green-islands" (Zug and Hammerstein 2015). Once the symbiont is removed, cytokinin concentrations are reduced in the mine and the green islands disappear, leading to increased moth mortality (Frago et al. 2012). In these systems, more researches should be performed for the detection of how symbionts influence the host plant physiology (Hansen and Moran 2014).

Population-Level Impacts of Endosymbionts

Symbionts have an ongoing impact on host population dynamics and demography as well as the genetic diversity of the host species (Feldhaar 2011). Several ecological models have explored the population dynamics in a host infected by vertically and horizontally transmitted microbes (Haine 2008). Endosymbionts, such as *Wolbachia, Arsenophonus, Spiroplasma*, and *Cardinium,* can sharply influence population dynamics via various ways, such as cytoplasmic incompatibility (CI), parthenogenesis induction (PI), feminization, and male-killing (Ishikawa 2006; Su et al. 2013). In the case of CI, uninfected eggs fertilized by sperm from infected males died, thus CI benefits infected females and favors the spread of symbiont through host populations (Zug and Hammerstein 2015). In PI, unfertilized haploid host eggs are converted into viable diploid female offspring, so it may result in a rapid declining of genetic diversity and increase the percentage of transmitting hosts (Su et al. 2013; Zug and Hammerstein 2015). In feminization, infected genetic males reproduce as females; and in the male killing, infected male embryos die while female embryos develop into infected females (Stouthamer et al. 1999). Feminization and male-killing distort the sex ratio of infected mothers toward females and thus directly increase the proportion of infected females (Su et al. 2013; Zug and Hammerstein 2015). All these sex-ratio distorters may decrease genetic diversity within a population and reduce effective population size, entailing strong negative effects, such as an increased rate of fixation of deleterious mutations and stronger genetic drift effects (Feldhaar 2011). The spread of these bacteria can drive small populations to extinction (Su et al. 2013) and in some cases enhance rates of speciation of hosts (Moran et al. 2008).

The impact of endosymbionts on traits relevant to dispersal may also influence the population dynamics of insect hosts (Feldhaar 2011). Pea aphids containing *R. insecticola*, for example, produced only half the number of winged offspring in response to crowding than those lacking this endosymbiont, and for two out of three aphid lineages, the timing of sexual reproduction was altered by the presence of *Regiella* symbiont (Leonardo et al. 2006). Thus, this facultative endosymbiont may limit gene flow (Feldhaar 2011). In contrast, a link between the presence of

Rickettsia in a pest-controlling money spider, *Erigoneatra,* and the tendency for the long-distance movement was demonstrated (Goodacre et al. 2009).

Symbionts Influence the Heat Sensitivity of the Insect Host

Insect symbionts offer valuable models to examine how microbes can influence the host's adaptation to a changing environment (Wernegreen 2012). Among different environmental factors, the temperature is the sole abiotic one that has received substantial attention so far. Temperature can either have direct effects on the insect hosts or indirect effects by changing the abundance of symbionts within the host or their efficiency of transmission to the offspring (Feldhaar 2011). These effects are sometimes in the direction that microbial partners expand the range of temperatures in which hosts can thrive. The well-known examples include cold-tolerant fungi as an obligate symbiont of leafcutter ants (Mueller et al. 2011), some heat-tolerant secondary symbionts of aphids (*Serratiasymbiotica* and *Hamiltonella defensa*) (Chen and Purcell 1997; Montllor et al. 2002; Russell and Moran 2006), and whitefly *Bemisiatabaci* (*Rickettsia*) (Brumin et al. 2011).

While some symbionts confer plasticity that accelerates adaptation, long-term obligate symbionts may be fragile in the face of a changing environment. These mutualisms are mostly heat-sensitive and can be depleted or lose entirely by temperatures that do not kill their hosts. Elevated temperature eliminates obligate bacterial mutualists in many insects like aphids (Buchner 1965b; Montllor et al. 2002), weevils (Heddi et al. 1999), and cockroaches (Sacchi et al. 1993). Aphids cannot tolerate high temperatures well because of limitations for both the host and symbiont. Aphid nymphs treated at 37 °C had slower development, and some of them lacked "normal" *Buchnera* and did not produce offspring (Ohtaka and Ishikawa 1991). The number of bacteriocytes and *Buchnera* symbionts has been shown to decrease dramatically at higher temperatures or heat shock (Montllor et al. 2002).

Likewise, carpenter ants are limited by the heat sensitivity of their obligate symbiont, *Blochmannia* (Fan and Wernegreen 2013). The high-temperature treatment reduced the reproductive rate and symbiont titer of *Cimexlectularius* (Chang 1974). In *Laodelphaxstriatellus,* high temperature, 35 °C, destroyed the yeast like in the mycetocytes. Under the continuous high temperature, no adults were obtained. The population of symbiotes in the fifth-instar nymphs, which were previously exposed to the high temperature for 3 days after hatching (heat treatment), was reduced (Noda and Saito 1979). The embryonic development and reproduction of *Nilaparvatalugens* were interrupted under heat treatment (Lee and Hou 1987).

Among stinkbugs, it has been shown that *Nezaraviridula* suffers from serious fitness deficiency under some simulated warming conditions (Prado et al. 2009; Musolin et al. 2010; Kikuchi et al. 2016;Moran 2016). The symbiont was detected in 100, 84, and 8.3% of the untreated control insects at 20, 25, and 30 °C, respectively, by using polymerase chain reaction (Prado et al. 2009). Heat treatment of egg capsules of plataspid stinkbug, *Megacoptapunctatissima*, disrupted symbiont transmission that resulted in the retarded host growth and development (Fukatsu and Hosokawa 2002). Elevation in rearing temperature for 5 °C resulted in symbiont eliminations of *Acrosternumhilare,* A. heegeri, A. arabicum, *Brachynemagermari,* and *Murgantiahistrionica* (Prado et al. 2010; Kashkouli et al. 2018). Specifically, under the high temperature (30 °C) condition, the highest level of *B. germari* preadult mortality was obtained and also, constant heat treatment did not allow *B. germari* females to lay eggs (Kashkouli et al. 2018).

Although the mechanisms underlying heat vulnerability of obligate mutualists of insects are unclear, microbial advances propose that some aspects of the susceptibility of symbionts to high temperature are related to the consistent patterns of degenerative evolution of their genomes. Low stability of AT-rich DNA, structural RNAs, and GroELunder heat stress and also loss of cell wall and its surface proteins were suggested as possible heat-sensitivity mechanisms of many symbionts (Moran 1996; Rathnayaka and Rakshit 2010; Wernegreen 2012).

Symbiont-Mediated Degradation of Pesticides

The development of insecticide resistance to a diverse range of insecticides has been a serious concern worldwide. The mechanisms underlying insecticide resistance were thought to be encoded by the insects' own genomes, which include target-site mutation, toxin avoidance behavior, up-regulation of degrading enzymes, and enhancement of drug excretion. However, classic culture-dependent methods conjugated with recent omics analyses have revealed the ability of symbiotic microbes to degrade pesticides, establishing insecticide resistance in the insect pests (Itoh et al. 2018).

A growing number of studies documented symbiont-mediated detoxification of insecticides by symbionts. In the early 1990s, detoxification of the organophosphate parathion by a symbiotic yeast was reported in the cigarette beetle, *Lasiodermaserricorne* (Shen and Dowd 1991). Then, organophosphate detoxification by *Pseudomonas melophthora*, a symbiont of *Rhagoletispomonella,* was reported (Boush and Matsumura 1967). Later on, a high *Wolbachia* density in insecticide-resistant populations of *Culexpipiens* was observed (Berticat et al. 2002). In 2012, it was demonstrated that the gut symbiotic bacteria of the genus *Burkholderia* are the causative agents of fenitrothion degradation in the bean bug, *Riptortuspedestris* (Kikuchi et al. 2012; Werren 2012).

More recent studies also suggest the symbiont-mediated resistance to pesticides in some other pests, including *Spodopterafrugiperda* (De Almeida et al. 2017), *Anopheles stephensi* (Soltani et al. 2017), Bactrocera dorsalis (Cheng et al. 2017), *Plutellaxylostella* (Xia et al. 2018), *Blatellagermanica* (Pietri et al. 2018), *Nilaparvatalugens* (Pang et al. 2018; Tang et al. 2020), and *Callosobruchusmaculatus* (Akami et al. 2019). All these findings indicate that symbionts can confer detoxification of chemical pesticides in crop pest insects more than previously thought (Van Den Bosch and Welte 2017).

Some microbial genes involved in the detoxification have been identified using genome and transcriptome analyses, in conjunction with physiological confirmations. The genes involved in producing organophosphate-hydrolyzing enzymes have gained more attention because they not only reduce pesticide toxicity but also clean up the pesticides from the environment (Itoh et al. 2018). Using the knowledge of symbiont-mediated detoxification will provide novel insight into the evolution and function of insect microbial symbiosis, and it may also lead to improved control strategies for insecticide-resistant pests (Pietri and Liang 2018) by, for example, developing symbiont-targeted pesticides (Blanton and Peterson 2020).

Protection Against Natural Enemies

Defensive symbionts can protect their insect host against biological threats, including predators, pathogens, and parasitoids (McLean 2019). Several natural examples of symbiont-mediated protections have been reported in insects. Here, the symbiont-induced anti-pathogenic effect is mentioned briefly because this fascinating subject is explained in detail in the next section ("symbionts mediate protection against natural enemies").

Protection of herbivorous insects by endosymbionts has been well established, particularly in aphids (Hansen and Moran 2014). Alongside *Buchnera*, some different secondary endosymbionts have been identified that occur regularly in aphid hosts, namely *Hamiltonelladefensa, Regiellainsecticola, Fukatsuiasymbiotica,* and *Serratiasymbiotica* as well as *Rickettsia, Rickettsiella, Spiroplasma,* and *Arsenophonus* (Feldhaar 2011; Hansen and Moran 2014; Patel et al. 2019; Ayoubi et al. 2020). Some strains of the endosymbiont species *H. defensa* and to a lesser extent *F. symbiotica,* S. symbiotica, and *R. insecticola* protect several aphid species, e.g., Fig. 5.2, against endoparasitic wasps (Oliver et al. 2003, 2010; Ferrari et al. 2004; Haine 2008; Von Burg et al. 2008; Vorburger et al. 2010; Cayetano et al. 2015). Recent studies strongly suggest that *H. defensa* has a dynamic genome, exhibiting evidence of recombination, phage-mediated gene uptake, and horizontal gene transfer and containing virulence and toxin-encoding genes (Oliver et al. 2010).

In addition, secondary symbionts may affect the defense against parasitoids in *Drosophila* fruit flies. *Spiroplasma* can defend Drosophila hydei against a common parasitoid wasp *Leptopilinaheteroma* (Xie et al. 2010). Other facultative endosymbionts have the opposite effect on parasitoids: *Wolbachia* infection in *Drosophilasimulans* leads to reduced encapsulation ability and therefore increased susceptibility to the wasp *L. heterotoma* (Ferrari and Vavre 2011). The discrepancy between different facultative symbionts in the resistance induction against parasitoids can be concluded as a result of the different strategies adopted by symbionts in order to invade host populations (Fytrou et al. 2006).

Aside from the enhanced resistance toward parasitoids, insect hosts may benefit from the production of toxin compounds by endosymbiotic bacteria that protect their hosts from predators. In several species of paederine beetles, an inherited *Pseudomonas* symbiont synthesizes the polyketide toxin, named pederin, that the beetle can use as a defense against wolf spiders (Kellner 2003; Piel et al. 2004; Brownlie and Johnson 2009). Another intracellular symbiont "Ca. Profftellaarmatura" synthesizes a novel polyketide toxin, named diaphorin, in Asian citrus psyllid, *Diaphorinacitri* (Nakabachi et al. 2013; Ramsey et al. 2015). Diaphorin is structurally very similar to onnamides and pederin and is found to be a toxin against yeast and cultured mammalian and insect cells (Ramsey et al. 2015; Oliver and Perlman 2020).

Symbiotic microbes have also the capacity to protect their hosts from different pathogenic microorganisms. The role of facultative symbionts in the defense against pathogenic fungi has also been studied in aphids. While *Hamiltonella* appears to have no effect on aphid susceptibility to fungal pathogens, at least four other secondary symbionts (*Rickettsia, Rickettsiella, Regiella,* and *Spiroplasma*) protect pea aphid against the fungal entomopathogen *Pandoraneoaphidis* (Fig. 5.2). In addition, symbionts provide protection against fungal pathogens in the beewolf digger wasps (Kaltenpoth et al. 2005, 2006; Goettler et al. 2007), attine ants (Currie et al. 1999a; Currie and Stuart 2001; Vieira et al. 2012), fungus-farming termites (Visser et al. 2012), bark/ambrosia beetles (Scott et al. 2008; Oh et al. 2009; Blodgett et al. 2010; Hulcr et al. 2011), and *Lagria* beetles (Flórez and Kaltenpoth 2017; Flórez et al. 2017).

Little is known about symbiont-mediated defense against nematodes. One reported case is in *Drosophila neotestacea*, in which *Spiroplasma* symbionts can protect the insect against a sterilising parasitic nematode, *Howardulaaoronymphium* (Jaenike et al. 2010). Inherited *Spiroplasma* strains have also been implicated in protection against trypanosome in tsetse flies (Schneider et al. 2019).

Wolbachia-induced anti-pathogenic effects in different flies have received substantial attention so far. The protections have been reported against several RNA viruses, different *Plasmodium* species, fungi, bacteria, and nematodes. Antiviral effects, in particular, have been observed frequently and across different *Wolbachia* strains, multiple hosts, and diverse viral families (Zug and Hammerstein 2015). Known cases of *Wolbachia*-different fly combinations and their mechanisms are discussed in the subheading "*Wolbachia*-different flies" (under the heading "Symbionts mediate protection against natural enemies" and in the part of "Pathogens") and their anti-pathogenic potential to control insect pests and insect-borne diseases is discussed in "Potential application of symbionts in pest control" in detail.

5.3 Symbionts Mediate Protection Against Natural Enemies

In biological control programs, natural enemies, including parasitoids, predators, and pathogens, are usually applied to curb populations of herbivorous insect pests. However, variable outcomes of biological control programs, which range from successful to unexpectedly ineffective, are usually achieved. Some of these variations have been shown to be regulated by the cryptic microbial defenders in a number of insects and an understanding of such mechanism is steadily increasing. In support of the theoretical predictions, several natural examples of symbiont-mediated protection have been reported recently in insects, and these are discussed below.

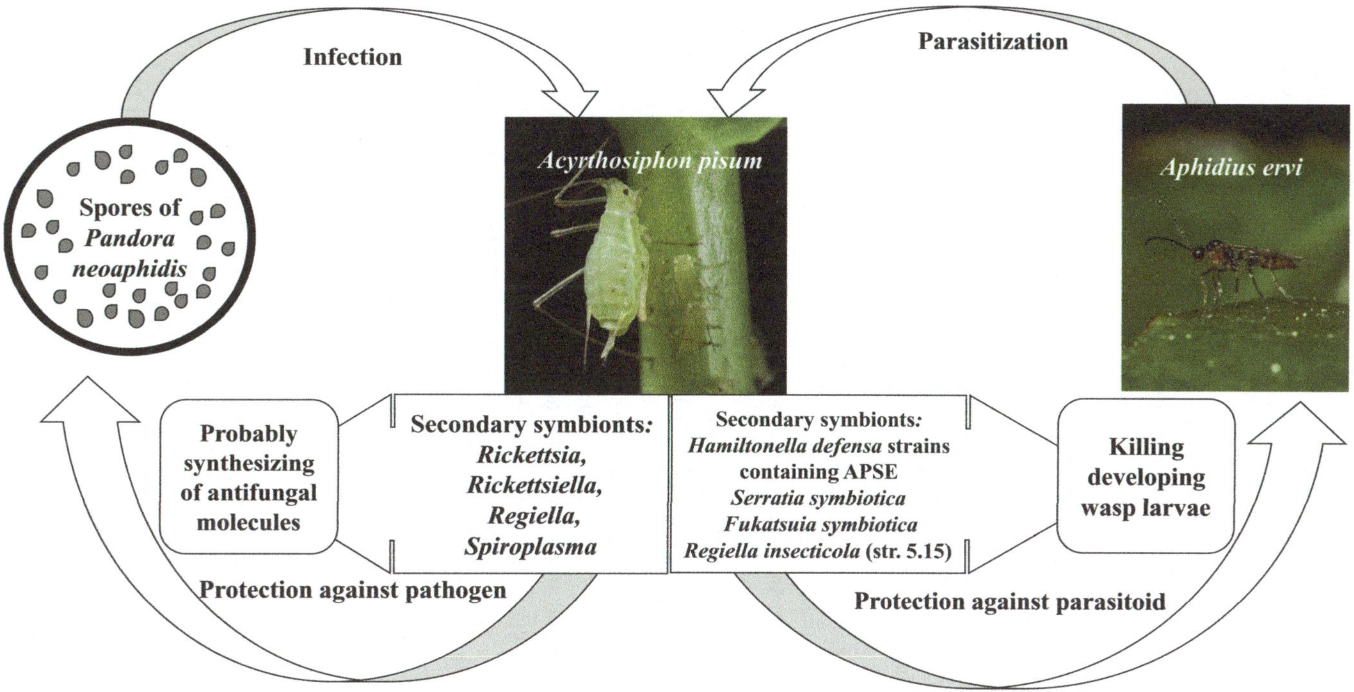

Fig. 5.2 Defensive symbionts protect *Paederus* spp. and Diaphorina citri against predators

5.3.1 Protection Against Predators

Aside from the enhanced resistance toward parasitoids, insect hosts may benefit from the production of toxins or body color differentiation by endosymbiotic bacteria to reduce the probability of predation. The most well-known examples related to this protection are discussed below.

Pseudomonas Sp. and *Paederus* Beetles

Rove beetles have long been of medical interest as they cause painful lesions called *Paederusdermatitis* (Oliver and Perlman 2020). In addition to these severe effects for humans, the beetles are known to have defensive value for some invertebrates. In several species of the genus *Paederus*, a polyketide toxin, pederin, is used as a chemical defense against predators (Piel et al. 2004). Experiments showed that pederin does not alter the behavior of insect predators, but strongly deters several species of wolf spiders and one jumping spider by reducing palatability of the beetles as prey (Kellner and Dettner 1996; Feldhaar 2011; Oliver and Perlman 2020). The primary mode of action of pederin is the inhibition of protein synthesis, leading to cytotoxic effects (Oliver and Perlman 2020) (Fig. 5.2).

Pederin was subsequently confirmed to originate from a bacterial symbiont of the beetles, rather than the beetles themselves (Kellner 2002; Piel 2002) (Fig. 5.2). PCR-based analysis of *16S rDNA* reveals that the symbiont is a member of the γ subdivision of the Proteobacteria that is clustered within the genus *Pseudomonas* (Kellner 2002). Genome sequencing indicates that the symbiont is a very close relative of *Pseudomonas aeruginosa* (Piel et al. 2004). These bacteria appear as the hypothesized common producers of pederin (Kellner 2002).

The transmission of defensive compound pederin and the endosymbiont are discussed in several studies. Females accumulate large amounts of the pederin and transfer in the eggs, while the endosymbionts are transmitted via the eggshell (Kellner 2003). Larvae-containing pederin survives the attacks of spiders without damage, whereas larvae descended from females that do not transfer pederin into their eggs are often killed and eaten (Kellner and Dettner 1996). Aposymbionts, which occur both naturally and in laboratory-reared specimens, can be induced to accumulate pederin and the symbiont if fed with symbiont-containing eggs during larval development (Kellner 2001). Successful colonization by endosymbionts depends on the number of eggs consumed and the age/stage of the feeding (Kellner 2003).

Polyketide toxins like pederin are widespread in unrelated animals from diverse habitats (Piel et al. 2004). Pederin is structurally very similar to the theopederins and onnamides isolated from bacteria inhabiting marine sponges and nosperin from lichens (Oliver and Perlman 2020). Another polyketide produced by the bacterial symbiont *Endobugula sertula* deters fishes from preying its bryozoan host and a similar symbiont-mediated mechanism is thought to protect marine isopods from fish predation (Haine 2008). A horizontal transfer of genes for the biosynthesis of protective substances could explain the widespread occurrence of pederin-type compounds among distantly related organisms (Piel et al. 2004).

"Ca. Profftellaarmatura" and Diaphorina citri

Asian citrus psyllid, Diaphorina citri, harbors intracellular symbiont "Ca. Profftellaarmatura" in a bacteriome (Nakabachi et al. 2013). Metagenomics revealed that *Profftella* is unique to D. citri and found in all insect populations analyzed worldwide (Ramsey et al. 2015; Hosseinzadeh et al. 2018). An experiment revealed that *Profftella* is a defensive symbiont with an extremely streamlined genome at 0.54 Mb (Nakabachi et al. 2013). Approximately 15% of its genome encoded complete gene cluster for the synthesis of a novel polyketide toxin (Ramsey et al. 2015). The toxin was extracted, pharmacologically and structurally characterized, and designated diaphorin (Nakabachi et al. 2013). The presence of *Profftella* and the production of diaphorin are observed without exception among individuals within and across geographically distant psyllid populations (Nakabachi et al. 2013; Flórez et al. 2015).

Diaphorin was found to have cytotoxicity to yeast and cultured mammalian and insect cells, but its toxicity on microbial cells has not been investigated (Ramsey et al. 2015; Oliver and Perlman 2020). Laboratory studies show that diaphorin is also toxic to aphids and ladybird beetles when injected (Fig. 5.2). However, it is not yet known that diaphorin is effective against which specific natural enemies in nature (Oliver and Perlman 2020).

Rickettsiellaviridis and Pea Aphid

Color variation within populations of the pea aphid affects aphid interactions with higher trophic levels including predators (and parasitoids, discussed previously in "Facultative endosymbionts and Acyrthosiphon pisum") (Tsuchida et al. 2010). The infection with the endosymbiont "Ca. Rickettsiellaviridis" increased amounts of blue-green polycyclic quinones (Tsuchida et al. 2010). Genomics revealed that the green pigments are produced not by the symbiont but by the host aphid and suggests the possibility that the symbiont may upregulate the host's production of polycyclic quinone pigments via cooption of secretion machinery and effector molecules for pathogenicity (Nikoh et al. 2018).

The increased amounts of blue-green polycyclic quinones change the aphid color from red to green in natural populations (Frago et al. 2012). Previous ecological studies demonstrate that ladybird beetles preferentially attack red aphids on green plants (Nikoh et al. 2018), so the infection with symbiont is expected to influence prey–predator interactions (Su et al. 2013).

5.3.2 Protection Against Parasitoids

Because the parasitoid wasp is used as a biological control agent, there is considerable interest in determining how resistance to parasitization is achieved in insect pests. Symbiont-mediated protection against parasitoids has been reported in several insects and there are some examples that support this (Table 5.1). In the following section, some well-known examples are discussed.

Table 5.1 Symbiont-mediated protection against parasitoids and predators

Symbiont(s)	Host(s)	Natural enemy(ies)	Protection mechanism(s)	References
Hamiltonelladefensa *Serratiasymbiotica* *Fukatsuiasymbiotica* *Regiellainsecticola*	*Acyrthosiphonpisum*	*Aphidiuservi*	Killing wasp larvae	Oliver et al. (2003), Ferrari et al. (2004), Haine (2008), Oliver and Perlman (2020)
R. Insecticola 5.15	*Myzuspersicae*	*Aphidiuscolemani* and *Diaeretiellarapae*	–	Vorburger et al. (2010)
H. Defense *R.insecticola*5.15	Aphis fabae	*Lysiphlebusfabarum* and *Aphidiuscolemani*	–	Vorburger et al. (2010), Cayetano et al. (2015)
H. defensa	*Rhopalosiphumpadi*	*Aphidiuscolemani*	–	Leybourne et al. (2020)
H. defensa	Aphis craccivora	*Binodoxyscommunis* and B. koreanus	–	Asplen et al. (2014)
Spiroplasma	D. hydei	*Leptopilinaheteroma*	–	Flórez et al. (2015), Xie et al. (2010)
“Ca. Rickettsiellaviridis”	Pea aphid	Ladybird, parasitoid wasps	Changing body color	Tsuchida et al. (2010), Nikoh et al. (2018)
Pseudomonas sp.	*Paederus* spp.	Predatory wolf spider	Secretion of pederin toxin	Kellner and Dettner (1996), Kellner (1999, 2001, 2002, 2003); Piel (2002), Piel et al. (2004)
“Ca. Profftellaarmatura”	*Diaphorinacitri*		Secretion of diaphorin toxin	Nakabachi et al. (2013), Ramsey et al. (2015)
Rickettsia, Rickettsiella, Regiella, and *Spiroplasma* symbionts	Pea aphid	*Pandora neoaphidis*	Synthesizing of antifungal molecules	Łukasik et al. (2013); Parker et al. (2013)

The symbiont, host, and natural enemy in each interaction and the known mechanism for the protection are listed

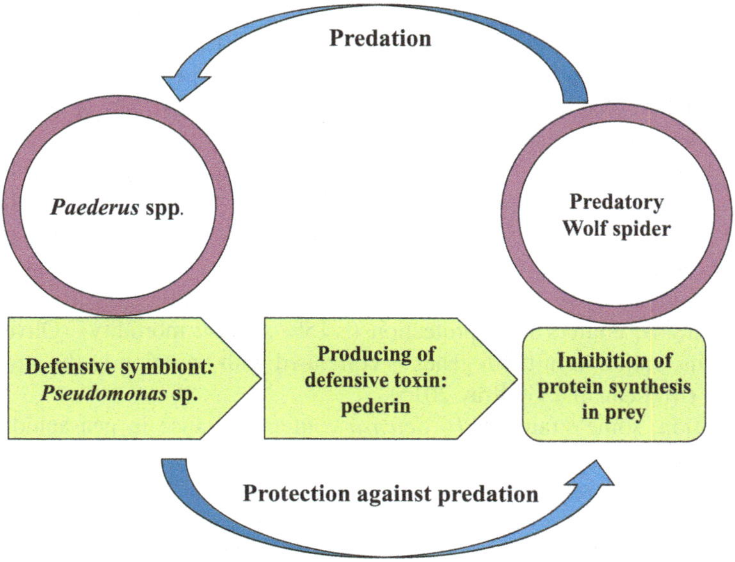

Fig. 5.3 Secondary symbionts protect pea aphid against parasitoid, *Aphidiuservi*, and pathogen, *Pandora neoaphidis*

Facultative Endosymbionts and *Acyrthosiphonpisum*

The pea aphid, *Acyrthosiphonpisum,* is the first insect in which many components of a diverse assemblage of bacterial symbionts have been studied (Oliver et al. 2010). A. pisum harbors facultative bacterial endosymbionts, including *Rickettsia, Spiroplasma,* and different γ-proteobacteria: *Hamiltonelladefensa, Fukatsuiasymbiotica, Serratiasymbiotica,* and *Regiellainsecticola* (Brownlie and Johnson 2009; Patel et al. 2019). Some strains of the endosymbiont species *H. defensa* and to a lesser extent S. symbiotica, *F. symbiotica,* and *R. insecticola* protect aphid against its dominant parasitoid natural enemy, *Aphidiuservi* (Fig. 5.3) (Oliver et al. 2003; Ferrari et al. 2004; Haine 2008; Oliver and Perlman 2020). The symbionts do not reduce oviposition of the wasp eggs within the aphid but the success of the wasps following parasitism was reduced by killing developing wasp larvae (Brownlie and Johnson 2009). Furthermore, parasitized aphids bearing *H. defensa*, but not S. symbiotica, produced significantly more offspring than parasitized uninfected aphids, indicating direct fitness benefits to *H. defensa* infection when under attack by parasitoids (Oliver et al. 2010).

Genome sequencing of *H. defensa* identified a bacteriophage (Oliver et al. 2009), called APSE (A. pisum secondary endosymbiont), that codes for several toxin genes and lyses symbiont cells (Brownlie and Johnson 2009; Hansen and Moran 2014; Oliver and Perlman 2020). Different strains of *H. defensa* are associated with different bacteriophage variants (named APSE-1, APSE-2, etc.). These variants encode homologs of different toxins that are known, or suspected, to target eukaryotic cells (Oliver et al. 2010). APSE-1 genome carries a gene homologous to the Shiga toxin (Stx) encoding genes found in prophages residing in important human and animal pathogens (Dale and Moran 2006). Phage variant APSE-2- contains a homolog of cytolethal distending toxin (cdtB), and aphids infected with *H. defensa* strains that carry APSE-2 are moderately protected from parasitism (~40% *A. ervi* mortality). APSE-3 encodes a tyrosine–aspartic acid repeat (YD-repeat)–containing protein, which also appears to be a toxin, and aphids infected with APSE-3–carrying *H. defensa* are highly protected from parasitism (>85% *A. ervi* mortality) (Oliver et al. 2009). In contrast, *H. defensa* strain lacking APSE, but identical at 10 chromosomal loci to the highly protective APSE-3–carrying strains, confers little protection (∼15% *A. ervi* mortality) (Oliver et al. 2010). Thus, variation in toxin genes is correlated with variation in the strength of protection (McLean and Godfray 2015).

In addition, some strains of *H. defensa* confer resistance in pea aphid against attack by a specialist parasitoid, *A. eadyi* (Ferrari et al. 2001, 2004) and an aphelinid parasitoid *Aphelinusabdominalis* (McLean and Godfray 2015). These studies suggest that different strains of the protective symbiont *H. defensa* may be adapted to defend their host against different parasitoid families (McLean and Godfray 2015).

Some strains of the sister species of *H. defensa*, *R. insecticola* 5 AU and LSR1, do not confer resistance to A. pisum (Oliver et al. 2003; Hansen et al. 2012), but another strain *R. insecticola* 5.15 significantly reduce parasitoid success and increase survivorship of A. pisum. To address the potential genetic basis of protection conferred by *R. insecticola* 5.15, genome sequencing of this strain was compared with nonprotective strain *R. insecticola* LSR1. Striking differences in gene sets were identified, which revealed that the strain *R. insecticola* 5.15 encoded five categories of pathogenicity factors including O-antigen biosynthetic pathway, an intact Type 1 Secretion System and its secreted RTX toxins, an intact SPI-1 Type 3 Secretion System and its effectors, hemin transport, and the two-component system PhoPQ. These gene sets are missing or inactivated in *R. insecticola* LSR1. Thus, resistance to parasitization is associated with the presence of *R. insecticola* 5.15 and is not dependent on aphid genotype (Hansen et al. 2012). Although *R. insecticola* 5.15 "behaves" virtually like *H. defensa*, the protection provided by *R. insecticola* strain R5.15 appears to rely on different mechanisms because APSE is not present in the genome of *R. insecticola* strain R5.15 (Hansen et al. 2012).

The symbiont *F. symbiotica* has also been reported to provide defense against parasitoids (Heyworth and Ferrari 2015). As with *Regiella*, *Fukatsuia* also does not contain APSEs but does have an intact cdtB homolog encoded on the main chromosome (Oliver and Perlman 2020).

Another endosymbiont "Ca. Rickettsiellaviridis" changes the pea aphids' body color from red to green (Tsuchida et al. 2010). The symbiont-induced body color change influences the rate at which insects are parasitized as green aphids suffer higher rates of parasitoid wasps attack (Nikoh et al. 2018).

Regiellainsecticola **and** ***Myzuspersicae***

In green peach aphid, *Myzuspersicae*, a set of 17 clones of the insect, was evaluated for susceptibility to two of their common parasitoids: *Aphidiuscolemani* and *Diaeretiellarapae.* It was demonstrated that one clone, the only one harboring a facultative endosymbiotic bacterium, *R. insecticola*, was entirely resistant to both parasitoids (Von Burg et al. 2008). Yet with just a single, naturally infected clone, it was not possible to infer whether the high resistance was a genetic effect or conferred by the endosymbiont. In another study, endosymbiont-induced resistance was confirmed by comparing naturally infected with cured M. persicae clones and by comparing uninfected with artificially infected clones in M. persicae. The results show clearly that unlike other strains of the symbiont, *R. insecticola* 5.15 strongly increases resistance to parasitoids (Vorburger et al. 2010).

Also, it was demonstrated that the resistance to parasitoids in M. persicae is not only derived by symbiont, but also host conferred resistance toward parasitization, indicating that aphids generally use a variety of mechanisms to aid in their defense (Martinez et al. 2014).

Hamiltonelladefensa, Regiellainsecticola, **and Aphis fabae**

The black bean aphids, Aphis fabae, was infected by the secondary symbionts *H. defensa* and *R. insecticola* 5.15, and exposed to the parasitoids *Lysiphlebusfabarum* and *Aphidiuscolemani*, respectively (Vorburger et al. 2010; Cayetano et al. 2015). These symbionts retained their capacity to protect against parasitoids in the new hosts (Jamin and Vorburger 2019). It was shown that uninfected A. fabae was being mummified at a significantly higher rate than the infected one with *R. insecticola* 5.15 (Vorburger et al. 2010). Another study confirmed that in black bean aphids, the strength of protection provided by different isolates of the *H. defensa* depends on the genotype of the attacking parasitoid. In other words, genotype-by-genotype interactions between the parasitoid and the host's heritable endosymbiont rather than the host itself were observed (Cayetano et al. 2015). Also, it was demonstrated that the resistance of black bean aphid to its common parasitoid *L. fabarum* encode by the defensive symbiont *H. defensa* and the host genome (Martinez et al. 2014).

Hamiltonella Defense **and** ***Rhopalosiphumpadi***

Recently, it was investigated that the bird cherry-oat aphid, *Rhopalosiphumpadi,* is infected by the facultative endosymbiont *H. defensa*. In parasitism assays, the survival of *H. defensa*-infected nymphs following attack by the parasitoid wasp *Aphidiuscolemani* was assessed. The experiment showed that the survivality of symbiont-infected nymphs are five-fold higher than of uninfected nymphs. In addition, aphid mortality after parasitoid attack was significantly lower for aphid lines harboring the facultative endosymbiont *H. defense* (Leybourne et al. 2020).

Hamiltonella Defense and Aphis craccivora

Symbiont protection mediated by *H. defensa* in the cowpea aphid, Aphis craccivora, against four parasitoid species: *Binodoxyscommunis,* B. koreanus, *Lysiphlebusorientalis,* and *Aphidiuscolemani* was investigated. Infection by *H. defensa* almost completely made resistance toward B. communis and B. koreanus, but had no effect on parasitism by L. orientalis and A. colemani (Asplen et al. 2014). This indicates that the outcome of a *H. defensa*–A. craccivora parasitoid relationship is species-specific.

Spiroplasma and _Drosophilahydei_

Symbiont-conferred protection against parasitoids has also been reported in D. hydei, in which *Spiroplasma* can defend the larvae against a common parasitoid wasp *L. heteroma* (Flórez et al. 2015). Experiments revealed that artificial infection with *Spiroplasma* enhances the survival of *Spiroplasma*-infected flies significantly (Xie et al. 2010).

Wolbachia and _Drosophilasimulans_

Wolbachia symbiont in D. simulans has the opposite effect on the parasitoid *L. heterotoma* (Fytrou et al. 2006). The infection leads to reduced encapsulation ability and therefore increased susceptibility to the wasp (Ferrari and Vavre 2011).

The discrepancy between different facultative symbionts in the resistance induction against parasitoids can be concluded as a result of the different strategies adopted by symbionts in order to invade host populations. While symbionts, such as those in the pea aphid, may spread by increasing their host's fitness, *Wolbachia* relies on manipulating host reproduction. The decrease in parasitoid resistance conferred by *Wolbachia* reduces the fitness of both the host and the symbiont, although for *Wolbachia* this cost is potentially offset by the effects of reproductive parasitism (Fytrou et al. 2006).

5.3.3 Protection Against Pathogens

Variation in resistance toward pathogens has been shown to be associated with the presence/absence of secondary symbionts in a number of insects (Feldhaar 2011). Symbiotic microbes have the capacity to protect their hosts from pathogenic microorganisms, and there is evidence for symbiont protection in diverse arthropods, like aphids and mosquitoes (Table 5.2) (Haine 2008).

Facultative Endosymbionts and Acyrthosiphon pisum

Pea aphids are protected against the fungal entomopathogen *Pandora* (*Erynia*) *neoaphidis* by at least four facultative symbionts, including *Rickettsia, Rickettsiella, Regiella,* and *Spiroplasma* (Fig. 5.3) (Łukasik et al. 2013; Parker et al. 2013; Flórez et al. 2015). These symbionts are capable of decreasing the mortality of aphids exposed to the fungi and also decreasing fungal sporulation on dead aphids by

Table 5.2 Symbiont-mediated protection against pathogens (Excluding *Wolbachia*, please see Fig. 5.5 for *Wolbachia*-different fly examples)

Symbiont(s)	Host(s)	Natural enemy(ies)	Mechanism(s)	References
Rickettsia, Rickettsiella, Regiella, and *Spiroplasma* symbionts	Pea aphid	*Pandora neoaphidis*	Synthesizing of antifungal molecules	Łukasik et al. (2013), Parker et al. (2013)
"Ca. Streptomyces philanthi"	Beewolf digger wasps	Fungui and bacteria in the soil	Antibiotic production (A complex cocktail of antibiotics)	Kaltenpoth et al. (2005, 2006, 2010), Goettler et al. (2007)
Pseudonocardia Streptomyces and *Amycolatopsis.*	Attine ants	*Escovopsis* and endophytic fungi	Antibiotic production	Currie et al. (1999a, b), Currie et al. (1999b, 2003b); Currie (2001a, b), Currie and Stuart (2001), Currie et al. (2003a), Cafaro et al. (2011), Vieira et al. (2012)
Actinobacterial isolates	Fungus-farming termites	*Pseudoxylaria*	Antibiotic production	Visser et al. (2012)
Streptomyces sp.	Bark/ambrosia beetle	*Ophiostoma minus*	Antibiotic production (mycangimycin)	Scott et al. (2008), Oh et al. (2009); Blodgett et al. (2010); Hulcr et al. (2011)
B. Gladioli	*Lagriavillosa*	Antagonistic fungi	Production of antifungal polyketid, lagriamide	Flórez and Kaltenpoth (2017), Flórez et al. (2017, 2018), Waterworth et al. (2020)
Spiroplasma sp.	*D. Neotestacea*	*Howardulaaoronymphium*	Toxins production, including a novel putative ribosome inactivating protein (RIP)	Jaenike et al. (2010), Jaenike and Brekke (2011); Hamilton et al. (2014)
Spiroplasma	Tsetse flies	Trypanosome	–	Schneider et al. (2019)

The symbiont, host, and natural enemy along with the known mechanism for the protection are listed

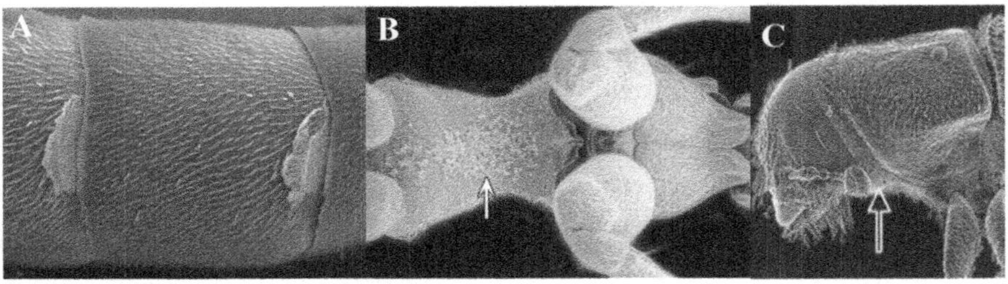

Fig. 5.4 Scanning electron micrographs showing the location of actinobacterial symbionts. (**a**) Symbiotic *Streptomyces* bacteria being secreted from the antennal glands of a female European beewolf, *Philanthus triangulum* (Kaltenpoth et al. 2006). (**b**) *Streptomyces* (arrow) under the forelegs of fungus-growing ant, *Apterostigma* sp. (Currie et al. 1999b). (**c**) Actinomycetous bacterium as well as a fungal food source, *Entomocorticium* sp. in mycangium of Dendroctonus frontalis (Scott et al. 2008)

probably synthesizing antifungal molecules (Brownlie and Johnson 2009; Flórez et al. 2015). For example, aphids that are infected with *R. insecticola* are at least five times more resistant to infection by the fungus than uninfected aphids. If the fungus does successfully infect and kill an aphid infected by *R. insecticola*, up to ten-fold fewer spores are produced (Brownlie and Johnson 2009). This may be adaptive for the aphids by reducing the spread of infection among groups of clonal aphids, thereby enhancing the inclusive fitness of not only individual insects but also the host population (Flórez et al. 2015).

It is demonstrated that *R. insecticola* protects pea aphids from *Zoophthoraoccidentalis*, a highly aphid-specific entomopathogen but not from *Beauveriabassiana* (Ascomycota: Hypocreales), a generalist that has been found in a variety of hosts. This finding highlights the complex influence of fungi on the dynamics of this economically important agricultural pest (Parker et al. 2013).

"*Candidatus* Streptomyces philanthi" and Beewolf Digger Wasps

Insects in three genera in the subfamily Philanthinae ("beewolves") are solitary digger wasps that engage in a highly specialized symbiotic association with "*Candidatus* Streptomyces philanthi" that protect the wasp offspring against pathogenic microorganisms (Kaltenpoth et al. 2005, 2006; Goettler et al. 2007). Transmission and scanning electron microscopy revealed that the bacterial symbionts are cultivated in large gland reservoirs in five antennomeres of female beewolves (Kaltenpoth et al. 2006; Goettler et al. 2007) (Fig. 5.4a). The antennal symbionts are transmitted vertically by an unusual mechanism of post-hatch transfer (Kaltenpoth et al. 2005). Female beewolves dig underground nests in soil, mass provision individual progeny in brood cells with insect prey (Kaltenpoth et al. 2010). After feeding on the provisioned prey, larvae spin a cocoon in which they usually overwinter (Kaltenpoth et al. 2006). During cocoon spinning, the bacteria are taken up by the larva and transfer to the walls of the cocoon (Kaltenpoth et al. 2005). For the following 2 weeks, the bacteria produce a cocktail of antibiotics, consisting primarily of streptochlorin and piericidin derivatives, which are distributed all over the surface of the cocoon and provide protection against detrimental fungi until the larval emergence several months later (Kaltenpoth et al. 2005; Flórez et al. 2015). Upon emergence, the adult female wasp picks up bacteria from the cocoon, which colonizes her antennal glands, continuing the cycle anew (Kaltenpoth et al. 2010).

The symbiotic association between beewolves and *Streptomyces* is categorized as a mutualistic relationship. Experimental removal of bacteria resulted in over 90% insect mortality, because bacteria enhance the survival probability of the larva, possibly by producing antibiotics (Kaltenpoth et al. 2005). On the other hand, the bacteria certainly benefit from the association by obtaining an ecological niche, a reliable route of transmission into the next generation, and nutrients from the beewolf (Kaltenpoth et al. 2006).

Pseudonocardia/Streptomyces/Amycolatopsis **and Fungus-Farming Ants**

Attine ants have long been an interesting model for symbiosis researchers because of their unique ability to "farm" fungi as a food source, providing the fungus with fresh plant material, and protecting it against competitors and pathogens (Currie 2001a; Currie and Stuart 2001; Brownlie and Johnson 2009). The successful fungus cultivation is threatened by specialized *Escovopsis* fungal pathogens and endophytic fungi, brought in by the ants with the plant substrate supplying the cultivars with nutrition (Currie et al. 1999b). To counteract the threats, ant workers combine continuous fungus grooming and weeding behaviors with the application of antimicrobial secretions from their metapleural glands as well as antimicrobials produced by symbiotic Actinobacteria (Fig. 5.4b) (Currie et al. 1999b; Currie and Stuart 2001; Vieira et al. 2012).

The Actinobacteria are typically maintained on the cuticle of workers and comprise vertically and occasionally horizontally transmitted *Pseudonocardia* symbionts (Currie et al. 1999b, 2003b; Poulsen et al. 2005; Cafaro et al. 2011) as well as environmentally acquired members of the genera *Streptomyces* and *Amycolatopsis* (Flórez et al. 2015). Infection experiments, bioassay challenges, and chemical analyses support the role of *Pseudonocardia* in defense against the specialized *Escovopsis* cultivar pathogens through antibiotic production (Currie et al. 1999a,2003a Currie 2001b). Antibiotic assays suggest that despite *Escovopsis* being generally susceptible to inhibition by diverse Actinobacteria, the ant-derived *Pseudonocardia* inhibit *Escovopsis* more strongly than they inhibit other fungi (Cafaro et al. 2011). *Streptomyces* and *Amycolatopsis*, on the other hand, defend mostly against endophytic fungi in the leaf substrate using antibiotics with broad-spectrum activities (Flórez et al. 2015).

Apart from leaf-cutter ants, *Allomerus* ants, which lie outside the tribe Attini, are associated with antifungal-producing Actinobacteria. These symbionts were isolated from the cuticle of *Allomerus* ants, and were hypothesized to play a role in the protection of the galleries against non-cultivar fungi isolated from their ant-plants (Seipke et al. 2012).

Streptomyces Sp. and Bark/Ambrosia Beetles

A similar ecological system, again involving fungal and Actinobacterial symbionts, was identified in the bark and ambrosia beetle. In a study on Dendroctonus frontalis bark beetles, it was found that *Streptomyces thermosacchari* is present in the insect oral secretions, galleries, and on the beetle's mycangia, a pair of sub-cuticular cavities located on their pronotum that are adapted for the transport of fungal symbiont and actinomycetous bacterium (Scott et al. 2008; Hulcr et al. 2011) (Fig. 5.4c).

The symbiont specifically protects fungal food source, *Entomocorticium* sp., from competing fungi, *Ophiostoma minus*, by producing a polyene peroxide, which was named mycangimycin (Scott et al. 2008). Successful maintenance of the D. frontalis symbiosis with fungal cultivar is likely mediated by the actinomycetous bacterium (Scott et al. 2008) as the bioassays revealed that mycangimycin selectively inhibits the beetle's antagonistic fungus but only slightly affects the beneficial fungus (Scott et al. 2008; Oh et al. 2009). However, *Streptomyces* are not consistently present in D. frontalis nests and are generally isolated at very low frequencies from some other species of bark and ambrosia beetles (Hulcr et al. 2011).

Another *Streptomyces* strain, SPB78, displayed no activity in Petri dish competition assays with associates of D. frontalis, but produces two antifungals, frontalamides A and B, under certain culture conditions (Blodgett et al. 2010).

Actinobacteria and Fungus-Farming Termites

Much less is known about the role of defensive bacterial symbionts in the gardens of the fungus-farming termites, compared to fungus-growing ants (Flórez et al. 2015). Fungus-growing termites live in mutualistic symbiosis with *Termitomyces*, but other fungi species, *Pseudoxylaria*, are latently present in the nests and appear to compete with *Termitomyces* for the substrate provided by the termites. The termites are thus predicted to have evolved strategies to suppress *Pseudoxylaria* within nests (Visser et al. 2012). As in leaf-cutter ants, Actinobacteria have been isolated from termite nests (Flórez et al. 2015). Antibiotic-activity screening of 288 Actinobacterial isolates revealed antifungal activity against fungal competitors (*Pseudoxylaria*) as well as the termite fungus (*Termitomyces*). A more detailed bioassay on 53 isolates showed that the termite fungus generally is more susceptible to inhibition than the competitor (Visser et al. 2012). This indicates that antifungals are either applied to specific areas by the termites, or unspecific Actinobacteria were isolated that do not act as defensive symbionts in fungus-farming termites (Visser et al. 2012; Flórez et al. 2015).

Instead, another study suggests that a strain of antibiotic-producing *Bacillus* may be a defensive symbiont involved in the cultivar protection of the *Macrotermesnatalensis* fungus-growing termite (Um et al. 2013). The bacterial symbiont produces a single major compound, bacillaene A, which specifically in vitro inhibits known (*Pseudoxylaria* and *Trichoderma*) and potentially competitive or antagonistic (*Coriolopsis*, *Umbelopsis,* and *Fusarium*) fungi (Um et al. 2013).

Previous work in *Odontotermesformosanus* fungus-growing termites has also suggested that gut- and fungus comb-residing *Bacillus* sp. produce a secretion that in vitro inhibits antagonistic fungi (*Trichoderma*), but not termite fungus (Mathew et al. 2012).

Burkholderia-Lagria Beetles

Another protective symbiosis was recently discovered between multiple strains of *Burkholderia* bacteria and tenebrionid beetles in the subfamily Lagriinae. Symbiotic bacteria are housed extracellularly in adult female accessory glands in the reproductive tract. Although at least some *Burkholderia* are vertically transmitted to offspring, phylogenetic analysis and the lack of symbiont monophyly indicate that these symbionts are predominantly horizontally transmitted (Flórez and Kaltenpoth 2017; Flórez et al. 2017).

In laboratory assays, the bacteria were found to protect eggs and larvae from pathogenic fungi (Flórez et al. 2017). Experiments in *Lagriavillosa* beetles demonstrate that the symbionts inhibit the growth of antagonistic fungi on the eggs of the insect host (Flórez and Kaltenpoth 2017; Flórez et al. 2017). *B. gladioli* inhibit the growth of the soil fungus *Trichoderma harzianum* and the entomopathogen *Beauveriabassiana* in vivo. In addition, aposymbiotic eggs suffer more frequently from infestation by *Purpureocilliumlilacinum* as a natural enemy of insect adults and larvae. These investigations revealed generalized antifungal protection made by *Burkholderia* symbionts (Flórez et al. 2017).

Burkholderia gladioli Lv-StB, which is unculturable and dominant in field-collected L. villosa, has a reduced genome that encodes antifungal polyketide lagriamide (Flórez et al. 2018; Waterworth et al. 2020). In spite of genome reduction and gene loss, horizontal acquisition of the lagriamide *lga* biosynthetic gene cluster has been occurred in *B. gladioli*. Interestingly, lagriamide is structurally similar to bistramides, defensive compounds found in marine tunicates, and are similar to poisons found in ascidians (Flórez et al. 2018; Waterworth et al. 2020).

On the other hand, the only culturable strain, *B. gladioli* Lv-StA, produces a wide range of poisons. HPLC-MS-based metabolic profiling of *B. gladioli* Lv-StA culture extracts confirmed the production of the azapteridinetoxoflavin and the polyynecaryoynencin as well as a polyketide structurally related to etnangien, which we named lagriene (Flórez et al. 2017). These compounds are likely responsible for the protective activity of this strain in situ (Flórez et al. 2018).

Spiroplasma Sp. and *Drosophila Neotestacea*

Little is known about symbiont-mediated defense against nematodes. The reported case is in the mushroom-feeding woodland fly, *Drosophila neotestacea*, in which *Spiroplasma* symbionts can protect the insect against a sterilising parasitic nematode, *Howardulaaoronymphium* (Jaenike et al. 2010). The symbiont significantly enhances the reproductive output of flies that are parasitized by the *H. aoronymphium* both in laboratory and wild populations (Jaenike et al. 2010; Hamilton et al. 2014). The presence of *Spiroplasma* results in reduced growth of the adult female nematodes within the host and ultimately in impaired fertility of the parasite as well as reduced virulence against the host (Jaenike et al. 2010). In addition, the nematodes appear sickly and small, and produce virtually no infective juveniles in *Spiroplasma*-infected flies (Jaenike et al. 2010).

In experimental populations, *Spiroplasma* spreads in the presence of nematodes, but declines in their absence (Jaenike and Brekke 2011). Although the findings show little support for exploitative competition or immune priming to mediate defense, the protection is proposed to be linked to the production of the putative toxins, including novel putative toxins called ribosome-inactivating proteins (RIPs) (Hamilton et al. 2014). RIPs target a highly conserved adenine residue in the α-sarcin/ricin loop of eukaryotic 28S ribosomal RNA (Oliver and Perlman 2020). This effect can explain why the recent spread of *Spiroplasma* in natural populations of *D. neotestacea* has coincided with a decline in the prevalence of *Howardula* parasitism in the wild (Jaenike and Brekke 2011).

Spiroplasma and Tsetse Flies

Inherited *Spiroplasma* strains have been implicated in protection against trypanosome in tsetse flies, with only 2% of *Spiroplasma* infected flies harboring trypanosome co-infections. In controlled laboratory infections, *Spiroplasma*-positive flies were less likely to be infected with trypanosome gut parasites. These results indicate that *Spiroplasma* infections may have important effects on vector control approaches to reduce trypanosome infections (Schneider et al. 2019). More work is needed to determine whether protection against trypanosome is mediated by symbiont-encoded toxins (esp. RIPs) (Oliver and Perlman 2020).

Wolbachia and Different Flies

During the last few years, numerous studies have reported that *Wolbachia* infection has an anti-pathogenic effect in different flies including *Drosophila*, *Aedes*, *Anopheles*, and *Culex* species (Fig. 5.5). The protections have been reported against several RNA viruses, different *Plasmodium* species, fungi, bacteria, and nematodes. Most of the best described protections are described here.

In *Aedesaegypti*, infections with some viruses, a pathogenic nematode, some bacteria as well as the malaria-causing protozoan parasite *Plasmodium*, are affected when the insect is infected with *Wolbachia* (Kambris et al. 2009; Moreira et al. 2009; van den Hurk et al. 2012; Ye et al. 2013). The *Wolbachia*'s antiviral and anti-*Plasmodium* properties in *Ae. aegypti* are proposed to achieve through one or more afterward mechanisms: activation of the host's immune system, competition for host resources, manipulation of the host defense pathways, such as the microRNA, and interference with pathogen replication (Moreira et al. 2009; van den Hurk et al. 2012; Ye et al. 2015). Whole-genome microarrays revealed that *Wolbachia* strains activated the expression of some immune genes including anti-microbial peptides, Toll pathway genes, and genes involved in melanization (Rancès et al. 2012). In addition, the upregulation of the mosquito's innate immune system inhibits the development of filarial nematodes in the mosquito (Kambris et al. 2009). Another work shows a correlation between the levels of innate immune priming induced by different *Wolbachia* strains in *Ae. aegypti* with the degree of protection conferred

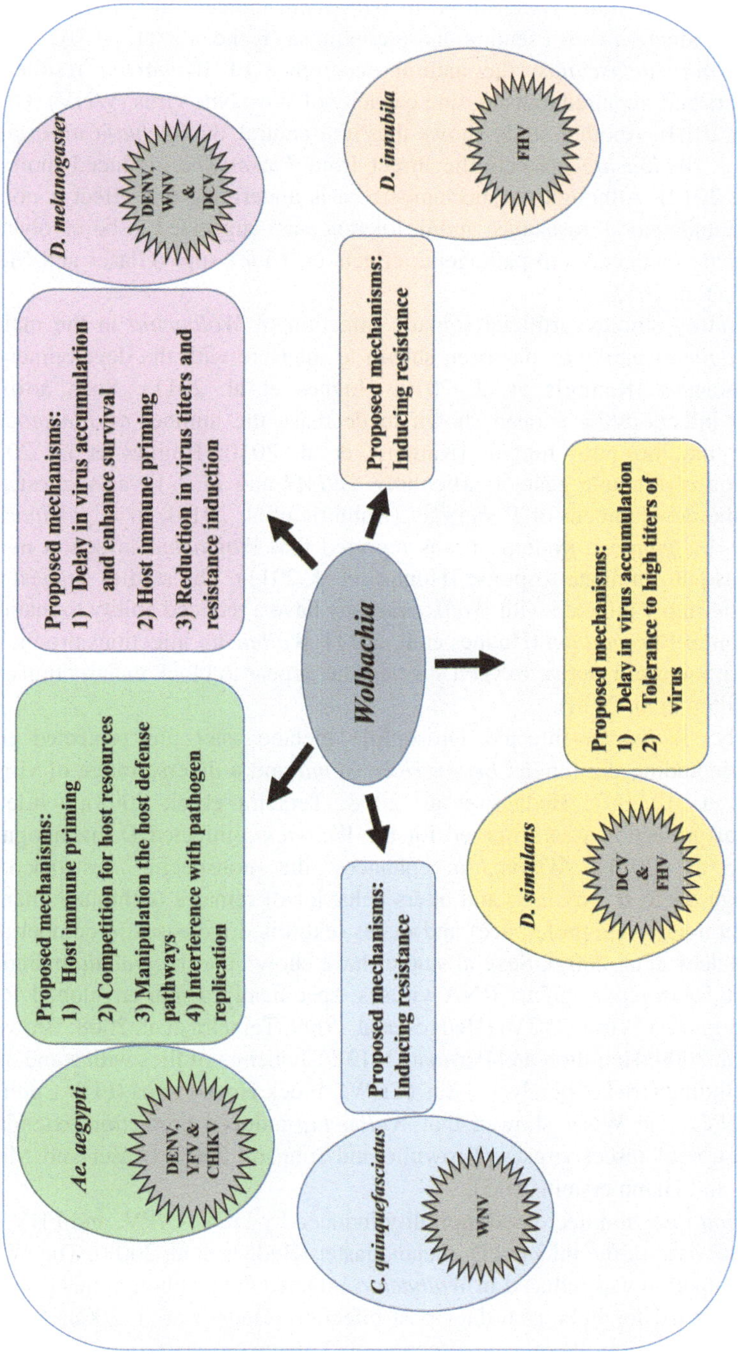

Fig. 5.5 *Wolbachia*-mediated protection against different pathogens of flies: *Aedes*(*Ae.*), *Drosophila* (*D.*), *Culex* (*C.*) *Drosophila* C viruses (DCV), Cricket paralysis virus (CrPV), Flock House virus (FHV), dengue virus (DENV), Yellow Fever Virus (YFV), chikungunya virus (CHIKV), and West Nile virus (WNV)

against bacterial pathogens (Ye et al. 2013). Co-expression of antimicrobial peptides, Cecropin A and Defensin A, in *Wolbachia*-infected *Ae. aegypti* induces resistance to infection with Pseudomonas aeruginosa (Kokoza et al. 2010).

In *Culexquinquefasciatus*, the natural occurrence of *Wolbachia* resulted in reduced titers and impaired transmission capacity of West Nile virus (WNV) (Glaser and Meola 2010). Another study shows that in a natural *Wolbachia–Culexpipiens* combination, *Wolbachia* protects the insect from *Plasmodium*-induced mortality (Zélé et al. 2012). Although the mechanistic basis underlying this effect is not yet completely understood, resistance and/or tolerance are suggested to be involved in the *Wolbachia*-mediated anti-pathogenic effects on *Culex* spp. (Glaser and Meola 2010; Zélé et al. 2012).

In laboratory studies, artificial somatic infection of *Wolbachia* in the malaria vector, *Anopheles gambiae,* has been shown to interfere with the development of malaria parasites (Kambris et al. 2010; Hughes et al. 2011). Such artificial *Wolbachia* infections have been shown to decrease the number of *Plasmodium* oocysts in the mosquito midgut (Kambris et al. 2010; Hughes et al. 2011). Upregulation of immune genes (particularly, *LRIM1* and *TEP1*) was suggested to influence the development of *P. berghei* (Kambris et al. 2010). In *P. falciparum-Wolbachia-An. gambiae* system, it was reported that *Wolbachia* infection modulate the mosquito immune response (Hughes et al. 2011). The studies suggest that *An. gambiae* stably infected with *Wolbachia* may have a reduced ability to maintain transmission of *Plasmodium* (Hughes et al. 2011). *Wolbachia* infections produce an enormous amount of reactive oxygen species that appear to block malaria transmission (Hamilton et al. 2014).

Naturally, *Wolbachia*-infected Drosophila melanogaster are protected from pathogens including the fungus *Beauveriabassiana* and a diverse range of viruses (Panteleev et al. 2007; Hedges et al. 2008; Teixeira et al. 2008), while no antibacterial protection was observed for the *Wolbachia*-infected D. melanogaster (Wong et al. 2011). *Wolbachia* enhances the nonspecific resistance of D. melanogaster to *B. bassiana* and alters behavior of females (exhibited changes in oviposition substrate preference) and males (exhibited increases in competitiveness) (Panteleev et al. 2007). Several studies have shown that *Wolbachia* infections can protect *Drosophila* against RNA viruses, specifically a non-enveloped RNA virus, *Drosophila* C virus (DCV) (Hedges et al. 2008; Teixeira et al. 2008; Brownlie and Johnson 2009; Hamilton and Perlman 2013). Challenge of flies with some other viruses including Cricket paralysis virus (CrPV), Flock House virus (FHV), dengue virus (DENV), and WNV showed that *Wolbachia*-induced protection extends to diverse groups of insect viruses (Brownlie and Johnson 2009; Glaser and Meola 2010; Zug and Hammerstein 2015).

Wolbachia infection decreased mortality induced by DCV, CrPV, and FHV and increased survival of the infected D. melanogaster (Hedges et al. 2008). The rate of DCV accumulation was reduced in *Wolbachia*-infected flies, although this decrease was not identified for FHV at 6 days post-infection (Hedges et al. 2008; Teixeira et al. 2008). In addition, *Wolbachia* infection of D. melanogaster reduced virus titers and induced strong resistance to WNV infection (Glaser and Meola 2010). The enhanced resistance in dengue-infected flies may result from the host's innate immune system being primed by *Wolbachia* (Rancès et al. 2012).

In addition, the natural strains of *Wolbachia* have been shown to defend D. simulans, and D. innubila against multiple RNA viruses (Osborne et al. 2009; Unckless and Jaenike 2012). Specifically, the experiments showed that *Wolbachia* increases the survival of D. innubila infected with flock house virus (FHV) and induces resistance to virus infection (Unckless and Jaenike 2012). Antiviral protection against DCV and FHV was observed for some *Wolbachia* strain–D. simulans line combinations with a delay in virus accumulation and a tolerance to high titers of virus (Osborne et al. 2009). These antipathogenic effects may be of importance in controlling vector-transmitted viral diseases.

5.4 The Potential Application of Symbionts in Pest Control

Insect symbionts offer an opportunity to deal with the anticipated elevated demand for novel pest management strategies created by growing human populations and global climate change (Engel and Moran 2013; Marzieh Kashkouli et al. 2018). Four main approaches have particular potential: (i) utilizing heterologous microorganisms, (ii) paratransgenesis, (iii) insect incompatibility technique (IIT), and (iv) the disruption of microbial symbionts required by insect pests (Fig. 5.6). These strategies are under development, particularly targeting the disease agent in the insect vectors of human disease agents. Here, we reviewed the well-known examples and the progress of using these strategies for the management of vector-borne diseases, in the context of the demand for novel methods of insect pest control that are both durable and environmentally benign.

5.4.1 Heterologous Associations

The presence of natural enemies is a situation that might reveal possible host benefits provided by symbionts. During the last few years, numerous studies have reported that *Wolbachia* infection has an anti-pathogenic effect in the host (Zug and Hammerstein 2015). A key element in the use of *Wolbachia* for the control of insect-borne disease has been the discovery that some *Wolbachia* strains can confer enhanced resistance toward various insect viruses in *Drosophila* and some other dipteran vectors, such as C. quinquefasciatus, *Ae. aegypti*, and *An. gambiae* (see "*Wolbachia*-different flies" section). Thus, the use of microbial symbionts is a promising research area to control the incidence of numerous devastating diseases, including malaria, dengue, yellow fever, and Chagas. To limit vector-borne diseases, one of the most active research areas include utilizing heterologous microorganisms that shorten life span and lower fertility of the insect host or that reduce its susceptibility to pathogens or parasites.

Heterologous associations are generated by the experimental transfer of microorganisms from one species into another species (naturally does not harbor

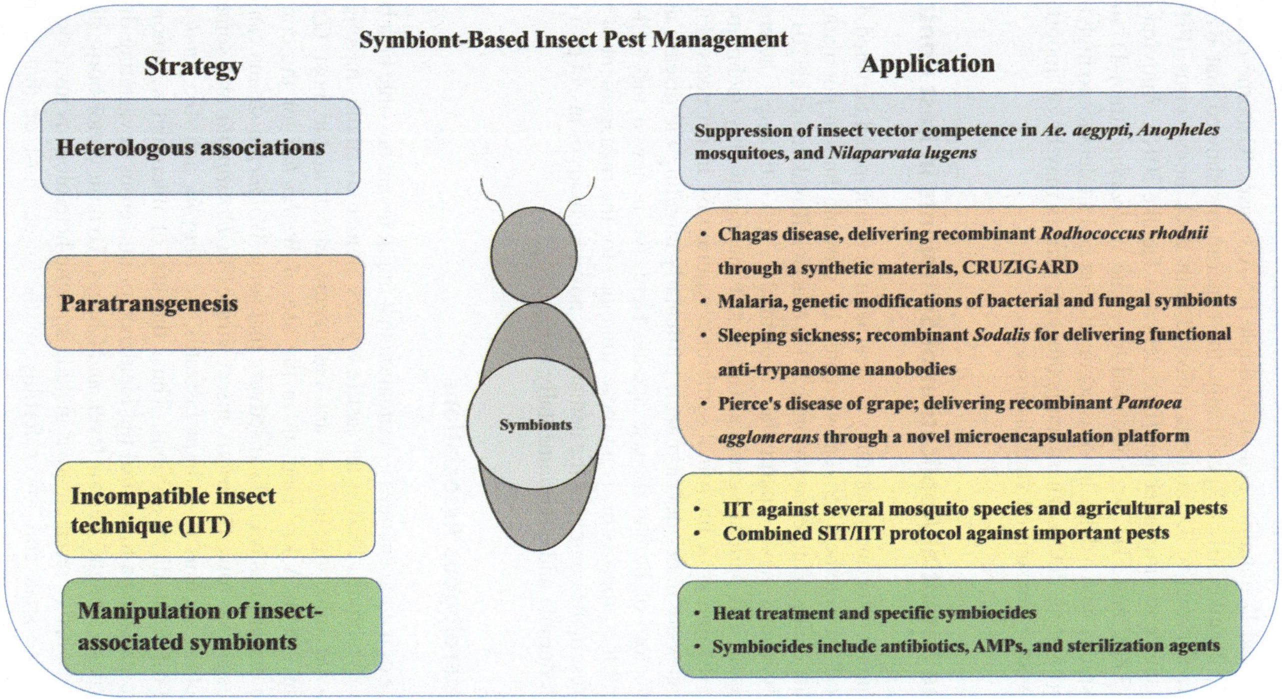

Fig. 5.6 Symbiont-based insect pest management strategies and applications. Sterile insect technique (SIT)

the microorganism) (Arora and Douglas 2017). Major vectors of human diseases, such as *An.* species and *Ae. Aegypti,* do not harbor *Wolbachia* naturally, so the transinfection of *Wolbachia* into mosquitoes is required (Iturbe-Ormaetxe et al. 2011;Walker and Moreira 2011). The presence of *Wolbachia* in transinfected mosquitoes interferes with pathogen transmission and with a wider range of pathogens (Moreira et al. 2009; Kambris et al. 2010) maybe because *Wolbachia* usually grows to high densities and more broadly inhibit the development of diverse parasites and pathogens in heterologous hosts compared to native host species (Glaser and Meola 2010). To facilitate the transfer of *Wolbachia* strains between distantly related insect species, mosquito cellline adaptation to new intracellular environments appears to be critical for transinfection success (Braig et al. 1994).

The life-shortening strain of *Wolbachia*, *w*MelPop, was predicted to have potential use in the control of mosquito-borne diseases by transferring into mosquito populations to reduce mosquito life span (McMeniman et al. 2009; Guruprasad et al. 2014). Although mosquito-borne pathogens can differ intrinsically, mosquito age is a critical factor for their transmission. Pathogens need to replicate in the body of the mosquito before reaching the salivary glands, which undergo an extrinsic incubation period (EIP) as the incubation period required from the ingestion of the pathogen until it is transmitted into a new human host by a bite (usually 10–14 days for dengue fever) (Iturbe-Ormaetxe et al. 2011; Walker and Moreira 2011). Since the EIP of many pathogens is quite long relative to the natural insect life span, female mosquitoes that are older than the EIP are vectors of epidemiological importance (Hughes et al. 2011; Iturbe-Ormaetxe et al. 2011). Transinfection of *w*MelPop into mosquito's population actually removes the older individuals and has the potential to decrease disease transmission (Sinkins and O'Neill 2000).

Several non-mutually exclusive hypotheses have been proposed to explain the antiviral effects of *Wolbachia* in heterologous associations in mosquitoes. In these novel hosts, both metabolic and immunological processes are likely involved: these include priming of the immune system such as the Toll pathways, and the production of toxic reactive oxygen species (Hamilton and Perlman 2013). The dual effect of life-shortening and pathogen interference can act synergistically, enhancing the prospects for *Wolbachia*-based disease control strategies in the transinfected hosts (Hughes et al. 2011). Here, heterologous associations in *Ae. Aegypti*, *Anopheles* mosquitoes, and *Nilaparvatalugens* are discussed.

Wolbachia Transinfection of *Ae. Aegypti*

The most known and developed insect pest management strategy based on the construction of a heterologous association relates to the introduction of *Wolbachia* (from *Drosophila* or other mosquitoes, e.g., Aedes albopictus and *Culexquinquefasciatus*) into *Ae. Aegypti* (Arora and Douglas 2017; Qadri et al. 2020). The *w*MelPop was introduced from D. melanogaster to *Ae. aegypti* by a two-step process comprising serial passage in mosquito cell culture to allow "adaptation" to the mosquito intracellular environment for over 3 years and then the mosquito cellline-adapted *Wolbachia* strain, *w*MelPop-CLA, was stably introduced into naturally uninfected embryos of *Ae. aegypti* using embryo microinjection, leading to stable colonization and vertical transmission (McMeniman et al. 2008, 2009).

Strain *w*MelPop halved the life span of transinfected *Ae. aegypti*, negatively affected mosquito survival, with no differences in fecundity (McMeniman et al. 2009). The heterologous *Wolbachia* can also cause CI with native *Wolbachia*-free *Ae. aegypti*, resulting in the displacement of the native *Wolbachia*-free populations and providing a ready mechanism for the drive of *Wolbachia* and defensive traits into vector populations (McMeniman et al. 2009; Hamilton and Perlman 2013; Ferguson et al. 2015). This effect is distinct from the related technology of IIT (insect incompatibility technique), which does not involve the construction of heterologous associations and relies on *Wolbachia*-mediated cytoplasmic incompatibility (Arora and Douglas 2017).

Infection of the *w*MelPop-CLA strain in *Ae. aegypti* results in approximately 50% reduction in the adult life span, thereby the ability of the insect to transmit dengue virus to a mammalian host decreased as much as 70% (McMeniman et al. 2009; Yeap et al. 2011; Ferguson et al. 2015). Remarkably, the acquisition of dengue virus by *Ae. aegypti* females bearing the heterologous *Wolbachia* was substantially reduced, both under laboratory conditions and in the field (Hoffmann et al. 2011; Walker et al. 2011). Laboratory and field examinations revealed the recapitulation of CI in mosquitoes and also the fact that the bacterial symbiont is able to approach fixation in mosquito populations within a few generations (Hoffmann et al. 2011; Hamilton and Perlman 2013; Berasategui et al. 2016). These experiments created a basis for the Eliminate Dengue Program (http://www.eliminatedengue.com), which has released *Wolbachia*-transinfected *Ae. aegypti* into several dengue-endemic regions (Arora and Douglas 2017).

In 2011, *Ae. aegypti* carrying the *w*Mel strain were released into the wild near Cairns, Australia, marking the first trial of microbiome manipulation of a wild insect population for the purpose of reducing vector competence (Hoffmann et al. 2011). A follow-up investigation in late 2019 indicated a 96% reduction in dengue incidence in *Wolbachia*-treated populations (Ryan et al. 2019). Subsequent releases have established *Wolbachia* throughout Indonesia (Tantowijoyo et al. 2020), Malaysia (Nazni et al. 2019), Vietnam, and Brazil (Arora and Douglas 2017).

Stable introduction of *Wolbachia* into *Ae. aegypti* mosquitoes and cell lines can suppress vector competence for some other pathogens, such as ZIKV, YFV, WNV, CHIKV, and Mayaro virus, as well as avian malaria parasites and filarial nematodes (Moreira et al. 2009; Walker et al. 2011; van den Hurk et al. 2012; Hussain et al. 2013; Ye et al. 2015; Aliota et al. 2016; Dutra et al. 2016; Parry et al. 2019). The strain *w*MelPop-CLA was able to diminish YFV replication and dissemination in transinfected *Ae. Aegypti* (van den Hurk et al. 2012). Oral feeding experiments of *w*MelPop-CLA transinfected *Ae. aegypti* revealed antiviral efficacy against CHIKV (Moreira et al. 2009; Walker et al. 2011). *Plasmodium* oocyst load was reduced by 67–88% in mosquitoes transinfected with *w*MelPop-CLA compared to untreated individuals 7 days after feeding on an infected chicken (Moreira et al. 2009). These studies provide excellent examples of how heterologous associations could affect a wide range of pathogens related to an insect vector (Berasategui et al. 2016).

Risk assessments have concluded that there is a negligible risk of the release of *Wolbachia*-infected A. aegypti, mostly related to a possible horizontal transfer of *Wolbachia* DNA into mosquito genomes, but this event takes place over evolutionary timescales and are extremely rare (Iturbe-Ormaetxe et al. 2011). Also, it is suggested that the potential of *Wolbachia*-infected A. aegypti to invade populations and persist will depend on interactions with environmental conditions, notably fluctuating temperature (Ross et al. 2017). Thus, increasing our understanding of virus/*Wolbachia* ecology and distribution is essential to understand the impact of the *Wolbachia*-mediated antiviral protection on natural *Ae. aegypti* populations.

Wolbachia Transinfection of *Anopheles* Mosquitoes

The effect of *Wolbachia* introduction on mosquito vector competence is not restricted to *Ae. aegypti*–RNA virus interactions (Arora and Douglas 2017). *Wolbachia* have also been introduced into the anopheline vectors of malaria, which they do not naturally infect. Transient somatic infections of *w*MelPop were created by intra-thoracic inoculation in *An. gambiae* females. In these insects, the mean *P. berghei* levels were reduced by 75–84% (Kambris et al. 2010). An additional study shows that transient somatic infections of two diverse *Wolbachia* strains (*w*MelPop and *w*AlbB) can substantially suppress the development of P. falciparum in *An. gambiae.* Both *Wolbachia* strains significantly inhibit *Plasmodium* oocyst levels in the mosquito midgut (Hughes et al. 2011). Other studies present that the transinfected *An. Stephensi* (with a *Wolbachia* strain native to *Ae. albopictus*) displays suppressed vector competence for *Plasmodium* and also induces CI, providing the basis for the spread of the vector-incompetent mosquitoes horizontally in natural populations (Bian et al. 2013; Joshi et al. 2017).

Several studies present evidence that the use of *Wolbachia* for malaria control will require induction of the CI phenotype and a stable infection through the vertical transmission to successfully invade wild populations (Walker and Moreira 2011). Thus an unstable threshold infection level (depends on the negative selection imposed by fitness costs of *Wolbachia* infection and positive selection associated with CI induction) is defined to assess the ability of any *Wolbachia* strain to successfully invade wild mosquito populations (Turelli 2010). Furthermore, the complexity of *Anopheles* mosquitoes populations would be a major complicating factor in the applied use of *Wolbachia* for malaria control (Walker and Moreira 2011).

Although the combination of vector/parasite does not occur in nature, the results suggest that if stable transinfections act in a similar manner to somatic infections, *Wolbachia* could potentially be used as part of a strategy to control the *Anopheles* mosquitoes in some malaria-endemic areas in which a single vector species is present. Further studies on this finding can greatly open the potential use of this methodology for malaria control.

Wolbachia Transinfection of *Nilaparvatalugens*

Although some progresses have been made in employing heterologous associations as a tool for protecting humans from mosquito-borne diseases, this approach has been poorly developed for the protection of plants from insect pests and their associated diseases. Recently, stable introduction of *Wolbachia* strain *w*Stri (from its native host, *Laodelphaxstriatellus*) into the brown planthopper, *Nilaparvatalugens* (cured of their native *w*Lug infection), was reported. The strain *w*Stri maintained perfect maternal transmission and induced high levels of CI, enabling rapid invasion of laboratory populations. Furthermore, *w*Stri inhibited both infection and transmission of Rice ragged stunt virus (RRSV) and mitigated virus-induced symptoms in rice plants. This study opens up the development of heterologous associations strategy against major agricultural pests and their transmitted pathogens (Gong et al. 2020).

5.4.2 Paratransgenesis

Paratransgenesis is the alteration of insect traits by genetic manipulation of associated microorganisms (Beard et al. 1998). The potential of this technology in insect pest control was proposed back in the early 1990s, especially in relation to mosquito vectors of human disease agents and several hemipteran crop pests (Beard et al. 1993; Arora and Douglas 2017). In this strategy, engineered gut bacteria may be used as vehicles to produce proteins impairing pathogen development and insect fitness (Olson et al. 1996; Coutinho-Abreu et al. 2010; Caragata and Walker 2012; Engel and Moran 2013).

Paratransgenesis is widely perceived as an alternative to transgenesis (direct genetic manipulation of the insects), despite the existence of various transgenic germlines of disease-transmitting insects (Coutinho-Abreu et al. 2010). Many challenges still exist to the application of transgenesis to control vector-borne diseases in the environment, such as reduced fitness of transgenes outside the laboratory (Coutinho-Abreu et al. 2010). Furthermore, the inefficiency of current methods for introducing and propagating transgenes in natural vector populations, and the negative public opinion could be mentioned (Engel and Moran 2013). Paratransgenesis could help to overcome the disadvantages of transgenesis without a genetic drive system required for insect transgenes and transgene instability in insect genomes (Arora and Douglas 2017; Qadri et al. 2020).

The key requirements for paratransgenesis are that the microbial partner can be cultured, transformed, and readily reintroduced into the insect hosts to facilitate the dissemination of the desired trait (Beard et al. 2002). In addition, the microbes should ideally be specific to target insects or harmless to nontarget hosts (Qadri et al. 2020). Gut bacteria are suitable for the development of paratransgenic applications because of having the aforementioned features. The large diversity of gut bacteria can easily be re-introduced into the host insect by, for example, fecal-oral transmission and disseminated in the environment via horizontal transfer to acquire gut microorganisms from conspecifics. Furthermore, most pathogens specifically colonize the guts of their vectors, which directly expose them to the effector proteins of the genetically modified gut symbionts. Thus, genetically amenable gut bacteria, i.e., *Sodalisglossinidius, Asaia* sp., and *Pantoeaagglomerans*, represent a valuable resource for paratransgenesis (Engel and Moran 2013). Applications of several genetically modified gut symbionts against Chagas disease, malaria, sleeping sickness, and pierce's disease of grape are fully discussed below.

Paratransgenesis Against Chagas Disease

Paratransgenesis was first carried out on the triatomine *R. prolixus*, the vector of the Chagas-causing protozoan *Trypanosoma cruzi* through the manipulation of the insect gut flora. A member of its microbial community, *Rodhococcusrhodnii*, co-localizes in the midgut with T. cruzi and is essential for the growth and development of the host (Beard et al. 2001, 2002). The transmission ecology of this symbiont and its amenability for in vitro cultivation and genetic transformation has presented the system as a useful platform to apply paratransgenesis as a way to limit the transmission of Chagas disease (Durvasula et al. 1999).

R. rhodnii has been transformed to produce anti-trypanosomal effector molecules, cecropin A and related pore-forming molecules, which were shown to exhibit effective inhibitory activity against T. cruzi in vivo (Beard et al. 1992, 1993; Qadri et al. 2020). In insects carrying the transformed bacteria, elimination or reduction in the number of T. cruzi, often to undetectable levels was observed (Durvasula et al. 1997). Furthermore, the transformation of the symbiont with an anti-trypanosome single-chain antibody, the first description of a functional mammalian antibody fragment expressed in an insect, was carried out and showed a significant reduction in parasite load (Durvasula et al. 1999).

The promising results from laboratory studies led to subsequent field trials in testing the dispersal efficiency of transformed symbionts. Naturally, the first instar nymphs of *R. prolixus* acquire the extracellular symbiont by probing the symbiont-contaminated feces of adults (coprophagy). In the field and semi-field trials, the insect progeny can be infected by adding engineered symbionts to synthetic insect-fecal materials called CRUZIGARD (Durvasula et al. 1999). CRUZIGARD consists of an inert guar gum matrix dyed with India ink and dispersed as droplets in the environment (Durvasula et al. 1999; Hurwitz et al. 2011).

More recently, a study has integrated paratransgenesis with RNA interference (RNAi) technology to control *R. prolixus*. In this system, engineered Escherichia coli strain HT115 expressing dsRNA for heme-binding protein and catalase was successfully introduced into the gut of *R. prolixus*. This combination results in serious fitness consequences for the bug including poor development of nymphs and reduced fecundity of females (Taracena et al. 2015).

Paratransgenesis Against Malaria

Paratransgenesis has also been applied to the anopheline mosquito vectors of *Plasmodium* through the manipulations of bacteria and fungi isolated from the mosquito. Early reports were based on the recombinant E. coli expressing a single-chain immunotoxin (Yoshida et al. 1999). However, E. coli could be used only as a laboratory model in paratransgenesis studies because they survive poorly in the mosquito midgut (Chavshin et al. 2013; Wang and Jacobs-Lorena 2013). Other bacteria, including *Pantoeaagglomerans, Asaiabogorensis*, and *Serratia* sp., have been transformed and applied to target *Plasmodium* persistence and development.

The Gram-negative bacteria of the genus *Asaia* may be suitable for paratransgenesis because they can be cultivated, genetically manipulated, and can recolonize the insect host (Favia et al. 2007, 2008). This symbiont is stably associated with midgut, salivary glands, and reproductive organs of *An. stephensi* (Favia et al. 2007). In the first use of *Asaia* sp. for paratransgenesis against *P. berghei,* genetically modified *Asaia* strains were constructed by fusing the siderophore receptor gene with anti-plasmodial effector genes. These genes included the scorpine antimicrobial peptide and a synthetic anti-Pbs21 scFv-Shiva1 immunotoxin composed of a single-chain antibody (scFv) against *P. berghei* ookinete surface protein 21-Shiva1 fusion protein. The development of *P. berghei* was significantly inhibited by effector proteins secreted from transformed *Asaia* (Bongio and Lampe 2015). Another study confirmed the ability of genetically modified strains of *Asaia* to colonize *An. gambiae* and suggest that *Asaia* has potential for use in the paratransgenic control of malaria transmitted by *An. Gambiae* (Damiani et al. 2010).

P. agglomerans, a bacterium commonly found in laboratory-reared *An. stephensi, An. gambiae, and An. albimanus*, easily grows in culture and can be engineered to express anti-*Plasmodium* effector proteins using Type I hemolysin secretion system derived from E. coli. These recombinant bacteria were found to strongly inhibit the development of P. falciparum and *P. berghei* in the midgut of *Anopheles* mosquitoes. Inhibition varied from 85% for the effector mPLA2 to 98% for scorpine (Wang et al. 2012; Wang and Jacobs-Lorena 2013).

Serratia sp. AS1, isolated from *Anopheles* ovaries, stably colonizes the mosquito midgut, female ovaries, and male accessory glands. The AS1 strain spreads rapidly using sexual and vertical transmissions, persisting for at least three generations. This strain was genetically engineered for the secretion of five different anti-*Plasmodium* effector proteins, and the recombinant strains inhibit the development of P. falciparum in mosquitoes and reduce the oocyte load by 93% (Wang et al. 2017).

The investigation of the midgut microbiota associated with *An. Stephensi* and *An. maculipennis* revealed that the majority of the identified bacteria belonged to the γ-proteobacteria class, including *Pseudomonas* sp. and *Aeromonas* sp. (Dinparast Djadid et al. 2011). Identification of culturable bacteria from wild *An. culicifacies* revealed 12 bacterial genera and predominantly the genus *Pseudomonas* (Chavshin et al. 2014). These studies could be helpful for the selection of a paratransgenesis candidate and the development of a paratransgenesis-based approach for the control of malaria.

Among fungi, the entomopathogen *Metarhiziumanisopliae* has also been manipulated to secrete the anti-plasmodial peptide SM1, a single-chain antibody that agglutinates sporozoites, and antimicrobial toxin scorpine. Application of these three types of *M. anisopliae* recombinants reduced sporozoite counts by 71%, 85%, and 90%, respectively. The results revealed that genetically modified *M. anisopliae* is capable of inhibiting the development of the parasite and could be a powerful weapon for combating malaria (Fang et al. 2011).

Paratransgenesis Against Sleeping Sickness

Similarly, paratransgenic approaches have been tested on tsetse flies, vector of T. brucei, the etiological agent of sleeping sickness. The ability to culture one of tsetse's commensal symbiotic microbes, *Sodalis* in vitro, has allowed for the development of a genetic transformation system for this organism (Aksoy et al. 2008). In the paratransgenesis studies, *Sodalis* was tested for its ability to deliver functional anti-trypanosome nanobodies, which demonstrated to release of significant amounts of these nanobodies in different tissues of the tsetse fly to block the transmission of the disease (De Vooght et al. 2012, 2014).

Paratransgenesis Against Pierce's Disease of Grape

The potential of paratransgenesis in crop protection is demonstrated against a pest and pathogen vector: the glassy-winged sharpshooter, *Homalodiscavitripennis*. The insect is a pest of grapes and citrus that spreads the pathogen *Xylellafastidiosa*, which causes Pierce's disease in grapes. An endophytic bacteria of grapes *P. agglomerans* E325 (an EPA-approved agent for managing fire blight in pears and apples) was genetically engineered and successfully harbored by the insect through an artificial feeding system. The genetically modified bacteria colonized in the foregut and persisted over a 15-day period. Furthermore, a novel microencapsulation platform for delivering the engineered bacteria to the gut of the insect under simulated field conditions has been established. Microencapsulation strategy may be useful for field application as it could decrease the environmental spread of foreign genetic material, horizontal gene transfer, and competition with native species by acting as a barrier between recombinant bacteria and the environment (Arora et al. 2015).

5.4.3 Incompatible Insect Technique (IIT)

Another potential *Wolbachia*-based approach to control vectors and other insect pests is the incompatible insect technique (IIT) (Stouthamer et al. 1999). The IIT is based on the mechanism of *Wolbachia*-induced CI, which manipulates natural populations of arthropod pests through embryonic lethality (for the definition, see the subheading "Population-level impacts of endosymbionts") (Bourtzis et al. 1998). The CI can be induced unidirectionally (crosses between infected males and uninfected females) or bidirectionally (crosses between individuals that are infected with different *Wolbachia* strains) (Werren 1997). Bidirectional CI (or unidirectional CI if the target population is uninfected) has been used for "population suppression" traditionally. In this method, the repeated releases of incompatible males lead to the gradual suppression of the target population, locally and temporally. On the other hand, "population replacement" is based on unidirectional CI (Bourtzis et al. 2014). In this method, infected females are introduced in the targeted population, which can establish and spread. In this scenario, the *Wolbachia* plays as a tool to limit pathogen transmission, directly or indirectly (Bourtzis et al. 2014).

The traditional IIT method (repeated releases of incompatible males) is analogous to the sterile insect technique (SIT) (Knipling 1955). SIT comprises the mass-rearing of the target species, sterilization (mainly through a gamma or x-ray irradiation), and inundative releases of the male insects sequentially into the target population (Zhang et al. 2015). Both methods are species-specific and environment-friendly techniques. In *Wolbachia*-based IIT, the genotype of IIT insects, the consequences of *Wolbachia* transinfection on host fitness (esp. mating competitiveness), and the stability of the association need to be critically assessed before field applications (Saridaki and Bourtzis 2010; Qadri et al. 2020).

IIT/SIT and Mosquito Species

Extensive research has been carried out to use IIT against several mosquito species. The first successful application of IIT in the field was achieved in Myanmar, where the target population of C. pipiens was almost eliminated (Laven 1967). There were trials to control *C. fatigens* in India by means of cytoplasmic incompatibility (Curtis and Adak 1974). Recently, transinfection of C. quinquefasciatus with the *w*Pip (Is) strain, naturally infecting C. pipiens, leads to the production of LR[*w*Pip(Is)] line. In addition to 100% embryo lethality from matings between LR[*w*Pip(Is)] males and all tested field females, most crosses between LR[*w*Pip(Is)] females and field males were incompatible (Atyame et al. 2011). In another study, two *Wolbachia* transinfections (from *Aedesalbopictus*) were generated in C. quinquefasciatus: a *w*AlbB single infection, and a wPip plus wAlbA superinfection. The *w*Pip*w*AlbA superinfection reached over 400-fold higher densities in the salivary glands (compared to the native strain, *w*Pip) results in complete unidirectional CI. These results support the feasibility of an IIT program using transinfected C. quinquefasciatus and stimulate the implementation of field tests for designing a control strategy (Ant et al. 2020).

Feasibility studies for the use of a combined SIT/IIT protocol to control populations of the mosquito species have provided encouraging results (Zhang et al. 2015). A new triple *Wolbachia*-infected strain of *Ae. albopictus* (i.e., infected with *w*AlbA, *w*AlbB, and *w*Pip), known as HC and expressing strong CI in appropriate matings, was recently developed (Zhang et al. 2015). Based on the fitness defects induced by *Wolbachia*, the combination of SIT with IIT (low-dose irradiation of HC strain) to control natural populations of *Ae. albopictus* was suggested for area-wide vector control (Zhang et al. 2015; Dimopoulos 2019). Also, it was demonstrated that *Wolbachia*-transinfected *Ae. aegypti* lines required a lower irradiation dose (combined SIT/IIT population suppression programs) to achieve complete female sterility than the uninfected ones (Carvalho et al. 2020). Another study establishes a combined SIT/IIT protocol (transinfected insects irradiated at lower doses) for *D. suzukii* management (Nikolouli et al. 2020).

IIT and Agricultural Pests

IIT has also been successfully tested against agricultural pests. In *Ephestiacautella*, the mass production and release of incompatible males (US strain was reproductively incompatible with an Iranian strain) were made into simulated warehouses for evaluating the population suppression. The results indicated that the insect populations were greatly reduced and suggested the use of reproductive incompatibility as the potential means of population suppression (Brower 1980). Another attempt is related to the Mediterranean fruit fly (medfly), *Ceratitiscapitata*, which has been transinfected with a *Wolbachia* strain from the cherry fruit fly *Rhagoletiscerasi* (Zabalou et al. 2004). This transinfection caused both unidirectional as well as bidirectional CI, suppressing the insect pest by single releases of infected males, in the laboratory. This study opens the possibility of using *Wolbachia*-induced CI as a novel environment-friendly tool for the control of medfly populations (Zabalou et al. 2004).

5.4.4 Manipulation of Insect-Associated Symbionts

The last approach is particularly valuable to control insect pests, by eliminating the microorganisms required for sustained insect growth, reproduction, and survival and/or disrupting the symbiont's transmission to the next host generation (Baumann 2005; Salem et al. 2015; Berasategui et al. 2016). For these purposes, the applied methods are using heat treatment and specific symbiocides (Douglas 1998; Arora et al. 2015). Several examples of how heat treatment could affect the insects indirectly by manipulating the inhabitant microbial partners are discussed in the abovementioned part "Symbionts influence the heat sensitivity of the insect host" (under the subheading "The importance of the symbionts in insect ecology" and the main heading "Insect symbionts, different types and roles"). Here, the applications of symbiocides, as effectors for perturbing/eliminating the symbionts, their interactions with the insect, and their transmissions, are discussed.

The potential of using specific symbiocides is greatest for insects, which are dependent on obligate bacteria that are transmitted vertically to the next generation (for the regain avoidance mainly happens in horizontal transmission). Proof of concept comes from the routine use of antibiotics to eliminate prokaryotic microorganisms from a wide range of insect species in important insect orders including Diptera, Hemiptera, Hymenoptera, Coleoptera, and Lepidoptera (Douglas 1998; Wilkinson 1998). In this method, the possible toxicity of antibiotics to other biotic creatures, like the insect host itself, is a serious limitation (Douglas 1998; Wilkinson 1998). In addition, the use of antibiotics is not affordable for large-scale applications. Last, there are some concerns associated with antibiotic resistance in environmental microorganisms (Arora and Douglas 2017).

The use of antimicrobial peptides (AMPs) has also been explored to manipulate insect symbionts (Qadri et al. 2020). This method is improved by heterologous expression of AMPs in some agricultural crops, which is cost-effective. But, like antibiotics, the serious limitation and problem is the possible toxicity to some other biotic creatures (Berasategui et al. 2016). Now, the strong motivation is to develop methods disrupting the insect–symbiont associations, specifically nutrient translocation between the insect and microbial partners, which are cost-effective and more specific (Price et al. 2014; Douglas 2015).

For disrupting the symbiont's transmission to the next host generation, egg surface sterilization might be helpful. In stinkbugs, the symbiotic bacteria are present on the surface of the egg mass for the duration of the preincubation period (Prado and Zucchi 2012; Taylor et al. 2016). To this end, a sterilizing agent, e.g., bleach or formaldehyde, can be applied on the egg mass to remove the bacteria from the surface and prevent the newly hatched nymphs from acquiring them. This method was applied against several stinkbugs including *Acrosternumhilare* and *Murgantiahistrionica* (Prado and Almeida 2009), A. heegeri (Kashkouli et al. 2020a, b), A. arabicum (Kashkouli et al. 2020a, b), *Adomerus* stinkbugs (Hosokawa et al. 2013), *Eurygasterintegriceps* (Kafil et al. 2013), *Graphosomalineatum* (Karamipour et al. 2016), *Brachynemagermari* (Kashkouli et al. 2019), and *Halyomorphahalys* (Taylor et al. 2014). Recently, this method has been used as a pest control strategy in the field using several antimicrobials and surfactants to decrease the population of H. halys, through affecting its symbiont (Taylor et al. 2016).

5.5 Conclusions

Insect-associated microbial communities are attracting growing interest today, mainly because of their ecological and economic importance. This chapter outlines various types of interactions between insects and microorganisms, which could result in important practical applications for the development of strategies for the management of insect-related problems. For example, knowing the influences of the symbionts on the insect–plant interactions, pesticide detoxification, as well as natural enemy protection will provide novel insight into the design of the pest management strategies. In addition, detailed studies and recent advances in the control of pests and vectors found on manipulation of microbial partners are fully described here. Using heterologous microorganisms and genetic manipulation of microbial symbionts are of interest mostly to control of insect vectors of human diseases. The first trial of microbiome manipulation of a wild insect population to reduce vector competence was recorded for *Ae. Aegypti* in Australia. Recently, stable introduction of *Wolbachia* strain *w*Stri (from its native host, Laodelphax striatellus) into the brown planthopper, Nilaparvata lugens (cured of their native *w*Lug infection), was reported. The strain *w*Stri maintained perfect maternal transmission, induced high levels of CI, and inhibited both infection and transmission of Rice ragged stunt virus (RRSV). Paratransgenesis is widely perceived as an alternative to transgenesis and was first carried out on *R. prolixus*, the vector of the Chagas-causing protozoan *Trypanosoma cruzi* through the manipulation of the insect gut flora. The potential of paratransgenesis in crop protection is demonstrated against a pest and pathogen vector: the glassy-winged sharp shooter, *Homalodisca vitripennis*. Endophytic bacteria of grapes *Pantoea agglomerans* E325 were genetically engineered, successfully harbored by the insect through an artificial feeding system, and colonized in the insect foregut. Furthermore, a novel microencapsulation platform for delivering the engineered bacteria to the gut of H. vitripennis under simulated field conditions has been established. The combination of SIT with IIT method, mainly through using transinfected insects irradiated at lower doses, is under investigation for improving the application and output of both methods. Targeting essential symbionts required by the insect can also be considered as a control strategy, resulting in the mortality or suppression of growth or fecundity of host insects. Appropriate choices of these strategies for the target pest species, along with sustained research on the different insect–microbe interactions and the related mechanisms, are the key to the success of each strategy. Further research will facilitate the implementation of these novel insect pest control strategies for managing the vectors of human diseases as well as agricultural insect pests. With the increasing interest and understanding of the insect–symbiont associations and their ecological features, they will be the matters of applicable field studies and time before pest/vector control programs utilize this information and technique.

5.6 Points to Remember

(i) The associations of insects with symbionts range from mutually beneficial (mutualism) to neutral (commensalism) or parasitic (parasitism) associations with respect to the effect of the symbiont on the host.

(ii) The symbionts play a prominent role in insect ecology by aiding in the digestion of food or providing nutrients, influencing insect–plant interactions, host population, heat tolerance, and pesticide detoxification, as well as protection from natural enemies.

(iii) *Wolbachia* induce anti-pathogenic effects against several RNA viruses, different *Plasmodium* species, fungi, bacteria, and nematodes. Thus, using *Wolbachia*, which particularly targets the disease agents, has received substantial attention so far.

(iv) Knowing the insect–symbiont interactions could result in important practical applications for the development of strategies for the management of insect-related problems.

(v) Main approaches for novel pest management strategies include (i) utilizing heterologous microorganisms, (ii) paratransgenesis, (iii) insect incompatibility technique (IIT), and (iv) the disruption of microbial symbionts required by insect pests.

(vi) Heterologous associations are generated by the experimental transfer of microorganisms from one species into another species.

(vii) Genetically amenable gut bacteria, i.e., *Sodalisglossinidius, Asaia* sp., and *Pantoeaagglomerans*, represent a valuable resource for the development of paratransgenic applications.

(viii) The IIT is based on the mechanism of *Wolbachia*-induced CI, unidirectionally or bidirectionally as means for the "population suppression," "population replacement" strategies. The combination of SIT/IIT programs against several insects especially mosquito species was suggested.

(ix) The microbial symbionts required by insect pests were disrupted, as a control strategy, resulting in insect mortality or suppression of insect growth or fecundity.
(x) Further research will facilitate the implementation of these novel insect pest control strategies for managing the vectors of human diseases as well as agricultural insect pests.

Acknowledgments The technical support of this research by the Department of Entomology, Tarbiat Modares University, is greatly appreciated.

References

Akami M, Njintang NY, Gbaye OA, Andongma AA, Rashid MA, Niu CY, Nukenine EN (2019) Gut bacteria of the cowpea beetle mediate its resistance to dichlorvos and susceptibility to Lippia adoensis essential oil. Sci Rep 9:1–13

Akman L, Yamashita A, Watanabe H, Oshima K, Shiba T, Hattori M, Aksoy S (2002) Genome sequence of the endocellular obligate symbiont of tsetse flies, Wigglesworthia glossinidia. Nat Genet 32:402–407

Aksoy S, Weiss B, Attardo G (2008) Paratransgenesis applied for control of tsetse transmitted sleeping sickness. In: Transgenesis and the management of vector-borne disease. Springer, New York, pp 35–48

Aliota MT, Walker EC, Uribe Yepes A, Dario Velez I, Christensen BM, Osorio JE (2016) The wMel strain of Wolbachia reduces transmission of chikungunya virus in *Aedes aegypti*. PLoS Negl Trop Dis 10:e0004677

Ant TH, Herd C, Louis F, Failloux AB, Sinkins SP (2020) Wolbachia transinfections in *Culex quinquefasciatus* generate cytoplasmic incompatibility. Insect Mol Biol 29:1–8

Arora AK, Douglas AE (2017) Hype or opportunity? Using microbial symbionts in novel strategies for insect pest control. J Insect Physiol 103:10–17

Arora AK, Forshaw A, Miller TA, Durvasula R (2015) A delivery system for field application of paratransgenic control. BMC Biotechnol 15:59

Asplen MK, Bano N, Brady CM, Desneux N, Hopper KR, Malouines C, Oliver KM, White JA, Heimpel GE (2014) Specialisation of bacterial endosymbionts that protect aphids from parasitoids. Ecol Entomol 39:736–739

Atyame CM, Pasteur N, Dumas E, Tortosa P, Tantely ML, Pocquet N, Licciardi S, Bheecarry A, Zumbo B, Weill M, Duron O (2011) Cytoplasmic incompatibility as a means of controlling *Culex pipiens* quinquefasciatus mosquito in the islands of the South-Western Indian Ocean. PLoS Negl Trop Dis 5:e1440

Ayoubi A, Talebi AA, Fathipour Y, Mehrabadi M (2020) Coinfection of the secondary symbionts, *Hamiltonella defensa* and Arsenophonus sp. contribute to the performance of the major aphid pest, *Aphis gossypii* (Hemiptera: Aphididae). Insect Science 27:86–98

Baumann P (2005) Biology of bacteriocyte-associated endosymbionts of plant sap-sucking insects. Annu Rev Microbiol 59:155–189

Beard CB, Mason PW, Aksoy S, Tesh RB, Ritchards FF (1992) Transformation of an insect symbiont and expression of a foreign genein the chagas' disease vector rhodnius prolixus. Am J Trop Med Hygiene:195–200

Beard CB, O'Neill SL, Tesh RB, Richards FF, Aksoy S (1993) Modification of arthropod vector competence via symbiotic bacteria. Parasitol Today 9:179–183

Beard CB, Durvasula RV, Richards FF (1998) Bacterial symbiosis in arthropods and the control of disease transmission. Emerg Infect Dis 4:581–591

Beard CB, Dotson EM, Pennington PM, Eichler S, Cordon-Rosales C, Durvasula RV (2001) Bacterial symbiosis and paratransgenic control of vector-borne Chagas disease. Int J Parasitol 31:621–627
Beard CB, Cordon-Rosales C, Durvasula RV (2002) Bacterial symbionts of the triatominae and their potential use in control of Chagas disease transmission. Annu Rev Entomol 47:123–141
Berasategui A, Shukla S, Salem H, Kaltenpoth M (2016) Potential applications of insect symbionts in biotechnology. Appl Microbiol Biotechnol 100:1567–1577
Berticat C, Rousset F, Raymond M, Berthomieu A, Weill M (2002) High Wolbachia density in insecticide-resistant mosquitoes. Proc Royal Soc B: Biol Sci 269:1413–1416
Bian G, Joshi D, Dong Y, Lu P, Zhou G, Pan X, Xu Y, Dimopoulos G, Xi Z (2013) Wolbachia invades Anopheles stephensi. Science 340:748–751
Blanton AG, Peterson BF (2020) Symbiont-mediated insecticide detoxification as an emerging problem in insect pests. Front Microbiol 11:2422
Blodgett JAV, Oh DC, Cao S, Currie CR, Kolter R, Clardy J (2010) Common biosynthetic origins for polycyclic tetramate macrolactams from phylogenetically diverse bacteria. Proc Natl Acad Sci U S A 107:11692–11697
Bongio NJ, Lampe DJ (2015) Inhibition of plasmodium berghei development in mosquitoes by effector proteins secreted from Asaia sp. bacteria using a novel native secretion signal. PLoS ONE 10:e0143541
Bourtzis K, Dobson SL, Braig HR, O'Neill SL (1998) Rescuing Wolbachia have been overlooked. Nature 391:852–853
Bourtzis K, Dobson SL, Xi Z, Rasgon JL, Calvitti M, Moreira LA, Bossin HC, Moretti R, Baton LA, Hughes GL, Mavingui P, Gilles JRL (2014) Harnessing mosquito-Wolbachia symbiosis for vector and disease control. Acta Trop 132:S150–S163
Boush MG, Matsumura F (1967) Insecticidal degradation by pseudomonas melophthora, the bacterial symbiote of the apple Maggot1. J Econ Entomol 60:918–920
Braig HR, Guzman H, Tesh RB, O'Neill SL (1994) Replacement of the natural Wolbachia symbiont of *Drosophila simulans* with a mosquito counterpart. Nature 367:453–455
Brower JH (1980) Reduction of almond moth/populations in simulated storages by the release of genetically incompatible males1. J Econ Entomol 73:415–418
Brownlie JC, Johnson KN (2009) Symbiont-mediated protection in insect hosts. Cell 17:348–354
Brumin M, Kontsedalov S, Ghanim M (2011) Rickettsia influences thermotolerance in the whitefly *Bemisia tabaci* B biotype. Insect Sci 18:57–66
Buchner P (1952) Endosymbiose der Tiere mit pflanzlichen Mikroorganismen. Verlag Birkhauser, Basel, Switzerland
Buchner P (1965a) A symbiosis in animals which suck plant juices. In: Endosymbiosis of animals with plant microorganisms, New York, pp 210–432
Buchner P (1965b) B endosymbiosis of animals with plant microorganisms. John Wiley, New York
Cafaro MJ, Poulsen M, Little AEF, Price SL, Gerardo NM, Wong B, Stuart AE, Larget B, Abbot P, Currie CR (2011) Specificity in the symbiotic association between fungus-growing ants and protective Pseudonocardia bacteria. Proc Royal Soc B: Biol Sci 278:1814–1822
Caragata EP, Walker T (2012) Using bacteria to treat diseases. Expert Opin Biol Ther 12:701–712
Carvalho DO, Torres-Monzon JA, Koskinioti P, Dilrukshi Wijegunawardana NDA, Liang X, Pillwax G, Xi Z, Bourtzis K (2020) *Aedes aegypti* lines for combined sterile insect technique and incompatible insect technique applications: the importance of host genomic background. Entomol Exp Appl 168:560–572
Casteel CL, Hansen AK, Walling LL, Paine TD (2012) Manipulation of plant defense responses by the tomato psyllid (Bactericerca cockerelli) and its associated endosymbiont Candidatus Liberibacter Psyllaurous. PLoS One 7:1–10
Cayetano L, Rothacher L, Simon JC, Vorburger C (2015) Cheaper is not always worse: strongly protective isolates of a defensive symbiont are less costly to the aphid host. Proc Royal Soc B Biol Sci 282:20142333

Chang KP (1974) Effects of elevated temperature on the mycetome and symbiotes of the bed bug *Cimex lectularius* (Heteroptera). J Invertebr Pathol 23:333–340

Chavshin AR, Oshaghi MA, Vatandoost H, Yakhchali B, Raeisi A, Zarenejad F (2013) *Escherichia coli* expressing a green fluorescent protein (GFP) in Anopheles stephensi: a preliminary model for paratransgenesis. Symbiosis 30:17–24

Chavshin AR, Oshaghi MA, Vatandoost H, Pourmand MR, Raeisi A, Terenius O (2014) Isolation and identification of culturable bacteria from wild anopheles culicifacies, a first step in a paratransgenesis approach. Parasites Vectors 7:419

Chen DQ, Purcell AH (1997) Occurrence and transmission of facultative endosymbionts in aphids. Curr Microbiol 34:220–225

Cheng D, Guo Z, Riegler M, Xi Z, Liang G, Xu Y (2017) Gut symbiont enhances insecticide resistance in a significant pest, the oriental fruit fly *Bactrocera dorsalis* (Hendel). Microbiome 5:1–12

Coutinho-Abreu IV, Zhu KY, Ramalho-Ortigao M (2010) Transgenesis and paratransgenesis to control insect-borne diseases: current status and future challenges. Parasitol Int 59:1–8

Currie CR (2001a) A a community of ants, fungi, and bacteria: a multilateral approach to studying symbiosis. Ann Rev Microbiol 55:357–380

Currie CR (2001b) B prevalence and impact of a virulent parasite on a tripartite mutualism. Oecologia 128:99–106

Currie CR, Stuart AE (2001) Weeding and grooming of pathogens in agriculture by ants. Proc Royal Soc B: Biol Sci 268:1033–1039

Currie CR, Mueller UG, Malloch D (1999a) The agricultural pathology of ant fungus gardens. Proc Natl Acad Sci U S A 96:7998–8002

Currie CR, Scottt JA, Summerbell RC, Malloch D (1999b) Fungus-growing ants use antibiotic-producing bacteria to control garden parasites. Nature 398:701–704

Currie CR, Bot ANM, Boomsma JJ (2003a) Experimental evidence of a tripartite mutualism: bacteria protect ant fungus gardens from specialized parasites. Oikos 101:91–102

Currie CR, Scottt JA, Summerbell RC, Malloch D (2003b) Corrigenda: fungus-growing ants use antibiotic-producing bacteria to control garden parasites. Nature 423:5002

Curtis CF, Adak T (1974) Population replacement in Culex fatigans by means of cytoplasmic incompatibility. I Laboratory experiments with non overlapping generations. Bull World Health Org 51:249–255

Dale C, Moran NA (2006) Molecular interactions between bacterial symbionts and their hosts. Cell 126:453–465

Damiani C, Ricci I, Crotti E, Rossi P, Rizzi A, Scuppa P, Capone A, Ulissi U, Epis S, Genchi M, Sagnon N, Faye I, Kang A, Chouaia B, Whitehorn C, Moussa GW, Mandrioli M, Esposito F, Sacchi L, Bandi C, Daffonchio D, Favia G (2010) Mosquito-bacteria symbiosis: the case of *Anopheles gambiae* and *Asaia*. Microb Ecol 60:644–654

De Almeida LG, De Moraes LAB, Trigo JR, Omoto C, Consoli FL (2017) The gut microbiota of insecticide-resistant insects houses insecticide-degrading bacteria : a potential source for biotechnological exploitation. PLoS One 12:1–19

De Bary A (1879) Die erscheinung der symbiose. Verlag von Karl J, Trübner, Strassburg

De Vooght L, Caljon G, Stijlemans B, De Baetselier P, Coosemans M, Van Den Abbeele J (2012) Expression and extracellular release of a functional anti-trypanosome Nanobody® in Sodalis glossinidius, a bacterial symbiont of the tsetse fly. Microb Cell Factories 11:23

De Vooght L, Caljon G, De Ridder K, Van Den Abbeele J (2014) Delivery of a functional anti-trypanosome nanobody in different tsetse fly tissues via a bacterial symbiont, sodalis glossinidius. Microb Cell Factories 13:156

Dicke M, Cusumano A, Poelman EH (2020) Microbial symbionts of parasitoids. Annu Rev Entomol 65:171–190

Dimopoulos G (2019) Combining sterile and incompatible insect techniques for *Aedes albopictus* suppression. Trends Parasitol 35:671–673

Dinparast Djadid N, Jazayeri H, Raz A, Favia G, Ricci I, Zakeri S (2011) Identification of the midgut microbiota of an. Stephensi and *C. maculipennis* for their application as a paratransgenic tool against malaria. PLoS One 6:e28484

Douglas AE (1998) Nutritional interactions in insect-microbial symbioses: aphids and their symbiotic bacteria buchnera. Annu Rev Entomol 43:17–37

Douglas AE (2009) The microbial dimension in insect nutritional ecology. Funct Ecol 23:38–47

Douglas AE (2015) Multiorganismal insects: diversity and function of resident microorganisms. Annu Rev Entomol 60:17–34

Durvasula RV, Gumbs A, Panackal A, Kruglov O, Aksoy S, Merrifield RB, Richards FF, Beard CB (1997) Prevention of insect-borne disease: an approach using transgenic symbiotic bacteria. Proc Natl Acad Sci U S A 94:3274–3278

Durvasula RV, Kroger A, Goodwin M, Panackal A, Kruglov O, Taneja J, Gumbs A, Richards FF, Beard CB, Cordon-Rosales C (1999) Strategy for introduction of foreign genes into field populations of Chagas disease vectors. Ann Entomol Soc Am 92:937–943

Dutra HLC, Rocha MN, Dias FBS, Mansur SB, Caragata EP, Moreira LA (2016) Wolbachia blocks currently circulating Zika virus isolates in Brazilian *Aedes aegypti* mosquitoes. Cell Host Microbe 19:771–774

Eichler S, Schaub GA (2002) Development of symbionts in triatomine bugs and the effects of infections with trypanosomatids. Exp Parasitol 100:17–27

Engel P, Moran NA (2013) The gut microbiota of insects—diversity in structure and function. FEMS Microbiol Rev 37:699–735

Fan Y, Wernegreen JJ (2013) Can't take the heat: high temperature depletes bacterial endosymbionts of ants. Microb Ecol 66:727–733

Fang W, Vega-Rodríguez J, Ghosh AK, Jacobs-Lorena M, Kang A, Leger RJS (2011) Development of transgenic fungi that kill human malaria parasites in mosquitoes. Science 331:1074–1077

Favia G, Ricci I, Damiani C, Raddadi N, Crotti E, Marzorati M, Rizzi A, Urso R, Brusetti L, Borin S, Mora D, Scuppa P, Pasqualini L, Clementi E, Genchi M, Corona S, Negri M, Grandi G, Alma A, Kramer L, Esposito F, Bandi C, Sacchi L, Daffonchio D (2007) Bacteria of the genus Asaia stably associate with Anopheles stephensi, an Asian malarial mosquito vector. Proc Natl Acad Scie U S A 104:9047–9051

Favia G, Ricci I, Marzorati M, Negri I, Alma A, Sacchi L, Bandi C, Daffonchio D (2008) Bacteria of the genus Asaia: a potential paratransgenic weapon against malaria. In: Transgenesis and the management of vector-borne disease. Springer, New York, pp 49–59

Feldhaar H (2011) Bacterial symbionts as mediators of ecologically important traits of insect hosts. Ecol Entomol 36:533–543

Ferguson NM, Kien DTH, Clapham H, Aguas R, Trung VT, Chau TNB, Popovici J, Ryan PA, O'Neill SL, McGraw EA, Long VT, Dui LT, Nguyen HL, Van Chau NV, Wills B, Simmons CP (2015) Modeling the impact on virus transmission of Wolbachia-mediated blocking of dengue virus infection of *Aedes aegypti*. Sci Transl Med 7:279ra37

Ferrari J, Vavre F (2011) Bacterial symbionts in insects or the story of communities affecting communities. Philos Trans Royal Soc B: Biol Sci 366:1389–1400

Ferrari J, Müller CB, Kraaijeveld AR, Godfray HCJ (2001) Clonal variation and covariation in aphid resistance to parasitoids and a pathogen. Evolution 55:1805–1814

Ferrari J, Darby AC, Daniell TIMJ, Godfray HCJ, Douglas AE (2004) Linking the bacterial community in pea aphids with host-plant use and natural enemy resistance. Ecol Entomol 29:60–65

Flórez LV, Kaltenpoth M (2017) Symbiont dynamics and strain diversity in the defensive mutualism between Lagria beetles and Burkholderia. Environ Microbiol 19:3674–3688

Flórez LV, Biedermann PHW, Engl T, Kaltenpoth M (2015) Defensive symbioses of animals with prokaryotic and eukaryotic microorganisms. Nat Prod Rep 32:904–936

Flórez LV, Scherlach K, Gaube P, Ross C, Sitte E, Hermes C, Rodrigues A, Hertweck C, Kaltenpoth M (2017) Antibiotic-producing symbionts dynamically transition between plant pathogenicity and insect-defensive mutualism. Nat Commun 8:1–9

Flórez LV, Scherlach K, Miller IJ, Rodrigues A, Kwan JC, Hertweck C, Kaltenpoth M (2018) An antifungal polyketide associated with horizontally acquired genes supports symbiont-mediated defense in Lagria villosa beetles. Nat Commun 9:1–10

Frago E, Dicke M, Godfray HCJ (2012) Insect symbionts as hidden players in insect-plant interactions. Trends Ecol Evol 27:705–711

Fukatsu T, Hosokawa T (2002) Capsule-transmitted gut symbiotic bacterium of the Japanese common plataspid stinkbug, megacopta punctatissima. Appl Environ Microbiol 68:389–396

Fytrou A, Schofield PG, Kraaijeveld AR, Hubbard SF (2006) Wolbachia infection suppresses both host defence and parasitoid counter-defence. Proc Royal Soc B: Biol Sci 273:791–796

Glaser RL, Meola MA (2010) The native Wolbachia endosymbionts of *Drosophila melanogaster* and *Culex quinquefasciatus* increase host resistance to west nile virus infection. PLoS One 5: e11977

Goettler W, Kaltenpoth M, Herzner G, Strohm E (2007) Morphology and ultrastructure of a bacteria cultivation organ: the antennal glands of female European beewolves, Philanthus triangulum (hymenoptera, Crabronidae). Arthropod Struct Dev 36:1–9

Gong JT, Li Y, Li TP, Liang Y, Hu L, Zhang D, Zhou CY, Yang C, Zhang X, Zha SS, Duan XZ, Baton LA, Hong XY, Hoffmann AA, Xi Z (2020) Stable introduction of plant-virus-inhibiting wolbachia into planthoppers for rice protection. Curr Biol 30:1–9

Goodacre SL, Martin OY, Bonte D, Hutchings L, Woolley C, Ibrahim K, Thomas CFG, Hewitt GM (2009) Microbial modification of host long-distance dispersal capacity. BMC Biol 7:1–8

Guruprasad NM, Jalali SK, Puttaraju HP (2014) Wolbachia-a foe for mosquitoes. Asian Pacific J Trop Dis 4:78–81

Haine ER (2008) Symbiont-mediated protection. Proc Royal Soc B: Biol Sci 275:353–361

Hamilton PT, Perlman SJ (2013) Host defense via symbiosis in drosophila. PLoS Pathog 9:1–4

Hamilton PT, Leong JS, Koop BF, Perlman SJ (2014) Transcriptional responses in a drosophila defensive symbiosis. Mol Ecol 23:1558–1570

Hansen AK, Moran NA (2014) The impact of microbial symbionts on host plant utilization by herbivorous insects. Mol Ecol 23:1473–1496

Hansen AK, Vorburger C, Moran NA (2012) Genomic basis of endosymbiont-conferred protection against an insect parasitoid. Genome Res 22:106–114

Heddi A, Grenier A-M, Chaque K, Charles H, Nardon P (1999) Four intracellular genomes direct weevil biology: nuclear, mitochondrial, principal endosymbiont, and Wolbachia. Proc Natl Acad Sci U S A 96:6814–6819

Hedges LM, Brownlie JC, Neill SLO, Johnson KN (2008) Wolbachia and virus protection in insects. Science 322:702

Heyworth ER, Ferrari J (2015) A facultative endosymbiont in aphids can provide diverse ecological benefits. J Evol Biol 28:1753–1760

Hoffmann AA, Montgomery BL, Popovici J, Iturbe-Ormaetxe I, Johnson PH, Muzzi F, Greenfield M, Durkan M, Leong YS, Dong Y, Cook H, Axford J, Callahan AG, Kenny N, Omodei C, McGraw EA, Ryan PA, Ritchie SA, Turelli M, O'Neill SL (2011) Successful establishment of Wolbachia in Aedes populations to suppress dengue transmission. Nature 476:454–459

Hosokawa T, Hironaka M, Inadomi K, Mukai H, Nikoh N, Fukatsu T (2013) Diverse strategies for vertical symbiont transmission among subsocial stinkbugs. PLoS One 8:4–11

Hosseinzadeh S, Shams-bakhsh M, Mann M, Fattah-hosseini S (2018) Distribution and variation of bacterial endosymbiont and B Candidatus Liberibacter asiaticus ^ titer in the huanglongbing insect vector, *Diaphorina citri* Kuwayama. Microbial Ecol 78:206–222

Hughes GL, Koga R, Xue P, Fukatsu T, Rasgon JL (2011) Wolbachia infections are virulent and inhibit the human malaria parasite *Plasmodium falciparum* in *Anopheles gambiae*. PLoS Pathog 7:e1002043

Hulcr J, Adams AS, Raffa K, Hofstetter RW, Klepzig KD, Currie CR (2011) Presence and diversity of streptomyces in dendroctonus and sympatric bark beetle galleries across North America. Microb Ecol 61:759–768

Hurwitz I, Fieck A, Read A, Hillesland H, Klein N, Kang A, Dur-vasula R (2011) Paratransgenic control of vector borne diseases. Int J Biol Sci 7:1334–1344

Hussain M, Lu G, Torres S, Edmonds JH, Kay BH, Khromykh AA, Asgari S (2013) Effect of Wolbachia on replication of west nile virus in a mosquito cell line and adult mosquitoes. J Virol 87:851–858

Ishikawa H (2006) Insect symbiosis: an introduction. In: Insect symbiosis, vol 1. CRC Press, pp 1–21

Itoh H, Tago K, Hayatsub M, Kikuchi Y (2018) Detoxifying symbiosis: microbe-mediated detoxification of phytotoxins and pesticides in insects. Nat Prod Rep 35:434–454

Iturbe-Ormaetxe I, Walker T, O'Neill SL (2011) Wolbachia and the biological control of mosquito-borne disease. EMBO Rep 12:508–518

Jaenike J, Brekke TD (2011) Defensive endosymbionts: a cryptic trophic level in community ecology. Ecol Lett 14:150–155

Jaenike J, Unckless R, Cockburn SN, Boelio LM, Perlman SJ (2010) Adaptation via symbiosis: recent spread of a drosophila defensive symbiont. Science 329:212–215

Jamin AR, Vorburger C (2019) Estimating costs of aphid resistance to parasitoids conferred by a protective strain of the bacterial endosymbiont Regiella insecticola. Entomol Exp Appl 167:252–260

Joshi D, Pan X, McFadden MJ, Bevins D, Liang X, Lu P, Thiem S, Xi Z (2017) The maternally inheritable Wolbachia wAlbB induces refractoriness to plasmodium berghei in Anopheles stephensi. Front Microbiol 8:366

Kafil M, Bandani AR, Kaltenpoth M, Goldansaz SH, Alavi SM (2013) Role of symbiotic bacteria in the growth and development of the Sunn pest, Eurygaster integriceps. J Insect Sci 13:99

Kaltenpoth M, Go W, Herzner G, Strohm E (2005) Symbiotic bacteria protect wasp larvae from fungal infestation. Curr Biol 15:475–479

Kaltenpoth M, Goettler W, Dale C, Stubblefield JW, Herzner G, Roeser-Mueller K, Strohm E (2006) "Candidatus Streptomyces philanthi", an endosymbiotic streptomycete in the antennae of Philanthus digger wasps. Int J Syst Evol Microbiol 56:1403–1411

Kaltenpoth M, Goettler W, Koehler S, Strohm E (2010) Life cycle and population dynamics of a protective insect symbiont reveal severe bottlenecks during vertical transmission. Evol Ecol 24:463–477

Kambris Z, Cook PE, Phuc HK, Sinkins SP (2009) Immune activation by life-shortening wolbachia and reduced filarial competence in mosquitoes. Science 326:134–136

Kambris Z, Blagborough AM, Pinto SB, Blagrove MSC, Godfray HCJ, Sinden RE, Sinkins SP (2010) Wolbachia stimulates immune gene expression and inhibits plasmodium development in *Anopheles gambiae*. PLoS Pathog 6:e1001143

Karamipour N, Mehrabadi M, Fathipour Y (2016) Gammaproteobacteria as essential primary symbionts in the striped shield bug, Graphosoma Lineatum (Hemiptera: Pentatomidae). Sci Rep 6:33168

Kashkouli M, Fathipour Y, Mehrabadi M (2018) Potential management tactics for pistachio stink bugs, Brachynema germari, *Acrosternum heegeri* and *Acrosternum arabicum* (Hemiptera : Pentatomidae): high temperature and chemical surface sterilants leading to symbiont suppression. J Econ Entomol 112:244–254

Kashkouli M, Fathipour Y, Mehrabadi M (2019) Heritable gammaproteobacterial symbiont improves the fitness of Brachynema germari Kolenati (Hemiptera : Pentatomidae). Environ Entomol 48:1079–1087

Kashkouli M, Castelli M, Floriano AM, Bandi C, Epis S, Fathipour Y, Mehrabadi M, Sassera D (2020a) Characterization of a novel Pantoea symbiont allows inference of a pattern of convergent genome reduction in bacteria associated with Pentatomidae. Environ Microbiol 23 (1):36–50

Kashkouli M, Fathipour Y, Mehrabadi M (2020b) Habitat visualization, acquisition features and necessity of the gammaproteobacterial symbiont of pistachio stink Bug, *Acrosternum heegeri* (Hem.: Pentatomidae). Bull Entomol Res 110:22–33

Kellner RL (1999) What is the basis of pederin polymorphism in Paederus riparius rove beetles? The endosymbiotic hypothesis. Entomol Exp Appl 93:41–49

Kellner RLL (2001) Suppression of pederin biosynthesis through antibiotic elimination of endosymbionts in Paederus sabaeus. J Insect Physiol 47:475–483

Kellner RLL (2002) Molecular identification of an endosymbiotic bacterium associated with pederin biosynthesis in Paederus sabaeus (Coleoptera: Staphylinidae). Insect Biochem Mol Biol 32:389–395

Kellner RLL (2003) Stadium-specific transmission of endosymbionts needed for pederin biosynthesis in three species of Paederus rove beetles. Entomol Exp Appl 107:115–124

Kellner RLL, Dettner K (1996) Differential efficacy of toxic pederin in deterring potential arthropod predators of Paederus (Coleoptera: Staphylinidae) offspring. Oecologia 107:293–300

Kenyon LJ, Meulia T, Sabree ZL (2015) Habitat visualization and genomic analysis of "candidatus pantoea carbekii," the primary symbiont of the brown marmorated stink bug. Genome Biol Evol 7:620–635

Kikuchi Y, Hayatsu M, Hosokawa T, Nagayama A, Tago K, Fukatsu T (2012) Symbiont-mediated insecticide resistance. Proc Natl Acad Sci U S A 109:8618–8622

Kikuchi Y, Tada A, Musolin DL, Hari N, Hosokawa T, Fujisaki K (2016) Collapse of insect gut symbiosis under simulated climate change. MBio 7:1–8

Knipling EF (1955) Possibilities of insect control or eradication through the use of sexually sterile males1. J Econ Entomol 48:459–462

Koch A (1960) Intracellular symbiosis in insects. Annu Rev Microbiol:121–140

Kokoza V, Ahmed A, Shin SW, Okafor N, Zou Z, Raikhel AS (2010) Blocking of plasmodium transmission by cooperative action of Cecropin a and Defensin a in transgenic *Aedes aegypti* mosquitoes. Proc Natl Acad Sci U S A 107:8111–8116

Laven H (1967) Eradication of Culex pipiens fatigans through cytoplasmic incompatibility. Nature 216:383–384

Lee YH, Hou RF (1987) Physiological roles of a yeast-like symbiote in reproduction and embryonic-development of the brown planthopper, *Nilaparvata lugens* Stal. J Insect Physiol 33:851–860

Leonardo TE, Mondor EB, Leonardo TE, Mondor EB (2006) Symbiont modifies host life-history traits that affect gene flow symbiont modifies host life-history traits that affect gene flow. Proc Royal Soc B: Biol Sci 273:1079–1084

Leybourne DJ, Bos JIB, Valentine TA, Karley AJ (2020) The price of protection: a defensive endosymbiont impairs nymph growth in the bird cherry-oat aphid, *Rhopalosiphum padi*. Insect Science 27:69–85

Li Q, Fan J, Sun J, Zhang Y, Hou M, Chen J (2019) Anti-plant defense response strategies mediated by the secondary symbiont hamiltonella defensa in the wheat aphid *Sitobion miscanthi*. Front Microbiol 10:1–12

Łukasik P, van Asch M, Guo H, Ferrari J, Godfray HCJ (2013) Unrelated facultative endosymbionts protect aphids against a fungal pathogen. Ecol Lett 16:214–218

Mala M, Weldegergis BT, Yang C, Mclean A, Godfray HCJ, Dicke M (2017) Symbionts protect aphids from parasitic wasps by attenuating herbivore-induced plant volatiles. Nat Commun 8:1–9

Martinez AJ, Ritter SG, Doremus MR, Russell JA, Oliver KM (2014) Aphid-encoded variability in susceptibility to a parasitoid. BMC Evol Biol 14:1–10

Mathew GM, Ju YM, Lai CY, Mathew DC, Huang CC (2012) Microbial community analysis in the termite gut and fungus comb of Odontotermes formosanus: the implication of bacillus as mutualists. FEMS Microbiol Ecol 79:504–517

McCutcheon JP, Moran NA (2012) Extreme genome reduction in symbiotic bacteria. Nat Rev Microbiol 10:13–26

McLean AH (2019) Cascading effects of defensive endosymbionts. Cur Opin Insect Sci 32:42–46
McLean AHC, Godfray HCJ (2015) Evidence for specificity in symbiontconferred protection against parasitoids. Proc Royal Soc B: Biol Sci 282:1–8
McMeniman CJ, Lane AM, Fong AWC, Voronin DA, Iturbe-Ormaetxe I, Yamada R, McGraw EA, O'Neill SL (2008) Host adaptation of a Wolbachia strain after long-term serial passage in mosquito cell lines. Appl Environ Microbiol 74:6963–6969
McMeniman CJ, Lane RV, Cass BN, Fong AWC, Sidhu M, Wang YF, O'Neill SL (2009) Stable introduction of a life-shortening Wolbachia infection into the mosquito *Aedes aegypti*. Science 323:141–144
Miller KE, Inward DJ, Gomez-Rodriguez C, Baselga A, Vogler AP (2019) Predicting the unpredictable: how host specific is the mycobiota of bark and ambrosia beetles? Fungal Ecol 42:100854
Montllor CB, Maxmen A, Purcell AH (2002) Facultative bacterial endosymbionts benefit pea aphids *Acyrthosiphon pisum* under heat stress. Ecol Entomol 27:189–195
Moran NA (1996) Accelerated evolution and Muller's rachet in endosymbiotic bacteria. Proc Natl Acad Sci 93:2873–2878
Moran NA (2016) When obligate partners melt down. mBio 7:e01904–e01916
Moran NA, McCutcheon JP, Nakabachi A (2008) Genomics and evolution of heritable bacterial symbionts. Annu Rev Genet 42:165–190
Moreira LA, Iturbe-Ormaetxe I, Jeffery JA, Lu G, Pyke AT, Hedges LM, Moreira LA, Rocha BC, Hall-mendelin S, Day A, Riegler M, Hugo LE, Johnson KN, Kay BH, Mcgraw EA, Van Den Hurk AF, Ryan PA, Neill SLO (2009) A Wolbachia symbiont in *Aedes aegypti* limits infection with dengue, Chikungunya , and plasmodium. Cell 139:1268–1278
Mueller UG, Mikheyev AS, Hong E, Sen R, Warren DL, Solomon SE, Ishak HD, Cooper M, Miller JL, Shaffer KA, Juenger TE (2011) Evolution of cold-tolerant fungal symbionts permits winter fungiculture by leafcutter ants at the northern frontier of a tropical ant-fungus symbiosis. Proc Natl Acad Sci U S A:108, 4053–4056
Musolin DL, Tougou D, Fujisaki K (2010) Too hot to handle? Phenological and life-history responses to simulated climate change of the southern green stink bug *Nezara viridula* (Heteroptera: Pentatomidae). Glob Chang Biol 16:73–87
Nakabachi A, Ueoka R, Oshima K, Teta R, Mangoni A, Gurgui M, Oldham NJ, Van Echten-Deckert G, Okamura K, Yamamoto K, Inoue H, Ohkuma M, Hongoh Y, Miyagishima SY, Hattori M, Piel J, Fukatsu T (2013) Defensive bacteriome symbiont with a drastically reduced genome. Curr Biol 23:1478–1484
Nazni WA, Hoffmann AA, NoorAfizah A, Cheong YL, Mancini MV, Golding N, Kamarul GMR, Arif MAK, Thohir H, NurSyamimi H, ZatilAqmar MZ, NurRuqqayah M, NorSyazwani A, Faiz A, Irfan FRMN, Rubaaini S, Nuradila N, Nizam NMN, Irwan SM, Endersby-Harshman NM, White VL, Ant TH, Herd CS, Hasnor AH, AbuBakar R, Hapsah DM, Khadijah K, Kamilan D, Lee SC, Paid YM, Fadzilah K, Topek O, Gill BS, Lee HL, Sinkins SP (2019) Establishment of Wolbachia strain wAlbB in Malaysian populations of *Aedes aegypti* for dengue control. Curr Biol 29:4241–4248
Nikoh N, Hosokawa T, Oshima K, Hattori M, Fukatsu T (2011) Reductive evolution of bacterial genome in insect gut environment. Genome Biol Evol 3:702–714
Nikoh N, Tsuchida T, Maeda T, Yamaguchi K, Shigenobu S, Koga R, Fukatsu T (2018) Genomic insight into symbiosis-induced insect color change. mBio 9:1–16
Nikolouli K, Sassù F, Mouton L, Stauffer C, Bourtzis K (2020) Combining sterile and incompatible insect techniques for the population suppression of Drosophila suzukii. J Pest Sci 93:647–661
Noda H, Saito T (1979) Effects of high temperature on the development of *Laodelphax striatellus* (Homoptera: Delphacidae) and on its intracellular yeastlike symbiotes. Appl Entomol Zool 14:64–75
Oh DC, Scott JJ, Currie CR, Clardy J (2009) Mycangimycin, a polyene peroxide from a mutualist Streptomyces sp. Org Lett 11:633–636

Ohtaka C, Ishikawa H (1991) Effects of heat treatment on the symbiotic system of an aphid mycetocyte. Symbiosis 11:19–30

Oliver KM, Perlman SJ (2020) Toxin-mediated protection against natural enemies by insect defensive symbionts. In: Mechanisms underlying microbial symbiosis. Elsevier Ltd., p 277

Oliver KM, Russell JA, Moran NA, Hunter MS (2003) Facultative bacterial symbionts in aphids confer resistance to parasitic wasps. Evolution 100:1803–1807

Oliver KM, Degnan PH, Hunter MS, Moran NA (2009) Bacteriophages encode factors required for protection in a symbiotic mutualism. Science 325:992–994

Oliver KM, Degnan PH, Burke GR, Moran NA (2010) Facultative symbionts in aphids and the horizontal transfer of ecologically important traits. Annu Rev Entomol 55:247–266

Olson KE, Higgs S, Gaines PJ, Powers AM, Davis BS, Kamrud KI, Carlson JO, Blair CD, Beaty BJ (1996) Genetically engineered resistance to dengue-2 virus transmission in mosquitoes. Science 272:884–886

Osborne SE, Leong YS, Neill SLO, Johnson KN (2009) Variation in antiviral protection mediated by different Wolbachia Strains in *Drosophila simulans*. PLoS Pathog 5(11):e1000656

Oulhen N, Schulz BJ, Carrier TJ (2016) English translation of Heinrich Anton de Bary's 1878 speech, 'die Erscheinung der Symbiose' ('De la symbiose'). Symbiosis 69:131–139

Pang R, Chen M, Yue L, Xing K, Li T, Kang K, Liang Z, Yuan L, Zhang W (2018) A distinct strain of Arsenophonus symbiont decreases insecticide resistance in its insect host. PLoS Genet 14: e1007725

Panteleev DY, Goryacheva II, Andrianov BV, Reznik NL, Lazebny OE, Kulikov AM (2007) The endosymbiotic bacterium Wolbachia enhances the nonspecific resistance to insect pathogens and alters behavior of *Drosophila melanogaster*. Russ J Genet 43:1066–1069

Parker BJ, Spragg CJ, Altincicek B, Gerardo NM, Parker BJ, Spragg CJ, Altincicek B, Gerardo NM (2013) Symbiont-mediated protection against fungal pathogens in pea aphids: a role for pathogen specificity? Appl Environ Microbiol 79:2455–2458

Parry R, Bishop C, De Hayr L, Asgari S (2019) Density-dependent enhanced replication of a densovirus in Wolbachia-infected Aedes cells is associated with production of piRNAs and higher virus-derived siRNAs. Virology 528:89–100

Patel V, Chevignon G, Manzano-Marın A, Brandt JW, Strand MR, Russell JA, Oliver KM (2019) Cultivation-assisted genome of candidatus fukatsuia symbiotica; the enigmatic "X-type" symbiont of aphids. Genome Biol Evol 11:3510–3522

Piel J (2002) A polyketide synthase-peptide synthetase gene cluster from an uncultured bacterial symbiont of Paederus beetles. Proc Natl Acad Sci U S A 99:14002–14007

Piel J, Höfer I, Hui D (2004) Evidence for a symbiosis island involved in horizontal acquisition of pederin biosynthetic capabilities by the bacterial symbiont of Paederus fuscipes beetles. J Bacteriol 186:1280–1286

Pietri JE, Liang D (2018) The links between insect symbionts and insecticide resistance: causal relationships and physiological tradeoffs. Ann Entomol Soc Am 111:92–97

Pietri JE, Tiffany C, Liang D (2018) Disruption of the microbiota affects physiological and evolutionary aspects of insecticide resistance in the German cockroach, an important urban pest. PLoS One 13:e0207985

Poulsen M, Cafaro M, Boomsma JJ, Currie CR (2005) Specificity of the mutualistic association between actinomycete bacteria and two sympatric species of Acromyrmex leaf-cutting ants. Mol Ecol 14:3597–3604

Prado SS, Almeida RPP (2009) Role of symbiotic gut bacteria in the development of Acrosternum hilare and *Murgantia histrionica*. Entomol Exp Appl 132:21–29

Prado SS, Zucchi TD (2012) Host-symbiont interactions for potentially managing heteropteran pests. Psyche 2012:1–9

Prado SS, Golden M, Follett PA, Daugherty MP (2009) Demography of gut symbiotic and Aposymbiotic *Nezara viridula* L. (Hemiptera : Pentatomidae). Environ Entomol 38(1):103–109

Prado SS, Hung KY, Daugherty MP, Almeida RPP (2010) Indirect effects of temperature on stink bug fitness, via maintenance of gut-associated symbionts. Appl Environ Microbiol 76:1261–1266

Price DRG, Feng H, Baker JD, Bavan S, Luetje CW, Wilson ACC (2014) Aphid amino acid transporter regulates glutamine supply to intracellular bacterial symbionts. Proc Natl Acad Sci U S A 111:320–325

Qadri M, Short S, Gast K, Hernandez J, Wong AC-N (2020) Microbiome innovation in agriculture: development of microbial based tools for insect pest management. Front Sustain Food Syst 4:547751

Ramsey JS, Johnson RS, Hoki JS, Kruse A, Mahoney J, Hilf ME, Hunter WB, Hall DG, Schroeder FC, MacCoss MJ, Cilia M (2015) Metabolic interplay between the asian citrus psyllid and its profftella symbiont: an achilles' heel of the citrus greening insect vector. PLoS One 10:1–21

Rancès E, Ye YH, Woolfit M, McGraw EA, O'Neill SL (2012) The relative importance of innate immune priming in Wolbachia-mediated dengue interference. PLoS Pathog 8:e1002548

Rathnayaka USK, Rakshit SK (2010) The stability of rRNA in heat-killed *Salmonella enterica* cells and its detection by fluorescent in situ hybridisation (FISH). Trop Life Sci Res 21:47–53

Richards G, Brooks MA (1958) Internal symbiosis in insects. Annu Rev Entomol:37–56

Ross PA, Wiwatanaratanabutr I, Axford JK, White VL, Endersby-Harshman NM, Hoffmann AA (2017) Wolbachia infections in *Aedes aegypti* differ markedly in their response to cyclical heat stress. PLoS Pathog 13:e1006006

Russell JA, Moran NA (2006) Costs and benefits of symbiont infection in aphids: variation among symbionts and across temperatures. Proc Royal Soc B: Biol Sci 273:603–610

Ryan PA, Turley AP, Wilson G, Hurst TP, Retzki K, Brown-Kenyon J, Hodgson L, Kenny N, Cook H, Montgomery BL, Paton CJ, Ritchie SA, Hoffmann AA, Jewell NP, Tanamas SK, Anders KL, Simmons CP, O'Neill SL (2019) Establishment of wMel Wolbachia in *Aedes aegypti* mosquitoes and reduction of local dengue transmission in Cairns and surrounding locations in northern Queensland, Australia. Gates Open Research 3:1547

Sacchi L, Grigolo A, Biscaldi G, Laudani U (1993) Effects of heat treatment on the symbiotic system of blattoidea: Morphofunctional alterations of bacteriocytes. Ital J Zool 60:271–279

Salem H, Florez L, Gerardo N, Kaltenpoth M (2015) An out-of-body experience: the extracellular dimension for the transmission of mutualistic bacteria in insects. Proc Royal Soc B: Biol Sci 282 (1804):20142957

Saridaki A, Bourtzis K (2010) Wolbachia: more than just a bug in insects genitals. Curr Opin Microbiol 13:67–72

Sasaki T, Kawamura M, Ishikawa H (1996) Nitrogen recycling in the Brown Planthopper, Nilapawata lugens: involvement of yeast-like endosymbionts in uric acid metabolism. J Insect Physiol 42:125–129

Schneider DI, Saarman N, Onyango MG, Hyseni C, Opiro R, Echodu R, O'Neill M, Bloch D, Vigneron A, Johnson TJ, Dion K, Weiss BL, Opiyo E, Caccone A, Aksoy S (2019) Spatio-temporal distribution of Spiroplasma infections in the tsetse fly (Glossina fuscipes fuscipes) in northern Uganda. PLoS Negl Trop Dis 13:1–22

Scott JJ, Oh DC, Yuceer MC, Klepzig KD, Clardy J, Currie CR (2008) Bacterial protection of beetle-fungus mutualism. Science 322:63

Seipke RF, Barke J, Ruiz-Gonzalez MX, Orivel J, Yu DW, Hutchings MI (2012) Fungus-growing Allomerus ants are associated with antibiotic-producing actinobacteria. Antonie Van Leeuwenhoek 101:443–447

Shen SK, Dowd PF (1991) Detoxification spectrum of the cigarette beetle symbiont Symbiotaphrina kochii in culture. Entomol Exp Appl 60:51–59

Shigenobu S, Watanabe H, Hattori M, Sakaki Y, Ishikawa H (2000) Genome sequence of the endocellular bacterial symbiont of aphids Buchnera sp. APS Nat 407:81–86

Sinkins S, O'Neill S (2000) Wolbachia as a vehicle to modify insect populations. In: Insect transgenesis: methods and applications. CRC Press, Boca Raton, pp 271–288

Smith TE, Moran NA (2019) Coordination of host and symbiont gene expression reveals a metabolic tug-of-war between aphids and Buchnera. Proc Natl Acad Sci 117:2113–2121

Soltani A, Vatandoost H, Oshaghi MA, Enayati AA, Chavshin AR (2017) The role of midgut symbiotic bacteria in resistance of Anopheles stephensi (Diptera: Culicidae) to organophosphate insecticides. Pathogens Global Health 111:289–296

Stouthamer R, Breeuwer JAJ, Hurst GDD (1999) Wolbachia pipientis: microbial manipulator of arthropod reproduction. Annu Rev Microbiol 53:71–102

Su Q, Zhou X, Zhang Y (2013) Symbiont-mediated functions in insect hosts. Commun Integr Biol 6:e23804

Sudakaran S, Kost C, Kaltenpoth M (2017) Symbiont acquisition and replacement as a source of ecological innovation. Trends Microbiol 25:375–390

Tang T, Zhang Y, Cai T, Deng X, Liu C, Li J, He S, Li J, Wan H (2020) Antibiotics increased host insecticide susceptibility via collapsed bacterial symbionts reducing detoxification metabolism in the brown planthopper, *Nilaparvata lugens*. J Pest Sci 94:757–767

Tantowijoyo W, Andari B, Arguni E, Budiwati N, Nurhayati I, Fitriana I, Ernesia I, Daniwijaya EW, Supriyati E, Yusdiana DH, Victorius M, Wardana DS, Ardiansyahid H, Ahmad RA, Ryan PA, Simmons CP, Hoffmann AA, Rancès E, Turley AP, Johnson P, Utarini A, O'neill SL (2020) Stable establishment of WMEL wolbachia in *Aedes aegypti* populations in Yogyakarta, Indonesia. PLoS Negl Trop Dis 14:e0008157

Taracena ML, Oliveira PL, Almendares O, Umaña C, Lowenberger C, Dotson EM, Paiva-Silva GO, Pennington PM (2015) Genetically modifying the insect gut microbiota to control Chagas disease vectors through systemic RNAi. PLoS Negl Trop Dis 9:1–14

Taylor CM, Coffey PL, DeLay BD, Dively GP (2014) The importance of gut symbionts in the development of the brown marmorated stink bug, *Halyomorpha halys* (Stål). PLoS ONE 9(3): e90312

Taylor C, Johnson V, Dively G (2016) Assessing the use of antimicrobials to sterilize brown marmorated stink bug egg masses and prevent symbiont acquisition. J Pest Sci 90:1–8

Teixeira L, Ferreira Á, Ashburner M (2008) The bacterial symbiont Wolbachia induces resistance to RNA viral infections in *Drosophila melanogaster*. PLoS Biol 6:2753–2763

Tsuchida T, Koga R, Horikawa M, Tsunoda T, Maoka T, Matsumoto S, Simon J-C, Fukatsu T (2010) Symbiotic bacterium modifies aphid body color. Science 330:1102–1104

Turelli M (2010) Cytoplasmic incompatibility in populations with overlapping generations. Evolution 64:232–241

Um S, Fraimout A, Sapountzis P, Oh DC, Poulsen M (2013) The fungus-growing termite Macrotermes natalensis harbors bacillaene-producing bacillus sp. that inhibit potentially antagonistic fungi. Sci Rep 3:1–7

Unckless RL, Jaenike J (2012) Maintenance of a male-killing wolbachia in *Drosophila innubila* by male-killing dependent and male-killing independent mechanisms. Evolution 66:678–689

Van Den Bosch TJM, Welte CU (2017) Detoxifying symbionts in agriculturally important pest insects. Microb Biotechnol 10:531–540

van den Hurk AF, Hall-Mendelin S, Pyke AT, Frentiu FD, McElroy K, Day A, Higgs S, O'Neill SL (2012) Impact of Wolbachia on infection with chikungunya and yellow fever viruses in the mosquito vector *Aedes aegypti*. PLoS Negl Trop Dis 6:e1892

Vieira AS, Morgan ED, Drijfhout FP, Camargo-Mathias MI (2012) Chemical composition of Metapleural gland secretions of fungus-growing and non-fungus-growing ants. J Chem Ecol 38:1289–1297

Visser AA, Nobre T, Currie CR, Aanen DK, Poulsen M (2012) Exploring the potential for actinobacteria as defensive symbionts in fungus-growing termites. Microb Ecol 63:975–985

Von Burg S, Ferrari J, Müller CB, Vorburger C (2008) Genetic variation and covariation of susceptibility to parasitoids in the aphid *Myzus persicae*: no evidence for trade-offs. Proc Royal Soc B: Biol Sci 275:1089–1094

Vorburger C, Gehrer L, Rodriguez P (2010) A strain of the bacterial symbiont Regiella insecticola protects aphids against parasitoids. Biol Lett 6:109–111

Walker T, Moreira LA (2011) Can wolbachia be used to control malaria? Memorias do Instituto Oswaldo Cruz 106:212–217

Walker T, Johnson PH, Moreira LA, Iturbe-Ormaetxe I, Frentiu FD, McMeniman CJ, Leong YS, Dong Y, Axford J, Kriesner P, Lloyd AL, Ritchie SA, O'Neill SL, Hoffmann AA (2011) The wMel Wolbachia strain blocks dengue and invades caged *Aedes aegypti* populations. Nature 476:450–455

Wang S, Jacobs-Lorena M (2013) Genetic approaches to interfere with malaria transmission by vector mosquitoes. Trends Biotechnol 31:185–193

Wang S, Ghosh AK, Bongio N, Stebbings KA, Lampe DJ, Jacobs-Lorena M (2012) Fighting malaria with engineered symbiotic bacteria from vector mosquitoes. Proc Natl Acad Sci U S A 109:12734–12739

Wang S, Dos-santos ALA, Huang W, Liu KC, Oshaghi MA, Wei G, Agre P, Jacobs-lorena M (2017) Driving mosquito refractoriness to *Plasmodium falciparum* with engineered symbiotic bacteria. Science 357:1399–1402

Wang Q, Yuan E, Ling X, Zhu-Salzman K, Guo H, Ge F, Sun Y (2020) An aphid facultative symbiont suppresses plant defence by manipulating aphid gene expression in salivary glands. Plant Cell Environ 43(9):2311–2322

Waterworth SC, Flórez LV, Rees ER, Hertweck C, Kaltenpoth M, Kwan JC (2020) Horizontal gene transfer to a defensive symbiont with a reduced genome in a multipartite beetle microbiome. mBio 11:1–23

Wernegreen JJ (2012) Mutualism meltdown in insects : bacteria constrain thermal adaptation. Curr Opin Microbiol 15:255–262

Werren JH (1997) Biology of Wolbachia. Annu Rev Entomol 42:587–609

Werren JH (2012) Symbionts provide pesticide detoxification. Proc Natl Acad Sci U S A 109:8364–8365

Wilkinson TL (1998) The elimination of intracellular microorganisms from insects: an analysis of antibiotic-treatment in the pea aphid (*Acyrthosiphon pisum*). Comp Biochem Physiol: Mol Integr Physiol 119:871–881

Wong ZS, Hedges LM, Brownlie JC, Johnson KN (2011) Wolbachia-mediated antibacterial protection and immune gene regulation in drosophila. PLoS One 6:e25430

Xia X, Sun B, Gurr GM, Vasseur L (2018) Gut microbiota mediate insecticide resistance in the Diamondback Moth , Plutella xylostella (L.). Front Microbiol 9:25

Xie J, Vilchez I, Mateos M (2010) Spiroplasma bacteria enhance survival of *Drosophila hydei* attacked by the parasitic wasp Leptopilina heterotoma. PLoS One 5:e12149

Ye YH, Woolfit M, Rancès E, O'Neill SL, McGraw EA (2013) Wolbachia-associated bacterial protection in the mosquito *Aedes aegypti*. PLoS Negl Trop Dis 7:e2362

Ye YH, Carrasco AM, Frentiu FD, Chenoweth SF, Beebe NW, van den Hurk AF, Simmons CP, O'Neill SL, McGraw EA (2015) Wolbachia reduces the transmission potential of dengue-infected *Aedes aegypti*. PLoS Negl Trop Dis 9:e0003894

Yeap HL, Mee P, Walker T, Weeks AR, O'Neill SL, Johnson P, Ritchie SA, Richardson KM, Doig C, Endersby NM, Hoffmann AA (2011) Dynamics of the "popcorn" Wolbachia infection in outbred *Aedes aegypti* informs prospects for mosquito vector control. Genetics 187:583–595

Yoshida S, Matsuoka H, Luo E, Iwai K, Arai M, Sinden RE, Ishii A (1999) A single-chain antibody fragment specific for the plasmodium berghei ookinete protein Pbs21 confers transmission blockade in the mosquito midgut. Mol Biochem Parasitol 104:195–204

Zabalou S, Riegler M, Theodorakopoulou M, Stauffer C, Savakis C, Bourtzis K (2004) Wolbachia-induced cytoplasmic incompatibility as a means for insect pest population control. Proc Natl Acad Sci U S A 101:15042–15045

Zélé F, Nicot A, Duron O, Rivero A (2012) Infection with Wolbachia protects mosquitoes against plasmodium-induced mortality in a natural system. J Evol Biol 25:1243–1252

Zhang D, Zheng X, Xi Z, Bourtzis K, Gilles JRL (2015) Combining the sterile insect technique with the incompatible insect technique: I-impact of Wolbachia infection on the fitness of triple- and double-infected strains of *Aedes albopictus*. PLoS One 10:e0121126

Zucchi TD, Prado SS, Cônsoli FL (2012) The gastric caeca of pentatomids as a house for actinomycetes. BMC Microbiol 12:1–7

Zug R, Hammerstein P (2015) Bad guys turned nice? A critical assessment of Wolbachia mutualisms in arthropod hosts. Biol Rev 90:89–111

Metagenomic Approaches for Insect Symbionts

6

Mani Chellappan and M. T. Ranjith

Contents

M. Chellappan (✉) · M. T. Ranjith
AINPVPM, Department of Agricultural Entomology, College of Agriculture, Vellanikkara, Kerala Agricultural University, Thrissur, Kerala, India
e-mail: mani.chellappan@kau.in

Omkar (ed.), *Microbial Approaches for Insect Pest Management*,
https://doi.org/10.1007/978-981-16-3595-3_6

Abstract

Insects, the most successful groups in animal kingdom, harbor diverse groups of microbes, such as bacteria, archaea, fungi, protozoa, and viruses, which profoundly influence their survival and adaptations over a wide range of ecological niches. These microbes are associated with their host insects permanently or transiently and such associations may be beneficial or harmful to the host insect under various instances. Attempts were made earlier to characterize insect microbiome by isolation and cultivation techniques and polymerase chain reaction (PCR)-based cloning methods that resulted in identification of a few groups of microbes. The metagenomic approaches under the next-generation sequencing platforms provide unparallel opportunities to understand the composition of the microbiome and their functional role in the biology of the insects, thus expanding our understanding from a single microbial species to the whole community. These approaches provide an ample opportunity to understand the components of the microbiome that can potentially and collectively affect the behavior and physiological traits of insects through genetic and metabolic interactions. For instance, endosymbionts (*i.e.*, microbes that live inside host cells or tissues) depend on the insect hosts for obtaining nutrients, provide fitness advantages to their insect hosts in terms of the breakdown of plant cell wall components, viz, cellulose, lignocelluloses, and xylan, supplying essential amino acids and vitamins to host insects, thereby upgrading the nutrient status of their diet, detoxification of lethal insecticide molecules, plant defensive compounds such as phenolics, and production of anti-microbial peptides against insect pathogens. However, in some instances, the microbes may also be pathogenic to the insect hosts by producing insecticidal toxins, which reduce viability and cause morbidity.

The culture-independent metagenomic approach allows us to characterize a variety of genes that microbes possess or are expressing, which signifies 'what they are doing' within the host. It also enables us to compare the performance of insect with changes in their microbiome composition. These approaches have a wide range of applications apart from the study of insect's microbial ecology. The microbiota associated with wood-feeding beetles can be exploited as source of novel enzymes in industrial bioprocesses. The information on microbial genes and enzymes involved in cellulose hydrolysis, vitamin production, and nitrogen fixation can be useful in improving the reliability and efficiency of industrial processes. Furthermore, insect–microbe relationships could be manipulated to improve pest control, by decreasing pest's fitness or by increasing the efficacy of pest management programs.

Keywords

Metagenome · Next Generation Sequencing · Microbial community · 16S rRNA gene

Learning Objectives

1. Insects harbour diverse group of microbes belonging to various taxa, which profoundly influence their survivability and adaptations over a wide range of ecological niches.
2. Over the past two decades, the insect microbiome analyses were carried out primarily by the classical approach involving the isolation cum cultivation techniques and polymerase chain reaction (PCR)-based cloning methods that resulted in identification of only a few groups of microbes.
3. However, the recent advances in metagenomic strategies and sequencing techniques revolutionized the study of insect microbiome and provided unparalleled opportunities to understand the composition and functional diversity of insect microbiome.
4. The insect metagenome analysis under next-generation sequencing (NGS) platforms offers valuable information that could be helpful in formulating novel approaches in pest management by manipulating the insect–microbe relationships.

6.1 Introduction

Approximately 4–6 × 10^{30} microorganisms are present on earth (Sleator et al. 2008). Out of the total, nearly 99% are not amenable for culture plate, but play an important role in a variety of environment, namely soil, water, atmosphere, plant and animal systems. Metagenomics (also called as ecogenomics or environmental genomics or community genomics) is the scientific study of DNA sequences collected directly from an environment to know the diversity and ecology of microorganisms of that specific environment. According to Chen and Pachter (2005), it is the application of modern genomic techniques for the study of microbial communities from an environment directly without actually culturing them (on earth remain uncultured). The study may help in our understanding on the microbial diversity in a specified environment, interaction between the communities and higher animals, and the biology, as a whole. Studies on the uncultured microorganisms will not only give in-depth details about their ecology, it also helps in the identification of novel enzymes, signal mimics, smart molecules and new generation antibiotics (Rajagopal 2009; Krishnan et al. 2014).

Insects, which represent more than half of all the biodiversity in the world, are one of the most diverse and successful organisms in the history of life on earth. The remarkable success is due to the abilities of insects to colonize highly diverse niches and the metamorphosis in its biology. A poikilothermic form, insects cannot regulate their body temperature, but with varied adaptations insects can survive both hot as well as sub-zero conditions (Finn et al. 2015).

Insect's digestive system harbours numerous microorganisms, which dictate the growth, development, adaptation and general fitness of the host. Alimentary canal of insects contains approximately 10 times more microorganisms than the total body cells of the insect (Rajagopal 2009). Microbes get into the digestive system of the insect through the food and reside inside as commensals or parasites or symbionts. The gut microbiota influences all aspects of insect physiology, ecology and evolution (the beneficial microbes help the insect in digestion and metabolism like cellulose and xylan hydrolysis, vitamin production, nitrogen fixation, insecticide resistance, antibiotic resistance, signal molecules like quorum sensors, etc.). The gut microbiota of insects also involves in food digestion, pesticide detoxification, growth and development of the organism, pathogen resistance, intra-specific communication and general physiology (Engel and Moran 2013; Douglas 2015; Jing et al. 2020). The contributions of these gut microorganisms in relation to insect functions are highly relevant in the field of public health and veterinary medicine, agriculture and ecology.

6.2 History and Milestones in the Metagenomic Research

Microorganisms occur in almost all habitats in nature, even in extreme environments, namely polar regions, desert, hot geysers, deep sea and inhospitable rocks. They play crucial roles in biology, palaeontology, soil health etc. The study of microorganisms is based on morphological features, growth and selection of some biochemical profiles in vogue for the past 300 years since the invention of microscope by Antonie Philips van Leeuwenhoek in 1676 (Schierbeek 1959; Roszak et al. 1984). Over the years, microbiologists realized that bulk of microorganisms (99%) cannot be cultured by routine culture media. The proposal to use ribosomal RNA genes as molecular markers for biological classification (Woese and Fox 1977) and automated sequencing method invented by Sangers et al. (1977a, b), in fact, revolutionized the study and classification of microorganisms in the late 1970s.

Stahl et al. (1984) demonstrated the direct analysis of 5S and 16S rRNA genes to describe the microbial diversity without culturing the microorganisms per se. This led to the subsequent isolation and cloning of DNA from environmental samples. Begon et al. (1986) proposed the microbial community concept as the set of microorganisms coexisting in the same space and time. During this period, the microbiologists conclusively learnt that the number of observed microorganisms in a microscope did not correspond with number of microorganisms obtained in culture plates (Staley and Konopka 1985). Several advances have been made in the ensuing decade, like polymerase chain reaction (PCR), rRNA genes cloning and sequencing, fluorescent in situ hybridization (FISH), denaturing gradient gel electrophoresis (DGGE and TGGE), restriction-fragment length polymorphism and terminal restriction-fragment length polymorphism (T-RFLP). Handelsman and co-workers in the late 1990s defined the study and characterization of uncultivable microorganism as metagenomics; it is the theoretical collection of all genomes from members in a microbial community from a specific environment (Handelsman et al. 1998). Metagenomics laid the foundation of the '-omics' techniques and revolutionized research in microbial ecology (Handelsman 2004).

Venter et al. (2004a, b) carried out the first ever large-scale metagenomic project by sequencing samples from Sargasso Sea close to Bermuda, which yielded a whopping 1.6 billion base pairs of DNA and led to the discovery of 1.2 million new genes. Following this, many studies on the microbial community structure in varied environment had been carried out to uncover several unknown facts as microbes continue to play a crucial part in those habitats.

6.3 Insect Microbiome and its Functional Role

Insect gut microbiome, the collective genome of the native microbiota of the gut has multiple relationships with their insect hosts, which range from the obligate mutualism to pathogenic (Dillon and Dillon 2004). The gut microbes play vital roles in the digestion of food ingested, produce essential vitamins, make the host to survive on suboptimal food sources, resisting hostile pathogen infection, aid in the detoxification of xenobiotics compounds, growth and development of overall physiology of the host insect (Jing et al. 2020). Acquisition, colonization and transmission of microbiome determine the success of the insect in an environment (Gupta and Nair 2020). In a mutual relationship, insects present distinctive habit for a variety of microorganisms to colonize and the gut microbes in turn provide numerous benefits to their insect hosts (Douglas 2015). In rare conditions, established symbionts can become opportunistic pathogen also, if conditions become so. A number of factors, namely digestive enzymes, pH, redox potential of the gut, type of food the host has ingested and the secondary plant compounds present in the food, dictate the microbial density and diversity in the insect gut (Dillon and Dillon 2004).

6.3.1 Nutritional Symbioses of Gut Microbes

Symbiosis is essential for the survival of insects in extreme environmental conditions (Gupta and Nair 2020). The insect host can be primarily (obligate) or secondarily (facultative) dependant on the microbial symbionts to get nutrition and protection from their natural enemies, respectively (Moran et al. 2008). Major role of gut microbes is to provide the essential nutrients to the host (Jing et al. 2020). For instance, spirochetes through acetogenesis and nitrogen fixation provide required carbon, nitrogen and energy requirements of termite (Breznak 2004). The symbionts also enable the host to overcome barriers, like plant allelochemicals (Dowd and Shen 1990), nutritionally poor diets and recalcitrant food resources (Slaytor 1992).

6.3.2 Protection against Natural Enemies of the Insect Host

Another important beneficial function of gut microbiota in insect is to provide a buffering action to help prevent the proliferation of pathogens (Kodama and Nakasuji 1971; Charnley et al. 1985; Dillon et al. 2002). Resident gut microbiota protect their insect hosts against invaders by multiple mechanisms including restricting nutrients or space, production of toxins and activation of insect immune system functions that are more deleterious to the invader than the resident (Douglas 2015).

6.3.3 Gut Microbes in Detoxification of Xenobiotics

Insect resistance or tolerance to xenobiotics is mostly mediated by the insect genome rather than the gut microbes (Dillon and Dillon 2004; Douglas 2015). However, a compelling evidence of gut microbiota, *Burkholderia*-mediated fenitrothion resistance emerged in *Riptortus pedestris* (Kikuchi et al. 2012).

6.3.4 Gut Microbes in Insect Communication and Mating

Inter- and intra-communication of insect may be mediated by the microorganisms associated with insects (Ezenwa et al. 2012; Gupta and Nair 2020). Gut microbe activities result in the production of some compounds, which may act as kairomones or pheromones. Aggregation pheromone in grasshopper, *Schistocerca gregaria,* is produced by the gut microbe *Pantoea agglomerans* (Dillon et al. 2002). In *Drosophila melanogaster*, mating preference is dictated by the gut microbiota where the flies mate preferentially with individuals harbouring similar microbiota (Sharon et al. 2010, 2011).

6.3.5 Trophic Interactions

The insect gut microbiota is involved in the behavioural aspects of the interactions between insects, their natural enemies and the host (Campbell 1990). In course of evolution, insects have evolved many strategies to feed on plants mediated by mutualistic symbionts (Frago et al. 2012). Insect symbionts have been reported to benefit their hosts; the best-known example is ambrosia beetles and their mutualistic fungi of bark, which make wood digestible for their host (Paine et al. 1997).

6.3.6 Interaction of Gut Microbiota in Productive Insects

Gut bacteria promote populations of beneficial insects by improving the fitness of productive insects, pollinators and biocontrol agents. In irradiated sterile male flies of the Mediterranean fruit fly, *Cerratitis capitata,* mating competence can be improved by feeding diet enriched in *Klebsiella oxytoca* (Lance et al. 2000; Ami et al. 2010) as irradiation causes shift in the microbial community and results in fitness decrease. Feeding on fortified diet significantly increased the sexual competitiveness of irradiated males, enhanced their survival and inhibited sexual receptivity of female flies (Gavriel et al. 2011). Pollinators, like bumble bees, are prone to the attack of parasitoid *Crithidia bombi* and depend upon the gut microbiota (Koch and Schmid-Hempel 2012).

6.4 Insect Microbiome Analysis: From Genomics to Metagenomics

Over the past two decades, the insect microbiome analysis was carried out primarily by the classical approach involving isolation of microorganisms from the various insect physiological systems, culturing them on solid or liquid growth medium containing appropriate sources of carbon, energy and electron acceptor and phenotypic characterization of isolates. This approach solely depended on the physiological conditions under which the organism isolated and sometimes the optimal conditions provided in the laboratory might impose selection pressure, thereby inhibiting the growth of a large number of microorganisms (Staley and Konopka 1985). The media used to isolate insect gut microorganisms were frequently the same as those employed in medical studies. However, some bacteria that were found to be numerically dominating in these media may be physiologically insignificant (Dillon and Dillon 2004). Thus, the focus was on developing the media for culturing of microorganisms, which satisfy the environmental factors, such as pH and available nutrients encountered in the insect physiological system. Further, both the simple morphological and physiological traits in most of the microbes provide only a few identification clues (Pace et al. 1986) and have revealed a large discrepancy between the relatively few culturable microorganisms and the significant diversity present in insect gut (Pace 1997; Head et al. 1998). It was also recognized that approximately 99% of microbes in the environment cannot be cultured (Amann et al. 1995) and due to limitations of culture methods, it was envisioned that most of the microbes associated with the insect gut were still to be identified (Stokes et al. 2001).

From late 1970s onwards, remarkable works were carried out to exploit the ribosomal RNA genes as molecular markers for classifying the life system (Woese and Fox 1977) and this approach in association with the Sanger automated sequencing (Sanger et al. 1977a, b) method revolutionized the study and classification of microorganisms. However, with the advancement in molecular techniques, three traditional molecular approaches, namely gene targeting PCR, molecular fingerprinting techniques, such as DGGE (denaturing gradient gel electrophoresis) and oligonucleotide probe-based hybridization techniques, such as FISH (fluorescent in situ hybridization) have been employed to investigate the insect gut microbial communities (Stokes et al. 2001), whereas the recent advances in sequencing techniques and the metagenomic strategies replaced the above techniques and revolutionized both gene discovery and biodiversity analysis of the insect gut symbiotic microbiota (Fig. 6.1).

6.4.1 Traditional Molecular Approaches in Microbiome Analysis

Gene-Specific PCR

This technique employs gene-specific primers to specifically amplify the target genes, such as conserved *16S rRNA* gene or a gene of specific functional interest from insect gut symbionts. Kane and Pierce (1994) were among the first to conceive the idea of using PCR-based ribosomal DNA sequencing to explore gut microbial communities of termites. Further, McKillip et al. (1997) analysed the composition of the microbiome in the midgut of leaf roller, *Pandemis pyrusana* Kearfott, using both PCR and culturing techniques. Later on, Lilburn et al. (1999) sequenced 98 clones of near-full-length *16S rDNA* of *Spirochaetes* in the gut of termite species, *Reticulitermes flavipes* and observed a substantial phylogenetic diversity in the termite gut. Schmitt Wagner et al. (2003) carried out phylogenetic analysis of *16S rRNA* genes recovered from the hindgut of soil-feeding termites and revealed an enormous diversity of bacteria in the different gut compartments, whereas Ohkuma and Kudo (1996) did the PCR targeting of *16S rRNA* in gut of termite species, *Reticulitermes speratus* and found that most of the gut microbial *16S rRNA*s amplified were unknown. Most of the earlier studies targeting *16S rRNA* gene analyses revealed a significant number of unknown bacterial species at the time.

Apart from *16S rRNA* gene analysis, gene-specific PCR has also been widely applied to identify genes from microbial communities, which are involved in various metabolic pathways. Gene targeting method was followed to clone a number of cellulases belonging to glycosyl hydrolase family 45 from the flagellates *Koruga bonita* and *Deltotrichonympha nana*, which are associated with termite gut (Li et al. 2003). Further, Inoue et al. (2005) identified a cellulase gene from lower termite hindgut using PCR with gene-specific primers and in situ hybridization.

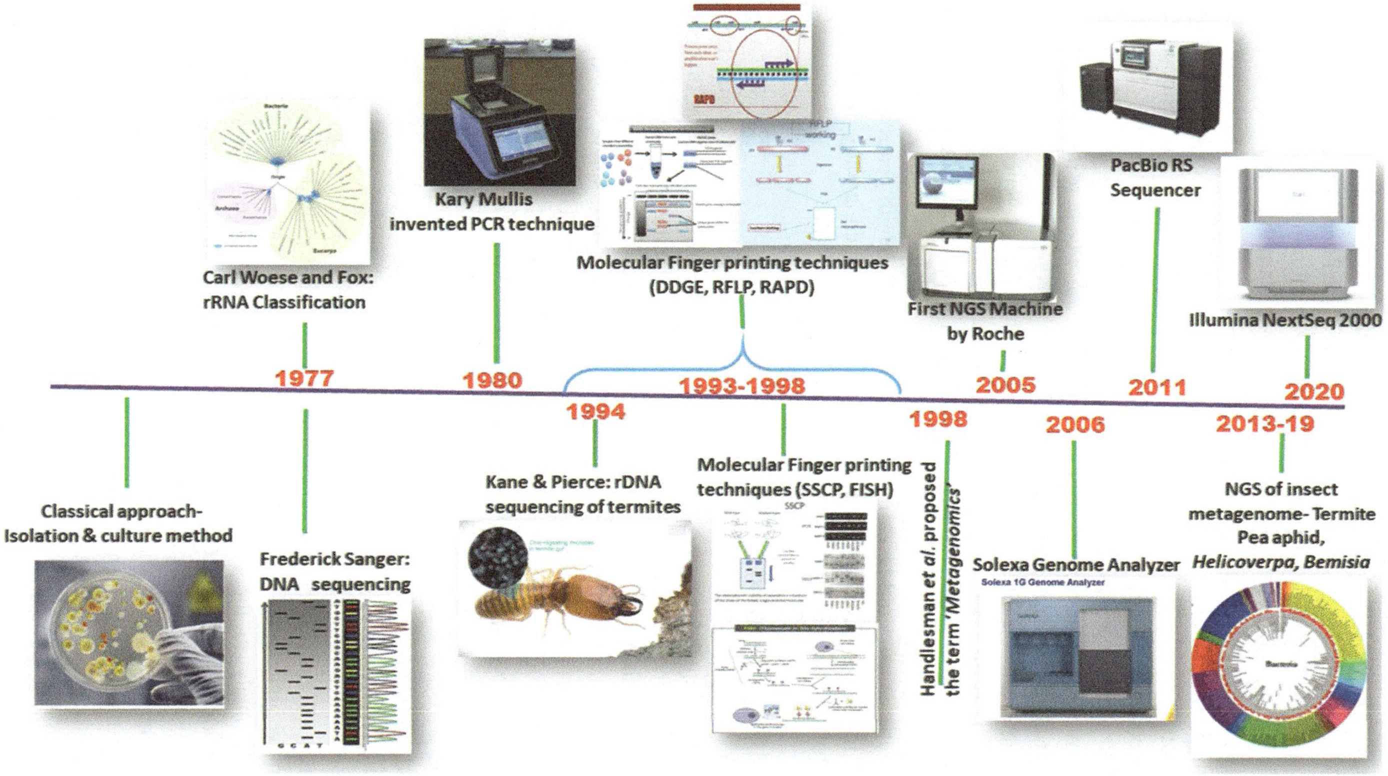

Fig. 6.1 Insect metagenomics—timeline and milestones

In addition to gene-targeting PCR of DNA samples, reverse transcriptase PCR (RT-PCR) from RNA has also been employed to clone genes from environmental samples (Manefield et al. 2002). Casu et al. (1996) identified a major excretory/secretory protease from *Lucilia cuprina* larvae by combining both RT-PCR and immune blotting technique. Later on, Noda et al. (1999) amplified a nitrogen fixing gene from microbial RNA in the gut of the termite, *Neotermes koshunensis* by RT-PCR method. Nakashima et al. (2002) carried out RT-PCR experiments and revealed that five GHF9 EG (Glycosyl Hydrolase Family 9 Endoglucanase) homologs were expressed in the salivary glands and the midgut of termites. Further, the RT-PCR technique was also employed for identifying the genes from gut bacterial communities of *Helicoverpa armigera* and *Manduca sexta,* respectively (Chougule et al. 2005; Brinkmann et al. 2008).

Though gene-specific PCR was proven to be effective for gene discovery and microbial diversity analysis, two major limitations have restricted the application of this technique. The gene-targeting PCR techniques depend on existing sequence information to design primers for PCR amplification and normally only partial sequence of the genes could be cloned; the cloning of full-length genes would have to involve further PCR-based chromosome walking, which greatly limited the application of this technique (Cowan et al. 2005).

Molecular Fingerprinting Techniques

Apart from the sequence, library-based gene targeting PCR and some other PCR-based techniques have also been widely used to analyse microbial diversity in various environmental samples. The molecular fingerprinting techniques used for microbiome analysis include denaturing or temperature gradient gel electrophoresis (DGGE or TGGE), restriction fragment length polymorphisms, single strand conformation polymorphism and random amplified polymorphic DNA (Muyzer et al. 1993; Lee et al. 1996; Liu et al. 1997; Kauppinen et al. 1999). These techniques are used to analyse the sequence of *16S rRNA* gene from different microbial species, where both molecular fingerprints and phylogenetic affiliation of microbial species can be generated. These above-mentioned molecular fingerprinting techniques have been found useful in exploring the microbial diversity associated with insect gut also (Smalla 2004).

Among the molecular fingerprinting techniques, DGGE is the most commonly used method to study insect gut microbial diversity that provides a much more comprehensive understanding of insect symbionts (da Mota et al. 2005). Reeson et al. (2003) analysed the microbial communities associated with wasp larva, *Vespula germanica* based on DGGE profiling and found that the wasp larvae are not solely dependent on one particular type of mutualist. Further, analysis of gut bacterial communities in Mediterranean fruit fly using both culture-dependent and culture-independent approaches, such as DGGE, revealed that the family Enterobacteriaceae is the most dominant species in the gut of fruit fly (Behar et al. 2005). The DGGE method is also used to analyse gut microorganisms in wood feeding termites (Hayashi et al. 2007), soil-feeding termites and their mounds (Fall et al. 2007), hindguts of scarab beetle larvae (Pittman et al. 2008; Vasanthakumar et al. 2006), gut of grasshoppers and locusts (Dillon et al. 2008), and diamond back moth (Raymond et al. 2008).

Restriction fragment length polymorphism (RFLP) analysis differentiates the homologous DNA sequences based on the distinct DNA fragment patterns generating from the sequence specificity toward restriction enzymes (Esumi et al. 1982). Harada and Ishikawa (1993) used RFLP to analyse *16S rRNA* gene from the group of prokaryote microbes in the gut of the pea aphid and the result suggested that gut microbes have a close relationship with aphid intracellular symbionts. However, due to the technical limitations and low resolution of traditional RFLP technique, terminal restriction fragment length polymorphism (T-RFLP) has been employed to study microbial diversity in insect gut (Shinzato et al. 2005). T-RFLP separates homologous DNA based on the length and sequence of the end sequence generated from restriction enzyme digestion of *16S rRNA*, which makes it much more efficient in revealing microbial diversity as in the case of bacterial *16S rRNA* genes analysis of the midguts of European cockchafer (*Melolontha melolontha*) larvae (Egert et al. 2005), soil-feeding termites (Kohler et al. 2008) and fungus-growing termites (Shinzato et al. 2007).

Another traditional molecular fingerprinting technique is random amplified polymorphic DNA (RAPD) and the analysis is based on amplification of genomic DNA using random primers. RAPD-PCR was carried out to compare the gut microbial composition between different generations of Western flower thrips, *Frankliniella occidentalis,* and the results revealed that some bacteria in the thrips could be passed from generation to generation for up to 50 generations (de Vries et al. 2001a, b). However, application of RAPD is very limited in gut microbiota analysis due to technical complexity and low reproducibility of the technique.

Single-strand conformation polymorphism (SSCP) is another technique that uses electrophoresis to separate single-strand DNA to differentiate the homologous sequences (Yandell 1991). Mohr and Tebbe (2006) used SSCP to study the diversity and phylogenetic relationship of bacteria in the guts of three bee species at the same oilseed rape field, whereas Brinkmann et al. (2008) used combination of PCR-SSCP, RT-PCR SSCP and stable isotope probing (SIP) to study the diversity of metabolically active bacteria in the larval gut of *Manduca sexta.*

Fluorescent in Situ Hybridization

Fluorescent in situ hybridization (FISH) is one of the most common techniques used in microbial ecology studies to visualize the symbiotic bacteria in the gut (Aminov et al. 2006; Cheung et al. 1977). The application of FISH in insect gut microbiota studies often involves fluorescently labelled probes targeting *16S rRNA* genes with sequences specific for a bacterial species or genus (Turroni et al. 2008). FISH has been used to detect, visualize and characterize the intracellular symbiotic bacteria of insects, such as aphids (Fukatsu et al. 1998), crickets (Domingo et al. 1998) and termites (Berchtold et al. 1999). The approach has been shown to be particularly useful in studying uncultivated microbes to observe the dynamics of microbiota (Santo Domingo et al. 1998). However, the analysis of complex bacterial communities from environmental samples by FISH with rRNA-targeted probes often encounters several technical problems and thus the detailed composition of the microbiota cannot be revealed. In addition, bacteria lives in less nutrient-rich environments with low ribosome content, could affect the sensitivity of detection (Smalla 2004).

To complement to FISH, DAPI (4,6-diamidino-2- phenylindole) and GFP (green fluorescent protein) can been used to visualize microbial communities inhabiting insect gut. DAPI staining of bacterial cells highlights the significant differences in the number of bacterial cells among different insect species, when reared under the similar conditions (Cazemier et al. 1997). Similarly, GFP can be used to track target microbial species in the host. Hurst and Jackson (2002) used GFP to show that the colonization of bacterium, *Serratia entomophila* in the gut of the host, *Costelytra zealandica,* is not confined to a specific site in the gut.

6.4.2 Metagenomics

Though the various traditional molecular techniques have greatly advanced our understanding of insect gut microbial communities, due to the inherent limitations of these techniques, they cannot provide detailed information regarding the gene and pathway for different biological processes and a comprehensive coverage of microbial taxonomy in the insect gut. In order to understand the biological processes involved in biomass degradation, a detailed understanding on the biocatalysts, pathways and compositions of insect gut symbionts is required. However, the high throughput 'metagenomic' approaches allow us to understand the complex properties of the microbiota, their dynamics and function in the natural system. Various metagenomic approaches answer fundamental questions, such as which organisms are present? (Taxonomic diversity), and what roles they play? (Functional metagenomics) (Vieites et al. 2008).

The term 'metagenomics' was coined in 1998 (Handelsman et al. 1998). It helps us to investigate complex microbial communities sampled directly from the environment, without culturing or isolating a single organism. The so-called 'metagenomics' often involves sequence-based, compositional and/or functional analyses of the combined microbial genomes contained within an environmental sample, such as the insect gut (Handelsman et al. 1998). The amplification of specific targeted genes, such as (V1toV9) of *16S rRNA, 18S rRNA,* ribosomal *ITS, NifH,* among others, by PCR before sequencing permit diversity analysis (Morgan and Huttenhower 2012). The diversity, composition and dynamics of a microbial community largely define its effectiveness, specificity and reactivity for a certain function related to life, biogeochemical cycles and environmental mitigation (Allen and Banfield 2005; Falkowski et al. 2008). In the past two decades, significant works have been carried out to explore the components of microbial communities from different niches at the molecular, organismic and ecological levels to reveal novel enzymes, functional pathways and requisite organisms for various applications (Green et al. 2008; Roussel et al. 2008).

Metagenome sequencing has also become important approaches for exploring biomass degrading mechanisms in wood-feeding insects. Several works have been carried out to explore the diversity of microbiota inhabiting the mid- and hindgut of higher (Warnecke et al. 2007) and lower termites (Todaka et al. 2007). However, some studies revealed that symbiotic bacteria and protozoa in the hindgut of the termite play an important role in the hydrolysis of cellulose and hemicellulose (Nakashima et al. 2002; Tokuda and Watanabe 2007; Warnecke et al. 2007; Zhou et al. 2007). Recently, numerous studies have been carried out using metagenomic approach under next-generation sequencing platform to explore the microbial communities associated with insects and their role in its survival and host–insect relationship. In addition to metagenomic approach, metatrascriptomics (refers to sequencing analysis of mRNA from a microbial population) and metaproteomics (refers to the quantification and identification of all the proteins in a microbial community) approaches have also been employed for better understanding of microbial diversity and function in the environment.

6.5 Types and Approaches in Metagenomics

One of the most significant developments in the field of microbial ecology in the past decade has been the advent of metagenomics and it is the explicit method of direct analysis of genomes present in an environmental sample. The field initially started with the cloning of environmental DNA, followed by functional expression screening (Handelsman et al. 1998) and was then quickly complemented by direct random shotgun sequencing of DNA from various environmental samples (Tyson et al. 2004; Venter et al. 2004a, b). Metagenomics provides an insight into the composition of functional genes present in microbial communities and gives a much broader description than usual phylogenetic surveys, which are based only on the diversity analysis of one gene, i.e., *16S rRNA* gene. It provides the valuable genetic information on potentially novel biocatalysts or enzymes involved in various metabolic pathways, genomic linkages between function and phylogeny of uncultured organisms, and evolutionary relationship of community function and structure. The types and approaches in metagenome analysis followed in various environmental fields are discussed hereunder.

6.5.1 Types of Metagenome Analysis

Two types of metagenomic analysis commonly used to unravel the microbial identity and their composition for high throughput sequencing data are: (i) amplicon-based analysis, which includes 16S ribosomal RNA for bacteria, internal transcribed spacer (ITS) and 18S region for fungi and eukaryotes, respectively, and (ii) whole metagenomic shotgun sequencing.

Amplicon-Based Analysis

16S sequencing is a widely used technique that relies on the variable regions (V1-V9) of the bacterial *16S rRNA* gene to make community-wide taxonomic assignments (Chakravorty et al. 2007). It is also used for microbial diversity analysis and for various environmental samples, such as soil (Chong et al. 2012), human gut (Dethlefsen et al. 2008) and various insect gut specimens (Hirsch et al. 2012a, b; Malathi et al. 2018). Some degree of divergence is allowed during the sequence similarity assessment stage of the analysis; typically, nearly identical sequences (>97%) are clustered into Operational Taxonomical Units (OTU) (Morgan and Huttenhower 2012). The limitation of this method is that if any two organisms have the same *16S rRNA* gene sequence, they may be classified as the same species in a 16S analysis, even if they are from different species. Because 16S analysis is based on the *16S rRNA* gene and OTUs are designated as taxa, it is difficult to discriminate between strains and, in some circumstances, closely related species. For example, 16S analysis cannot distinguish *Escherichia coli* O157:H7 from *E. coli* K-12 (Weinstock, 2012) but it can separate *Shigella flexneri* from *E. coli* (Hilton et al., 2016). Similarly, the *18S rRNA* is mainly used for taxonomic studies of fungi, while the *ITS* region is widely adopted for analysing fungal diversity in environmental samples (Bromberg et al. 2015).

Shotgun Metagenome Analysis

Shotgun metagenomic analysis has the ability to identify the majority of the organisms (culturable and unculturable bacteria) in the environmental sample. It helps to create a community biodiversity profile, which can be further utilized for functional composition analysis of organism lineages (i.e., genera or taxa) (Tringe et al. 2005). Before initiating a whole metagenomic study, an understanding of the potential microbial diversity and the relative abundance of species in the environmental sample is very important. Chen et al. (2018) carried out comparative shotgun metagenome analysis of silkworm, *Bombyx mori,* and the sequence datasets not only provide first insights into all bacterial genes in silkworm guts, but also help to generate hypotheses for subsequent analysis of functional traits of gut microbiota. A higher sequencing depth is required to detect a rare taxa from the given environmental sample (Sharpton 2014). This makes shotgun metagenomic sequencing much more expensive than 16S sequencing (Quail et al. 2012).

6.5.2 Approaches in Metagenome Analysis

There are two principal approaches in metagenome analysis: (i) the sequence-based metagenomics, and (ii) functional metagenomics. Sequence-based metagenomics involves metagenome sequencing and downstream data analysis, whereas functional metagenomics involves screening of DNA or cDNA library for gene discovery.

Sequence-Based Metagenomics

Sequence-based analysis of metagenomic DNA from insect gut has been well explored during the past decade to mine out the associated microbial communities.

However, metagenome analysis was first carried out with the conventional Sanger sequencing techniques, which are mainly used toward the *16S rRNA* library or metagenomic DNA library preparations (Smalla 2004). Warnecke et al. (2007) prepared the metagenomic DNA library of termite hindgut symbiotic microbiota with Sanger sequencing techniques, where approximately 71 million pairs of sequence data were generated and assembled, but they are highly fragmented in nature. In order to have a better understanding, 15 fosmids were selected for further sequencing and analysis through shotgun method. The data have led to a comprehensive coverage and quantification of the microbial composition in termite gut symbionts.

The advances in next-generation sequencing technology have offered the potential to revolutionize metagenome analysis (Marusina 2006). When next-generation sequencing is used, the approach can be the direct shotgun sequencing of metagenomic DNA. 454 sequencing technology is the first available next-generation sequencing technique and the platform is based on 'pyrosequencing' and emulsion PCR amplification (Margulies et al. 2005). The sequence read length for 454 sequencing can be up to 400 bases and the throughput is relatively lower at 400 million bases per run. The advantage of the 454 sequencing is the read length, which makes it easier for the sequence assembly in de novo sequencing (Shendure and Ji 2008; Yuan et al. 2008).

Illumina MiSeq, formerly known as Solexa, is based on the concept of 'sequencing by synthesis' (SBS) (Mardis 2008; Adams et al. 2009). With the latest development of the technology, Illumina genome analyser can generate pairwise end sequencing of 100 base pairs and 40 gigabase sequences per run. The two NGS platforms are ABSOLiD and Helocus, both of which have similar sequencing throughput and less sequence read-length (Mardis 2008). Thus, 454 and Illumina have been the major approaches for metagenome sequencing, where 454 offer the longer read length, while the strength of Illumina is the sequence throughput (Stangier 2009). Recently, next-generation sequencing-based metagenome analysis was carried out to explore microbial communities associated with major insect pest of global importance (Hirsch et al. 2012a, b; Scully et al. 2013; Ranjith et al. 2016; Jones et al. 2019; Harish et al. 2019).

Functional Metagenomics

Functional metagenomics involves screening for target genes in a library constructed with metagenomic DNA or RNA (Allen et al. 2009). Generally, metagenomic DNA can be stored stably as a DNA library for further investigation. Similarly, RNA can be reverse transcribed to build a cDNA library. The information available within a DNA or cDNA library can be used to determine community diversity and search for the enzymes with a particular activity (Steele and Streit 2005).

In order to construct metagenomic DNA library, the basic steps include the extraction of metagenomic DNA, the generation of suitably sized DNA fragments, and the cloning of these fragments into an appropriate vector (Cowan et al. 2005). For the construction of metagenomic cDNA library, total RNA will be extracted and cDNA will be synthesized for building into a proper vector. Both types of libraries can be screened for genes of interest via DNA hybridization technique using the probes of target genes or homologous genes (Demaneche et al. 2009). The approach has been widely used to search for various genes from insect guts. Shen and Jacobs-Lorena (1997) were the first to clone the chitinase gene from a cDNA library through screening and showed that it got expressed exclusively in the midgut of *Anopheles gambiae* adult females using Northern Blot techniques.

One of the major limitations of the traditional screening strategy is the need for specific probes to a certain gene. The sensitivity and reproducibility often also depend on the probe design. The combination of library screening with gene expression and/or enzyme activity assay has been developed to overcome such limitations. The method has been successfully applied to discover new genes and enzymes with different activities.

One of the recent developments in the functional metagenomics is the use of biosensor technology for gene discovery from the insect symbionts. Guan et al. (2007) constructed a metagenomic DNA library of midgut microbiota of gypsy moth, and analysed it using an intracellular screen named as METREX. The biosensor detects compounds that induce the expression of GFP from a bacterial quorum promoter by fluorescence microscopy or fluorescence-activated cell sorting (Williamson et al. 2005). Further, they identified an active metagenomic clone encoding a monooxygenase homologue that mediates a pathway of indole oxidation. Further, the metagenomic analysis of whole gut microbiota in four subspecies of termites revealed that they have shared conserved functional and carbohydrate-active enzyme profile and specialized in cellulose and chitin degradation (Grieco et al. 2019).

The functional metagenomics based on the cDNA library allows us to identify the novel enzymes and the genes encoding for particular enzymes; however, the analysis is limited often by the availability of probes for screening the cDNA library and the assay for determination of function of specific protein (Moran et al. 2008; Chaves et al. 2009). A more comprehensive approach is required to sequence the metatranscriptome of microbial communities and annotate them to discover the novel genes.

6.6 Steps Involved in Metagenomic Studies

Metagenomics is the study of collective genomes and genes from the members of a microbiota residing at a particular environment. This collection is obtained through sequencing of DNA extracted from an environmental sample followed by annotating the sequence data in silico, thereby increasing the understanding of the dynamics of the microbial community understudy. The various steps involved in insect metagenomic studies (Fig. 6.2), and tools and techniques used in metagenome sequence data analysis are discussed hereunder.

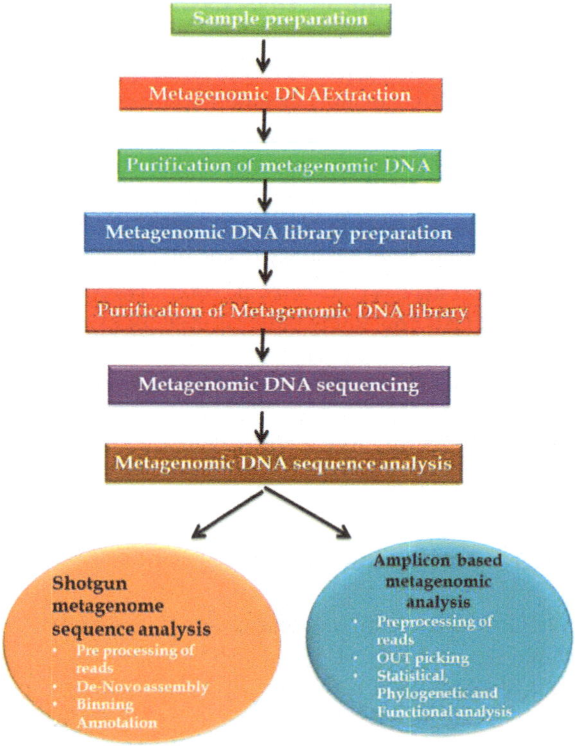

Fig. 6.2 Schematic diagram of steps involved in metagenomic studies

6.6.1 Sample Preparation

The sample preparation step for metagenomic analysis is crucial and must be carefully designed, with immediate analysis or freezing of samples for late analysis. Proper care must be taken to avoid multiple freeze-thaw cycles, which can alter the profile of the microbial community under investigation (Quince et al. 2017). The insect specimens collected from field should be kept overnight for starving and further immobilize them by treating with chloroform (100%). In order to make the surface of insect free from other extraneous microbial communities, which may interfere during the downstream analysis of gut microbiota, it should be surface sterilized with an antibacterial agent, such as streptomycin (0.05%) for approximately 1 h. Further, the antibacterial agent should be removed by surface washing with sufficient quantity of sterile water. As per the objective of study, the gut regions of the insects should be dissected out with utmost care under aseptic conditions.

6.6.2 Metagenomic DNA Extraction

Most of the insect gut metagenomic DNA extraction procedure has been adopted from soil DNA isolation methods (Zhou et al. 1996) with slight modifications. In metagenomic DNA isolation, two major strategies have been employed viz., cell recovery method and the direct lysis method (Roose-Amsaleg et al. 2001). In the cell recovery method, prior to cell lysis, the intact microorganisms associated with insect gut will be isolated by following either frequent homogenization or differential centrifugation or by gradient centrifugation in media, such as percoll or sucrose (Hopkins et al. 1991; Robe et al. 2003). However, in the direct lysis method, either SDS or CTAB containing buffer is used for extracting the metagenomic DNA. Some commercially available kits can also be used for the extraction of metagenomic DNA from uncultured organisms. However, the extraction protocol must be standardized under laboratory conditions, since most of these kits are not designed specifically for insect metagenomic DNA isolation.

6.6.3 Purification of Metagenomic DNA

If the total content of microorganisms from samples may not be efficiently extracted, it may lead to loss of DNA diversity (Josefsen et al. 2015). Since the extracted metagenomic DNA is prone to degradation by nucleases from the external environment, its integrity needs to be protected by inhibiting those enzymes with denaturing agents, which are commonly available in commercial kits. It is also necessary to remove the metal ions to avoid interference with DNA purification steps based on ion exchange. Silica-based columns are also used to bind DNA under high pH and salt concentrations, which helps to remove metal ion interferents (Bag et al. 2016). The DNA interference from dead microbial cells may be eliminated by treatment with propidium monoazide (PMA) or ethidium monoazide (EMA) before DNA extraction. These are DNA intercalating agents that pass only through ruptured membranes and after exposure of the treated cells to ultraviolet light; these agents prevent PCR amplification of the DNA of dead cells (Mayo et al. 2014).

6.6.4 Metagenomic DNA Library Preparation

After metagenomic DNA extraction and purification, the DNA fragmentation and insertion of adapters into the end regions of fragments will be carried out according to various protocols, depending on the sequencing platform (Van Djick et al. 2014). The DNA fragmentation can be performed with physical methods (i.e., ultrasonication), chemical reagents and enzymes with or without transposase activity (Head et al. 2014). Enzymes with transposase activity are highly advantageous because they perform both fragmentation and insertion of labelled or unlabelled sequencing adapters simultaneously, depending on the protocol of choice.

The sequencing adapters will be inserted in the DNA fragments and are specific to each sequencing platform (Van Djick et al. 2014). The adapters are ligated to a support or solid surface to enable spatial separation of fragments. Each fragment will serve as a template for the synthesis of new fragments in the amplification phase and different samples can be sequenced simultaneously during the process (Metzker 2010). The use of DNA indexes allows the processing of a pool of samples and correlates a given fragment with its original sample. The Illumina® platform has a unique indexing process that combines both the adapter and the indexes (barcodes) instead of adding the indexes to the ends of the each mould molecule, as performed for other sequencing platforms (Meyer and Kircher 2010).

For the preparation of libraries, two different approaches can be adopted: (i) paired-end, and (ii) mate-pair. Libraries with short-sized inserts are called paired-end libraries, whereas libraries with long-sized inserts are called mate-pair libraries. Both libraries support the sequencing data to discriminate the physical distance between two reads aligned in the reference genome. According to Van Nieuwerburgh et al. (2012), the success of de novo assembly from short reads depends on the determination of physical distance of the fragment, which is very important to specify the order and orientation of a *contig* in the genome. Thus, the preparation of a paired-end library is highly recommended to complete the regions of the genome containing small gaps, because the short-sized fragments can easily fill empty spaces and provide confirmation for the closing of a draft genome.

6.6.5 Purification of Metagenomic DNA Library

The generated libraries need to be purified before sequencing by selecting appropriately sized fragments and removing free adapters, dimers of adapters and other possible artifacts. This step can be performed with magnetic beads or agarose gel. If dimers of adapters are not removed, they can form clusters in the flow cell and lead to the generation of unwanted sequencing data (Head et al. 2014).

6.6.6 Metagenomic DNA Sequencing

The first step in metagenomic DNA sequencing is to choose a sequencing platform of a particular manufacturer, with due attention given to the set of data generated from the platform in each run (output). Among the companies that market sequencing platforms, Illumina® currently stands out for offering a variety of highly compatible platforms (Goodwin et al. 2016). In addition, Illumina® platforms provide the highest high-throughput per run and the lowest cost per sequenced base among all companies (Van Djick et al. 2014).

Illumina® platform uses the Sequencing By Synthesis (SBS) technique coupled with bridge amplification process in the flow cell (Shokralla et al. 2012). SBS sequencing uses the enzymes, such as DNA polymerase or DNA ligase, for the massive parallel amplification of template DNA. During the operation of SBS platforms, DNA polymerase adds labelled dNTPs on real time uninterruptedly, which are easily distinguishable from nucleotides not incorporated into the template DNA with the aids of an optical reader (Fuller et al. 2009).

Either single-end (SE) or paired-End (PE) sequencing can be opted and this profoundly influences the downstream analysis. SE sequencing refers to sequencing from a single end of the library fragment, whereas PE sequencing refers to sequencing from both ends of the fragment in a two-way elongation process (Van Djick et al. 2014). Paired-end sequencing is the most common approach and it is cost-effective, because it generates two reads for the same fragment per run.

6.6.7 Metagenomic Sequence Data Analysis

Various pipelines are used for downstream analysis in different metagenomic methods and the requisite bioinformatics' tools (Table 6.1).

Shortgun Metagenome-Sequence Analysis

Pre-Processing of Sequence Reads

The raw reads generated from the next-generation sequencing platform are subjected to adapter trimming, quality filtration and de-replication. If the metagenomic sample is isolated from a host organism, then host contamination is typically removed by aligning to the reference genome of the host organism, using Bowtie2 or other short-read mapper (Oulas et al. 2015).

De Novo Assembly

Assembly is computationally expensive and it requires sophisticated algorithms based on de Bruijn graphs. Tools that are specifically designed for metagenomic applications are mainly built on de Bruijn graph algorithms. A few common metagenomics assembly tools include CLC workbench, Meta-Ray, MetaVelvet-SL, MetaVelvet, Meta-IDBASOAP and metaSPAdes (Nurk et al. 2017; Luo et al. 2012). If an appropriate reference metagenome is available in the database, a reference-based assembly may be performed (Nagarajan et al. 2010).

Binning

Binning is the process of clustering the reads or contigs into a highly similar groups, and assigning the groups to specific taxa, such as species, subspecies or genera. Two types of algorithms are available: (a) composition-based binning, and (b) similarity-based binning. Certain binning tools make use of hybrid approaches, which run both kinds of algorithms. In composition-based binning, the groups occur in a supervised or semi-supervised manner, where the DNA fragments are with similar composition, whereas in similarity-based binning, it aligns the DNA fragments to database or reference sequences (Leung et al. 2011).

Table 6.1 Bioinformatic tools used in metagenomic analysis

Sl. No.	Name	Application
1	**QIAGEN CLC Main workbench** digitalinsights.qiagen.com/products-overview/discovery-insightsportfolio/analysis-andvisualization/qiagen-clc-mainworkbench/	For gene expression analysis, primer design, molecular cloning, phylogenetic analyses and sequence data management
2	**MetaVelvet** metavelvet.dna.bio.keio.ac.jp	For de novo metagenomic assembly
3	**MaxBin2 online** https://kbase.us/applist/apps/kb_maxbin/run_maxbin2/release?gclid=EAIaIQobChMI8tY2IqH7AIVxm4qCh2B6Qr6EAAYASAAEgIxc_D_BwE	For Metagenome Assembly Binning
4	**Pfam** https://pfam.xfam.org/	The Pfam database is a large collection of protein families, each represented by *multiple sequence alignments* and *hidden Markov models (HMMs)*
5	**TIGRFAMs** http://tigrfams.jcvi.org/cgi-bin/index.cgi	For protein sequence classification, and associated information designed to support automated annotation of (mostly prokaryotic) proteins.
6	**WebMGA** http://weizhonglilab.org/webMGA/server/tigrfam/	Web service for metagenomic analysis
7	**Ribosomal database project (RDP)** http://rdp.cme.msu.edu/	For quality-controlled, aligned and annotated bacterial and archaeal 16S rRNA sequences, and fungal 28S rRNA sequences
8	**Phylosift** https://phylosift.wordpress.com/	Phylogeny-driven metagenomic classification and comparison
9	**metAMOS** https://github.com/marbl/metAMOS	Metagenomic assembly and classification pipeline
10	**MG-RAST** https://www.mg-rast.org/	Open source, open submission web application server that suggests automatic phylogenetic and functional analysis of metagenomes.
11	**MEGAN6** https://uni-tuebingen.de/fakultaeten/mathematisch-naturwissenschaftliche-	Comprehensive toolbox for interactively analysing microbiome data

(continued)

Table 6.1 (continued)

Sl. No.	Name	Application
	fakultaet/fachbereiche/informatik/lehrstuehle/algorithms-in-bioinformatics/software/megan6/	
12	**Qiime** http://qiime.org/	For performing microbiome analysis From raw DNA sequencing data
13	**UCLUST** https://www.drive5.com/usearch/manual/uclust_algo.html	It is an algorithm designed to cluster nucleotide or amino-acid sequences into clusters based on sequence similarity
14	**AmphoraNet** https://pitgroup.org/amphoranet/	For metagenomic and genomic phylotyping
15	**Silva** https://www.arb-silva.de/	A comprehensive on-line resource for quality checked and aligned ribosomal RNA sequence data
16	**PICRUSt** http://picrust.github.io/picrust/	To predict metagenome functional content from marker gene (e.g., 16S rRNA) surveys and full genomes.
17	**PyNAST** https://biocore.github.io/pynast/	For adding new 16 s rRNA sequences to existing 16 s rRNA alignments.
18	**SRA (sequence read archive)** https://www.ncbi.nlm.nih.gov/sra	It is the largest repository of high throughput sequencing data (including metagenome sequencing data)
19	**Greengenes** https://greengenes.secondgenome.com/	It is a full-length 16S rRNA gene database that provides a curated taxonomy based on de novo tree inference
20	**Krona tools** https://hpc.nih.gov/apps/kronatools.html	It allows hierarchical data to be explored with zooming, multi-layered pie charts

Annotation

Annotation is the prediction of CDS (coding DNA sequences) of the genome, followed by its functional assignment based on similarity searches of query sequences against databases containing a known functional and/or taxonomic information. The taxonomic information can be displayed using Krona, which plays hierarchical data as an interactive multi-layered pie-chart (Ondov et al. 2011). The predicted genes are annotated to identify homologous genes using Gene ontology terms, KEGG pathways, protein families using Pfam or TIGRfams, clusters of orthologous genes (COGs/KOGs) or orthologous families and functional motifs using Inter Pro. Some tools, such as Kaiju, assign taxonomy status using a reference database, and also integrate the Krona tool for visualization of taxonomic composition, whereas COGNIZER can be used for functional annotation, which applies a new approach of search strategy that helps in reducing the computational requirements (Gosh et al. 2018).

Amplicon-Based Metagenomic Analysis

Pre-Processing of Reads for Amplicon Analysis

During this process, the raw files generated from the next-generation sequencing platform will be subjected to de-multiplexing, adapter trimming, and quality filtration (Plummer et al. 2015) and the detection of PCR chimera and its removal will be carried out using UCHIME algorithm (Sinclair et al. 2015).

OTU Picking and Taxonomic Assignment

OTU picking groups are the similar sequences by clustering or a similarity-based method. OUT picking in the most popular tool QIIME is performed using the UCLUST programme. The UCLUST program uses the algorithm USEARCH to assign the sequences to clusters (Edgar 2010). Each OUT represents a cluster of sequences with similarity greater than a threshold, typically 97–98%, which is then assigned to a corresponding taxonomic group. There are various OUT picking strategies: (1) De novo, wherein the reads are clustered without reference to known sequences; (2) Closed-reference, where the reads are clustered based on the alignment to a reference database; or (3) Open reference method, where clusters read against a reference database and also clusters unaligned reads using a de novo approach. All these methods are incorporated in the tool, QIIME (Oulas et al. 2015).

Statistical Analysis

The taxonomic tree in Newick format can be obtained from QIIME tool and it can be visualized using any tree display tool, such as FigTree. The alpha diversity measures the variability within a single population, which measures the richness, dominance and evenness. Rarefaction analysis is used to assess the coverage of the microbial community contained in the sample and the resultant rarefaction curves plot the sample size versus the estimated number of genera (Jaenicke et al. 2011).

Beta diversity measures the diversity across many samples or populations, which is calculated using various matrices, such as weighted and unweighted UniFrac and

PCoA (Principal Coordinate Analysis). It includes the absolute or relative overlap between the samples for estimating the taxa shared among them. The calculation of both the alpha and beta diversity is well supported by QIIME tool.

Phylogenetic Analysis
The phylogenetic analysis helps in identifying the species and its lineages at taxonomic levels. The various tools used for analysing the phylogenetic relationship in metagenomes are AmphoraNet, TIPP (taxonomic identification and phylogenetic profiling) and Phylosift (Gosh et al. 2018).

Functional Analysis
In order to predict the functional composition of microbial communities from the 16S profile, the tool PICRUSt can be used. It employs an extended ancestral-state reconstruction algorithm, which predicts the gene families and further combines the gene families to estimate the composite metagenome. The annotation of the predicted gene family counts can be obtained from orthologous groups of gene families, KOGs, COGs, NOGs, or Pfam families (Langille et al. 2013).

6.7 Metagenome Analysis of Insect Pests: An Overview

Nowadays, the insect-associated microbial communities are attracting wide attention mainly because of their ecological and economic importance. Microorganisms have been investigated for their profound influence on their host partner by directly mediating interactions with other species or indirectly by impacting the host genetic diversity. Moreover, microorganisms can help insects to counteract the defence mechanisms offered by the host plants, provide protection against natural enemies, influence the reproductive ability and help to survive on nutritionally marginal diets (Ferrari and Vavre 2011). Recently, the study of host-microorganism interactions has attracted a wide attention with the introduction of metagenomic techniques. A wide range of research described the insect-associated microbial community using metagenomic tools and the glimpse on metagenome analysis of insect pests of global importance are discussed hereunder.

6.7.1 Termites

Termites pose serious threat to a wide range of agricultural crops, structures, especially wooden materials and prove themselves a major insect-pest to human-kind. The gut of termite is a rich reservoir of microbes, belongs to Bacteria, Archea and Euckarya and the higher termites are capable of digesting the lignocellulose in various stages of humification with the help of an array of symbiotic prokaryotic microbiota housed in their compartmented intestinal tract. The metagenomic profiling of hindgut pouches of wood (*Amitermes wheeleri*) and dung (*Nasutitermes corniger*) feeding termites based on *16S rRNA* pyro-sequencing revealed that

Firmicutes and *Spirochaetes* were the most abundant phyla in *A. wheeleri* in contrast to *N.corniger* where *Spirochaetes* and *Fibrobacteres* dominated. Further, functional analysis revealed that the microbiota associated with *A. wheeleri* involved in hemicellulose breakdown and fixed-nitrogen utilization, whereas, those associated with *N. corniger* possess glycoside hydrolases attacking celluloses and nitrogen fixation genes (He et al. 2013).

Later on, metagenomic profiling of highly compartmented hindgut of six wood or soil feeding termite reveals that P1 compartment of the most of termite species is dominated by Firmicutes, whereas P4 is generally more diverse when compared to other compartments and displayed an increasing abundance of Bacteroidetes (Rossmassler et al. 2015). Metagenomic analysis of whole gut microbiota in seven species of termites (Termitidae) with different feeding habits from four locations at Brazil reveals that in termite species feeding on litter, the bacteria belong to the phylum Firmicutes are abundant, whereas in humus feeding termite species, the bacteria belonging to the phylum Proteobacteria are abundant. The gut microbiota of all four examined subfamilies of termites shared a conserved functional carbohydrate-active enzyme profile specialized for cellulose and chitin degradation (Grieco et al. 2019).

6.7.2 Pea Aphid, *Acyrthosiphon pisum*

The pea aphid, *Acyrthosiphon pisum*, a pest of legume crops represents a well-studied case of symbiotic associations. The 454 pyro-sequencing of pea aphid resulted in a range of 2838–16,637 sequence reads with a median of 4199 reads per sample. In total, *Buchnera* sequences comprised an average of 88.4% of the sequence reads followed by *Serratia symbiotica* with an average sequence read abundance of 4.3%. The X-type, Rickettsia, *H. defensa* and *R. insecticola* were next in read abundance, ranging from 1.2 to 1.7% of sequence reads, on average, across all samples (Russell et al. 2013).

The diversity analysis of bacterial communities associated with nine biotypes of the pea aphid complex using pyro-sequencing of *16S rRNA* genes reveals that *Spiroplasma* was the most dominant taxon in number of sequences (48%) followed by *Rickettsia* (25%) and *Buchnera* (21%) (Gauthier et al. 2015). Cariou et al. (2018) compared both *16S rDNA* amplicon sequencing and hybridization capture for pea aphid microbiota diversity analysis and found that both the methods provide description of 8 bacterial taxa, namely *Buchnera aphidicola, Hamiltonella defensa, Rickettsiella viridis, Rickettsia* sp., *Regiella insecticola, Fukatsuia, Serratia symbiotica* and *Spiroplasma* sp. and considered as qualitatively and quantitatively robust on such a sample with low microbial complexity.

6.7.3 Boll Worm, *Helicoverpa Armigera*

Helicoverpa armigera (Hübner), commonly known as American bollworm or gram caterpillar or tomato fruit borer, is a polyphagous insect pest known to infest many economically important crops throughout the world. The dreaded nature of this pest is attributed to number of factors, among which the gut microbiota also play a major role to thrive in various crop ecosystem. T-FRLP analysis of the gut bacterial community associated with *H. armigera* from tomato, chickpea and cotton crops at different locations showed that among the 12 bacterial phylotypes detected, *Enterococcus faecalis* and *Enterobacter* sp. were the major phylotypes found in all the larvae regardless of the crop or location of samples collected including artificial diets (Priya et al. 2012).

Further, Ranjith et al. (2016) analysed the composition and diversity of gut bacterial communities associated with *H. armigera* based on Illumina Next-Generation Sequencing (NGS) of 16S ribosomal RNA. The NGS dataset consisted of 864,813 high-quality paired end sequences with mean length of 150 base pairs. A highly diverse groups of bacteria were present in the sample with an approximate of 2303 operational taxonomic units (OTUs). A total of 17 bacterial phyla, 34 classes, 84 orders, 173 families, 334 genera, and 707 species were deduced from the sequence analysis. *Actinobacteria* was the most dominant taxon, followed by *Proteobacteria* and *Firmicutes*. Dar et al. (2018) identified cellulose degrading bacteria *Klebsiella* sp. MD21 from the gut of cotton bollworm, *Helicoverpa armigera* based on *16S rRNA* gene sequencing and demonstrated that *H. armigera* can be used as source of cellulolytic bacteria, which can be utilized in both biorefinery and pulp industries.

6.7.4 Whitefly, *Bemisia tabaci*

Silverleaf whitefly, *Bemisia tabaci* (Gennadius), is one of the polyphagous sucking insect pests, infesting more than 900 species of plants and serve as a vector for spreading more than 200 viral diseases. Harish et al. (2019) studied the composition of bacterial communities associated with whitefly infesting cassava from two different zones (zone P: plains; zone H: high ranges) of Kerala, India, using the next-generation sequencing of *16S rDNA*. Sequence analysis revealed a marked difference in the relative abundance of gut inhabiting bacteria present in the populations. In the P population, the taxonomic status of bacteria identified were 16 phyla, 27 classes, 56 orders, 91 families, 236 genera and 409 species, whereas in H population, it was earmarked as 16, 31, 60, 88, 225 and 355, respectively. The most dominant bacterium present in P population was *Arsenophonus* sp. (Enterobacteriaceae), which aids in virus transmission, whereas in the H population, *Bacillus* sp. was found relatively abundant. This study pinpoints the association between whitefly biotypes and secondary symbionts and the role of bacteria in modifying the host characteristics, such as transmission of various virus groups, expanding the host range, imparting the insecticide resistance and speciation.

The comparative analysis of endosymbionts present in 21 globally collected species in the *B. tabaci* complex, and two samples of *B. afer* using PacBio sequencing of full-length bacterial *16S rRNA* gene amplicons revealed the new putative bacteria and one among them was *Halomonas,* first confirmed to be present in MED *B. Tabaci* (Indiragandhi et al. 2010). Similarly, new secondary endosymbiotic strains of *Rickettsia* and *Arsenophonus* were also found associated with the whitefly samples collected from different locations (Wang et al. 2019).

Shah et al. (2020) characterized bacterial communities present in wild adult *B. tabaci* infesting cotton plants in eight major cotton growing districts of southern Punjab, Pakistan based on 16S rDNA next-generation sequencing and identified 50 known and 7 unknown genera of bacteria belonging to 10 phyla, 20 classes, 30 orders and 40 families. Proteobacteria was the most abundant phylum followed by Bacteriodetes, Firmicutes and Actinobacteria.

6.7.5 Diamond Back Moth, *Plutella xylostella (L.)*

The diamond back moth, *Plutella xylostella* (L.) (Lepidoptera: Plutellidae), is one of the most destructive insect pests infesting the cruciferous vegetables, such as cabbage, broccoli and cauliflower across the globe. The first report of high-throughput DNA sequencing of the entire microbiota of DBM reveals that more than 97% of the bacteria were from three orders, namely Enterobacteriales, Vibrionales and Lactobacillales. Both chlorpyriphos and fipronil resistant lines used in the study had more Lactobacillales and the much scarcer taxa Pseudomonadales and Xanthomonadales with fewer Enterobacteriales compared with the susceptible strain and this is consistent with the hypothesis that Lactobacillales or other scarcer taxa play a role in conferring DBM insecticide resistance (Xia et al. 2013).

Metagenomic analysis of diamond back moth reveals that the phylum, Proteobacteria was the dominant taxon in the *P. xylostella* gut microbiota, followed by Firmicutes. Functional metagenome analysis reveals the role of gut bacteria in metabolic activities associated with glycans, carbohydrates, amino acids, vitamins, xenobiotics and terpenoids, which are linked to digestion, nutrition and detoxification. The most enriched functions within these activities were carbohydrate metabolism and amino acid metabolism (nutrition), followed by xenobiotic degradation and terpenoid metabolism (detoxification of plant defensive compounds) (Xia et al. 2017).

6.8 Application of Metagenomics in Insect Pest Management

Metagenomics has wide range of application from clinical to environmental samples, from food safety to industrial waste and also in identifying the pathogens, which can infest various hosts including humans and animals. Metagenome analysis provides the information on both the diversity and function of microbiota associated with insect pest. These inherent gut microbiota play a crucial role in its insect survival by upgrading the nutrient status of diet, aids in digestion of recalcitrant food, protection from parasites, pathogens, lethal insecticidal molecules and development and maintenance of host immune system (Gill et al. 2004; Wernegreen 2002). The metagenomics offers new technologies and conceptual approaches to the entomologists by facilitating the study of impacts of microbes on insect function and to adopt various pest control strategies based on manipulation of microbial partners.

The application of insect metagenomic studies in formulating various novel approaches in pest management and few are discussed hereunder.

6.8.1 Improve Biosurveillance Programme

Metagenomics techniques can be used to improve bio-surveillance programmes, as a tool to detect the arrival, origin, invasion pathways and adaptation traits of invasive species (Roe et al. 2018) in different ecosystems. It can be employed in the monitoring of critical areas viz., port of entry where massive trapping is the most common practice followed to identify the arrival of invasive insect species (Rassati et al. 2018; Poland and Rassati 2018; Rassati et al. 2015). This regular mass trapping is time-consuming and laborious process, often requires extensive taxonomic knowledge of different systematic groups. However, the metagenome analysis simplifies the process by analysing the entire genetic pool of single traps, and detecting not only the arrival of an invasive insect species, but also likely plant pathogens (Malacrinò et al. 2017; Roe et al. 2018).

6.8.2 Suppression of Vector Competence of Insects

On the basis of comparative metagenome analysis, Hajeri et al. (2014) developed a novel method of RNAi mediated control for the Asian citrus psyllid, *Diaphorina citri*, a vector for multiple citrus diseases, by modifying the genome of the citrus tristeza virus into a stable vector producing dsRNA. The progeny of *D. citri* individuals feeding on plants infected with the modified virus showed increased mortality.

6.8.3 Manipulation of Host Range of Insect Pests

Microbial symbionts play a major role in determining the host range of phytophagous insects, but persuasive evidence is rare (Hansen and Moran 2014). However, exceptional phenomenon has been noticed in two plataspid stinkbugs, *Megacopta punctatissima* and *M. cribraria*. In their native range in Japan, *M. punctatissima* is an agricultural pest, especially of soybean crops, but *M. cribraria* performs very poorly on soybean. Metagenome profiling reveals that the capacity of *M. punctatissima* to utilize soybean is mediated by the bacterial symbiont *Ishikawaella* localized to the distal portion of the insect gut. When the newly hatched nymphs of *M. punctatissima* were administered with *Ishikawaella* symbiont of the other species, *M. cribaria* leads to poor consumption with high mortality on soybean (Hosokawa et al. 2007). The *Megacopta* association has great potential for manipulation to suppress the infestation of soybean crops because heterologous associations can be generated very easily by feeding neonate nymphs on symbionts from a different insect species and the acquired partner is then transmitted vertically with high fidelity (Hosokawa et al. 2005).

6.8.4 Heterologous Symbionts those Are Insecticidal

Many insect-microbial associations are co-evolved, with the implication that certain microorganisms that are benign in their native insect host may be deleterious when introduced to a different insect. Incompatibility can occur naturally on hybridization between two related insect species with maternally inherited symbionts. This phenomenon has been demonstrated very clearly through metagenomic profiling of two species of jewel wasps, *Nasonia vitripennis* and *N. girauldi*, with genetic evidence that interspecific crosses yield incompatibilities between a maternally inherited 'factor' and the nuclear genome of the hybrid (Breeuwer and Werren 1995). Identification of the microbial symbionts as the 'factor' comes from the finding that antibiotic treatment protects against hybrid lethality and that lethality is revived by adding back specific gut bacteria, *Providencia* sp. and *Proteus mirabilis*, derived from each of the two parental jewel wasp species (Brucker and Bordenstein 2013). The high populations of *P. mirabilis* in hybrid insects are indicative of immunological dysfunction even though the underlying mechanisms are not understood fully (Chandler and Turelli 2014).

For application to control insect pests, there are two key requirements: (1) the pest and non-pest species hybridize under field conditions, yielding viable progeny; and (2) an association can be constructed between the non-pest species and microorganism(s) that are benign in the non-pest species but lethal to hybrids between the non-pest and pest species, as well as posing no risk to other species or the wider environment. Under these conditions, implementation would have many parallels to the sterile insect technique, but with the mass release of the non-pest species bearing the microorganisms instead of sterile conspecific male insects.

6.8.5 Paratransgenesis and Induced Lethality in Insect Pests

The most developed application of genetically modified microorganisms in insect pest control is paratransgenesis, which can be defined as the alteration of insect traits by genetic manipulation of associated microorganisms (Beard et al. 1998). The potential of this technology in insect pest control has been appreciated for more than twenty years (Beard et al. 1993), especially in relation to mosquito vectors of human disease agents. The key requirements for paratransgenesis are that the microbial partner is culturable under ideal condition and amenable to genetic manipulation as well as readily transmitted among insects to facilitate the transfer of the desired trait (Beard et al. 2002). The possibility of using this technique to manage the agriculturally important insect pests is to be thoroughly explored with metagenome analysis.

6.8.6 Genetically Modified Microorganisms as Insecticides

Genetic technologies can be applied to modify microorganisms to express traits that are virulent to the insect. The use of microorganisms for delivery of dsRNA relates to the promise of RNA-interference (RNAi) to target insect pests by suppressing the expression of essential insect genes. *In planta* RNAi is now used widely in research on herbivorous insects, with an insecticidal RNAi against the Western corn rootworm *Diabrotica virgifera virgifera* reported to be close to commercial release in transgenic corn. In addition, the environmental release of dsRNA against insect pests found associated with soils, water and other natural habitats are being promoted by advanced encapsulation technologies (Scott et al. 2013).

6.8.7 Elimination of Vertically Transmitted Obligate Microbial Partner

The goal of targeting microbial partners is to control insect pests by eliminating the microorganisms required for sustained insect growth, reproduction and survival. Unlike the use of heterologous or genetically modified microorganisms, which involve the administration of microorganisms to insects, this strategy involves the use of specific symbiocides, i.e., effectors that perturb the resident microbial partners and their interactions with the insect.

The insect systems ideally suited to this strategy involve bacteria that are localized to specialized insect cells known as bacteriocytes. Because the bacterial partners are obligately vertically transmitted and unknown apart from their insect hosts (Buchner 1965; Douglas 1989) a treated insect cannot regain the association horizontally from other insects or the environment.

A strong motivation to develop methods that target the bacteriocyte symbioses comes from the expectation of specific molecular targets linked to the coevolutionary interactions between the participating insect and microbial lineages (Douglas 2015). This can be explicitly studied based on metagenome analysis of pest taxa potentially amenable to this strategy include sap feeding hemipterans (aphids, whiteflies, planthoppers, leafhoppers etc.), and various xylophagous and stored product coleopteran pests (many curculionids and chrysomelids, the anobids and bostrychids).

Grape plants transformed with constructs coding the anti-microbial peptide, cecropin B, with either melittin or elastase, reduced the *Xylella* abundance and disease symptoms in the plants (Dandekar et al. 2012; Li et al. 2015). The antimicrobials circulating in the xylem sap of these plants are presumably ingested by the xylem-feeding insects, including the leafhoppers that vector *Xylella,* but their activity against the obligate bacterial symbionts in the leafhoppers (Wu et al. 2006) needs to be investigated. The feasibility of selective symbiocides is supported further by the relative ease with which orally delivered antibiotics and antibodies can cross the gut wall to the hemocoel and internal organs of insects (Bonning and Chougule 2014; Jeffers and Roe 2008).

6.8.8 Elimination of Horizontally Transmitted Obligate Microbial Partner

This strategy combines the use of genetically modified microorganisms as a delivery vehicle to target obligate microbial partners. Gut metagenome analysis reveals that the lower termites are absolutely dependent on cellulose degrading trichomonad and hypermastigote protists in their hindgut, providing an opportunity to control the pest by targeting the protist symbionts. Various antimicrobial peptides, including melittin, cecropin, and the synthetic product Hecate have been demonstrated to lyse these protists, but their application has been constrained by challenges in their delivery to the hindgut. This limitation has been overcome by using microorganisms as the delivery vehicle to target the obligate symbiont of the insect pest instead of the insect.

The commercially available yeast, *Kluyveromyces lactis,* engineered to express melittin was used as microbial delivery vehicle and administrated to the termite *Coptotermes formosanus* was found effective in eliminating the protists without detectable direct damage to the insect gut (Husseneder et al. 2016). It was also demonstrated that a bacterial isolate from *C. formosanus*, *Trabulsiella odontotermitis*, is genetically transformable and transmitted efficiently among termites (Tikhe et al. 2016) and offers a route to use a natural symbiont as the delivery vehicle for the toxic peptides.

6.9 Future Perspective

The study of insect metagenome analysis yields valuable information on role of different groups of microbes in insect physiology, pest management, evolutionary relationships and the tritrophic interactions existing in the nature. It also gives an insight into the various microbe derived novel biocatalysts, which can be used for various applications, including pest management and biorefinery development. In particular, the gut systems of many herbivore insects can be considered as effective bioreactors, where biomass material can be deconstructed for the synthesis of various bioproducts important for insect growth and development (Breznak 2004). The coordinative function of both host insect and symbiont derived enzymes plays an important role in biomass processing and degradation. Thus, study of insect gut symbiotic microbiota at the systems level will enable us to design the next-generation biorefinery for various levels of industrial applications. Similarly, the insect microbiome analysis provides the role of different microbiota in insect survival and development in a particular environment.

The insect metagenome analysis has experienced dramatic changes during the past two decades. The initial studies of insect gut microbes were based on culture-dependent platforms, which provided a very limited information on the diversity and functions. The culture-dependent analysis was quickly replaced and complemented by the advancement in molecular techniques, which is a key partner in culture independent methods. The molecular fingerprinting techniques, like DGGE, SSCP, RFLP and FISH, allow us to better explore the complexity of natural microbial communities present in an ecosystem. However, the development of metagenomic approaches and the advancements in next-generation sequencing techniques allow us to explore the metagenomes from insect gut symbiotic microbiota to an unprecedented depth and comprehensiveness.

In addition to metagenomics, metatranscriptomic and metaproteomic profilings are also providing important information regarding the function of insect hosts and symbionts from different perspectives. The integration of information will lead to a systems-level understanding of insect gut as the system for biomass deconstruction, nutrient biosynthesis and to formulate various novel approaches in pest management. Despite significant progresses, several aspects of research need to be emphasized to better exploit insect gut systems for various biotechnology applications.

Though the metagenomic approach provided a thorough knowledge on the microbial census in the insect gut, identification of novel genes and the development of potential biotechnological applications is a great challenge due to the presence of both diverse microbial communities and the variability existing in their genomes. Most of the bioinformatics programmes are designed for collecting and depositing of the metagenomic sequence composition, and their respective data management. However, more sophisticated bioinformatics tools are yet to be developed to analyse the *hitherto* unexplored microbial genes of insect gut metagenomics. Though the new high throughput next sequencing technologies enable us for identifying a novel candidate gene, the assay for protein function exemplify one of the most important and inimitable tools for identifying their target genes. Thus, the development of high throughput functional screening methods will also be necessary to assess the functional role in particular system.

Most of the insect metagenomic studies are focusing merely on exploring the gut microbial composition and their functional diversity. However, metagenome analysis also provides valuable information to formulate various pest management strategies based on manipulation of insect-associated microorganisms. The status of the various strategies varies from generalized concepts and experimental proof-of-principle under defined laboratory conditions to products suitable for field application and ongoing field trials in multiple countries. Furthermore, knowledge of the molecular basis of most strategies on microbial manipulations offers the opportunity for modification of the product in response to a novel insect pest and resistance evolution in insect pests.

6.10 Conclusions

Insect-associated microbial communities are attracting increasing interest nowadays, mainly because of their ecological and economic importance. They play essential roles in the growth, development, pathogenesis and environmental adaptation of host insects. At present, we are capable of exploring the microbial communities associated with insects, their composition, diversity and interaction with their hosts. In particular, the modern molecular techniques, metagenomics revolutionized the field with enormous data to enable unprecedented understanding of insect gut symbiotic microbiota and their interactions with hosts. The metagenome approaches together with the recent advancements in next-generation sequencing provide enormous sequencing information, allowing in-depth microbial diversity analysis and modelling of pathways for biological processes, such as biomass degradation. In addition, insect gut metagenome analysis data also provide conceptual approaches to the plant protection specialists to formulate various novel pest management strategies based on manipulation of insect-associated microorganisms. Certainly, metagenomics in combination with metaproteomic and metatranscriptomic approaches and modern bioinformatics tools enable us to retrieve pivotal information that can effectively be used in combating ravages of insect pests.

6.11 Points to Be Remember

- Insects, the most successful groups in animal kingdom, harbour diverse groups of microbes viz., bacteria, archaea, fungi, protozoa and viruses.
- These microbes are associated with their host insects permanently or transiently and such associations may be beneficial or harmful to the host insects under various instances.
- The initial studies of insect microbiome were based on culture-dependent platforms, which provided very limited information for the diversity and functions of insect gut symbiotic microbiota.
- This classical approach was quickly replaced and complemented by traditional molecular approaches, like gene specific PCR, molecular fingerprinting techniques (DGGE or TGGE, RFLP, SSCP and RAPD) and FISH allowed us to better explore the complexity of natural microbial communities.
- The recently developed metagenome sequencing techniques, in particular, the advancements in next-generation sequencing techniques allow us to explore the metagenomes from insect gut symbiotic microbiota to an unprecedented depth and comprehensiveness.
- Two types of metagenomic analysis are commonly used to unravel the identity and composition of microbes for a high throughput sequencing data are amplicon-based analysis and shotgun metagenome analysis and the two principal approaches in metagenome analysis are sequence-based metagenomics and functional metagenomics.

- The major steps involved in metagenomic studies include sample preparation, extraction and purification of metagenomic DNA, metagenomic DNA library preparation, sequencing under various next-generation platforms, sequence data analysis and interpretation.
- Recently, a number of researches have been carried out to explore metagenome of insect pest of global importance, viz. termite, pea aphid, cotton bollworm, silver leaf whitefly, diamond back moth, etc. reveals the crucial role played by the microbes in insect nutrition, protection from parasites, pathogens, lethal insecticidal molecules and development and maintenance of immune system.
- Insect metagenomic research will aid in formulation of various novel pest management approaches viz, improved bio-surveillance programme, suppression of vector competence of insects, manipulation of insect host range, use of heterologous symbionts, paratransgenesis and induce lethality in insects, genetically modified microorganisms as insecticides and elimination of both vertically and horizontally transmitted obligate microbial partners.

References

Adams IP, Glover RH, Monger WA, Mumford R, Jackeviciene E, Navalinskiene M, Samuitiene M, Boonham N (2009) Next-generation sequencing and metagenomic analysis: a universal diagnostic tool in plant virology. Mol Plant Pathol 10(4):537–545

Allen EE, Banfield JF (2005) Community genomics in microbial ecology and evolution. Nat Rev Microbiol 3(6):489–498

Allen HK, Moe LA, Rodbumrer J, Gaarder A, Handelsman J (2009) Functional metagenomics reveals diverse beta-lactamases in a remote Alaskan soil. ISME J 3(2):243–251

Amann RI, Ludwig W, Schleifer KH (1995) Phylogenetic identification and in situ detection of individual microbial cells without cultivation. Microbiol Rev 59(1):143–169

Aminov RI, Walker AW, Duncan SH, Harmsen HJ, Welling GW, Flint HJ (2006) Molecular diversity, cultivation, and improved detection by fluorescent in situ hybridization of a dominant group of human gut bacteria related to *Roseburia* spp. or *Eubacterium rectale*. Appl Environ Microbiol 72(9):6371–6376

Ami E, Yuval B, Jurkevitch E (2010) Manipulation of the microbiota of mass-reared Mediterranean fruit flies Ceratitis capitata (Diptera: Tephritidae) improves sterile male sexual performance. ISME J 4:28–37. https://doi.org/10.1038/ismej.2009.82

Bag S, Saha B, Mehta O, Anbumani D et al (2016) An improved method for high qualitymetagenomics DNA extraction from human and environmental samples. Sci Rep 6. https://doi.org/10.1038/srep26775

Beard CB, O'Neill SL, Tesh RB, Richards FF, Aksoy S (1993) Modification of arthropod vector competence via symbiotic bacteria. Parasitol Today 9(5):179–183

Beard CB, Durvasula RV, Richards FF (1998) Bacterial symbiosis in arthropods and the control of disease transmission. Emerg Infect Dis 4(4):581–591

Beard CB, Cordon-Rosales C, Durvasula RV (2002) Bacterial symbionts of the triatominae and their potential use in control of Chagas disease transmission. Annu Rev Entomol 47:123–141

Begon M, Harper JL, Townsend CR (1986) Ecology: individuals, populations and communities. Blackwell Scientific Publications, Oxford

Behar A, Yuval B, Jurkevitch E (2005) Enterobacteriamediated nitrogen fixation in natural populations of the fruit fly Ceratitis capitata. Mol Ecol 14(9):2637–2643

Berchtold M, Chatzinotas A, Schonhuber W, Brune A, Amann R, Hahn D, Konig H (1999) Differential enumeration and in situ localization of microorganisms in the hindgut of the lower termite *Mastotermes darwiniensis* by hybridization with rRNA-targeted probes. Arch Microbiol 172:407–416

Bonning BC, Chougule NP (2014) Delivery of intrahemocoelic peptides for insect pest management. Trends Biotechnol 32(2):91–98

Breeuwer JAJ, Werren JH (1995) Hybrid breakdown between two haplo diploid species: the role of nuclear and cytoplasmic genes. Evolution 49(4):705–717

Breznak JA (2004) Invertebrates-insects. In: Bull AT (ed) Microbial diversity and bioprospecting. ASM Press, Washington, D.C., pp 191–203

Brinkmann N, Martens R, Tebbc CC (2008) Origin and diversity of metabolically active gut bacteria from laboratorybred larvae of *Manduca sexta* (Sphingidae, Lepidoptera, Insecta). Appl Environ Microbiol 74(23):7189–7196

Bromberg JS, Fricke WF, Brinkman CC, Simon T, Mongodin EF (2015) Microbiota: implications for immunity and transplantation. Nat Rev Nephrol 11(6):342–353

Brucker RM, Bordenstein SR (2013) The hologenomic basis of speciation: gut bacteria cause hybrid lethality in the genus Nasonia. Science 341(6146):667–669

Buchner P (1965) Endosymbioses of animals with plant microoorganisms. John Wiley and Sons, Chichester, UK

Campbell BC (1990) On the role of microbial symbiotes in herbivorous insects. In: Bernays EA (ed) Insect-plant interactions. CRC Press p, Boca Raton, FL, pp 1–44

Cariou M, Ribière C, Morlière S et al (2018) Comparing 16S rDNA amplicon sequencing and hybridization capture for pea aphid microbiota diversity analysis. BMC Res Notes 11:461. https://doi.org/10.1186/s13104-018-3559-3

Casu RE, Eisemann CH, Vuocolo T, Tellam RL (1996) The major excretory/secretory protease from *Lucilia cuprina* larvae is also a gut digestive protease. Int J Parasitol 26(6):623–628

Cazemier AE, Hackstein JHP, Op den Camp HJM, Rosenberg J, van der Drift C (1997) Bacteria in the intestinal tract of different species of arthropods. Microb Ecol 33:189–197

Chakravorty S, Helb D, Burday M, Connell N, Alland D (2007) A detailed analysis of 16S ribosomal RNA gene segments for the diagnosis of pathogenic bacteria. J Microbiol Methods 69(2):330–339

Chandler JA, Turelli M (2014) Comment on "the hologenomic basis of speciation: gut bacteria cause hybrid lethality in the genus Nasonia". Science 345(6200):1011

Charnley AK, Hunt J, Dillon RJ (1985) The germ-free culture of desert locusts, Schistocerca gregaria. J Insect Physiol 31(6):477–485

Chaves S, Neto M, Tenreiro R (2009) Insect-symbiont systems: from complex relationships to biotechnological applications. Biotechnol J 4(12):1753–1765

Chen K, Pachter L (2005) Bioinformatics for whole-genome shotgun sequencing of microbial communities. PLoS Comput Biol 1(2):e24. 106–112

Chen B, Yu T, Xie S, Du K, Liang X, Lan Y, Sun C, Lu X, Shao Y (2018) Comparative shotgun metagenomic data of the silkworm Bombyx mori gut microbiome. Sci Data 5:180285. https://doi.org/10.1038/sdata.2018.285

Cheung SW, Tishler PV, Atkins L, Sengupta SK, Modest EJ, Forget BG (1977) Gene mapping by fluorescent in situ hybridization. Cell Biol Int Rep 1(3):255–262

Chong CW, Pearce DA, Convey P, Yew WC, Tan IKP (2012) Patterns in the distribution of soil bacterial 16S rRNA gene sequences from different regions of Antarctica. Geoderma 181:45–55

Chougule NP, Giri AP, Sainani MN, Gupta VS (2005) Gene expression patterns of *Helicoverpa armigera* gut proteases. Insect Biochem Mol Biol 35(4):355–367

Cowan D, Meyer Q, Stafford W, Muyanga S, Cameron R, Wittwer P (2005) Metagenomic gene discovery: past, present and future. Trends Biotechnol 23(6):321–329

da Mota FF, Gomes EA, Paiva E, Seldin L (2005) Assessment of the diversity of *Paenibacillus* species in environmental samples by a novel rpoB-based PCR-DGGE method. FEMS Microbiol Ecol 53(2):317–328

Dandekar AM, Gouran H, Ibanez AM, Uratsu SL, Aguero CB, McFarland S, Borhani Y, Feldstein PA, Bruening G, Nascimento R et al (2012) An engineered innate immune defense protects grapevines from Pierce disease. Proc Natl Acad Sci U S A 109(10):3721–3725

Dar MA, Shaikh AA, Pawar KD, Pandit RS (2018) Exploring the gut of *Helicoverpa armigera* for cellulose degrading bacteria and evaluation of a potential strain for lignocellulosic biomass deconstruction. Process Biochem 73:142–153

de Vries EJ, Breeuwer JAJ, Jacobs G, Mollema C (2001a) The association of western flower thrips, *Frankliniellaoccidentalis*, with a near *Erwinia* species gut bacteria: transient or permanent? J Invertebr Pathol 77(2):120–128

de Vries EJ, Jacobs G, Breeuwer JAJ (2001b) Growth and transmission of gut bacteria in the western flower thrips, Frankliniella occidentalis. J Invertebr Pathol 77(2):129–137

Demaneche S, David MM, Navarro E, Simonet P, Vogel TM (2009) Evaluation of functional gene enrichment in a soil metagenomic clone library. J Microbiol Methods 76(1):105–107

Dethlefsen L, Huse S, Sogin ML, Relman DA (2008) The pervasive effects of an antibiotic on the human gut microbiota, as revealed by deep 16S rRNA sequencing. PLoS Biol 6(11):e280

Dillon RJ, Dillon VM (2004) The gut bacteria of insects: non-pathogenic interactions. Annu Rev Entomol 49:71–92

Dillon RJ, Vennard CT, Charnley AK (2002) A note: gut bacteria produce components of a locust cohesion pheromone. J Appl Microbiol 92:759–763

Dillon RJ, Webster G, Weightman AJ, Dillon VM, Blanford S, Charnley AK (2008) Composition of Acridid gut bacterial communities as revealed by 16S rRNA gene analysis. J Invertebr Pathol 97(3):265–272

Domingo JWS, Kaufman MG, Klug MJ, Tiedje JM (1998) Characterization of the cricket hindgut microbiota with fluorescently labeled rRNA-targeted oligonucleotide probes. Appl Environ Microbiol 64(2):752–755

Douglas AE (1989) Mycetocyte symbiosis in insects. Biol Rev 64(4):409–434

Douglas AE (2015) Multiorganismal insects: diversity and function of resident microorganisms. Annu Rev Entomol 60:17–34

Dowd PF, Shen SK (1990) The contribution of symbiotic yeast to toxin resistance of the cigarette beetle (Lasioderma serricorne). Entomologia Experimentalis et Applicata 56:241–248

Edgar RC (2010) Search and clustering orders of magnitude faster than BLAST. Bioinformatics 26 (19):2460–2461

Egert M, Stingl U, Bruun LD, Pommerenke B, Brune A, Friedrich MW (2005) Structure and topology of microbial communities in the major gut compartments of *Melolonthamelolontha* larvae (Coleoptera: Scarabaeidae). Appl Environ Microbiol 71(8):4556–4566

Engel P, Moran NA (2013) The gut microbiota of insects – diversity in structure and function. FEMS Microbiol Rev 37:699–735

Esumi H, Takahashi Y, Sato S, Sugimura T (1982) Restriction fragment length polymorphism of the rat albumin gene in Sprague-Dawley rats and its application in genetic study of analbuminemia. Nucleic Acids Res 10(14):4247–4257

Ezenwa VO, Gerardo NM, Inouye DW, Medina M, Xavier JB (2012) Microbiology: animal behaviorand the microbiome. Science 338:198–199

Falkowski PG, Fenchel T, Delong EF (2008) The microbial engines that drive Earth's biogeochemical cycles. Science 320(5879):1034–1039

Fall S, Hamelin J, Ndiaye F, Assigbetse K, Aragno M, Chotte JL, Brauman A (2007) Differences between bacterial communities in the gut of a soil-feeding termite. Appl Environ Microbiol 73 (16):5199–5208

Ferrari J, Vavre F (2011) Bacterial symbionts in insects or the story of communities affecting communities. Philos Trans R Soc Lond Ser B Biol Sci 366(1569):1389–1400. https://doi.org/10.1098/rstb.2010.0226

Finn RN, Chauvigné F, Stavang JA, Belles X, Cerda J (2015) Insect glycerol transporters evolved by functional co-option and gene replacement. Nat Commun 6:7814

Frago E, Dicke M, Godfray HC (2012) Insect symbionts as hidden players in insect-plant interactions. Trends Ecol Evol 27:705–711

Fukatsu T, Watanable K, Sekiguchi Y (1998) Specific detection of intracellular symbiotic bacteria of aphids by oligonucleotide-probed in situ hybridization. Appl Entomol Zool 33(3):461–472

Fuller CW, Middendorf LR, Benner SA, Church GM, Harris T, Huang X, Jovanovich SB, Nelson JR, Schloss JA, Schwartz DC, Vezenov DV (2009) The challenges of sequencing by synthesis. Nat Biotechnol 27(11):1013–1023. https://doi.org/10.1038/nbt.1585

Gauthier JP, Outreman Y, Mieuzet L, Simon JC (2015) Bacterial communities associated with host-adapted populations of pea aphids revealed by deep sequencing of 16S ribosomal DNA. PLoS ONE 10(3):e0120664. https://doi.org/10.1371/journal.pone.0120664

Gavriel S, Jurkevitch E, Gazit Y, Yuval B (2011) Bacterially enriched diet improves sexual performance of sterile male Mediterranean fruit flies. J Appl Entomol 135:564–573

Gill R, Latorre A, Moya A (2004) Bacterial endosymbionts of insects: insights fromcomparative genomics. Environ Microbiol 6(11):1109–1122

Goodwin S, McPherson JD, McCombie R (2016) Coming of age: tenyears of next-generation sequencing technologies. Nat Rev Genet 17:333–351. https://doi.org/10.1038/nrg.2016.49

Gosh A, Mehta A, Khan AM (2018) Metagenomic analysis and its application. Encyclopedia Bioinform Comput Biol:1–9. https://doi.org/10.1016/B978-0-12-809633-8.20178-7

Green JL, Bohannan BJ, Whitaker RJ (2008) Microbial biogeography: from taxonomy to traits. Science 320:1039–1043

Grieco MB, Lopes FAC, Oliveira LS (2019) Metagenomic analysis of the whole gut microbiota in Brazilian Termitidae termites *Cornitermes cumulans, Cyrilliotermes strictinasus,Syntermes dirus, Nasutitermes jaraguae, Nasutitermes aquilinus,Grigiotermes bequaerti*, and *Orthognathotermes mirim*. Curr Microbiol 76(6):687–697

Guan CH, Ju JJ, Borlee BR, Williamson LL, Shen B, Raffa KF, Handelsman J (2007) Signal mimics derived from a metagenomic analysis of the gypsy moth gut microbiota. Appl Environ Microbiol 73(11):3669–3676

Gupta A, Nair S (2020) Dynamics of insect-microbiome interaction influence host and microbial symbiont. Front Microbiol 11:1357

Hajeri S, Killiny N, El-Mohtar C, Dawson WO, Gowda S (2014) Citrus tristeza virus-based RNAi in citrus plants induces gene silencing in *Diaphorina citri*, a phloem-sap sucking insect vector of citrus greening disease (Huanglongbing). J Biotechnol 176:42–49

Handelsman J, Rondon MR, Brady SF, Clardy J, Goodman RM (1998) Molecular biological access to the chemistry of unknown soil microbes: a new frontier for natural products. Chem Biol 5 (10):45–49

Handelsman J (2004) Metagenomics: application of genomics to uncultured microorganisms. Microbiol Mol Biol Rev MMBR 68(4):669–685. https://doi.org/10.1128/MMBR.68.4.669-685.2004

Hansen AK, Moran NA (2014) The impact of microbial symbionts on host plant utilization by herbivorous insects. Mol Ecol 23:1473–1496

Harada H, Ishikawa H (1993) Gut microbe of aphid closely-related to its intracellular symbiont. Biosystems 31(2-3):185–191

Harish ER, ManiChellappan K, Makeshkumar T, Mathew D, Ranjith MT, Girija D (2019) Next-generation sequencing reveals endosymbiont variability in cassava whitefly, Bemisia tabaci, across the agro-ecological zones of Kerala, India. Genome 62(9):571–584

Hayashi A, Aoyagi H, Yoshimura T, Tanaka H (2007) Development of novel method for screening microorganisms using symbiotic association between insect (Coptotermes formosanus Shiraki) and intestinal microorganisms. J Biosci Bioeng 103(4):358–367

He S, Ivanova N, Kirton E, Allgaier M, Bergin C, Scheffrahn RH et al (2013) Comparative metagenomic and Metatranscriptomic analysis of hindgut paunch microbiota in wood- and dung-feeding higher termites. PLoS One 8(4):e61126

Head IM, Saunders JR, Pickup RW (1998) Microbial evolution, diversity, and ecology:Adecade of ribosomal RNA analysis of uncultivated microorganisms. Microb Ecol 35:1–21

Head SR, Komori HK, LaMere SA, Whisenant T, Van Nieuwerburgh F, Salomon DR, Ordoukhanian P (2014) Library construction for nextgeneration sequencing: overviews and challenges. Biotech 56:61–64., n 66, 68, passim. https://doi.org/10.2144/000114133

Hilton SK, Castro-Nallar E, Perez-Losada M et al (2016) Metataxonomic and metagenomic approaches vs. culture-based techniques for clinical pathology. Front Microbiol 7:484. https://doi.org/10.3389/fmicb.2016.00484

Hirsch J, Strohmeier S, Pfannkuchen M, Reineke A (2012a) Assessment of bacterial endosymbiont diversity in Otiorhynchus spp. (Coleoptera: Curculionidae) larvae using a multitag 454 pyrosequencing approach. BMC Microbiol 12(Suppl 1):S6

Hirsch J, Strohmeier S, Pfannkuchen M et al (2012b) Assessment of bacterial endosymbiont diversity in *Otiorhynchus*spp. (Coleoptera: Curculionidae) larvae using a multitag 454 pyrosequencing approach. BMC Microbiol 12:S6. https://doi.org/10.1186/1471-2180-12-S1-S6

Hopkins DW, Macnaughton SJ, O'Donnell AG (1991) A dispersion and differential centrifugation technique for representatively sampling microorganisms from soil. Soil Biol Biochem 23 (3):217–225

Hosokawa T, Kikuchi Y, Meng XY, Fukatsu T (2005) The making of symbiont capsule in the plataspid stinkbug Megacoptan punctatissima. FEMS Microbiol Ecol 54(3):471–477

Hosokawa T, Kikuchi Y, Shimada M, Fukatsu T (2007) Obligate symbiont involved in pest status of host insect. Proc Biol Sci 274(1621):1979–1984

Hurst MRH, Jackson TA (2002) Use of the green fluorescent protein to monitor the fate of *Serratia entomophila* causing amber disease in the New Zealand grass grub, *Costelytrazealandica*. J Microbiol Methods 50(1):1–8

Husseneder C, Donaldson JR, Foil LD (2016) Genetically engineered yeast expressing a lytic peptide from bee venom (Melittin) kills symbiotic protozoa in the gut of Formosan subterranean termites. PLoS One 11(3):e0151675

Indiragandhi P, Yoon C, Yang JO et al (2010) Microbial communities in the developmental stages of B and Q biotypes of sweetpotato whitefly, Bemisia tabaci (hemiptera: Aleyrodidae). J Korean Soc Appl Biol Chem 53:605–617. https://doi.org/10.3839/jksabc.2010.093

Inoue T, Moriya S, Ohkuma M, Kudo T (2005) Molecular cloning and characterization of a cellulase gene from a symbiotic protist of the lower termite, Coptotermes formosanus. Gene 349:67–75

Jaenicke S, Ander C, Bekel T (2011) Comparative and joint analysis of two metagenomic data sets from a biogas fermenter obtained by 454-pyrosequencing. PLoS One 6(1):e14519

Jeffers LA, Roe RM (2008) The movement of proteins across the insect and tick digestive system. J Insect Physiol 54(2):319–332

Jing TZ, Qi FH, Wang ZY (2020) Most dominant roles of insect gut bacteria: digestion, detoxification, or essential nutrient provision? Microbiome 8:38

Jones AG, Mason CJ, Felton GW, Hoover K (2019) Host plant and population source drive diversity of microbial gut communities in two polyphagous insects. Sci Rep 9:2792. https://doi.org/10.1038/s41598-019-39163-9

Josefsen MH, Andersen SC, Christensen J, Hoorfar J (2015) Microbial food safety: potential of DNA extraction methods for use in diagnostic metagenomics. J Microbiol Methods 114:30–34. https://doi.org/10.1016/j.mimet.2015.04.016

Kane MD, Pierce NE (1994) Diversity within diversity: molecular approaches to studying microbial interactions withinsects. In: Schierwater B, Streit B, Wagner GP, DeSalle R (eds) Molecular methods in ecology and evolution. Birkhauser Verlag, Berlin, Germany, pp 509–524

Kauppinen J, Mantyjarvi R, Katil ML (1999) Mycobacterium malmoense-specific nested PCR based on a conserved sequence detected in random amplified polymorphic DNA fingerprints. J Clin Microbiol 37(5):1454–1458

Kikuchi Y, Hayatsu M, Hosokawa T, Nagayama A, Tago K, Fukatsu T (2012) Symbiont-mediated insecticide resistance. Proc Natl Acad Sci U S A 109:8618–8622

Koch H, Schmid-Hempel P (2012) Gut microbiota instead of host genotype drives the specificity in the interaction of a natural host-parasite system. Ecol Lett 15:1095–1103

Kodama R, Nakasuji Y (1971) Further studies on the pathogenic mechanism of bacterial diseases in gnotobiotic silkworm larvae. IFO Res Commun 5:1–9

Kohler T, Stingl U, Meuser K, Brune A (2008) Novel lineages of Planctomycetes densely colonize the alkaline gut of soil-feeding termites (*Cubitermes* spp.). Environ Microbiol 10(5):1260–1270

Krishnan M, Bharathiraja C, Pandiarajan J, Prasanna VA, Rajendhran J, Gunasekaran P (2014) Insect gut microbiome: an unexploited reserve for biotechnological application. Asian Pacific J Trop Biomed 4(1):516–522

Lance DR, McInnis DO, Rendon P, Jackson CG (2000) Courtship among sterile and wild Ceratitis capitata (Diptera: Tephritidae) in field cages in Hawaii and Guatemala. Ann Entomol Soc Am 93:1179–1185

Langille MG, Zaneveld J, Caporaso JG (2013) Predictive functional profiling of microbial communities using 16S rRNA marker gene sequences. Nat Biotechnol 31:814–821

Lee DH, Zo YG, Kin SJ (1996) Nonradioactive method to study genetic profiles of natural bacterial communities by PCR-single-strand-conformation polymorphism. Appl Environ Microbiol 62 (9):3112–3120

Leung HC, Yiu SM, Yang B (2011) A robust and accurate binning algorithm for metagenomic sequences with arbitrary species abundance ratio. Bioinformatics 27(11):1489–1495

Li L, Frohlich J, Pfeiffer P, Konig H (2003) Termite gut symbiotic archaezoa are becoming living metabolic fossils. Eukaryot Cell 2(5):1091–1098

Li ZT, Hopkins DL, Gray DJ (2015) Overexpression of antimicrobial lytic peptides protects grapevine from Pierce's disease under greenhouse but not field conditions. Transgenic Res 24 (5):821–836

Lilburn TG, Schmidt TM, Breznak JA (1999) Phylogenetic diversity of termite gut spirochaetes. Environ Microbiol 1(4):331–345

Liu WT, Marsh TL, Cheng H, Forney LJ (1997) Characterization of microbial diversity by determining terminal restriction fragment length polymorphisms of genes encoding 16S rRNA. Appl Environ Microbiol 63(11):4516–4522

Luo R, Liu B, Xie Y (2012) SOAP denovo2: an empirically improved memory-efficient short-readdenovoassembler. Gigascience 1(1):18

Malacrinò A, Rassati D, Schena L, Mehzabin R, Battisti A, Palmeri V (2017) Fungal communities associated with bark and ambrosia beetles trapped at international harbours. Fungal Ecol 28:44–52. https://doi.org/10.1016/j.funeco.2017.04.007

Malathi VM, More RP, Anandham R, Gracy GR, Mohan M, Venketesan T, Samaddar S, Jalali SK, Tongmin (2018) Gut bacterial diversity of insecticide-susceptible and –resistant nymphs of the Brown Planthopper Nilaparvata lugens Stål (Hemiptera: Delphacidae) and elucidation of their putative functional roles. J Microbiol Biotechnol 28(6):976–986

Manefield M, Whiteley AS, Griffiths RI, Bailey MJ (2002) RNA stable isotope probing, a novel means of linking microbial community function to phylogeny. Appl Environ Microbiol 68 (11):5367–5373

Mardis ER (2008) The impact of next-generation sequencing technology on genetics. Trends Genet 24(3):133–141

Margulies M, Egholm M, Altman WE et al (2005) Genome sequencing in microfabricated high-density picolitre reactors. Nature 437(7057):376–380

Marusina K (2006) The next generation of DNA sequencing. Genet Eng News 26:28–31

Mayo B, Rachid CTCC, Alegría A, Leite AMO, Peixoto RS, Delgado S (2014) Impact of next generation sequencing techniques in food microbiology. Curr Genomics 15:293–309. https://doi.org/10.2174/1389202915666140616233211

McKillip JL, Small CL, Brown JL, Brunner JF, Spence KD (1997) Sporogenous midgut bacteria of the leafroller, *Pandemis pyrusana* (Lepidoptera: Tortricidae). Environ Entomol 26 (6):1475–1481

Metzker ML (2010) Sequencing technologies- the next generation. Nat Rev Genet 11:31–46. https://doi.org/10.1038/nrg2626
Meyer M, Kircher M (2010) Illumina sequencing library preparation forn highlymultiplexed target capture and sequencing. Cold Spring Harb Protoc. https://doi.org/10.1101/pdb.prot5448
Mohr KI, Tebbe CC (2006) Diversity and phylotype consistency of bacteria in the guts of three bee species (*Apoidea*) at an oilseed rape field. Environ Microbiol 8(2):258–272
Moran NA, McCutcheon JP, Nakabachi A (2008) Genomics and evolution of heritable bacterial symbionts. Annu Rev Genet 42:165–190
Morgan XC, Huttenhower C (2012) Human microbiome analysis. PLoS Comput Biol 8(12): e1002808
Muyzer G, de Waal EC, Uitterlinden AG (1993) Profiling of complex microbial populations by denaturing gradient gel-electrophoresis analysis of polymerase chain reactionamplified genes coding for 16s rRNA. Appl Environ Microbiol 59(3):695–700
Nagarajan N, Cook C, DiBonaventura M (2010) Finishing genomes with limited resources: lessons from an ensemble of microbial genomes. BMC Genomics 11(1):242
Nakashima K, Watanabe H, Saitoh H, Tokuda G, Azuma JI (2002) Dual cellulose-digesting system of the wood-feeding termite, Coptotermes formosanus Shiraki. Insect Biochem Mol Biol 32 (7):777–784
Noda S, Ohkuma M, Usami R, Horikoshi K, Kudo T (1999) Culture-independent characterization of a gene responsible for nitrogen fixation in the symbiotic microbial community in the gut of the termite *Neotermes koshunensis*. Appl Environ Microbiol 65(11):4935–4942
Nurk S, Meleshko D, Korobeynikov A, Pevzner PA (2017) metaSPAdes:A new versatile metagenomica ssembler. Genome Res 27(5):824–834
Ohkuma M, Kudo T (1996) Phylogenetic diversity of theintestinal bacterial community in the termite *Reticulitermessperatus*. Appl Environ Microbiol 62(2):461–468
Ondov BD, Bergman NH, Phillippy AM (2011) Interactive metagenomic visualization in a web browser. BMC Bioinform 12:385
Oulas A, Pavloudi C, Polymenakou P (2015) Metagenomics:tools and insights for analysing next-generation sequencing data derived from biodiversity studies. Bioinform Biol Insights 9(9):75
Pace NR (1997) A molecular view of microbial diversity and the biosphere. Science 276 (5313):734–740
Pace NR, Stahl DA, Lane DJ, Olsen GJ (1986) The analysis of natural microbial populations by ribosomal RNA sequences. Adv Microb Ecol 9:1–55
Paine TD, Raffa KF, Harrington TC (1997) Interactions among Scolytid bark beetles, their associated fungi, and live host conifers. Annu Rev Entomol 42:179–206
Pittman GW, Brumbley SM, Allsopp PG, O'Neill SL (2008) "Endomicrobia" and other bacteria associated with the hindgut of *Dermolepida albohirtum* larvae. Appl Environ Microbiol 74 (3):762–767
Plummer E, Twin J, Bulach DM, Garland SM, Tabrizi SN (2015) A comparison of three bioinformatics pipelines for the analysis of preterm gut microbiota using 16S rRNA gene sequencing data. J Proteomics Bioinformatics 8(12):283
Poland TM, Rassati D (2018) Improved biosecurity surveillance of non-native forest insects: a review of current methods. J Pest Sci:1–13. https://doi.org/10.1007/s10340-018-1004-y
Priya NG, Ojha H, Kajla MK, Raj A, Rajagopal R (2012) Host plant induced variationin gut bacteria of *Helicoverpa armigera*. PLoS One 7:e30768
Quail MA, Smith M, Coupland P et al (2012) A tale of three next generation sequencing platforms: comparison of IonTorrent, Pacific biosciences and Illumina MiSeq sequencers. BMC Genomics 13:341. https://doi.org/10.1186/1471-2164-13-341
Quince C, Walker AW, Simpson JT, Loman NJ, Segata N (2017) Shotgun metagenomics, from sampling to analysis. Nat Biotechnol 35:833–844. https://doi.org/10.1038/nbt.3935
Rajagopal R (2009) Beneficial interactions between insects and gut bacteria. Indian J Microbiol 49: 114–119

Ranjith MT, Chellappan M, Harish ER, Girija D, Nazeem PA (2016) Bacterial communities associated with the gut of tomato fruit borer, *Helicoverpa armigera* (Hubner) (Lepidoptera: Noctuidae) based on Illumina next-generation sequencing. J Asia Pac Entomol 19(2):333–340

Rassati D, Faccoli M, Petrucco Toffolo E, Battisti A, Marini L (2015) Improving the early detection of alien wood-boring beetles in ports and surrounding forests. J Appl Ecol 52:50–58. https://doi.org/10.1111/1365-2664.123

Rassati D, Faccoli M, Marini L, Haack RA, Battisti A, Petrucco Toffolo E (2018) Exploring the role of wood waste landfills in early detection of non-native wood-boring beetles. J Pest Sci 88:563–572. https://doi.org/10.1007/s10340-014-0639-6

Raymond B, Lijek RS, Griffiths RI, Bonsall MB (2008) Ecological consequences of ingestion of *Bacilluscereus* on Bacillus thuringiensis infections and on the gut flora of a lepidopteran host. J Invertebr Pathol 99(1):103–111

Reeson AF, Jankovic T, Kasper ML, Rogers S, Austin AD (2003) Application of 16S rDNA-DGGE to examine the microbial ecology associated with a social wasp *Vespulagermanica*. Insect Mol Biol 12(1):85–91

Robe P, Nalin R, Capellano C, Vogel TM, Simonet P (2003) Extraction of DNA from soil. Eur J Soil Biol 39:183–190

Roe AD, Torson AS, Bilodeau G, Bilodeau P, Blackburn GS, Cui M, Cusson M, Doucet D, Griess VC, Lafond V et al (2018) Biosurveillance of forest insects: part I—integration and application of genomic tools to the surveillance of non-native forest insects. J Pest Sci:1–20. https://doi.org/10.1007/s10340-018-1027-4

Roose-Amsaleg CL, Garnier-Sillam E, Harry M (2001) Extraction and purification of microbial DNA from soil and sediment samples. Appl Soil Ecol 18(1):47–60

Rossmassler K, Dietrich C, Thompson C et al (2015) Metagenomic analysis of the microbiota in the highly compartmented hindguts of six wood- or soil-feeding higher termites. Microbiome 3:56. https://doi.org/10.1186/s40168-015-0118-1

Roszak DB, Grimes DJ, Colwell RR (1984) Viable but nonrecoverable stage of *salmonella enteritidis* in aquatic systems. Can J Microbiol 30:334–338

Roussel EG, Bonavita MAC, Querellou J, Cragg BA, Webster G, Prieur D, Parkes RJ (2008) Extending the sub-sea-floor biosphere. Science 320(5879):1046–1046

Russell JA, Weldon S, Smith AH, Kim KL, Hu Y, Łukasik P, Oliver KM (2013) Uncovering symbiont-driven genetic diversity across north American pea aphids. Mol Ecol 22 (7):2045–2059

Sanger F, Nicklen S, Coulson AR (1977a) DNAsequencing with chain- terminating inhibitors. Proc Natl Acad Sci U S A 74(12):5463–5467. https://doi.org/10.1073/pnas.74.12.5463

Sanger F, Nicklen S, Coulson AR (1977b) DNA sequencing with chain terminating inhibitors. Proc Natl Acad Sci U S A 74:5463–5467

Santo Domingo JW, Kaufman MG, Klug MJ, Holben WE, Harris D, Tiedje JM (1998) Influence of diet on the structure and function of the bacterial hindgut community of crickets. Mol Ecol 7 (6):761–767

Schierbeek A (1959) Measuring the invisible world: the life and works of Antoni van Leeuwenhoek. Abelard-Schuman, London. Available online at: https://www.questia.com/library/73684/measuring-the-invisible-world-thelife-and-works (Accessed Sep 12, 2020)

Schmitt Wagner D, Friedrich MW, Wagner, B.& Brune, A. (2003) Phylogenetic diversity, abundance, and axial distribution of bacteria in the intestinal tract of two soil-feeding termites (*Cubitermes* spp.). Appl Environ Microbiol 69(10):6007–6017

Scott JG, Michel K, Bartholomay LC, Siegfried BD, Hunter WB, Smagghe G, Zhu KY, Douglas AE (2013) Towards the elements of successful insect RNAi. J Insect Physiol 59(12):1212–1221

Scully ED, Geib SM, Hoover K, Tien M, Tringe SG et al (2013) Metagenomic profiling reveals lignocellulose degrading system in a microbial community associated with a wood-feeding beetle. PLoS One 8(9):e73827. https://doi.org/10.1371/journal.pone.0073827

Shah SHJ, Malik AH, Zhang B, Bao Y, Qazi J (2020) Metagenomic analysis of relative abundance and diversity of bacterial microbiota in Bemisia tabaci infesting cotton crop in Pakistan. Infect Genet Evol 84:104381

Sharon G, Segal D, Ringo JM, Hefetz A, Zilber-Rosenberg I, Rosenberg E (2010) Commensal bacteria play a role in mating preference of Drosophila melanogaster. Proc Natl Acad Sci U S A 107:20051–20056

Sharon G, Segal D, Zilber-Rosenberg I, Rosenberg E (2011) Symbiotic bacteria are responsible for diet-induced mating preference in Drosophila melanogaster, providing support for the hologenome concept of evolution. Gut Microbes 2:190–192

Sharpton TJ (2014) An introduction to the analysis of shotgun metagenomic data. Front Plant Sci 5: 209. https://doi.org/10.3389/fpls.2014.00209

Shen Z, Jacobs-Lorena M (1997) Characterization of a novel gut-specific chitinase gene from the human malaria vector Anopheles gambiae. J Biol Chem 272:28895–28900

Shendure J, Ji H (2008) Next-generationDNAsequencing. Nat Biotechnol 26:1135–1145

Shinzato N, Muramatsu M, Matsui T, Watanabe Y (2005) Molecular phylogenetic diversity of the bacterial community in the gut of the termite Coptotermes formosanus. Biosci Biotechnol Biochem 69(6):1145–1155

Shinzato N, Muramatsu M, Matsui T, Watanabe Y (2007) Phylogenetic analysis of the gut bacterialmicroflora of the fungus-growing termite *Odontotermes formosanus*. Biosci Biotechnol Biochem 71:906–915

Shokralla S, Spall JL, Gibson JF, Hajibabaei M (2012) Nextgenerationsequencing technologies for environmental DNA research. Mol Ecol 21(8):1794–1805. https://doi.org/10.1111/j.1365-294X.2012.05538.x

Sinclair L, Osman OA, Bertilsson S, Eiler A (2015) Microbial community composition and diversity via 16S rRNA gene amplicons: evaluating the illumina platform. PLoS One 10(2): e0116955

Slaytor M (1992) Cellulose digestion in termites and cockroaches: what role do symbionts play? Comp Biochem Physiol B: Comp Biochem 103(4):775–784

Sleator RD, Shortall C, Hill C (2008) Under the microscope metagenomics. Lett Appl Microbiol 47: 361–366

Smalla K (2004) Culture-independent microbiology. In: Bull AT (ed) Microbial diversity and bioprospecting. ASM Press, Washington, D.C., pp 88–99

Stahl DA, Lane GJ, Pace NR (1984) Analysis of hydrothermal vent-associated symbionts by ribosomal RNA sequences. Science 224(4647):409–411

Staley JT, Konopka A (1985) Measurement of *in situ* activities of nonphotosynthetic microorganisms in aquatic and terrestrial habitats. Annu Rev Microbiol 39:321–346

Stangier KA (2009) Next-generation sequencing: a short comparison. LaboratoriumsMedizin-J Lab Med 33:153–157

Steele HL, Streit WR (2005) Metagenomics: advances in ecology and biotechnology. FEMS Microbiol Lett 247(2):105–111

Stokes HW, Holmes AJ, Nield BS, Holley MP, Nevalaine KMH, Mabbutt BC, Gillings MR (2001) Gene cassette PCR: sequence-independent recovery of entire genes from environmental DNA. Appl Environ Microbiol 67(11):5240–5246

Tikhe CV, Martin TM, Howells A, Delatte J, Husseneder C (2016) Assessment of genetically engineered Trabulsiella odontotermitis as a 'Trojan Horse' for paratransgenesis in termites. BMC Microbiol 16:202. https://doi.org/10.1186/s12866-016-0822-4

Todaka MS, Saita K, Hondo T, Kiuchi I, Takasu H, Ohkuma M, Piero C, Hayashizaki Y, Kudo T (2007) Environmental cDNA analysis of the genes involved in lignocellulose digestion in the symbiotic protist community of *Reticulitermes speratus*. FEMS Microbiol Ecol 59(3):592–599

Tokuda G, Watanabe H (2007) Hidden cellulases in termites: revision of an old hypothesis. Biol Lett 3(3):336–339

Tringe SG, VonMering C, Kobayashi A et al (2005) Comparative metagenomics of microbial communities. Science 308(5721):554–557

Turroni F, Ribbera A, Foroni E, van Sinderen D, Ventura M (2008) Human gut microbiota and bifidobacteria: from composition to functionality. Antonie Van Leeuwenhoek 94(1):35–50

Tyson GW, Chapman J, Hugenholtz P, Allen EE, Ram RJ, Richardson PM, Solovyev VV, Rubin EM, Rokhsar DS, Banfield JF (2004) Community structure and metabolism through reconstruction of microbial genomes from the environment. Nature 428(6978):37–43

Van Djick EL, Auger H, Jaszczyszyn Y, Thermes C (2014) Ten years of next-generation sequencing technology. Trends Genet 30:418–426. https://doi.org/10.1016/j.tig.2014.07.001

Van Nieuwerburgh F, Thompson RC, Ledesma J, Deforce D, Gaasterland T, Ordoukhanian P, Head SR (2012) Illumina mate-paired DNA sequencing-library preparation using Cre-lox recombination. Nucleic Acids Res 40(3). https://doi.org/10.1093/nar/gkr1000

Vasanthakumar A, Delalibera I, Handelsman J, Klepzig KD, Schloss PD, Raffa KF (2006) Characterization of gut-associated bacteria in larvae and adults of the southern pine beetle, Dendroctonus frontalis Zimmermann. Environ Entomol 35(6):1710–1717

Venter JC, Remington K, Heidelberg JF et al (2004a) Environmental genome shotgun sequencing of the Sargasso Sea. Science 304(5667):66–74

Venter JC, Remington K, Heidelberg JF, Halpern AL, Rusch D et al (2004b) Environmental genome shotgun sequencing of the Sargasso Sea. Science 304:66–74

Vieites JM, Guazzaroni ME, Beloqui A, Golyshin PN, Ferrer M (2008) Metagenomics approaches in systems microbiology. FEMS Microbiol Rev 33(1):236–255

Wang H, Lei T, Xia W et al (2019) Insight into the microbial world of Bemisia tabaci cryptic species complex and its relationships with its host. Sci Rep 9:6568. https://doi.org/10.1038/s41598-019-42793-8

Warnecke F, Luginbuhl P, Ivanova N et al (2007) Metagenomic and functional analysis of hindgut microbiota of a wood-feeding higher termite. Nature 450(7169):560–565

Wernegreen JJ (2002) Genome evolution in bacterial endosymbionts of insects. Nat Rev Genetics 3 (11):850–861

Williamson LL, Borlee BR, Schloss PD, Guan CH, Allen HK, Handelsman J (2005) Intracellular screen to identify metagenomic clones that induce or inhibit a quorumsensing biosensor. Appl Environ Microbiol 71(10):6335–6344

Weinstock GM (2012) Genomic approaches to study human microbiota. Nature 489:250–256

Woese CR, Fox GE (1977) Phylogenetic structure of the prokaryotic domain: the primary kingdoms. Proc Natl Acad Sci U S A 74(11):5088–5090. https://doi.org/10.1073/pnas.74.11.5088

Wu D, Daugherty SC, Van Aken SE, Pai GH, Watkins KL, Khouri H, Tallon LJ, Zaborsky JM, Dunbar HE, Tran PL et al (2006) Metabolic complementarity and genomics of the dual bacterial symbiosis of sharpshooters. PLoS Biol 4(6):e188

Xia X, Zheng D, Zhong H, Qin B, Gurr GM, Vasseur L et al (2013) DNA sequencing reveals the midgut microbiota of diamondback moth, *Plutella xylostella* (L.) and a possible relationship with insecticide resistance. PLoS ONE 8(7): e68852. https://doi.org/10.1371/journal.pone.0068852

Xia X, Gurr GM, Vasseur L et al (2017) Metagenomic sequencing of diamondback moth gut microbiome unveils key Holobiont adaptations for herbivory. Front Microbiol 8:663. https://doi.org/10.3389/fmicb.2017.00663

Yandell DW (1991) SSCP (single strand conformation polymorphism) analysis of PCR fragments. Jpn J Cancer Res 82:1468–1469

Yuan JS, Galbraith DW, Dai SY, Griffin P, Stewart CN (2008) Plant systems biology comes of age. Trends Plant Sci 13(4):165–171

Zhou J, Bruns MA, Tiedje JM (1996) DNA recovery from soils of diverse composition. Appl Environ Microbiol 62(2):316–322

Zhou XG, Smith JA, Oi FM, Koehler PG, Bennett GW, Scharf ME (2007) Correlation of cellulase gene expression and cellulolytic activity throughout the gut of the termite Reticulitermes flavipes. Gene 395(1-2):29–39

Entomopathogenic Fungi

7

Amritesh C. Shukla and Karina Afzal

Contents

Abstract

With the rising need of switching over to sustainable agricultural practices, utilization of entomopathogenic fungi (EPF) as biocontrol agents provides better and safe substitute against chemical insecticides, which are associated with

A. C. Shukla (✉) · K. Afzal
Biocontrol Laboratory, Department of Botany, University of Lucknow, Lucknow, Uttar Pradesh, India

Omkar (ed.), *Microbial Approaches for Insect Pest Management*,
https://doi.org/10.1007/978-981-16-3595-3_7

several environmental and health hazards. Entomopathogenic fungi act as a parasite of insects and kill or critically disable the insects. These include different classes of fungi, viz., Oomycota, Chytridiomycota, Zygomycota, Ascomycota, Deuteromycota, Basidiomycota, and Entomophthoromycota that infect and kill the insects. Some of the merits related with the uses of entomopathogenic fungi as biocontrol agents are high host specificity, insignificant effect on the beneficial insects/nontarget organisms, and simple mass production. The EFP are reported to infect a very wide range of insects, such as lepidopterous larvae, aphids, and thrips, which are of enormous concern in the agriculture, globally. This approach of using EPF as biocontrol agent, instead of chemical pesticides, seems to be very effective and promising in the near future as it moves toward sustainable agricultural practices and protecting the environment, which is the need of the hour.

Keywords

Entomopathogenic fungi · Secondary metabolites · Biopesticides · Biocontrol agent · Insect pests

Learning Objectives

1. The entomopathogenic fungi (EPF) are amid pioneering microorganisms to be applied as biocontrol agent for pest management. EPF kill insect pests at several stages of life.
2. In addition, their incidences along with widespread occurrence make them suitable contenders for pesticides in integrated pest management strategies.
3. The knowledge of recent findings about EPF will signify new frontier contribution toward pathogenesis and multifunctional applications.
4. Therefore, in this chapter, we aim to foreground various aspects of EPF biology including their general characteristics, classification, mechanism of action, maintenance in culture, and some commercial formulations developed for field application.

7.1 Introduction

The fungi are eukaryotic organisms, ranging from unicellular to multicellular filamentous in form. They are chemo-organotrophic creatures without chlorophyll. They reproduce either asexually or sexually or by both means. Fungi often associate with other organisms, creating various biological associations. Entomopathogenic fungi (EPF) are parasitic microorganisms having the potential to cause infection and kill arthropods. Though they majorly belong to arthropod carcasses, yet they are found naturally occurring in soil (Behie and Bidochka 2014). The members of this group are designated under six classes, viz., Oomycetes, Chytridiomycota, Entomophtoromycota, Microsporidia, Basidiomycota, and majorly Ascomycota. Because of their cosmopolitan occurrence and extensive diversity, EPF considerably contribute as biocontrol agents for sustainable management of insect populations as they own a unique model of infection in various orders of insects. Progressive findings concerning the genomic biology of EPF have revealed that genetic makeup of such microorganisms is advanced for fungal adaptation with plenty of insect hosts. Recent investigations reveal that they also work as endophytes as well as biocontrol mediators of plant pathogens (Behie et al. 2013); moreover, they endorse plant development as rhizosphere fungi. Entomopathogens as biocontrol weapons impart numerous assets over conventional insecticides including high efficiency, low costs, safety for beneficial organisms, lessening of its remains in the surrounding environment, along with amplified biodiversity in the human-controlled community (Asi et al. 2013; Ortiz-Urquiza and Keyhani 2013; Gul et al. 2014). Considering the high enzymatic pursuit, the potential to synthesize secondary metabolites, along with virtuous growth in culture media, their probable application in other fields of biotechnology, like biosynthesis of nanoparticles, making them economically significant (Kozłowska et al. 2019; Dou et al. 2019). The knowledge about their mechanism of virulence and level of tolerance toward adverse conditions, along with application of genetic engineering will potentiate cost-efficient products of mycoinsecticides for pest management in agricultural fields. Moreover, the recent findings concerning exploitation of their genetic diversity, vast ecological occurrence, and wide functional sphere make these fungi highly applicable for integrated pest management.

7.2 Groups of Entomopathogenic Fungi

The kingdom Fungi is the main eukaryotic group with about 700 well-recognized EPF species, accounting for less than 1% of whole fungal species (McLaughlin et al. 2009). EPF do not form a single monophyletic group. EPF are found in three major groupings, viz., Blastocladiomycota, Entomophthoromycota, and Microsporidia and 12 classes under 6 phyla of fungi. Ascomycota, Chytridiomycota, Deuteromycota, Oomycota, and Zygomycota (Humber 1997) are the main divisions harboring these pathogenic fungi. So far, 12 species of Oomycetes, 65 species of Chytridiomycota, 474 species of Entomophtoromycota, 339 species of Microsporidia, 476 species of Ascomycota, and 238 species of Basidiomycota have been described (Araújo and Hughes 2016; Jaber and Enkerli 2017). Amid the diverse phyla, the species belonging to genus *Beauveria, Hirsutella, Verticillium, Nomuraea,* and *Metarhizium* of diverse environmental groups are most conspicuously significant EPF, which are commercially applied effectively at field levels. Further, biological and ecological features of EPF have also been well reported (Steinhaus 1964; Samson et al. 1988; Balazy 1993).

Fig. 7.1 Some common Entomopathogenic fungi

7.2.1 Classification of the Entomopathogenic Fungi

EPF are vital biological control mediators, undergoing rigorous research for more than 100 years. Further, based on the characteristics of the hyphae, composition of cell wall, mode of nutrition and reproduction, entomopathogenic fungi can be classified into different groups. The details of the morphological view of EPF are being summarized as follows (Figs. 7.1 and 7.2):

Phylum Oomycota

The most distinctive feature of Oomycetes includes production of zoospores in sporangia. The coenocytic hyphae of these fungi possess cellulose (without chitin) and their zoospores are biflagellate. Sexual reproduction takes place between gametangia occurring either on the same hyphae or on different hyphae. They reproduce by oospores, which are thick-walled, and they have mitochondria with tubular cristae at the cellular level. They are parasitic on animals and plants, yet a few species are saprophytes. This phylum includes orders *Lagenidiales,* and *Leptomitales*. *Lagenidium giganteum* and *Leptolegnia chapmanii* are parasitic on mosquito larvae, arthropods, crabs, and some aquatic crustaceans (Hatai et al. 2000).

Phylum Chytridiomycota

The most distinctive feature of members of this phylum includes production of motile zoospores having solitary whiplash flagellum, inserted posteriorly. The cell wall of the members of this fungal group is predominantly made up of chitin, their hyphae are coenocytic. This group of fungus is regarded as basal, as per their comparative rRNA phylogenetic analysis. This phylum includes orders Blastocladiales, Chytridiales, and Blastocladiales. Some common genus belonging

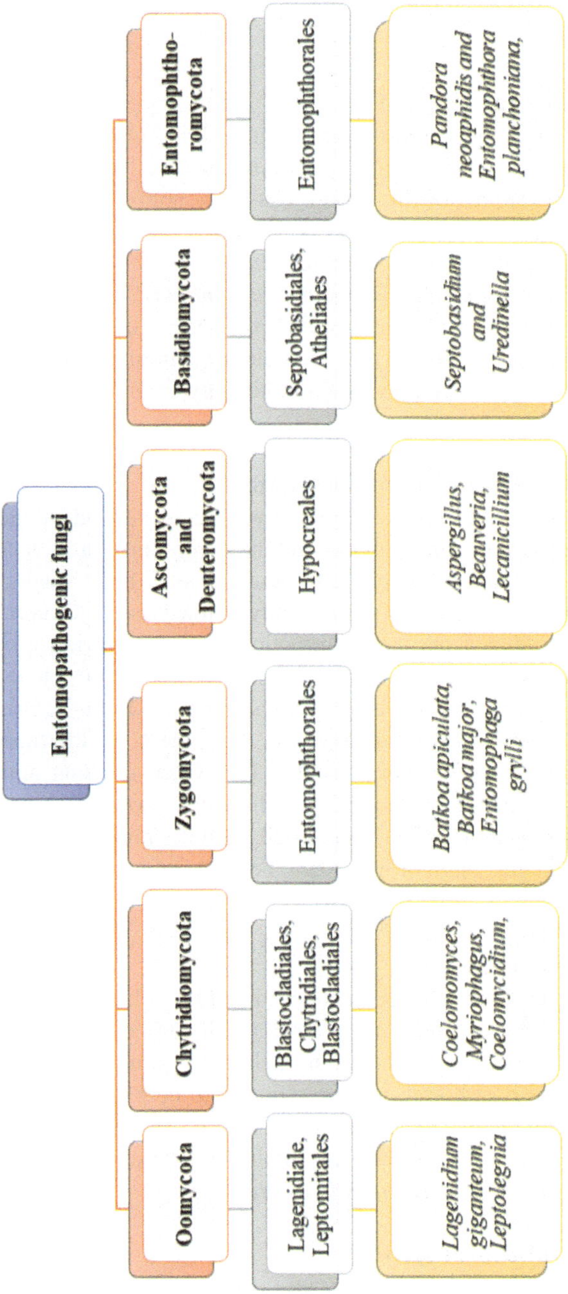

Fig. 7.2 Classification of the entomopathogenic fungus

to this phylum includes *Coelomomyces, Coelomycidium, Myriophagus, Coelomycidium (Blastocladiales),* and *Myriophagus* (Chytridiales). The most common insect hosts infected by them are members of Hemiptera and dipteran flies and mosquitoes.

Phylum Zygomycota

The members of this fungal group possess multicellular and nonseptate mycelium, and after fusion, the gametangia form zygospores (sexual stage), which is one of the most characteristic features of this group. Class Trichomycetes within the phylum consists mostly of species related to insects. The order Entomophthorales holds more than 200 insect-infecting species including *Batkoa apiculata, Entomophaga grylli, Entomophaga maimaiga, Conidiobolus thromboids, Pandora neoaphidis, Zoophthora radicans, Neozygites parvispora,* and various others. The most common insect hosts infected by them are hemipterans, homopterans, lepidopterans, grasshoppers, dipterans, leafhoppers Lepidoptera, gypsy moth larvae, psyllids, and others.

Phylum Ascomycota and Deuteromycota

The members of phylum Deuteromycota are well characterized by the presence of septate mycelium bearing conidiophores and they reproduce through conidia. However, members belonging to phylum Ascomycota bear distinctive feature "ascospores" that develop in the fruiting body, named ascus. Generally, there are eight ascospores produced per ascus formation of ascus bearing ascospores. The order Hypocreales belonging to Ascomycota includes various genera, including *Aspergillus, Aschersonia, Beauveria, Culicinomyces, Metarhizium, Hirsutella, Tolypocladium, Lecanicillium, Paecilomyces*, and others. Furthermore, more than 300 entomopathogenic species are present in Cordyceps. Among numerous insect hosts infected by them are whiteflies, grasshoppers, mosquitoes, potato beetles, Hymenoptera, Lepidoptera, coding moth, boll weevil, chinch bug, granary weevil, cowpea curculio, lygus bug, brown plant hopper, fire ants, termites, European cockchafer, sugarcane borer, and others.

Phylum Basidiomycota

One of the exclusive traits for the members includes "clamp connections." Their sexual spores called basidiospores are formed outside the reproductive cells known as basidia. The members of this group belong to orders Septobasidiales and Atheliales. Only few Basidiomycetes are reported to be pathogenic to insects. The common entomopathogenic genera include *Fibula rhizoctonia, Uredinella,* and *Septobasidium* (Samson et al. 1988), and the insect host infected by them includes termite eggs and scale insects (*Diaspididae, Hemiptera*).

Phylum Entomophthoromycota

The mycelium is well defined, coenocytic, or septate in a somatic state. The protoplast is changeable in shape, either amoeboid or hyphoid, and a few members form rhizoids or cystidia. Their conidiophores are branched or unbranched and the

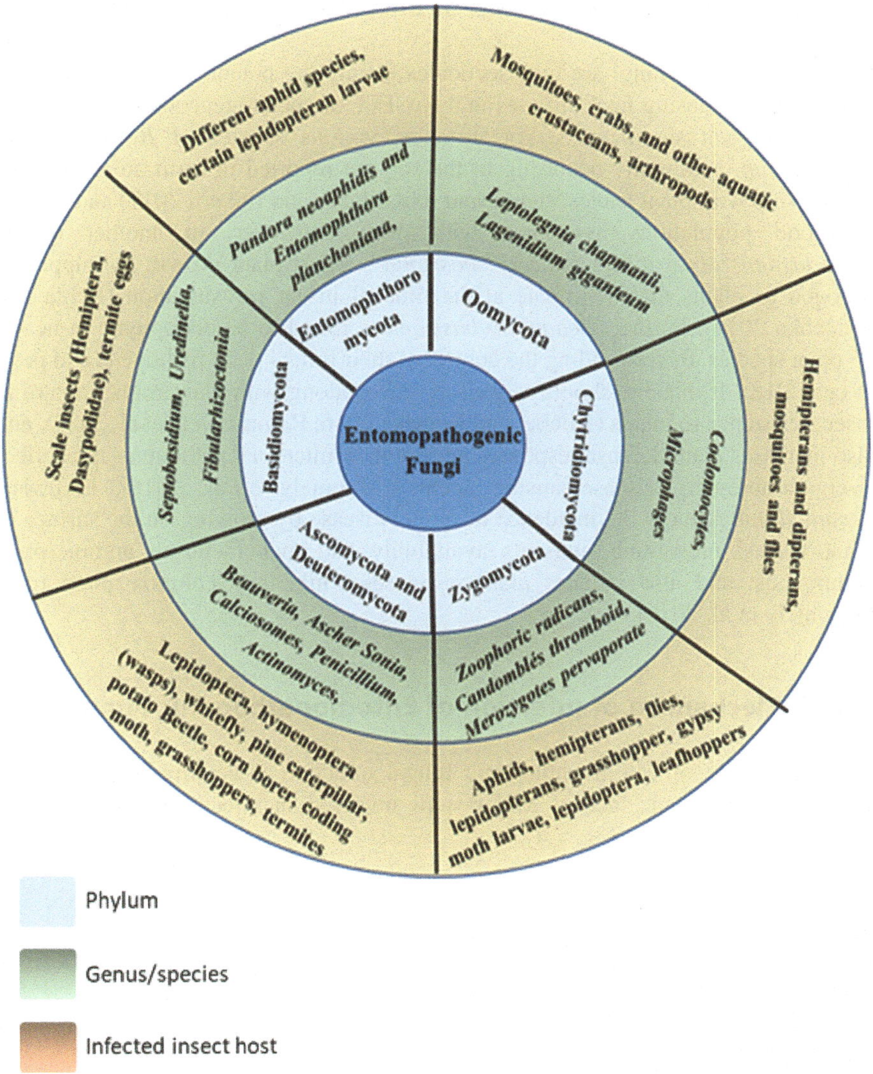

Fig. 7.3 Common Entomopathogenic fungi with their host insect

primary spores are true conidia with uni-, pluri-, or multinucleate conditions. The order Entomophthorales includes various species, e.g., *Entomophthora planchoniana, Pandora neoaphidis,* and *Entomophaga maimaiga,* which infect certain lepidopteran larvae and various aphid species (Fig. 7.3).

7.3 General Characteristics of Entomopathogenic Fungi

Entomopathogenic fungi are bioinsecticides having the potential to infect and kill arthropods by causing fatal diseases in them. EPF are heterogeneous microbes with huge ecological significance. For instance, various species of *Beauveria* and *Metarhizium*, commonly occurring in the soil are reported to form an endophytic relationship with plant leaves, stems, and roots (Jaber and Enkerli 2017) and control arthropod populations associated with them. However, in another report, *Metarhizium robertsii* and *Beaveria bassiana* support plant growth by supplying nitrogen to plants that assimilate at the time of insect parasitization (Behie and Bidochka 2014). Further, *Beauveria bassiana* is found to be endophytic in nearly 25 plant species, thereby aiding the control of their fungal plant pathogens and pests (Vega 2018). It inhabits shoots as well as leaves along with plant roots, imparting insect resistance to plants (Klieber and Reineke 2016; Ramakuwela et al. 2020), and also increases plant defense responses against their microbial pathogens, thus effectively suppressing disease-causing agents (Moonjely et al. 2016). Likewise, *Lecanicillium* reduces the incidence of fungal disease by growing on the surface of plant leaves, preventing nutrients availability and manufacturing antimicrobial compounds, and also induces plant responses while they colonize plant roots (Moonjely et al. 2016).

7.4 Mechanism of Infection of Entomopathogenic Fungi

Mode of action of entomopathogenic fungal infection is determined by various intrinsic and extrinsic factors. Major steps involved in infection process can be summarized as follows (Fig. 7.4):

- Adhesion of spores to cuticle of insect.
- Activation of some defensive biochemical process in insect foe defense.
- Germination of spore.
- Penetration of cuticle.
- Growth in the haemocoel.
- Death of insect and saprophytic feeding by fungi.
- Mummification.

Entomopathogenic fungi directly penetrate the cuticle of insects to cause infection. Rather, their ingestion by an insect is not necessary, as in the case of viruses or bacteria (Bilgo et al. 2018). During the initial phase of the infection process, the spores adhere to arthropod shells and then follow the two phases: the first one depending on the action of electrostatic and hydrophobic forces, while the other one involving the enzymatic activities and hydrophobins; the low-molecular-weight proteins (Skinner et al. 2014). Suitable environmental variables, such as satisfactory humidity and temperature, are necessary for spores to germinate on the insect's cuticle. Further, the presence of adequate energy and carbon sources and the optimal

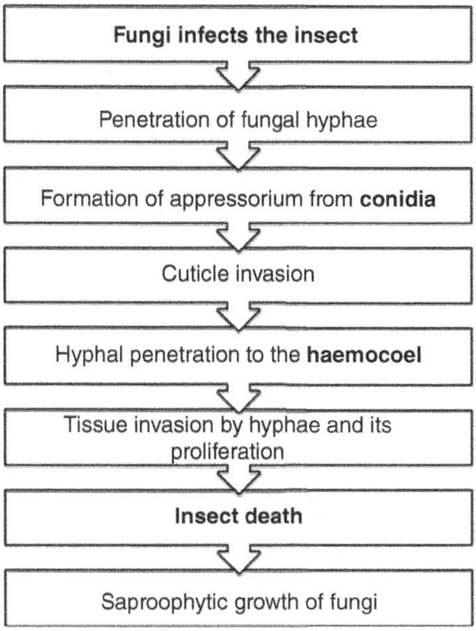

Fig. 7.4 Entomopathogenic fungi and it's mode of infestation

temperature (between 20 and 30 °C) are prerequisites for the growth and germination of EPF (Skinner et al. 2014). Thereafter, appressoria emerge by applying considerable mechanical pressure on the cuticle and start producing proteolytic, lipolytic, and chitinolytic enzymes that disrupt the insect's body shells (Lacey et al. 2015). The development of the fungal hyphae begins once inside the host's body cavity, i.e., hemocoel and in some cases, there is the production of blastospores that pass in the host's hemolymph where they begin to develop secondary hyphae inside the host's tissues. The process is followed by the synthesis and production of secondary metabolites by fungi soon after their commencement, thereby disrupting the host's physiological processes, immune responses, and causing paralysis (Donzelli and Krasnoff 2016). The development of infection leads to the destruction of the insect's body because it starts depleting the nutrient, even the internal organs are damaged by the developing hyphae (Fan et al. 2017). As the infection continues to develop, the insect's body subsequently becomes stiff because EPF continuously absorb the fluids from the initially soft insect's body. Sometimes, the corpse of insects destroyed by EPF may initially turn into a dark red color, as in the case of the genus *Beauveria*. The approximate duration of the entire infection process takes about 14 days after infection; however, the initial symptoms of the disease appear early, within 7 days after infection, which may differ depending on fungal species. Once the insect is dead and deprived of all nutrients, the hyphae of the EPF leave the cadaver of the insect body through intersegmental areas and openings, like mouth hole and anus. Formerly, infective spores or resting spores are formed, which promotes the spread and infection cycle of fungus, also known as mummification (Skinner et al. 2014) (Fig. 7.5).

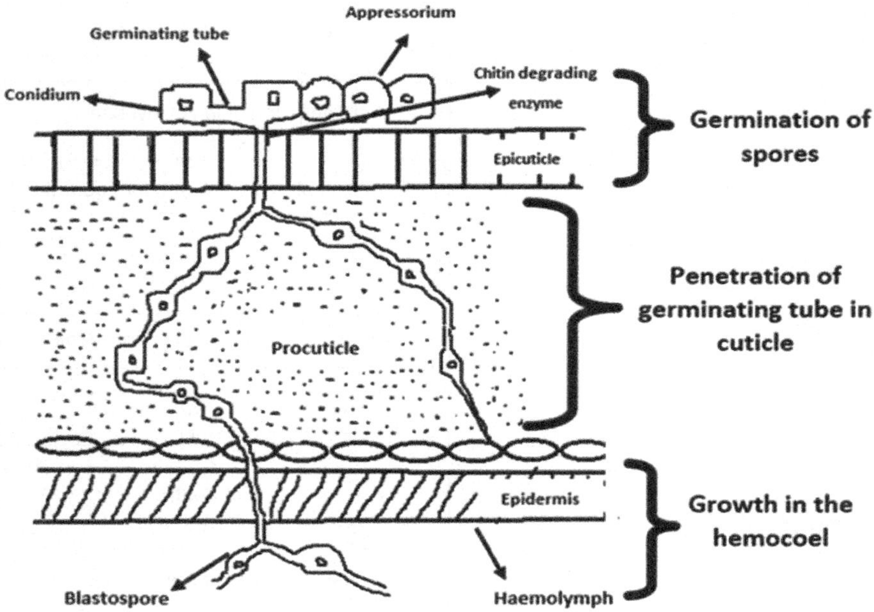

Fig. 7.5 Mechanism of action of Entomopathogenic fungi

Another significant progression involved in the process of infection by EPF is at biochemical levels involving infectious mediators. These **infectious agents** required by EPF for pathogenesis include several lytic enzymes, adhesion molecules, and secondary metabolites. A brief description of these agents is narrated below:

7.4.1 Adhesins

The phenomenon of adhesion of spores to the surface of the arthropod's body is the beginning of the spread of contagion. Two types of proteins, viz. adhesins (MAD1 and MAD2) and hydrophobins, aid in close union as well as recognition of the host by the fungal species (Greenfield et al. 2014).

7.4.2 Lytic Enzymes

Lytic enzymes are the utmost essential components of the infection process by EPF. These enzymes are synthesized and produced once the spore gets involved in the cuticle and starts to develop appressorium (Santi et al. 2010). The major function of these enzymes is the hydrolysis of the components of the insect cuticle and hence to promote the penetration of outer covers of arthropods by appressoria. Lipases, which hydrolyze the lipoproteins and lipids, are produced first, residing in the outer cuticle of the insect (Pedrini et al. 2007). These enzymes function by breaking the ester bonds of triacylglycerols and allowing the subsequent release of free fatty acids, glycerol, monoacylglycerols, and diacylglycerols (Silva et al. 2009). Also, lipases improve hydrophobic interactions among the fungus and the cuticle surface, hence, promoting adherence of germinating spores to insect cuticles (Santi et al. 2010).

Some proteolytic enzymes hydrolyze peptide bonds of insect cuticles and play crucial roles in developing virulence of EPF. Subtilisin (Pr1) is a serine endoprotease that modifies the surface of the cuticle by degrading some cuticle proteins, so as to facilitate the adhesion of spores. It is reported to be found in *B. bassiana* (Donatti et al. 2008) and O. sinensis (Zhang et al. 2012). Some other proteases, like cysteine Pr4 proteinase, trypsin-like acid Pr2 protease, and metalloprotease reported from *M. anisopliae*; serine elastase found in *Conidiobolus coronatus* and *B. bassiana*, and Pr1- and Pr2- like serine proteases occurring in *Aschersonia aleyrodids, Metarhizium rileyi,* and *Beaveria brongniartii* also contribute to the process of pathogenesis (Zhao et al. 2016).

The chitinases, another group of lytic enzymes, are grouped on the basis of their site of action on the chitin molecule. Endochitinases are enzymes that hydrolyze the b-1,4-glycosidic bonds present inside the chitin molecules, and exochitinases are the ones hydrolyzing N-acetylglucosamine oligomers derived from the cleavage by endochitinases. The complete degradation of insect chitin requires the combined action of endo- and exochitinases. As reported by Duo-Chuan (2006), chitinolytic enzymes have been uncovered in several EPF.

EPF also produce additional enzymes, along with proteases, chitinases, and lipases that are not promptly involved in the collapse of the cuticle, still contribute significantly in pathogenesis. Acid trehalase (ATM1) is one of them, which hydrolyzes the main disaccharide of host hemolymph, i.e., trehalose and results in the release of two glucose molecules, thereby supplying a nutrient for EPF. As reported by Jin et al. (2015), disruption of the ATM1 gene in *M. acridum* causes a notable lessening in the virulence of EPF, which was confirmed by their failure to grow inside the host's body.

7.4.3 Role of Secondary Metabolites in Infection

Secondary metabolites are low-molecular weight organic compounds secreted abundantly by entomopathogenic fungi in response to environmental conditions. These compounds are crucial for affirming the vital functions of the host's system under stress conditions and efficiently protecting against infecting pathogens by reducing insect resistance (Donzelli and Krasnoff 2016). They can be classified on the basis of their chemical structures into the following groups, viz., cyclic depsipeptides (including cyclic tetradepsipeptides and cyclic hexadepsipeptides), amino acid derivatives, peptides (including dipeptides, octadepsipeptides, and depsipeptides), peptide hybrids, polyketides, and terpenoids (Donzelli and Krasnoff 2016; Wang et al. 2018; Fig. 7.6).

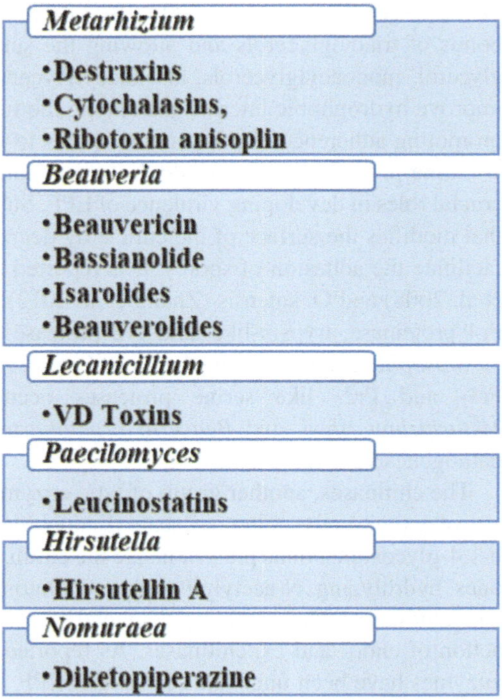

Fig. 7.6 Toxins produced by some Entomopathogenic fungi

Moreover, EPF belonging to deuteromycetes produce diverse fungal toxins that have numerous hostile effects in target insect at tissue level. Few vital toxins synthesized by EPF are summarized as follows:

Destruxins

Destruxins (dtxs), cyclic hexadepsipeptides found mainly associated with species of EPF genus *Metarhizium*. Dtxs are also found abundantly in potatoes (Carpio et al. 2016), strawberries, and maize (Taibon et al. 2015). Out of >40 types of dtxs reported so far, dtxs A, B, and E are of utmost significance for causing pathogenesis (Arroyo-Manzanares et al. 2017). Dtxs possess various biologically significant properties, such as antiproliferative, cytotoxic, antimicrobial, antiviral, and immunosuppressive properties. Dtxs are most prominently known for their phytotoxicity and insecticidal activity. Dtxs function by inhibiting the activity of V-ATPase proton pumps (Liu and Tzeng 2012).

Beauvericins

Beauvericins are commonly produced by EPF genera, such as *Isaria tenuipes*, *I. fumosorosea*, or *B. bassiana* (Weng et al. 2019), and possess various interesting properties. *Bassianolide* and other cyclic hexadepsipeptides are isolated from *Lecanicillum* sp. and *B. bassiana*. They hold robust cytotoxic abilities against tumor cells and were reported to restrict the activity of inhibitors of the cholesterol acyltransferase. Another, secondary metabolite produced by *C. militaris* and *Isaria cicadae* is Cordycepine, which belongs to octadepsipeptides. Several reports have demonstrated the insecticidal potential of cordycepin since a long time agoback, including Kim et al. (2002), affirming the larvicidal activity of cordycepin from the fruiting body of *C. militaris* against *Plutella xylostella*. Further, Rosa et al. (2013) showed insecticidal activity of cordycepin against *Trypanosoma evansi*. VIu et al. (2014) also illustrated the iInsecticidal effects of cordycepine on larvae of the Colorado potato beetle Leptinotarsa decemlineata.

Oosporein

Oosporein is another secondary metabolite, isolated chiefly from B. brongniartii, or *B. bassiana* is a very reactive oktadepsipeptide having an antiviral and insecticidal activity (Feng et al. 2015). Further, McNamara et al. (2019) reported that oosporein obtained from *B. caledonica* contributed toward faster mortality of insects due to its greatest immunosuppressive properties.

Since EPF harbor a pool of highly valuable secondary metabolites with diverse activities, researchers are greatly concerned about exploring such compounds, studying their structures, and investigating their biosynthesis mechanisms in detail.

7.5 Culture of Entomopathogenic Fungi

7.5.1 Maintenance of Culture

Regarding this, a single colony isolation of a fungus from the insect (host) is required. A representative strategy is to first isolate the fungus from the host so as to make primary "mother culture" that is genetically uniform. The subsequent re-culturing of the fungi is to be made from this mother culture. Next, the removal of some part of the prime stock from stored one is to be made on annual or semi-annual basis to develop various subcultures on agar medium that can be applied in mass production for use in the future. Although cautions should be taken, still the preserved fungus cultures must not be sub-cultured several times on artificial media because successive passages through artificial media increase the risk of virulence attenuation (Shah et al. 2007) as well as variations in morphogenesis (Butt et al. 2006). The consistent sub-culturing of a fungus is not recommended as it increases the chances of genetic variations, attenuation in sporulation, or virulence abilities through frequent sub-culturing (Ansari and Butt 2011). Preferably, the primary culture should not be passaged in vitro more than four or five times from an insect host. It must be ensured that the fungus remains viable during storage and at the same time preservation method should be such that it inhibits genetic variations. Storage at low-temperature in liquid nitrogen, or under desiccation conditions, such as storage of dry spores with silica desiccant, freeze-drying is the norm. Further, the sporulated fungus in stock cultures can be preserved by means of small agar pieces positioned in 10% glycerol and kept at −80 °C. In some laboratories, a commercial form of this practice (Microbank, Pro-Lab Diagnostics) is applied. Humber (2012) has described various methodologies for the appropriate preservation of such fungal cultures.

Some key information like site and date of collection and substratum or insect host as well as a notation of a code for every fungal isolate must be recorded and maintained related to fungal stock cultures.

7.5.2 Process Sterility

It is necessary to maintain sterility in order to prevent process contamination. Air, equipment, and the fermentation medium must be purified. Furthermore, it is compulsory to remove all the native microorganisms from raw materials and use apparatus at the beginning of the procedure. Various sterilization techniques used could be like the use of special filters, gamma-radiation, heat, and many more. If proper precautions are not taken, the contaminants could rapidly outstrip the desired strain, and the end product will be inadmissibly contaminated resulting in insignificant production.

7.5.3 Nutrients

Since nutrients are the building blocks, supporting fungal growth. They are key elements that provide co-factors and energy source for biochemical reactions. Depending upon the fungal species and strains under consideration, different concentrations of minerals, vitamins, oxygen, carbon, nitrogen, and hydrogen are required. As reported by Jackson (1997), fungal morphogenesis, specific growth rate, propagule formation, and propagule quality and fitness for application in biological control are all affected by the type and level of nutrient used.

Another key element under consideration is dissolved oxygen. During aerobic fermentation of the filamentous fungal entomopathogens, dissolved oxygen is often considered as the limiting factor. There is a specific requirement of an adequate supply of oxygen for the successful cultivation of EPF. The application of high-speed agitations in oxygen-enriched cultures is one of the means of improving the oxygenation of media and obtaining larger and faster biomass growth of EPF in liquid culture fermentation. Since cultures subjected to oxidative stress run the risk of limited growth and reduction in cell viability and biomass dry weight, therefore, high concentrations of oxygen in the atmosphere ($>21\%$ O_2) or dissolved oxygen in various cultures can be detrimental to fungal growth.

Different genotypes and strains of EPF are not equally receptive to the similar oxygen availability in the growing environment; therefore, specific studies on oxygen rate consumption are compulsory to obtain optimal oxygen requirements by a specific strain (Tlecuitl-Beristain et al. 2010; Garza-López et al. 2012). Recently established, response surface methodology (RSM) is variously used to proficiently determine the unsurpassed parameters.

7.6 Product Formulations

The substantial application of entomopathogenic fungi in diverse arenas make them of utmost importance to mankind. The myco-biocontrol of insects is among one of the best-applied fields of immense significance concerning environmental and food safety. Therefore, their mass production at low cost is among the most desirable targets. Numerous fungal-based products embracing *Metarhizium anisopliae, B. bassiana, Lecanicillium* spp., and *Isaria* spp. have been established for use contrary to a wide variety of pests of household, field, greenhouse, and forests. Currently, a wide range of commercial formulations obtained from such fungi is accessible to farmers in many parts of the world. Widespread research works have been conducted so far for improving fungal mass growth and production as well as to estimate the consequence of changes in additives, substrates, and additional aspects on the viability, virulence, and thermotolerance of spores of entomopathogenic fungal species (Machado et al. 2010; Kassa et al. 2004). Some commercial formulations of EPF used in different countries are given in Table 7.1 (Kaushal et al. 2016).

7.6.1 Mass Production

Miscellaneous approaches have been established for the practical application of EPF to control insect pests. Various approaches including direct application of collected cadavers of infected insects into fields in the same or upcoming season, or use of artificially infected insects in the laboratory and others are being practiced. However, a commonly used method includes the production of the EPF on artificial media. Vegetable materials, like cracked barley, rice or wheat bran, are the most exploited resources and used in the preparation of the medium for mass production schemes. Gouli et al. (2008) have established a comparatively simple millet-based fungal production system that provides nutritional support to growing fungi in the soil without an insect host. Another practice makes use of fungal-based baits to attract the target insects. Here, insects with a prominent feeding stage in the soil are used as targets for bait.

In Russia, large-scale production of conidia is being established by growing the EPF in a fermenter. The mycelium obtained is then placed in shallow pans of approximately 1.0 cm depth, and conidia are produced in the pans after some time. Production of the mycelium in submerged culture is another one. After culture, the fungus' usually filtered from the medium to produce thin mats, approximately 3–6 mm thick, which are treated with a sugar solution (10% maltose or sucrose) as a desiccation protectant. The mat is air-dried at room temperature and then milled to particles of 2 mm size. The application of these particles into the field initiates production of conidia in presence of moisture. The particles of mycelium can produce new conidia for several consecutive days in the field, thereby, providing fresh inocula for a considerable period after the introduction of the fungus. Mass

Table 7.1 Mercantile formulations of entomopathogenic fungal pesticides

S. N.	Fungi	Formulation	Crop	Target pest	Product and company
1	*Aeschersorzia aleyrodis*	Wettable powder	Cucumber, potato	Whitefly	Koppert/ Holland
2	*B. Bassiana*	Suspendible granules	Coffee	Coffee berry borer	Conidia, AgroEvo, Germany, and Columbia
3	*Beauveria bassiana*	Liquid formulation	Cotton and glasshouse crop	Sucking insects	Naturalis™, tray bioscience, USA
4	*B. Bassiana*	Wettable powder	Field crops	Whiteflies/ aphids/Thrips	Mycontrol-WP/ Mycotech Corp, USA
5	*B. Bassiana*	Suspendible granules	Coffee	Coffee berry borer	Conidia, AgroEvo, Germany, and Columbia
6	*B. Bassiana*	Microgranules of mycelium	Maize	Corn borer	Ostrinil/natural plant protection/ France
7	*B. brongniartii*	Barley kernels colonized with fungus	Pasture	Scarab beetle larvae	Engerlingspilz/ Andermatt/ biocontrol
8	*Beauveria brongniartii*	Microgranules of mycelium	Sugarcane	Scarab beetle larvae	Betel/natural plant protection/ France
9	*Metarhizium anisopliae*	Conidia on a mycelium placed in trap/ chamber	Houses	Termites	Bio-path™/ EcoScience/ USA
10	*M. Anisopliae*	Granules of mycelium	Glasshouse ornamental crops, nursery stock houses	Black vine weevil	Biologie Bio 1020/Bayer AG, Germany
11	*M. Anisopliae*	Conidia produced on grains.	Pasture/turf	Locusts, grasshoppers and red-headed cockchafer	Biogreen/ biocare technology Pvt. ltd./Australia
12	*P. Fumosoroseus*	Wettable powder	Wide range of crops	Mites	Priority/T. Stanes, India
13	*Pacilomyces fumosoroseus*	Wettable powder	Glasshouse crops	Whiteflies/ Thrips	PFR-21™/WR grace USA
14	*V. Lecanii*	Wettable powder	Glasshouse crops	Aphids, whiteflies and Thrips	Vertatec/ Koppert/ Netherlands

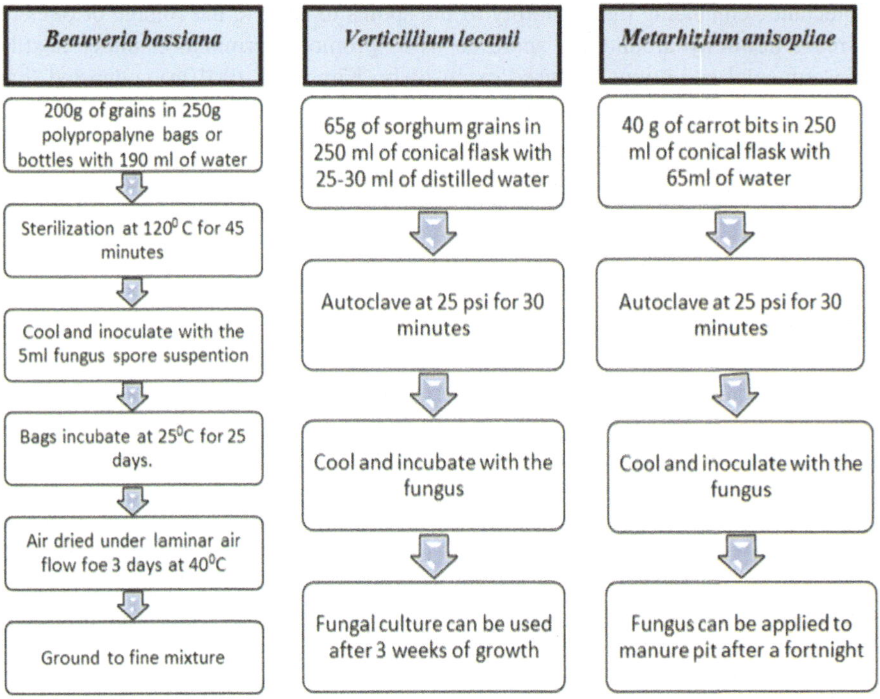

Fig. 7.7 Mass production of few Entomopathogenic fungi

production of some commercially useful entomopathogenic fungi are summarized in Fig. 7.7.

Mass-produced mycelium of EPF obtained after sieving from the medium is applied in making mycelial mats that are dried at room temperature. Mycopesticidal formulation is of major concern because of the brief half-life of the perishable conidia of entomopathogens in sunlight when applied on plant leaves. Another commonly exploited product is in the form of microencapsulation of dried mycelium, pregelatinized with starch. However, the utmost commercially established formulations of EPF are present in the form of wettable powders or emulsified oils.

7.6.2 Wettable Powders

The most commonly produced formulation of entomopathogenic fungi are wettable powder (WP). As reported by Burges (1998), WP is comprised 50–80% technical powder, 15–45% filler, 3–5% surfactant, and 1–10% dispersant. WP is assorted with water and applied to the verdure in the form of a standard insecticidal spray at ultra-low volume or hydraulic concentrations. Even these applications are efficient when applied to the soil as a drench. These formulations are being established using an extensive array of compounds, each with exclusive properties affecting particular aspects to enhance spore survival or efficacy. Additives, such as UV light protectants, enhancing the capability of the spores to stick to the foliage or ones to increase humidity around the spore so as to promote germination under hostile environmental conditions are used extensively. Kim et al. (2010b) suggested that phloxine B (0.005 g/m), a photoactive dye protects from phytotoxicity. Since conidial viability in storage is greatly affected by moisture, makes it an important factor concerning shelf life. Therefore, the probable use of moisture absorbents, such as silica gel, calcium chloride, white carbon, sodium sulfate, or magnesium sulfate have been recommended in 10% WP conidial powder formulations.

7.6.3 Oil Formulations

Various oils are generally added to many fungal products for improving their shelf life and increasing the field efficacy of such formulations in a dry climate. When oil is used as a carrier, it helps to wet the plant leaves and waxy surface of insects. Moore et al. (1995a, b) reported that addition of oil to spore powder improves the survival rate and viability of conidia. Oils also stimulate the germination of spores and simplify the adhesion of spores to the insect, also facilitating the penetration by disrupting the waxy layer of the cuticle. Paraffin oil, mineral oil, and various other isoparaffinic hydrocarbon solvents are mostly used as carriers for oil-based formulations. Kim et al. (2011) have suggested the use of some additional oils, like methyl oleate, vegetable oils, corn, and cotton seed oil for production of more formulations based out of entomopathogenic fungi.

7.7 Patents Granted on Entomopathogenic Fungi Formulations

Various patents on formulations of entomopathogenic fungi have also been granted. Some of them are summarized in Table 7.2.

7.8 Conclusion

The findings clearly unveil the powerful application of EPF as a biocontrol agent, owing to great significance because of the sanctuary amendment and ecofriendly mode of action. Yet, there is a lot of scope for probable manipulations in desirable traits of EPF toward better overall field activity. Moreover, novel research is needed to develop techniques and formulations/nanoformulations with improved and rapid integrated pest management mechanisms. Their possession of enzymes capable of minimizing toxic anthropogenic compounds reveals their prodigious role as environment protectors. Nonetheless, despite their enormous uses in biocontrol processes, they are undervalued due to lack of information on their proficiencies. Therefore, there is need for additional researches and industrial interests concerning the usage of entomopathogenic fungi with maximized efficacy, amended compatibility, and pliant investments.

Table 7.2 Patents granted on entomopathogenic fungi formulations

S. No.	Patent No.	Country	Inventor	Tittle
1.	WO2008087294A3	WIPO patent	Samantha Besse, Antoine Bonhomme	Use of entomopathogenic fungi as a means for the biological control of paysandisia archon
2.	CA2699272C	Canada patent	Mark A. Jackson, Stefan T. Jaronski	Composition of entomopathogenic fungus and method of production and application for insect control
3.	EP0738317A1	European patent	Clifford A. Bradley, James H. Britton	Formulations of entomopathogenic fungi for use as biological insecticides
4.	WO2009035925A2	WIPO patent	Mark A. Jackson, Stefan T. Jaronski	Composition of entomopathogenic fungus and method of production and application for insect control
5.	US20050095259A1	United States patent	Chien Liu	Manufacturing method for entomopathogenic fungi
6.	WO1995010597A1	WIPO patent	Clifford A. Bradley, James H. Britton	Formulations of entomopathogenic fungi for use as biological insecticides
7.	US8501207B2	United States patent	Stamets P	Mycoattractants and mycopesticides
8.	US20130156740A1	United States patent	Leland JE	Biopesticide methods and compositions
9.	US8226938 B1	United States patent	Miekle et al.,	Biocontrol of Varroa mites with Beauveria bassiana
10.	US008227224 B2	United States patents	Kalisz et al	Method of making moulded part comprising mycelium coupled to mechanical device

7.9 Points to Remember

(i) Since last few decades, synthetics for crop protection were used for management of insect pests; however, use of such chemical pesticides is now declining very drastically due to various health and environmental hazards.
(ii) Further, development of pesticides resistance properties of the insects forced to develop an eco-friendly biological way to manage such problem. So far, more than 500 arthropods have shown to develop resistance to such chemicals.

(iii) Plentiful microbial species have been progressively used for successful control of infections, but very few of them are effective and persistent in the market.
(iv) Most commonly categorized EPF fall under divisions viz. Ascomycota, Deuteromycota, and Zygomycota, and they comprise good applicant to be explored for ecofriendly management of pest and diseases associated with agricultural crops.
(v) Diverse extracellular enzymes and various secondary metabolites produced by EPF could be used for the development of mycopesticide, viz., entomopathogenic fungi-*Verticillium lecanii* have been used for their proteolytic, amylolytic as well as lipolytic enzymes.
(vi) Recent expansion of contemporary techniques in the field of biotechnology has significantly improved the efficacy of the entomopathogenic fungal species, using their genetic and biochemical manipulations, but still there are many hindrances that impede the advancements of EPF in the field of mycopesticide that have to be resolved, for developing an alternative to the synthetics.

References

Ansari MA, Butt TM (2011) Effects of successive subculturing on stability, virulence, conidial yield, germination and shelf-life of entomopathogenic fungi. J Appl Microbiol 110:1460–1469

Araújo JPM, Hughes DP (2016) Diversity of Entomopathogenic fungi which groups conquered the insect body? In: Lovett B, Leger RJS (eds) Advances in genetics, vol 94. Elsevier, Amsterdam, pp 1–39. https://doi.org/10.1016/bs.adgen.2016.01.001

Arroyo-Manzanares N, Diana Di Mavungu J, Garrido-Jurado I et al (2017) Analytical strategy for determination of known and unknown destruxins using hybrid quadrupole-Orbitrap high-resolution mass spectrometry. Anal Bioanal Chem 409:3347–3357. https://doi.org/10.1007/s00216-017-0276-z

Asi MR, Bashir MH, Afzal M, Zia K, Akram M (2013) Potential of entomopathogenic fungi for biocontrol of Spodoptera litura Fabricius (Lepidoptera: Noctuidae). J Anim Plant Sci 23 (3):913–918

Balazy S (1993) Entomophthorales, Flora of Poland (Flora Polska), fungi (Mycota). Polish Acad Sci W Szafer Inst Botany, Krakow 24:1–356

Behie SW, Bidochka MJ (2014) Ubiquity of insect-derived nitrogen transfer to plants by endophytic insect-pathogenic fungi: an additional branch of the soil nitrogen cycle. Appl Environ Microbiol 80:1553–1560. https://doi.org/10.1128/AEM.03338-13

Behie SW, Padilla-Guerrero IE, Bidochka MJ (2013) Nutrient transfer to plants by phylogenetically diverse fungi suggests convergent evolutionary strategies in rhizospheric symbionts. Commun Integr Biol 6:e22321. https://doi.org/10.4161/cib.22321

Bilgo E, Lovett B, Leger RJS et al (2018) Native entomopathogenic Metarhizium spp. from Burkina Faso and their virulence against the malaria vector anopheles coluzzii and non-target insects. Parasites Vectors 11:11–16. https://doi.org/10.1186/s13071-018-2796-6

Burges HD (1998) Formulation of mycoinsecticides. In: Burges HD (ed) Formulation of microbial pesticides: beneficial microorganisms, nematodes and seed treatments. Kluwer Academic Publishers, Dordrecht, The Netherlands, pp 132–185

Butt TM, Wang C, Shah FA, Hall R (2006) Degeneration of entomogenous fungi. In: Eilenberg J, Hokkanen (eds) An ecological and societal approach to biological control. Springer, Dordrecht, pp 213–226

Carpio A, Arroyo-Manzanares N, Ríos-Moreno A et al (2016) Development of a QuEChERS-based extraction method for the determination of destruxins in potato plants by UHPLC–MS/MS. Talanta 146:815–822. https://doi.org/10.1016/j.talanta.2015.06.008

Donatti AC, Furlaneto-Maia L, Fungaro MHP, Furlaneto MC (2008) Production and regulation of cuticle-degrading proteases from Beauveria bassiana in the presence of Rhammatocerus schistocercoides cuticle. Curr Microbiol 56:256–260. https://doi.org/10.1007/s00284-007-9071-y

Donzelli BGG, Krasnoff SB (2016) Molecular genetics of secondary chemistry in Metarhizium fungi. In: Lovett B, Leger RJS (eds) Advances in genetics, vol 94. Elsevier, Amsterdam, pp 365–436

Dou F, Wang Z, Li G, Dun B (2019) Microbial transformation of flavonoids by Isaria fumoso rosea ACCC 37814. Molecules 24:1028. https://doi.org/10.3390/molecules24061028

Duo-Chuan L (2006) Review of fungal chitinases. Mycopathologia 161:345–360. https://doi.org/10.1007/s11046-006-0024-y

Fan Y, Liu X, Keyhani NO et al (2017) Regulatory cascade and biological activity of Beauveria bassiana oosporein that limits bacterial growth after host death. Proc Natl Acad Sci U S A 114:E1578–E1586. https://doi.org/10.1073/pnas.1616543114

Feng P, Shang Y, Cen K, Wang C (2015) Fungal biosynthesis of the bibenzoquinone oosporein to evade insect immunity. Proc Natl Acad Sci 112:11365–11370. https://doi.org/10.1073/pnas.1503200112

Garza-López PM, Konigsberg M, Gómez-Quiroz LH, Loera O (2012) Physiological and antioxidant response by Beauveria bassiana Bals (Vuill.) to different oxygen concentrations. World J Ind Microbiol Biotechnol 28:353–359

Greenfield BPJ, Lord AM, Dudley E, Butt TM (2014) Conidia of the insect pathogenic fungus, Metarhizium anisopliae, fail to adhere to mosquito larval cuticle. R Soc Open Sci 1(2):140193. https://doi.org/10.1098/rsos.140193

Gul HT, Saeed S, Khan FZA (2014) Entomopathogenic fungi as effective insect pest management tactic: a review. Appl Sci Bus Econ 1:10–18

Hatai K, Roza D, Nakayama T (2000) Identification of lower fungi isolated from larvae of mangrove crab *Scylla serrata*, in Indonesia. Mycoscience 41(6):565–572

Humber RA (1997) Fungi: identification. In: Lacey LA (ed) Manual of techniques in insect pathology. Academic, London, pp 153–185

Humber RA (2012) Preservation of entomopathogenic fungal cultures. In: Lacey LA (ed) Manual of techniques in insect pathology, 2nd edn. Academic Press, San Diego, pp 317–327

Jaber LR, Enkerli J (2017) Fungal entomopathogens as endophytes: can they promote plant growth? Biocontrol Sci Tech 27:28–41

Jackson MA (1997) Optimizing nutritional conditions for the liquid culture production of effective fungal biological control agents. J Ind Microbiol Biotechnol 19:180–187

Jin K, Peng G, Liu Y, Xia Y (2015) The acid trehalase, ATM1, contributes to the in vivo growth and virulence of the entomopathogenic fungus, Metarhizium acridum. Fungal Genet Biol 77:61–67. https://doi.org/10.1016/j.fgb.2015.03.013

Kassa A, Vidal SD, Zimmermann G (2004) Production and processing of Metarhizium anisopliae var. acridum submerged conidia for locust and grasshopper control. Mycol Res 108:93–100

Kim JS, Je YH, Woo EO, Park JS (2011) Persistence of Isaria fumosorosea (Hypocreales: Cordycipitaceae) SFP-198 conidia in corn oil based suspension. Mycopathologia 171:67–75

Klieber J, Reineke A (2016) The entomopathogen Beauveria bassiana has epiphytic and endophytic activity against the tomato leaf miner Tuta absoluta. J Appl Entomol 140:580–589. https://doi.org/10.1111/jen.12287

Kozłowska E, Dymarska M, Kostrzewa-Susłow E, Janeczko T (2019) Cascade biotransformation of estrogens by Isaria fumosorosea KCh J2. Sci Rep 9:1–8. https://doi.org/10.1038/s41598-019-47225-1

Lacey LA, Grzywacz D, Shapiro-Ilan DI et al (2015) Insect pathogens as biological control agents: back to the future. J Invertebr Pathol 132:1–41. https://doi.org/10.1016/j.jip.2015.07.009

Liu BL, Tzeng YM (2012) Development and applications of destruxins: a review. Biotechnol Adv 30:1242–1254. https://doi.org/10.1016/j.biotechadv.2011.10.006

Mc Namara L, Dolan SK, Walsh JMD et al (2019) Oosporein, an abundant metabolite in Beauveria caledonica, with a feedback induction mechanism and a role in insect virulence. Fungal Biol 123:601–610. https://doi.org/10.1016/j.funbio.2019.01.004

McLaughlin DJ, Hibbett DS, Lutzoni F, Spatafora JW, Vilgalys R (2009) The search for the fungal tree of life. Trends Microbiol 17:488–497

Moonjely SS, Barelli L, Bidochka MJ (2016) Insect pathogenic fungi as endophytes. In: Lovett B, Leger RJS (eds) Advances in genetics. Elsevier, Amsterdam, pp 107–135

Moore O, Bateman RP, Carey M, Prior C (1995a) Long-ter m storage of Metarhizium flavoviride conidia in oil formulation for the control of locusts and grasshoppers. Biocontrol Sci Tech 5:193–199

Moore O, Bateman RP, Carey M, Prior C (1995b) Long-term storage of Metarhizium flavoviride conidia in oil formulation for the control of locusts and grasshoppers. Biocontrol Sci Tech 5:193–199

Ortiz-Urquiza A, Keyhani O (2013) Action on the surface: entomopathogenic fungi versus the insect cuticle. Insects 4:357–374

Pedrini N, Rosana C, Patricia Juarez M (2007) Biochemistry of insect epicuticule degradation by entomopathogenic fungi. Comp Biochem Physiol 146:124–137

Ramakuwela T, Hatting J, Bock C et al (2020) Establishment of Beauveria bassiana as a fungal endophyte in pecan (*Carya illinoinensis*) seedlings and its virulence against pecan insect pests. Biol Control 140:104102. https://doi.org/10.1016/j.biocontrol.2019.104102

Samson RA, Evans HC, Latg JP (1988) Atlas of Entomopathogenic fungi. Springer, Berlin Heidelberg, New York

Santi L, Beys da Silva WO, Berger M et al (2010) Conidial surface proteins of Metarhizium anisopliae: source of activities related with toxic effects, host penetration and pathogenesis. Toxicon 55:874–880. https://doi.org/10.1016/j.toxicon.2009.12.012

Shah FA, Allen N, Wright CJ, Butt TM (2007) Repeated in vitro subculturing alters spore surface properties and virulence of Metarhizium anisopliae. FEMS Microbiol 276:60–66

Silva WOB, Santi L, Berger M et al (2009) Characterization of a spore surface lipase from the biocontrol agent Metarhizium anisopliae. Process Biochem 44:829–834. https://doi.org/10.1016/j.procbio.2009.03.019

Skinner M, Parker BL, Kim JS (2014) Role of entomopathogenic fungi. In: Abrol DP (ed) Integrated pest management. Academic Press, Cambridge, pp 169–191

Steinhaus EA (1964) Microbial disease of insects. In: Debach P (ed) Biological control of insect pest and weeds. Chapman and Hall, London, pp 515–547

Tlecuitl-Beristain S, Viniegra-Gonzalez G, Diaz-Godinez G, Loera O (2010) Medium selection and effect of higher oxygen concentration pulses on Metarhizium anisopliae var. lepidiotum conidial production and quality. Mycopathologia 169:387–394

Vega FE (2018) The use of fungal entomopathogens as endophytes in biological control: a review. Mycologia 110(1):4–30

Wang X, Gong X, Li P et al (2018) Structural diversity and biological activities of cyclic depsipeptides from fungi. Molecules 23:169. https://doi.org/10.3390/molecules23010169

Weng Q, Zhang X, Chen W, Hu Q (2019) Secondary metabolites and the risks of Isaria fumosorosea and Isaria farinosa. Molecules 24(4):E664. https://doi.org/10.3390/molecules24040664

Zhang Y-J, Li E, Wang C, Li Y, Liu X (2012) Ophiocordyceps sinensis, the flagship fungus of China: terminology, life strategy and ecology. Mycology 3:2–10. https://doi.org/10.1080/21501203.2011.654354

Zhao H, Lovett B, Fang W (2016) Genetically engineering entomopathogenic fungi. In: Lovett B, Leger RJT (eds) Advances in genetics, vol 94. Elsevier, Amsterdam, pp 137–163

Entomopathogenic Protozoa

8

Anita Yadav, Neeshma Jaiswal, Shivji Malviya, and Sandeep K. Malhotra

Contents

A. Yadav
Department of Zoology, C.M.P.P.G. College, A Constituent College of the University of Allahabad, Prayagraj, Uttar Pradesh, India
e-mail: anitaydv997@gmail.com

N. Jaiswal (✉)
Department of Zoology, Babasaheb Bhimrao Ambedkar University (A Central University), Lucknow, Uttar Pradesh, India
e-mail: neeshversity@gmail.com

S. Malviya
Department of Zoology, Govt. H.N.B.(P.G.) College, Naini, Prayagraj, Uttar Pradesh, India

U.P. Higher Education Service Commission, Prayagraj, Uttar Pradesh, India

S. K. Malhotra
Parasitology Laboratory, Department of Zoology, University of Allahabad, Prayagraj, Uttar Pradesh, India
e-mail: sandimtec@gmail.com

Omkar (ed.), *Microbial Approaches for Insect Pest Management*,
https://doi.org/10.1007/978-981-16-3595-3_8

Summary

Entomopathogenic infections by protozoans predominantly, the microsporidians, are of paramount significance in being ineradicable that caused subtle pathologies to induce reproductive productivity decline and shortening of life spans. These have been considered incredibly fit as long-term regulators of pests for effective checks of disease outbreaks. Protozoans are an important constituent of biopesticides with a narrow target range, hence used for a specific problem; their slow-action characteristics rejuvenate suppression of pests with limited field persistence, and devoid of residues. The detection and diagnostic potential for the length of survival of spores in the habitat has been facilitated by the application of advanced biochemical techniques. The critical inputs regulate mutualism after invasion of microsporidia into intracellular environment of the host-cell and influence protein synthesis, growth and endomitosis during merogony. A determined impact of lessening noxious pest hosts with the assistance of non-chemical application techniques in the field attributed commercial significance to the protists (microsporidia) and a variety of other protozoa. Vertical transmission comes into effect once the microsporidians invade gonads of insect pests, followed by the contamination of eggs. Therefore, not only that spores adhere to eggshell surface, and egg contents comprise spores and developmental stages yet the larvae hatched from internally contaminated eggs are often not infected. Microsporidia with reduced mitochondria, that is mitosomes, are *sans* mitochondrial genome, and the energy for their functioning is extracted from the host's intracellular environment. Mitosomes, devoid of cristae, might be dependent on ATP produced by glycolytic pathway because oxidative phosphorylation was dysfunctional. Pest suppression characteristics of biocontrol agent are a factor dependent on host's ecology.

Keywords

Entomopathogenic · Protozoa · Microsporidia · Biocontrol · Molecular · Strategy · Transovarial transmission · Environmental

8.1 Introduction

The protozoans belonging to certain phyla are pathogenic to insects. Some of these insects are important pests of various agricultural crops. Thus, these protozoans may act as biocontrol agents and play a significant role to regulate the population of insect pests. The protozoans, unicellular eukaryotes, particularly under the Sub Kingdom Protozoa, viz., neogregarines and microsporidia, are relatively more important in this context. The potential of protozoans to invade the body of their hosts in huge numbers, and kill them by inducing dysfunction of organs or depleting the host of essential reserves, has earned these organisms a unique position in their own way. The refinement of the concept of management of pests incorporating conservation efforts in coherence with refurbishing enhanced populations of the existing natural control agents has been of primary interest in the application of protozoans in the biocontrol of insects. The issues of application of protozoan organisms as important elements of economic entomology have been frequently reviewed (Brooks 1974, 1979; Mclaughlin 1971, 1973; Tanada 1976). Initially it was believed that the actions by protozoans were neither rapid nor useful for a broader range of effectivity. The slower-paced performance of the pathogenic protozoan forms deprived them of their penetrative capabilities to cause mortality into higher abundance of host pest on their own. Their effective use is, therefore, implied in high threshold pest populations to cause damage or else when they are required as a component of pathogen-plus-chemical formulation or in a multiple pathogen combination to act upon pests. In addition, in *Lambornella clarki* Corliss and Coats, an endoparasitic ciliate of the tree-hole mosquito, *Aedes sierrensis* (Ludlow), exhibited remarkable potential as a biocontrol agent to keep effective check on it, to prevent outbreak of dangerous diseases. The armament of cuticular cysts is deployed by the ciliates of the genus *Lambornella* that penetrate to cause damage by attaching to the cuticle of culicine mosquitoes, particularly tree-hole breeding species, for a remarkable mosquito control. Clark and Brandl (1976) gave an illustrated account of the mechanism of dissolution of the cuticle of larval mosquitoes to facilitate penetration into epidermis for its arrival in haemocoele. Two mosquito-associated species have been reported, viz., *Lambornella stegomyiae* Keilin and *L. clarki* Corliss & Coats. Egeter et al. (1986) and Washburn and Anderson (1986) also emphasized its effective role as a potential biocontrol agent of the container breeding mosquitoes.

The typical cyst forming ability of this endoparasitic ciliate comes in handy for it to survive over the years to pass over unfavourable conditions. The property of resistance to the cystic framework is the typical characteristic of such protozoans that equipped these pathogenic organisms with appreciable resistance against desiccation. This unique ability of the cysts to withstand adversities during the stages of development enabled the organism encased within, to overwinter or cross over to the next weather circle stage of the following year. Taking a cue out of such apparent characteristics of the cysts of this endoparasite, Anderson et al. (1986a, b) advocated field trials and perused in vitro methods to determine the efficacy and feasibility of their use in field experiments. The association of a variety of protozoans, particularly microsporidians, with mosquitoes and other arthropods of medicinal significance, is known (Lacey and Undeen 1986). The exposition of their practical utility has been restricted because of their complex life cycle, as well as in vivo technology to be employed for maintenance. The application of novel innoculative technique to ensure the enhanced productivity oriented release highlighted its functional significance in the dynamics of mosquitoes, under specific habitat conditions. The report of Flint and Dreistadt (1998) compiled microorganisms, particularly entomophilic protozoans as pathogens in coordination with a variety of bacteria, fungi, nematodes and viruses that possessed the natural potential to infect and kill the hosts if the congregated pest populations flourished under conditions of high humidity. Certain protozoans, viz. the microsporidia, *Nosema pyrausta* (Fig. 8.5) and *Vairimorpha necatrix* (Fig. 8.6), are the typical examples that demonstrated regulation of lepidopteran pest populations. The abundance of pest populations would be monitored the moderate to low pathogenicity because the inverse relationship between their pathogenicity and prevalence was obvious. While *V. necatrix* operated through the insect gut wall invasion as well as disruption due to the bacterial septicaemia. Though this ensured survival and propagation of larval populations, yet adulthood was never attained. Thus, this microsporidian could rarely register its occurrence in natural populations as it was deprived of the transovarial transmission. Under the circumstances, therefore, its potential as an important constituent of being a microbial pesticide, on a short-term control operation, has been realized. On the other hand, the natural companion in the lepidopteran pests, *N. pyrausta* (Paillot 1927) that parasitized a single host, that is, the European corn borer, with acquired lower pathogenicity in the stressful environment, succeeded to kill larvae. Resultantly, therefore, the adulthood was achieved by most survivors but whose power to reproduce was adversely affected and length of life curtailed. This microsporidian was encountered with higher prevalence in natural precincts, equipped with the virtues of being transmitted transovarially. In such ways though not being an organism that showed severe pathogenicity, it still has been attributed a significant constituent of the mechanism regulating natural pest populations.

8.2 Protozoan Assessment as Biocontrol Agent

The microsporidia (Fig. 8.1) of various protozoan organisms encountered in a variety of insects in terrestrial as well as aquatic ecosystems have been summarized in Table 8.1. The occurrence of microsporidia in the insect pests has been listed by insect host taxonomic order in Table 8.2. The insect orderwise enumeration of microsporidian fauna is presented in Table 8.3. The reported species of insects frequently occurring in the insect pests in the forest areas have been compiled in Table 8.4. The common as well as specific names of insect pests, that occurred in forests and microsporidia harboured by them are listed in Table 8.5. The precise details of spore wall proteins of microsporidia and their subcellular location are summarily presented in Table 8.6.

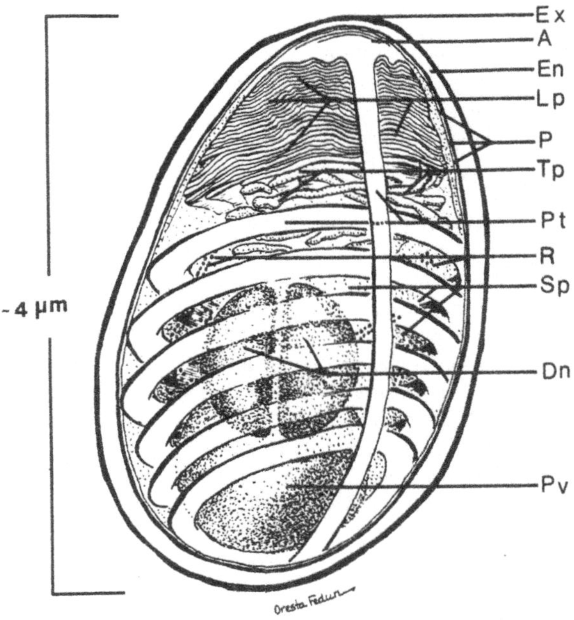

Fig. 8.1 Internal structure (Diagrammatic) of a microsporidian spore (Courtesy: Cali and Takvorian, 2014) (Reprinted from Han et al. 2020) (A, Anchoring disc; Dn, Diplokaryon; En, Endospore; Ex, Exospore; Lp, Lamellar Ploroplast; P, Unit membrane; Pt, Polar tube; Pv, Posterior vacuole; R, Ribosome; Sp, Sporoplasm; Tp, Tubular Ploroplast)

8.2.1 Phylum Ciliophora

Though not recognized as obligate endoparasitic forms, a few strains, like *T. pyriformis* of genus *Tetrahymena* spp. that otherwise are free-living forms, possessed the traits of parasitism. Corliss and Coats (1976) enlisted parasites of genus *Tetrahymena* Furgason and *Lambornella* Keilin of the haemocoele of culicine mosquitoes. The mortality-induced infectivity in the haemocoele of mosquito hosts by both *Tetrahymena* and *Lambornella* (Table 8.1) was taken into account, although the exact data on the rates of mortality were evading.

The evidence of culturability of *Tetrahymena* (Table 8.1) and *Lambornella* being available, the findings of Clark and Brandl (1976) to conclude prevalence of *L. clarki* in *Aedes sierrensis* to the level of 50% strengthened the contrary view of low control potential of ciliophorans. This was because of lower infectivity or abstained natural epizootics of ciliates underlined by McLaughlin (1971) and Henry (1981) that could respond to the likelihood of decimation of infected lot of individuals in the population due to severe pathogenicity. WHO released the document WHO/VBC/81.803 briefing the details of results to give evidence.

8.2.2 Phylum Sarcomastigophora

The evidence is in favour of the spread of these parasitic protozoa due to the faecal ingestion, which is not necessarily associated with the formation of cysts. But the

Table 8.1 Insect pest hosts and protozoan organisms harboured by them

Sl. No.	Protozoan organism (s)	Insect pest(s)	Site of infection	Author(s)[a]
1.	Gregarine, *Dirofilaria immitis*	*A. aegypti* and other culicine mosquitoes	Malpighian capsules	Sneller (1979)
2.	*Mattesia grandis* McLaughlin	Cotton boll weevil, *Anthonomis grandis*	Gonads	McLaughlin (1971)
3.	*Mattesia trogodermae* (Canning)	Stored grain pest, *Trogoderma glabrum*	Pheromones as tools	Shapas et al. (1977)
4.	*Adelina tribolii* and *Farino cystistribolii* (in formulations)	Flour beetles, *Tribolium* spp.	Pheromones as tools	Shapas et al. (1977)
5.	*Lambornella clarki* Corliss and coats, an endoparasitic ciliate	Tree hole mosquito, *Aedes sierrensis* (Ludlow)	Encysted stages	Wallace (1979)
6.	*Lambornella*	Culicine mosquitoes	The cuticle of culicine mosquitoes	Wallace (1979)
7.	Microsporidia of *Vairimor*	Lepidopteran host populations	–	–
8.	*Tetrahymena* Furgason and *Lambornella* Keilin	Culicine mosquitoes	Haemocoel	Corliss and Coats (1976)
9.	*L. clarki*	*Aedes sierrensis*	–	Clark and Brandl (1976)
10.	*Nosema pyrausta* (Paillot)	European corn borers, *Ostrinia nubilalis*	Malpighian tubules	Introduced into the United States of America in 1917
11.	Cyst forming protozoans of the genus *Lambornella*, free-swimming ciliates	The mosquito hosts through culture invasion into the aptly suited	Encysted stages	Wallace (1979)
12.	*N. Pyrausta*	In larvae of *O. Nubilalis*	–	–
13.	*Nosema sp.*	Borer population	–	–
14.	*N. Pyrausta*	*O. nubilalis*	–	Kramer (1959a, 1959b)
15.	Invasive spores of microsporidian organisms	In larvae, pupae as well as adult moths	In the yolk accompanying embryonated oocytes in the ovary	Zimmack and Brindley (1957)

(continued)

Table 8.1 (continued)

Sl. No.	Protozoan organism (s)	Insect pest(s)	Site of infection	Author(s)[a]
16.	Larval infestation of the dimorphic Diptera *Amblyospora* and *Thelohania*	–	–	Hazard and Oldacre (1976)
17.	*N. heliothidis* (Lutz and Splendore)	Corn earworms, *Heliothiszea*	Larval tissues	Gaugler and Brooks (1975)
18.	*N. fumiferanae* (Thomson)	Spruce budworm, *Choristoneura fumiferana*	In the actively feeding larvae post-emergence	Thomson (1958)
19.	3 species of microsporidia	Winter moth, *Operophtera brumata* (L.) (Canning & Barker, 1981)	Newly laid, non-embryonated eggs (Canning & Barker 1981; On the surface of egg shells & inside egg contents (Kramer 1959a; Brooks 1968)	Canning (1982)
20.	*V. Necatrix*	Seasonal pests; young larvae	Gut, malphighian tubules & salivary or silk glands	–
21.	The ciliate, *Lambornella*	Mosquitoes	Cuticle	Wallace (1979)
22.	Trypanosomatid flagellates	Insects	Contaminated faecal material	Wallace (1979)
23.	A large variety of protozoans	Insects		Fine (1981)
24.	*Vairimorpha necatrix* (Kramer); *N. pyrausta* and *V. necatrix* (Kramer)	Army worms, *Mythimna Mythimna (Pseudaletia) unipuncta* (Haworth)	a microbial insecticide	Tanada and Chang (1962); Tanada (1964)
25.	*Vairimorpha necatrix* (Kramer)	Forest insects; several major agricultural pests	A phytophagous Lepidoptera	Maddox et al. (1981); Maddox et al. (1981)

[a]References are species descriptions or research reports. (Reprinted from publications by various authors)

possibility of their being pathogenic is a rarer one. Therefore, in a review published by Wallace (1979), a large number of flagellates belonging to the group of kinetoplastic flagellates, commonly being harboured by a variety of invertebrates, including arthropods, were not considered to be biocontrol organisms.

Table 8.2 Genera of microsporidian type species isolated from insects; listed by insect host taxonomic order

Insect order	Microsporidian genus	Insect order	Microsporidian genus
Collembola	Auraspora	Diptera	Tricornia
Diptera	Aedispora		Tubilinosema
	Amblyospora		Vavraia
	Andreanna		Weiseria
	Anisofilariata		Pegmatheca
	Bohuslva		Pernicivesicula
	Campanulospora		Pilosporella
	Caudospora		Polydispyrenia
	Chapmanium		Ringueletium
	Coccospora		Scipionospora
	Crepidulospora		Semenovaia
	Crispospora		Senoma
	Cristulospora		Simuliospora
	Culicospora		Spherospora
	Culicosporella		Spiroglugea
	Cylindrospora		Striatospora
	Dimeiospora		Systenostrema
	Edhazardia		Tabanispora
	Evlachovaia		Toxoglugea
	Flabelliforma		Toxospora
	Golbergia		Trichoctosporea
	Hazardia	Coleoptera	Anncaliia
	Helmichia		Cannngia
	Hessea		Chytridiopsis
	Hirsutusporos		Endoreticulatus
	Hyalinocysta		Ovavesicula
	Intrapredatorus	Ephemeroptera	Mitoplistophora
	Janacekia		Pankovaia
	Krishtalia		Stempellia
	Merocinta		Telomyxa
	Napamichum		Trichoduboscqia
	Neoperezia	Hemiptera	Becnelia
	Octosporea	Hymenoptera	Antonospora
	Octotetraspora		Burenella
	Parapleistophora		Kneallhazia
	Parastempellia	Isoptera	Duboscquia
	Parathelohania		

Table 8.3 The total number of identified microsporidian genera distributed along various orders of insects

Sl. No.	Insect order	Microsporidian genera	Insect order	Microsporidian genera totals
1.		*Cystosporogenes*	Collembola	1
2.	*Archips cerasivoranus*	*Larssoniella*	Diptera	57
3.	*Choristoneura con flictana*	*Nosema*	Coleoptera	5
4.	*Choristoneura fumiferana*	*Orthosomella*	Ephemeroptera	5
		Vairimorpha	Hemiptera	1
		Nudispora	Hymeoptera	3
		Resiomeria	Isoptera	1
5.	*Dendroctonus species*	*Heterovesicula*	Lepidoptera	5
		Johenrea	Odonata	2
		Liebermannia	Orthoptera	4
6.	*Euproctis chrysorrhoea*	*Paranosema*	Syphonaptera	2
		Nolleria	Thysanura	1
		Pulicispora	Trichoptera	3
7.	*Hyphantria cunea*	*Buxtehudea*		
		Episeptum	Total	90
		Issia		
8.	*Ips* spp.	*Tardivesicula*		

[a]References are species descriptions or research reports. (Reprinted from Solter, L.F., Becnely, J.J. & Oiy, D.H., 2012. Microsporidian Entomopathogens, DOI: https://doi.org/10.1016/B978-0-12-384,984-7.00007-5)

8.2.3 Phylum Apicomplexa

At least three modes were employed to explore the role of gregarines (sub-class Gregarinia) and some adeleine coccidia (sub-class Coccidia, sub-order Adeleina) of the Apicomplexa in biological control of insects.

Laboratory Experiments

A gregarine, *D. immitis* exhibited the potential to damage Malpighian tubules of *A. aegypti* (Table 8.1) and other culicine mosquitoes in the laboratory experiments conducted by Sneller (1979).

Field Trials

Sneller (1979) also reported parasitization by *D. immitis* and damage in process, of the Malphigian tubules, by this gregarine. Similarly, the neogregarines and adeleine coccidia, the common intracellular parasitic forms sustained one or more cycles of

Table 8.4 Partial list of microsporidia infecting forest insect pests

Sl. No.	Host species	Microsporidian species	References[a]
1.	*Agrilus anxius*	*Cystosporogenes* sp.	Kyei-Poku et al. (2011)
2.	*Archips cerasivoranus*	*Endoreticulatus (Pleistophora) schubergi*	Wilson and Burke (1978)
3.	*Choristoneura conflictana*	*Nosema thomsoni*	Wilson and Burke (1971)
4.	*Choristoneura fumiferana*	*Cystosporogenes* sp. *(legeri?)*	Van Frankenhuyzen et al. (2004)
		Endoreticulatus (Pleistophora) schubergi	Wilson (1975)
		Nosema fumiferanae	Thompson (1955)
		Thelohania sp.	Wilson (1975)
5.	*Dendroctonus species*	*Nosema dendroctoni*	Weiser (1970)
		Chytidiopsis typographi	Knell and Allen (1978)
		Unikaryon minutum	
6.	*Euproctis chrysorrhoea*	*Nosema chrysorrhoeae*	Hylis et al. (2006)
		Nosema kovacevici	Purrini and Weiser (1975)
		Endoreticulatus sp.	Purrini (1975)
7.	*Hyphantria cunea*	*Endoreticulatus schubergi hyphantriae*	Weiser (1971)
		Nosema sp. (*bombycis*-type)	
		Vairimorpha sp.	–
8.	*Ips* spp.	*Chytidiopsis typographi*	Purrini and Weiser (1985)
		Larsoniella duplicate	Weiser et al. (2006)
		Nosema typographi	Weiser et al. (1997)
		Unikaryon montanum	Wegensteiner et al. (1996)
9.	*Lymantria dispar*	*Endoreticulatus schubergi*	McManus and Solter (2003)
		Nosema lymantriae	
		Nosema serbica	
		Vairimorpha disparis	
10.	*Malacosoma americanum*	*Nosema* sp. (*bombycis*-type)	Weiser and Veber (1975)
	Malacosoma disstria	*Nosema disstriae*	Thomson (1959)
	Operophtera brumata	*Cystosporogenes operophterae*	Canning et al. (1983)
11.	*Pristiphora erichsoni*	*Thelohania pristiphorae*	Smirnoff (1967)
	Tmicus piniperda	*Caningia tomici*	Kohlmayr et al. (2003)
	Tortrix viridana	*Osema tortricis*	Franz and Huger (1971)

[a]References are species descriptions or research reports. (Reprinted from Solter, L.F., Becnely, J.J. & Oiy, D.H., 2012. Microsporidian Entomopathogens, DOI: https://doi.org/10.1016/B978-0-12-384,984-7.00007-5)

Table 8.5 Partial list of forest insects from which microsporidia have been reported

Sl. No.	Host species		References[a]
	Common name	Specific name	
1.	Eastern tent caterpillar	*Malacosoma americanum*	–
2.	Fall webworm	*Hyphantria cunea*	Weiser and Veber (1975) Nordin and Maddox (1974)
3.	Forest tent caterpillar	*Malacosoma disstria*	Thomson (1959)
4.	Green tortrix	*Tortrix viridana*	Lipa (1976), Franz and Huger (1971)
5.	Larch sawfly	*Pristiphora erichsoni*	Smirnoff (1966), Quednau (1968)
6.	Large aspen tortrix	*Choristoneura conflictana*	Wilson and Burke (1971)
7.	Spruce bedworm	*Choristoneura fumiferana*	Thomson (1958)
8.	Uglynest caterpillar	*Archips cerasivoranus*	Wilson and Burke (1978)
9.	Winter moth	*Operophtera brumata*	Canning et al. (1983)

[a]References are species descriptions or research reports. (Reprinted from (Reprinted from Maddox et al. 1998. In: McManus, M.L. & Liebhold, A.M. (Eds.)0.1998. Proceedings: Population Dynamics, Impacts, & Integrated Management of Forest Defoliating Insects. USDA Forest Service General Technical Report NE-247, p.187–197)

schizogony, to produce them in enormous numbers. The destructive abilities of the latter and excessive consumption of reserves in the host body led to the greater number of death of host bodies, with decline in reproductive ability in the colonies of insects maintained in the laboratory. McLaughlin (1971) highlighted success of propagation of *Mattesia grandis* McLaughlin (Table 8.1), a parasite of cotton boll weevil, *Anthonomus grandis* as a biocontrol agent of insect pest, in field trials. Their application was successful on their engagement as supplement in the role of feed stimulant.

Stored Grain Pests

The use of pheromones in the stored grain pest, *Trogoderma glabrum*, which lured insects as a tool (Shapas et al. 1977), has partially resolved the problem of outreach of infective spores of *Mattesia trogodermae* (Canning). Several other protozoan mixed formulations have also been applied to utilize *Adelina tribolii* and *Farino cystistribolii* against flour beetles, *Tribolium* spp.

Table 8.6 The identified spore wall proteins of microsporidia (Reprinted from Han et al. 2020)

	Protein	Subcellular location	Function domain	Mw (kDa)	Amino acids/ GenBank ID	References (as in)
Encephalitozoon cuniculi	EcSWP1	Exospore	–	45.9	450aa ECU10_1660	Bohne et al. (2000)
	EcEnP1	Endospore	HBM	40.6	357a ECU10_0820	Peuvel-Fanget et al. (2006)
	EcEnP2/ EcSWP3	Endospore	Transmembrane	22.5	221 aaECUO1_1270	Peuvel-Fanget et al. (2006), Xu et al. (2006)
	EcCDA	Endospore and plasma membrane	Glycoside hydrolase and deacetylase	28.1	254aa ECU11_0510	Brosson et al. (2005)
Encephalitozoon intestinalis	EiSWP1	Exospore	–	41.5	388aa AF355750.1	Hayman et al. (2001)
	EiSWP2	Exospore	–	107.2	1002aa AF3557 49.1a	Hayman et al. (2001)
	EiEnP1	Exospore and endospore and polar membrane layer	HBM	39.1	348aa EF539266	Southern et al. (2007)
Encephalitozoon hellem	EhSWP1a EhSWP1b	Exospore Exospore	–	54.9 57.9	509aa FJ8770923 533aa FJ8770924	Pollonais et al. (2010)
Nosema bombycis	NbSWP5	Exospore and endospore	–	20.3	186aa EF683105	Lie et al. (2012)
	NbSWP7	Exospore and polar tube	–	32.8	287aa EOB13707.1	Yang et al. (2015)
	NbSWP9	Exospore, endospore and polar tube	Transmembrane helix region (TMHMM)	42.8	367aa EOB13793.1	Yang et al. (2014)
	NbSWP11	Exospore and endospore	DnaJ domain	52.3	446aa EF683111	Yang et al. (2014)

	NbSWP12	Exospore and endospore	BAR-2 domain	44.0	221aa EOFB683112	Chen et al. (2013)
	NbSWP16	Exospore	HBM	44	221aa EOB14338	Wang et al. (2015)
	NbSWP25	Endospore	HBM	30.7	268aa EF683102	Wu et al. (2009)
	NbSWP26	Exospore, endospore and plasma membrane	HBM	25.7	223aa EU677842	Li et al. (2009)
	NbSWP30	Endospore	–	32.1	278aa EU683101	Wu et al. (2008)
	NbSWP32	Exospore	–	37.4	316aa	Wu et al. (2008)
	EOB14572	Endospore and pollen tube	Four tandem repeats	37.0	316aa NBO_24g0018	Wang et al. (2015)
Enterocytozoon hepatopenaei	EHSWP1	Endospore and exospore	HBM, Bar-2	27.0	228aa MGO15710	Jarcentak et al. (2018)
Antonospora	AlocSWP2	Endospore and exospore	GPI, HBM	25	222aa	Chen et al. (2013)

8.2.4 Phylum Microspora

Microsporidia are among the largest group of successful biocontrol organisms under Protozoa. The widely spread out resistant spores initiate infections into pest populations. The spores hatch after ingestion, and penetration (Fig. 8.2) of its polar filament into a gut epithelial cell occurred, pushing through the peritrophic membrane barrier. However, the cytoplasm of the host cell received inoculation of sporoplasm in cases where polar filament was longer enough to have reached up to the cytoplasm of the host cell after passing through the gut wall and penetrating haemocytes directly (Malone 1990). The spores were resultantly produced by sporogony that was preceded by frequently occurring binary or multiple fissions, after initial invasion by the tube of polar filament. It is also known that the fatality, as well as the decline in reproductive potential of the host, and their life span dawned upon because the tissues of the host were replaced by the invasive infective spores in heavier numbers (Fig. 8.3).

It has been emphasized that the negativity features of greater time lag between initiation of infection by a microsporidia and the resultant tissue destruction, which might extend to several weeks, affected the potential of these infective agents that are detrimental to pest, to facilitate their usefulness in vector control. However, the

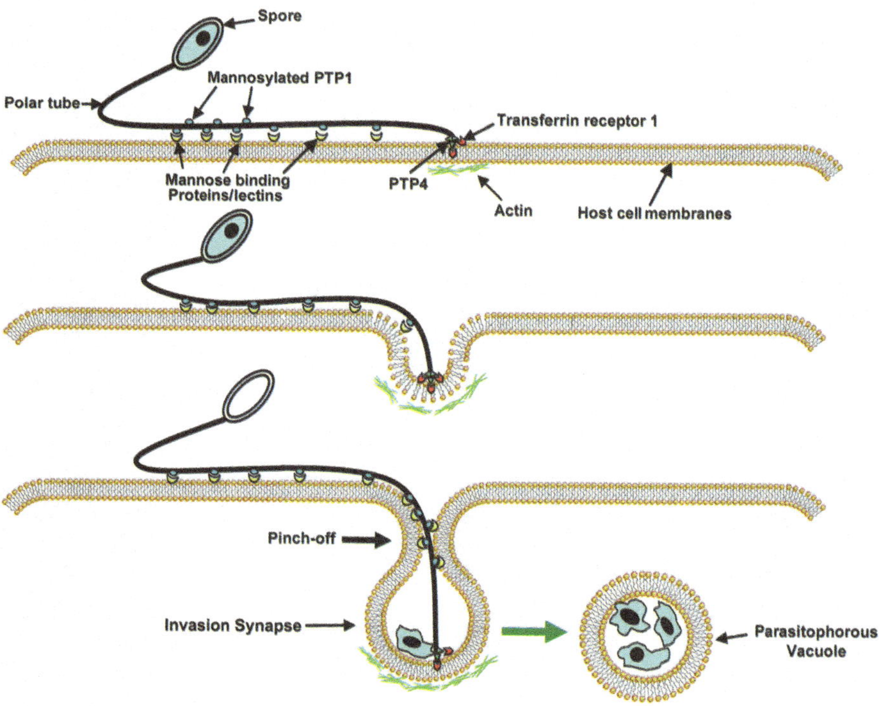

Fig. 8.2 Invasion by microsporidia into a cell of the host (a model) (Reprinted from Han et al. 2020)

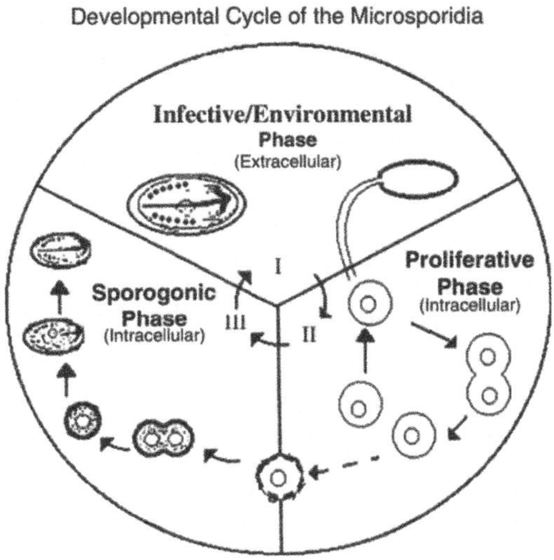

Fig. 8.3 A typical microsporidial developmental cycle (Fig. 8.3) can be divided into three phases. Phase I, the infective/environmental phase, is the only extracellular part of the cycle. It is represented by mature spores shed into the environment from previously infected hosts. Under appropriate conditions, the spores germinate (e.g., if the spores are ingested by an appropriate host). They are activated by the digestive tract environment; this results in the explosive expulsion of the polar filament (which 'everts' becoming a hollow tube). If the polar tube pierces a host cell, the spore contents, the sporoplasm, are injected into it and phase II begins. Phase II is the proliferative phase, the first phase of intracellular development. During this part of the microsporidian life cycle, organisms are usually in direct contact with the host cell cytoplasm or in a parasitophorous vacuole as they increase in number. The transition to phase III, the sporogonic phase, represents the organisms' commitment to spore formation. In many life cycles, this is morphologically indicated by parasite secretions through the plasmalemma producing a 'thickened' membrane (many also form a surrounding sporophorous vesicle, SPOV). The number of cell divisions that follow varies, depending on the genus in question, and results in sporoblast cells that develop into spores (Reprinted from Cali et al. 1999)

characteristic of transovarial transmission negates this ineptitude to overcome by attaining of the retention of the potential of spread full to capacity of the infective microsporidia.

The activated transferability, sustainability of infective stages, and severe ferocity of invasive penetrative capability equip a biocontrol agent to achieve success. A long-term strategic management scheme encompassed the manipulation of several alternative methods. These included field trials to test the methods to ensure availability of bulk volumes of microsporidia and neogregarines, incorporating tests for ensuring their safety during the process of production, as well as testing of bait formulations to suit appropriate mass production. This has resulted into the registration of a microsporidia as a microbial pesticide with the United States Environmental Protection Agency.

A lot of efforts have been put in to explore novel methods of introduction, keeping an eye on the evaluation of their pathogenicity and the potential of their wider spread. While, on the other hand, the groups like ciliates, flagellates, amoebae and the less pathogenic Apicomplexa have been under focus to investigate their prevalence during random surveys as well as the descriptions of their newer taxa, if any.

The concerns have been expressed to verify the viability of cyst formation activity under experimental conditions (Fig. 8.4), though the cyst forming protozoans of the genus *Lambornella*, the free-swimming ciliates were amply encountered in natural conditions. This is yet to be tested whether *Lambornella* organisms could be cultured in the laboratory, and promoted for invasion into the aptly suited mosquito hosts, under natural habitats, by forming cuticular cysts. Therefore, most of the studies illustrating appropriate protozoan organisms for biocontrol of insects depended heavily on microsporidians and neogregarines.

8.3 Pathogenicity

The manifestation of pathogenic effects of the cysts (Fig. 8.4) harboured by pests was illustrated by the declining pest population growth emanated as a result of reduced fecundity as well as their longevity. This was implied as the long-term control measure. However, the short-term measure to restrict pest population growth, that attained critical peaks, emerged to be the mortality of larvae that were to be employed to rescue and ensure relief measures against adult mosquitoes.

The efforts have also been made to reveal mystery of short-term and long-term population control of pests that comprised proven record of *N. pyrausta* (Paillot) (Tables 8.3, 8.4) and *V. necatrix* (Kramer) (Tables 8.3, 8.4) as the most suited candidates for being agents of biocontrol.

8.3.1 *N. Pyrausta* (Paillot) (Fig. 8.5)

The European corn borers, *Ostrinia nubilalis*, introduced into the United States of America in 1917, along with *N. pyrausta* (Paillot) (Fig. 8.5), parasitizing their Malpighian tubules, comprised one of the most extensively studied moth host-microsporidian systems. The European microsporidian introduced sporadically in 1951 into the USA (Steinhaus 1951) was the subject of investigation (Zimmack et al. 1954) conducted on its distribution in 1954 in seven of the North Central states. The studies dealing with relative fecundity status of female moths revealed striking differences between moths that showed marked differences in their infectivity status. Fewer eggs were laid by the infected moths, whose longevity was short and the non-infected moths, who lived longer, laid numerous eggs (Zimmack and Brindley 1957). Simultaneously, faster-paced growth was hampered in the former, coupled with dim survival prospects.

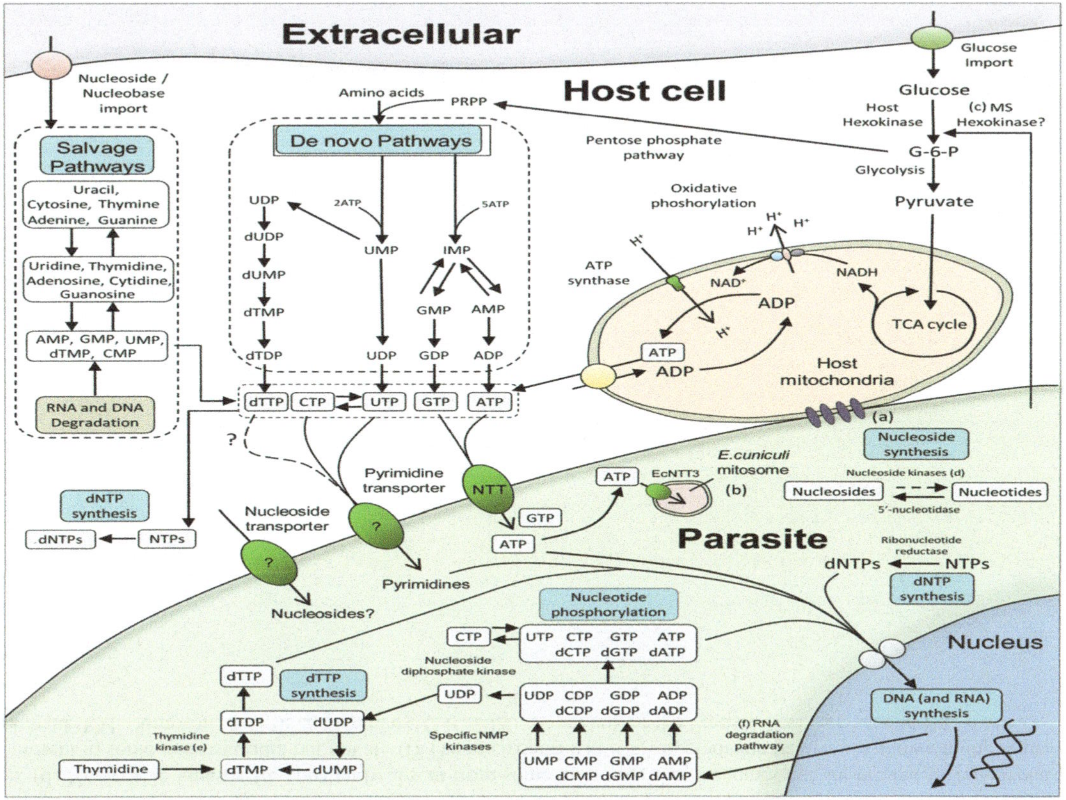

Fig. 8.4 Nucleotide acquisition and metabolism in host cells and microsporidia
Schematic illustration showing nucleotide metabolism in a typical microsporidian parasite within an infected host cell. Host cells can make nucleotides via de novo biosynthesis and regenerate ATP by oxidative phosphorylation – these pathways are absent in microsporidia. Host purine nucleotides can be stolen using

Fig. 8.4 (continued) microsporidia NTT transporters and then efficiently used and recycled by the parasites. Key: (**a**) *E. cuniculi* physically tethers mitochondria using an unidentified protein. (**b**) Only EcNTT3 has been found in the mitosome. (**c**) Nematocida may secrete a hexokinase into the host cell to stimulate host nucleotide production. (**d**) Nucleoside kinases are apparently absent from some microsporidian genomes but are present in *Trachypleistophora hominis*. (**e**) Thymidine kinase is present in some microsporidia but not all. (**f**) The microsporidian RNA degradation pathway is shown in the figure below.
(Reprinted from: Archibald, J.M., Simpson, A.G.B. & Slamovits, C.H. 2020. Handbook of the Protists. Springer. second Ed., p.1657)

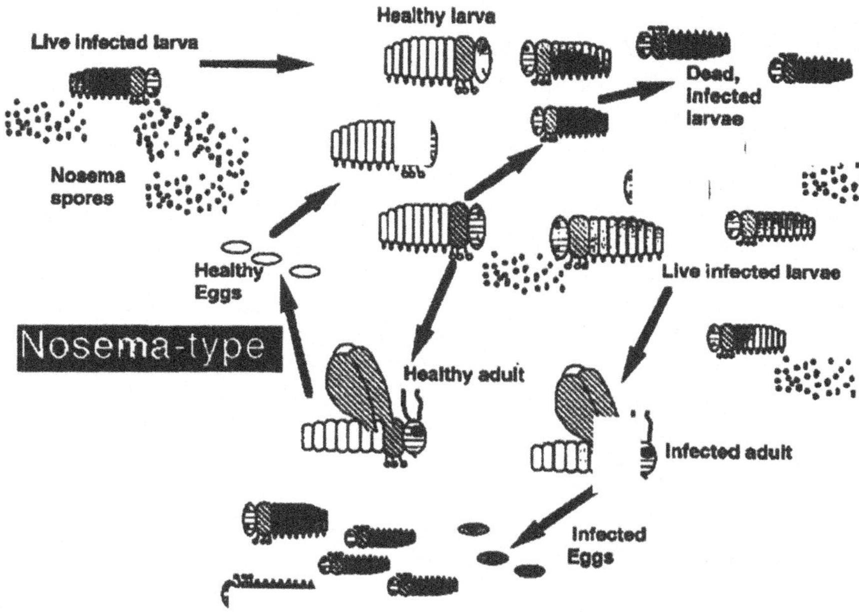

Fig. 8.5 Diagrammatic representation of the interactions between *Nosema*-type microsporidia and their hosts. Healthy larvae become infected by ingesting microsporidian spores which are present in the faeces and/or silk of infected individuals. Larvae infected by ingesting spores may die from the infection if they consume many spores at an early larval stage, but many infected larvae can develop into infected adults. Mortality may occur during pupation and emergence as adults. Much of the mortality caused by *Nosema*-type microsporidia (Tables 8.3, 8.4) occurs in the transovarially infected offspring in infected females. Transovarially infected larvae may be heavily infected and die in early larval stadia
(Reprinted from Maddox et al. 1998. Microsporidia affecting forest Lepidoptera. (In: (Eds.) Maddox, J.V., McManus, M.L. & Solter, L.F. Proceedings: Population Dynamics, Impacts, & Integrated Management of Forest Defoliating Insects.). USDA Forest Service General Technical Report NE-247.).

The evidence of invasive spores of microsporidian organisms was traced by Zimmack and Brindley (1957) in the yolk accompanying embryonated oocytes in the ovary. This was compiled in histological analysis. The infected female moths laid embryonated eggs with spores and the shallow foveae on the egg shell were found embedded with spores by Kramer (1959a). Several other reports have illustrated noticeable prevalence of infection in larvae, pupae as well as adult insects.

The concept of stress-induced effect of fatality to the insect pests infected with *N. pyrausta* was formalized by Kramer (1959b). The stress of winter was under investigation. The natural populations of *O. nubilalis* with reduced longevity and fecundity, under influence of the infection by *N. pyrausta*, were thus regulated, as reported by Van Denburgh and Burbutis (1962). The natural populations of *O. nubilalis* infected by the larvae of *N. pyrausta* in two counties of Nebraska between 1957 and 1972 were correlated with the density of parasites by Hill and

Gary (1979). The two epizootic waves of *Nosema* (Tables 8.4, 8.5, 8.6) occurred at the onset of crash in borer populations that co-occurred with peak in parasite prevalence. The event could thus be correlated with the introduction of mortality on augmentation of prevalence by *N. pyrausta* in larvae of *O. nubilalis*.

Hill and Gary (1979) provided evidence of the eventuality that the parasitic forms did not act directly to the detriment of pest populations, as those due to a combination of environmental factors that regulated fecundity. The intricacies of regulatory dynamics put forth by Anderson and May (1981) could explain that most of the features of the population cycles concluded could actually conform to the host/parasite association model, derived earlier by Anderson and May (1980).

On the basis of the review of long-term observations, the effect of environmental factors emerged to be the significant operator that induced a supportive role to the dynamics of infections by *N. pyrausta*, which on its own was unable to cause significant damage to the pest, but, in combination with environmental attributes, was much more destructive.

8.3.2 *V. Necatrix* (Kramer) (Fig. 8.6)

The utility of *V. necatrix* (Kramer) as a stronger pathogen than *N. pyrausta* and *V. necatrix* (Kramer) is established, and its application as a microbial insecticide has been illustrated effectively (Tanada and Chang 1962). Its initial recovery from the army worms, *Mythimna (Pseudaletia) unipuncta* (Haworth) in Hawaii from more than half of the dead larvae, was ensured from a decimated pest population. The alarming signals emanated from it due to its uninhibited substantial prevalence the following year, at the same site again, while the population density of the army worms declined (Tanada 1964). Though its prevalence in the routine field crops is minimal, its outbreak in certain forest insects has been substantial (Maddox et al. 1981). A noticeable occurrence of *V. necatrix* has been encountered in several major agricultural pests, particularly a variety of phytophagous Lepidoptera (Maddox et al. 1981).

As the ruptured gut due to polar filament extrusion to reach at haemocoele triggered bacterial septicaemia drawing source bacteria from the midgut lumen, cent percent mortality level was attained following infection. The domain of this disease is with a larval base, in which the sites of infections are primarily the fat bodies. Adult insects evade infections because of the non-survival of larvae under the influence of disease as well as non-occurrence of transovarial transmission. This disease primarily being fatal to larvae than adult insects, it has been recognized as a larvicide for quickened depletion of the pest population.

The entry of protozoan organisms into the body of their hosts is most of the time through ingestion of their cysts. It is very rare that any other route is employed by protozoans to attack the susceptible pest hosts; for instance, the penetrative cysts of the ciliate *Lambornella* invade the mosquito's cuticle. It is though intriguing that the life cycle of the flagellate, *Trypanosoma* sp., commonly parasitizing insects (Wallace 1979), does not comprise true cysts, but the transmittance of infection

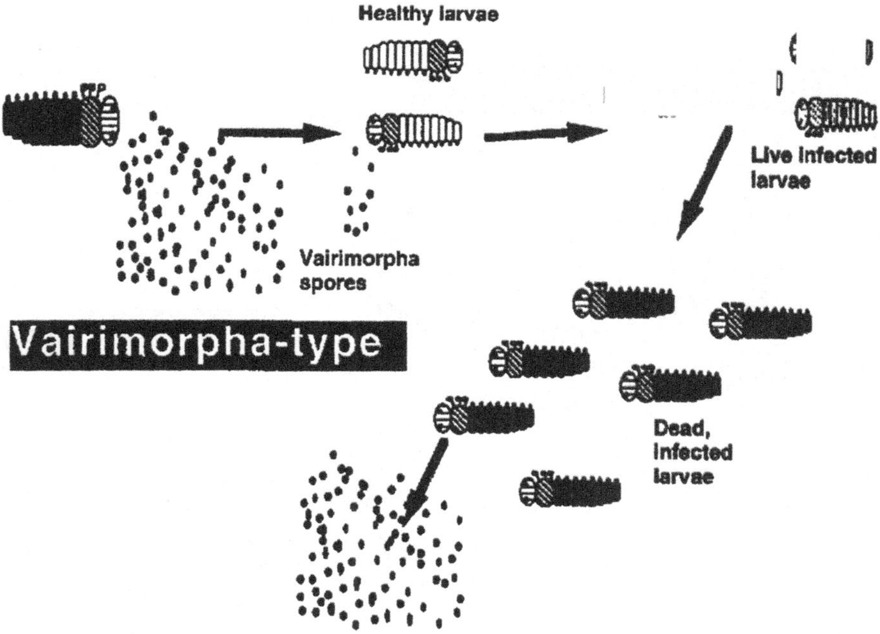

Fig. 8.6 Diagrammatic representation of the interactions between Vairimorpha-type microsporidia and their hosts. Healthy larvae become infected by ingesting microsporidian spores which are released when infected larvae die and disintegrate. Spores are seldom present in the faeces or silk of infected larvae. Most Vairimorpha-type microsporidia are very pathogenic; individuals infected during the larval stage seldom develop into infected adults. These infected individuals usually die as larvae
(Reprinted from Maddox et al. 1998. Microsporidia affecting forest Lepidoptera. (In: (Eds.) Maddox, J.V., McManus, M.L. & Solter, L.F. Proceedings: Population Dynamics, Impacts, & Integrated Management of Forest Defoliating Insects.). USDA Forest Service General Technical Report NE-247.).

was suggested to have occurred by ingestion of contaminated faecal material. This infection is never passed on from mother to offspring, but it is transmitted from individual to individual.

8.3.3 *Endoreticulatus Schubergi* (Zwolfer) (Fig. 8.7)

The protists (microsporidia) in *Endoreticulatus* group (Table 8.2) were entirely distinguished from the other two, namely *N. pyrausta* and *V. necatrix*, described as above. Restricted to midgut epithelial sites, their phylogenetic interrelationships segregated these from others in the other two groups, while simultaneously their life cycles and decimated pathogenicity were also the differentiating features. They were capable of producing only a uninucleate environmental spore, enveloped in a cover, that encompassed 16, 32 or more spores. The most commonly encountered

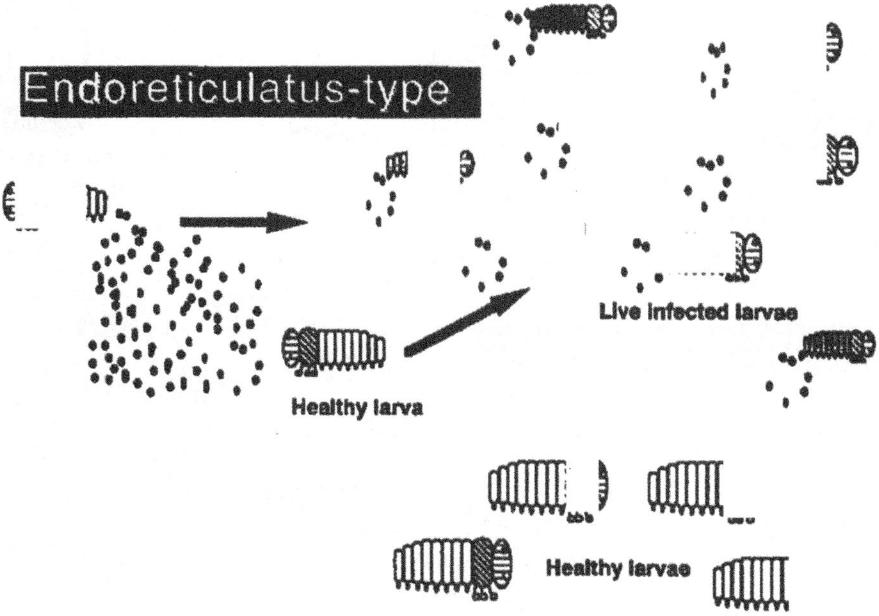

Fig. 8.7 Diagrammatic representation of the interactions between Endoreticulatus – type microsporidia and their hosts. Healthy larvae became infected by ingesting microsporidian spores which are present in the faeces of infected individuals. The mortality rate is very low in individuals infected with Endoreticulatus-type microsporidia. Infected larvae develop more slowly than healthy larvae and produce faeces contaminated with spores throughout their larval development (Reprinted from Maddox et al. 1998. In: (Eds.) McManus, M.L. & Liebhold, A.M. Proceedings: Population Dynamics, Impacts, & Integrated Management of Forest Defoliating Insects. USDA Forest Service General Technical Report NE-247, p.187–197.)

Endoreticulatus schubergi (Fig. 8.7) never adopted the transovarial pathway to reach at the site of infection in larval populations of forest Lepidoptera. It, however, had the potential to produce chronic infections. Though their vertical transmission on the surface of eggs existed (transovum transmission) to a minimal extent, this pathway was even a subdued one when compared to the actual vertical transmission within the environment of the egg (transovarial transmission). The mechanism of transmission of *Endoreticulatus schubergi* and associated protists are illustrated in Fig. 8.7, identifying the stages at which mortality occurred. A large variety of Lepidopteran pest hosts harboured *Endoreticulatus* spp. quite frequently, but the impact was felt to the maximum extent of a 'sublethal' type only, because of the chronic nature of infection by this species (Maddox et al. 1998).

8.4 Transmission

Protozoa normally enter their hosts when encysted stages are ingested and rarely by any other route (e.g. the use of invasion cysts of the ciliate *Lambornella* (Table 8.3) to penetrate the cuticle of mosquitoes). True cysts are unknown in some trypanosomatid flagellates of insects (Wallace 1979) but even in these cases the mode of infection is probably by ingestion of contaminated faecal material. Transmission is defined as horizontal when the pathogen is passed from individual to individual within and between generations but not directly from parent to offspring, whatever the mode of entry. When a parasite infects the gonads of its host, vertical transmission from parent to progeny is possible. This may be transovum, when the infective stages contaminate the external surface of the egg and are consumed by the larvae at eclosion, or transovarial, when the infection is located within the eggshell. Occasionally the infected parent is the male. The different explored routes summarizing transmission pathways of a large variety of protozoans in the insects were reviewed earlier by Fine (1975). Two specific modes of transmission, (i) Horizontal alone and the other (ii) Horizontal as well as Vertical, were concluded. Though the horizontal transfer of cysts has been a natural component of the life cycle's pathway, the transfer of protozoan organism or their cysts from parents to the offspring is likely facilitated only by the invasion by a parasite or its developing entities into the reproductive bodies, like gonads, or eggs. This could be termed transovarial transmission.

The uniformly adopted strategy by a majority of protozoans to effectively propagate individual to individual transmission essentially required maintenance of parasites throughout the year, including the potential of their reproductive entities, particularly cystic bodies to survive during the unfavourable environment. In such a manner, these bodies enabled overwintering of the developing stages, for example, cysts, diapausing larvae, pupae or non-feeding adults, so as to ensure continuation of the progeny.

8.4.1 Horizontal Alone

Most Protozoa are maintained this way in invertebrate host populations. There is no strong evidence that any amoebae, flagellates, ciliates or Apicomplexa are transmitted by the vertical route. Restriction of transmission to the horizontal route requires that the parasites are maintained in the living hosts throughout the year or that they can survive in the environment through periods when the host is quiescent and non-feeding, that is, as eggs, diapausing larvae, pupae or non-feeding adults. In seasonal pests, it is likely that the initial generation of the new season will have a low prevalence of disease because infected hosts may have a poor over-wintering survival rate, and the number of available infective stages may be low. The latter could have resulted due to loss of viability with time and their ultimate removal from the environment. Infections will multiply in young larvae, releasing infective stages at death to spread through the new generation. Spread will be enhanced if spores are released continuously from live hosts, as would occur with infections of the gut, Malpighian tubules and salivary or silk glands. In multivoltine host, prevalences would increase with each generation. The characteristics of *V. necatrix* are noticeably remarkable to have a rapid spread within a generation. But its virulence (effecting rapid removal of the parasitized sector from the population), coupled with the absence of a vertical route of transmission and low survival rate of spores in the environment, is sufficient to account for the low prevalences and rare sightings of this parasite in nature.

8.4.2 Horizontal Alone As Well as Vertical & Horizontal Routes

The other alternative mode of transmission could be that of vertical transmission in a way that the invasion of microsporidians into the contents of the eggs could result into hatching of larvae from the contaminated egg material (Kramer 1959a; Brooks 1968). The winter moth, *Operophtera brumata* (L.) (Canning and Lom 1986), population in an oakwood habitat recorded augmented prevalences of microsporidia. The spread of as many as three species of microsporidia was noticed (Canning 1982). The possibility of occupation of microsporidia on the surface of egg shells as well as inside the contents of the eggs (Kramer 1959b; Brooks 1968) opened up interesting avenues. The rising proportion of incidence of spores and developmental stages of the three species of microsporidia harboured by the winter moth, *O. brumata* (L.), were encountered in the newly laid, non-embryonated eggs examined by Canning and Lom (1986).

Prior to eclosion, the non-occurrence of parasites either in the larval tissues contained in eggs, or in larvae post 1 day of hatching necessitated the location of site of survival of spores during embryogenesis. The spores were shelved in polar plug, meconium, that represented remnants of yolk entering into the lumen of the gut, during late embryogenesis that facilitated merger of stomodaeum with proctodaeum to form midgut. This not only extended disinfecting protection to larvae but also ensured transovarial transfer for vertical transmission. Sections of larvae within the eggs showed that spores were present only in a plug of material, occupying the lumen of the gut. This plug, the meconium, represents the remnants of the yolk taken into the gut lumen at the time when the anterior and posterior parts of the gut, stomodaeum and proctodaeum, join up via the midgut late in embryogenesis. This suggested itself as a remarkable adaptation to ensure that the larvae did not succumb to overwhelming infections while still within the egg. If the larvae were to die, the purpose for which transovarial transmissions has evolved would be defeated. Therefore, it remained to be ascertained whether the infections in newly hatched larvae are initiated in the epithelial cells of gut due to sporoplasm inoculation after hatching of spores, or whether these spores are passed in the meconium and are eaten by, or accidentally contaminate, the larvae. Thomson (1958) confirmed the occurrence of a similar phenomenon for *N. fumiferanae* (Thomson)-spruce budworm, *Choristoneura fumiferana* (Table 8.3) system. *N. fumiferanae* were fatal only in the actively feeding larvae post-hatching.

Gaugler and Brooks (1975) emphasized that the failure of heavier egg infections of *N. heliothidis* (Lutz & Splendore) to reduce egg hatch in corn earworms, *Heliothis zea*, and account for the normal, embryonic development and lack of disease symptoms in newly hatched larvae apparently occurred due to the failure of the egg stages to penetrate larval tissues during embryogenesis. Therefore, the mystery of the mechanism of attained infections by eggs in larvae in the microsporidian—host systems remained to be cracked. The sex-wise differentiation in the larval infestation of the dimorphic diptera, namely *Amblyospora* (Table 8.2) and *Thelohania*, revealed heavier infections in males than females (Hazard and Oldacre 1976).

8.5 Protozoans as Biocontrol Agents to Restrict Grasshoppers and Locusts

The density of the pests, attributes of the environment, as well as the conditions of plants are, in fact, instrumental to attract the wrath of the grasshoppers that might be recognized as chronic pests (Watts et al. 1982). Though due to behavioural advantage to the grassland songbirds, utilizing feeding on grasshoppers resulted into extensive breeding of the former (Martin et al. 1997), yet colossal loss was caused to the crops and vegetation, due to chewing and clipping by even lower densities of pests that consumed the precious crops of farmers and poor people. Although the protozoans were considered to be the slow-acting control agents, which affected rapid breeding and feeding speed of acridid organisms, yet their potential for action fell short of comparably rapidly acting fungi and certain chemical insecticides. Bidochka and Khachatourians (1991), Brooks (1988), Mclaughlin (1971), Henry (1990), Maddox (1987), and Raina et al. (1987) put forth extensive reviews to succinctly account for the protozoan potential as biocontrol agents of insect pests. A number of earlier workers (King and Taylor 1936) have been assertive about the enfeebling effect of a variety of protozoa on grasshoppers. Microsporidia have spearheaded the targeted detriment of acridid grasshoppers, but besides Phylum *Microspora*, a great variety of Amoebida, and Eugregarinida, Phylum *Apicomplexa* of Neogregarinida as well as Phylum Ciliophora have succeeded as a biocontrol agent. The adverse influence of *Malameha locustae* (King and Taylor 1936) harboured by *Locusta pardalina* (Walker) resulted into effect on fecundity of these locusts. The proof of their ferocious attacks on the colonies too, of several susceptible species (Pickford and Randell 1969), maintained in the laboratory was obtained by Jackson et al. (1968). The common occurrence of the pathogen on the colonies of *Schistocerca gregaria* (Forskil) including the nondiapause strain of *Melanoplus sanguinipes* was noticed. Interestingly, simultaneous enhanced resistance against insecticides in the cyst-harboured Malpighian tubules (Hinks and Ewen 1986) co-occurred in certain extreme cases, whose account of life history and pathogenesis has been recorded by Braun et al. (1988). Several Orthoptera commonly harboured gregarines in nature as well as in certain poorly maintained colonies of *S. gregaria* (Harry 1970). Certain well-known protozoans that could succeed in making field crickets physiologically dysfunctional as well as brought disturbances in the ecology of field crickets (Zuk 1987) were brought on record. The taxonomic status of Gregarinida, placed under Apicomplexa, was reviewed by Levine et al. (1980). These were reportedly encountered in the gastro-intestinal tract of more than 30 species of grasshoppers. Though not thoroughly analyzed, gregarines existed in the relationship of symbionts with grasshoppers but their role as potential pathogens is being actively investigated world over (Johnson et al. 1997). However, serious efforts are being made to employ microsporidia as candidates for the potential control of grasshoppers and locusts (Henry and Oma 1981). Based on the characteristics of rapid transmission, fecundity and ease of formulation, *N. locustae* are the favoured candidates for use in the biocontrol of locusts and grasshoppers. A wide spectrum of *Nosema* spp. were encountered in the field populations of *Locusta migratoria migratorioides* (Reiche and Fairnaire) at the Imperial College, London, by Canning (1953, 1962b). Northern Great Plains and Prairies of North America were the native sites of *N. locustae*. Its spores were also frequently harboured by *M. bivittatus*, *M. mericanzs Saus.* (*M. sanguinipes*) and *M. dawsoni* (Scudder). This has commonly been recognized as a common inhabitant of Acrididae with more than 80 species having frequently harboured spores of *N. locustae* (Henry 1969; Henry 1985; van der Paauw et al. 1990; Bomar et al. 1993). The life cycle of *N. locustae* illustrated that unless its infective spores have arrived in the lumen of the gut through ingestion of *N. locustae*, they shall not germinate to release binucleate sporoplasm into the live cells of host pests by rapid extrusion of polar tubes. Therefore, through this mechanism the contact of sporoplasms with the epithelial cells of the midgut is successfully established or the sporoplasms reach directly into the gut cells (Raina and Ewen 1979). The invasion of the yellow organ, the principal fat body organ performing the function of storage of energy, by the sporoplasm enabled generation of developmental entities, like meronts, sporonts, sporoblasts and spores in succession (Canning 1962a). A starved energy reserve of the pest due to microsporidial infection and the ensuing hypertrophy, thereafter, disrupted the dynamicity of intermediate metabolism in the fat cells of the principal fat body, and weakened the body of the pest to induce debilitating conditions. In certain other cases, the virulent pathogeny of microsporidians, greater than *N. locustae*, induced mortality occurred in acridids. As for example, the pathogenicity by *N. cuneatum* Henry enhanced susceptibility to protozoan infections with effective hindrance to development resulting into frequent mortality at an even greater rate. The experiment enumerated by Mussgnug (1980) that the effective outcome on control of pathogenicity by application of *N. locustae* alone was at a relatively lower level than when pathogens combined in the formulations of Malathion- *N. locustae* were applied.

8.6 Forest and Tree Crop Pests

The typical example of vertical transmission of microsporidia of *N. fumiferanae* brought to the fore the primary pathway of disease transmission through consumption of its spores by the spruce bud worm, *Choristoneura fumiferana* (Table 8.3), in the forest and tree crops. Virtually, the fourth and fifth instars received a spore inoculum from the frass and cadavers that were generated after the death of larvae, which were naturally infected from parenteral infections, but the latter could not attain maturity. On the other hand, the larvae that did not receive infections from their parents, and instead their fourth and fifth instar stages received the spore inoculum from the frass and cadavers, attained maturity as well as harboured infections. Thus microsporidia are maintained in the pest hosts by vertical transmission, once the consumption of contaminated spores triggers the augmented incidence of infection.

The spores of *Nosema* were illustrated to be germinated in gut and live cells of gut, fat body and other tissues of Lepidoptera, beetles, locusts and a variety of Orthoptera. They multiply to trigger sporulation that culminated with destruction of organs resulting into being chronic and debilitating. The protozoans were transported by rotten food stuff, cannibalistic behaviour of the organisms through the transovarial pathway (Capinera and Hibbard 1987). The living insects happen to be the fertile ground for the spores of protozoans that sustain with appreciably reduced virulence. It was demonstrated that specifically *Nosema locustae* has had a restricted role in the biocontrol of grasshoppers. Certain formulations of other protozoans, like *Nosema* combined with *Bacillus thuringiensis*, were utilized as dry baits sprinkled over the surface of the soil against other pests, but their availability was restricted commercially.

8.7 Observations on the Sporoplasm

The transport of sporoplasm into the cells of the host pests preceded by the microsporidia infection of host cells involved expedited ejection of the polar tube (Fig. 8.2) (Weidner 1970; Frixione et al. 1992; Takvorian et al. 2005; Han et al. 2017). Weidner (1972) and Han et al. (2017) elucidated that the event of the onset of the infection triggered sticking of microsporidian spores to the cells of the host pests or the surrounding tissues. The indomitable association of the polar tube during infection of the host cells by microsporidia to target spore germination was propagated by Cali et al. (2017). It was also asserted by Takvorian et al. (2013) and Vavra and Larsson (2014) that the transfer of infection is possible by the outflow of infectious sporoplasm, to reach polar tube after its passage from spores. The mature spore formation occurred after initiation of a reproduction cycle, that terminates into meronts' emergence, once the sporoplasm invaded into the host cell (Cali et al. 2017; Han and Weiss 2017), and the formation of proliferative forms, meronts was preceded by the production of constituents of the cycle, like sporonts and sporoblasts. The microsporidia not only were devoid of functional mitochondria, but also did not comprise the full complement of genes capable of ATP generation, yet could contribute to the completion of glycolysis. The protein coding gene complement of the microsporidial genome was restricted to ~3000.

Three of the total four nucleotide transport proteins (NTT1–4), identified to date from certain microsporidia namely, *Encephalitozoon cuniculi* and *Trachipleistophora hominis*, have been known to be the constituents of the membrane of sporoplasm. It has been illustrated in the literature dealing with microsporidia evolution, enumerating the findings of Tsaousis et al. (2008), Heinz et al. (2014) and Dean et al. (2018) that the horizontal gene transfer was instrumental in the displacement of these nucleotide transport proteins from bacteria to microsporidia. Resultantly, these nucleotide transport proteins illustrated their potential to transport ATP, GTP, NAD+ and purine nucleotides from the cytoplasm of the host. The mechanism to transport energy as well as nucleotides from the cells of the hosts is also undertaken by microsporidia major facilitator superfamily (MFS), which is an alternative to the nucleotide transport proteins (NTT1–4), being a sporoplasm surface located protein family. These were ThMFS1–4, of which the location of ThMFS1 and ThMFS3 was traced to be in the sporoplasm plasma membrane during infection. These were also reported (Major et al. 2019) to be performing the similar function of transportation of ATP, GTP and purine, identical to those by NTTs. But it is remarkable that as yet the pyrimidine nucleotide import system of NTTs nor ThMFS remained undeciphered (Heinz et al. 2014; Dean et al. 2018; Major et al. 2019).

Microsporidia are supposed to possess appreciably reduced mitochondria, called mitosome, *sans* mitochondrial genome as well as ATP generation capacity (Williams et al. 2002; Goldberg et al. 2008). Typically, mitosomes constitute an essential constituent of Microsporidia, Diplomonads, Amoebozoa and Apicomplexa (Tovar et al. 1999, 2003; Williams et al. 2002; Keithly et al. 2005) that are a double membrane bound entity. The role of nucleotide transport proteins to ensure the import of nuclear encoded proteins for the functional activity of mitosomes, as well as the maintenance of organelles, to compensate the absence of cristae as well as their own DNA in the morphologically reduced sized mitosomes than mitochondria is well recognized (Burri et al. 2006; Hans-Peter Braun 2009; Tachezy 2019). The energy produced through glycolytic pathway becomes available for use by microsporidian mitosomes because of the process of oxidative phosphorylation to produce ATP being dysfunctional. However, the former process of energy generation operated only in spores, but became inactivated within the cytosol of pests, while intracellular growth and replication were initiated (Dolgikh et al. 2011; Heinz et al. 2012; and Williams et al. 2014). Han et al. (2019) demonstrated aggregation of mitochondria around microsporidia of the cells of their host pests, for energy supply, as confirmed in their experimentation on Encephalitozoonidae harboured by a parasitophorous vacuole of the cells of host pests.

8.8 Molecular Characterization of the Organisms of Biocontrol

The validity of the names of species of microsporidians was conducted in recent years by using chromosomal DNA comparisons (Munderloh et al. 1990). The control basis for vectors of *Nosema* spp. has been analyzed to highlight their molecular characteristics (Strett & Henry 1985). The *N. Locustae* spore structure profiling for polypeptide configuration was done to record pre-eminent polypeptides of exospores and spores post-buffer treatment comprising 4% sodium dodecyl sulphate and 0.02 M dithiothreitol (Heckmann 2020). An electron optically oriented recent approach to characterize a variety of biological control agents was introduced by Avery and Anthony (1983) to crack the mysteries of the molecular basis for biocontrol. In addition, the cell culture revelation could be helpful to elucidate the elements of biology of the candidates for microbial biocontrol. The biology of microsporidians in the in vitro cultures comprising continuous cell lines was demonstrated by Raina et al. (1987). The successful elements required to conduct such studies were aseptic spores in adequate number; applicable stimulus for germination; convenient accessibility to cell lines, along with a sensitive collaborators to conduct research (Kurtti et al. 1990; Heckmann 2020). The apparent proximity between the two microsporidians, *N. payrausta* and *N. furnacalis*, naturally parasitizing the European corn borer, *Ostrinia nubilalis*, and the Asian corn borer, *O. furnacalis*, respectively, was test cultured on experimentally reared caterpillars of *O. nubilalis* and centrifugally purified. This experimental illustration reassuringly proved the closeness of the two strains of *N. payrausta* and *N. furnacalis*, with even greater strength than it was previously demonstrated (Munderloh et al. 1990). Since the intricacies of distribution and interactions of microsporidia in the intracellular environment have long remained an enigma, certain advanced methods of detection, for example, ELISA (Enzyme-linked Immunosorbent Assay) of these spores, have been applied. These have helped to assess potential length of survival of spores in the habitat, in addition to their detection in the microhabitat. The facts that have emerged from the interactions of microsporidia of *N. fumiferana* infesting *Choristoneura fumiferana* (Table 8.3) revealed influence on the nutritional physiology of the hosts, which rendered them nutritionally deficient. This provided comparative data on nutritional efficiencies that revealed noticeable decline in Consumptive Index (CI), Nitrogen Consumptive Index (NCI), Relative Growth Rate (RGR) and gross (ECI) and net (ECD) production efficiencies in the infected than non-infected larvae. The approximate digestibility (AD), N utilization efficiency (NUE) and larval moisture content are taken into account to assess that of the healthy and diseased insects, reared on 2.5%N and 4.5% N diets, the enhanced CI in the former with deficient NCI than those reared on the latter diet, were the significant findings. However, no effect of dietary N was seen on the mortality of healthy insects. The outcome thus indicated sharing of products of host cell metabolism for parasitic metabolism and productivity requirements. A presumptive improvement in the growth and survival of host as well as biocontrol agent at the augmented levels of nitrogen was concluded (Heckmann 2020).

8.9 Molecular Mechanism in Pathogeny

The aspects of physiology, as well as biochemistry of microsporidians, have been scarcely worked upon, and information on the operative Kreb's cycle in these has been wanting. Isotopic labelling and analysis of intensities of radiolabelled bands have come in handy to crack the mysteries of metabolism in the microsporidia. The cell fractionation achieved by segregating molecular entities based on differentiating isoelectric points, electrophoretic analysis of the enzymes encased within microsporidia, HPLC of the routine precursors or labelled metabolites of biologically significant molecules, functional localization of the specific enzymes in the cellular environment by chemical applications could reveal successfully the details of metabolic pathways within microsporidia (Heckmann 2020). It was presumptively postulated that the pre-emptive regulation of cyclic events within the host cell provided alternative energy pathway for microsporidia, as mitochondria and endoplasmic reticulum aggregated intracellularly, in the vicinity of the former. The greater variety of species diversity and their origin could be deciphered by the knowledge on aspects of evolution of form within the built environment of microsporidian organism, but the information on these was conclusively unavailable (Heckmann 2020). The microsporidia has essentially to survive within the host cell itself after invasion into it, especially following the principle of mutualism by triggering critical inputs during host–cell protein synthesis, growth and endomitosis during merogony. Soon thereafter, at the onset of sporogony, the microsporidia attained the characteristics of a pathogen, after this organism enveloped as a separate entity within the environment of host environment, and finally lysed the cell (Bulla Jr. and Cheng 1976).

The microsporidians predominantly being intracellular parasitic organisms, it's only at the cellular level that their response could be manifested. But a variety in these manifestations was obvious; for instance, in the genus *Pleistophora*-type of event the interiors of the host cell are systematically engulfed completely. Once the invader started division to ensure growth, the manifestation into destruction of the cell was evident. Once the microsporidia occupied the interior of a cell, the lytic action to convert the cytoplasm into an array of vesicles, endoplasmic reticulum, cisternae and ribosomes jumbled into an unorganized mass (Bulla Jr. and Cheng 1976). No pathogenesis was encountered beyond the infected region in this event. However, in other genera, exorbitant hypertrophy in the infected cell under influence from the microsporidian resulted. The connective tissue cells and those of mesenchyme origin, reaching up to 14 mm size comprised the hypertrophic cells (Weiser 1969). These are termed xenoma. The physiological and morphological integration of the cell of the host whose internal environment and composition are totally altered, mainly because of the enhanced sized microsporidian, contributed significantly to the formation of xenoma, as a separate entity that was nurtured depleting natural constituents of the host cell. With the gradual growth of xenoma noticeable alterations in its wall triggered inflammation as well as proliferation. This resulted into replacement of xenoma by the granuloma, whose tissues invaded into the former, but with a diminished centrally placed spore mass. The molecular mechanism operating to express mitochondrial association with microsporidia has been a mystery even today (Lom and DyKova 1992). Though the secondary xenomas emanated from the reminiscent spores within, yet these never protruded or were liberated to the exterior. It is obvious that the granulation tissue invasion brought about changes into the wall of xenoma, and thus isolated the microsporidian with restricted auto-infection probability. The viable spores thus could only dehisce out if the host perished or else these were engulfed by the fibroblasts. Thus the mechanism of defence at the cellular level is strengthened with the central role of phagocytes that diminish the macrophage-engulfed spores, if their initial disintegration failed within granuloma. The gradual disintegration of the contents of the spore in the interior of phagocytes left an emptied folded membrane, that itself disintegrated ultimately. The evidence of chitinase enzyme activity was thus concluded by Heckmann (2020) confirming complete disintegration of the phagocytised spores that are redundant not to trigger any fresh infection activity episode further. The molecular level implications of the entry of microsporidians into the live cells were dealt with by Heckmann (2020) to emphasize that to analyze response of the host to the invader, such interactions would have to be understood at the level of cell.

The indications of interactions involving a surface protein 1 (EhSSP1) on the surface of the microsporidial sporoplasm of *E. hellem* with mitochondria from the cell of the host pest have been expressed (Han et al. 2019). All the three forms of voltage-dependent anion selective channels (VDAC1–3) being expressed in the cytoplasm of the outer membrane of mitochondria interact with EhSSP1. Han et al. (2019) asserted that the strength of the association of the mitochondria with microsprodian parasitophorous vacuole would weaken on account of this interaction being disrupted. Han et al. (2019) demonstrated that the energy retention encountered by the microsporidia in the cell of its host was presumably facilitated by the interaction of EhSSP1 with VDAC. It was peculiar to note that an unidentified cell protein within the host pest incorporated with the constituents of an invasion synapse (sporoplasm and polar tube in the cellular environment, form a synapse) also succeeded to interact with EhSSP1. This could trigger a novel response while sporoplasm is transported into the host cell cytoplasm. Further additional investigations could be conducted to explain this phenomenon. The flagellates, *Trypanosoma* sp., followed distinctively segregated life cycle pathway *sans* cysts, and instead, transfer of developmental stages occurred through direct ingestion of excretory matter. Naturally occurring DNA catenanes (Waraich et al. 2020) encountered primarily in the mitochondrial DNA isolated from HeLa cell lines (Hudson and Vinograd 1967) were assigned the task of the predominant method of organizing DNA in the mitochondria of trypanosomes. The network of the latter mitochondrial DNA, also called kinetoplast DNA (kDNA) (Englund et al. 1982), comprised two types of DNA circles that are interlinked to form an intricate network of maxicircles (20–40 kbp) and mini circles $\langle$ 2.5 kbp) (Lukes et al. 2002).

The soil-dwelling amoeba of genus *Dictyostelium* was the first free-living protozoan genome to be fully sequenced. It provided opportunity to examine activity of mutants of pathogens, as attenuation of their virulence resulted into depletion of their virulence in *Drosophila* as well (Han et al. 2020). This meant that *Dictyostelium* provided a valid tool to examine the virulence mechanism of pathogens of insect pests. Han et al. (2020) carried out genetic manipulation in *Dictyostelium* that could illustrate the application of autophagy at relative ease during infections by pathogens. The cysteine protease, Paracaspase, was found in *Dictyostelium.* But sufficient data is not available to examine molecular function of these caspase-like proteins or their role in programmed cell death. Such detailed investigations could also lead to development of treatment therapies against parasitic protozoans (Han et al. 2020).

8.10 Microsporidia Invasion

The transport of sporoplasm into the cells of the host pests preceded by the Microsporidia infection of host cells involved expedited ejection of the polar tube (Fig. 8.2) (Weidner 1970; Frixione et al. 1992; Takvorian et al. 2005; Han et al. 2017). Weidner (1972) and Han et al. (2017) elucidated that the event of the onset of the infection triggered sticking of microsporidian spores to the cells of the host pests or the surrounding tissues. The significant contribution to these initial steps during infection is those of spore wall proteins, that is, SWPs, as illustrated by Southern et al. (2007). The investigations revealed a number of SWPs interacting with host cells due to stronger affinity to the heparin-binding motif (HBM) and sulphated glycosaminoglycans (GAGs) associated with the surface of cells of the host pests on *Nosema bombycis*, (Table 8.6) *Encephalitozoon* spp. and *Antonospora locustae* (Table 8.6) (Hayman et al. 2001; Hayman et al. 2005: Southern et al. 2007; Li et al. 2009; Wu et al. 2009; Chen et al. 2017). A simultaneous report of the critical involvement of the integrin constituent of the host cell into the infection process by the microsporidium, *Encephalitozoon intestinalis. E. intestinalis* has also been on record in recent years (Leonard and Hayman 2017).

The canonical integrin-binding motif, arginine-glycine-aspartic acid (RGD), that regulated the binding of extracellular matrix (ECM) proteins with host cell integrins, were supposed to be present within a large number of hypothetical proteins, as revealed by the *E. intestinalis* genome analysis. A variety of pathogenic microbes, viz., several parasites, bacteria and viruses, that are reported to have stuck to the host cells were attributed to comprise proteins conducting interactions with host cell integrins (Patti et al. 1994; Bartlett and Park 2010). It was further demonstrated that the microsporidial spore attachment and host cell infection were barred by the incubation of host cells with RGD-peptides or recombinant alpha3 beta1 and alpha 5 beta 1 human integrin proteins (Leonard and Hayman 2017). It, therefore, conclusively suggested that, for the germination for production of spores and subsequent host cell invasion to succeed, the attachment of spores was an important part of the whole mechanism (Leonard and Hayman 2017).

8.11 Environmental Interactions

The body constituents of microsporidia are usually very soft and fragile that are usually susceptible to a variety of environmental attributes, but its spores can withstand adversities, externally to the cellular environment. The protozoans, with bacteria (decomposers), and ciliates (bacterivores) entered into effective interspecific and intraspecific interactions, in association with mosquito larvae (primary consumers) succeeded to influence characteristics, like cell size, cyst production and growth rate (TerHorst 2010, 2011). According to Maddox and Solter (1996), the survival period of life cycle stages of microsporidia, that belonged to the terrestrial category, distinct from the aquatic ones, was normally over 30 years, in liquid nitrogen storage conditions. But the usual survival time was from 1 month to a year, under natural environmental conditions, when these were under protection from UV radiation and other co-occurring microbes or degradation factors (Maddox 1973; Maddox et al. 1981; Brooks 1980, 1988; Goertz and Hoch 2008). However, microsporidia from aquatic environment were not that sturdy, and could easily be under the influence of environmental degradation (Becnel and Johnson 2000). It was quite an unusual characteristic of microsporidia that it could coordinate its presence or absence while their hosts are unavailable under variable conditions of seasonality or wide fluctuations in their population density. Subsequently, their unique potential to develop in a suitable host species, which was related to its original host species (Lange and Azzaro 2008) or in an alternate or intermediate host, when their routine hosts were not available, was an advantageous adaptation (Micieli et al. 2009).

The vertical transmission in certain types of infections by *Edharzardia aedis* in the host, *Aedis aegypti*, (Table 8.1) might not be so ferocious to the extent as to facilitate its inoculation to the local environment with spores for horizontal transmission among larvae of the mosquitoes (Koella et al. 1998). The most common instance of adoption of survival strategy by *N. pyrausta* microsporidia in diapause stages of the host, as well as to pass over the overwintering period through its fifth instar stage in *O. nubilalis* (Andreadis 1986; Siegel et al. 1988), and the relatively more common transovarial transmission through infected eggs of *N. Portugal* (Maddox et al. 1999) are available. This, of course, is quite imperative that the infection stress employed under additional physiological alterations during diapause could as well be fatal to the host; yet the protection of the infecting agent in the surviving host's environment is facilitated to enable these to remain available, till the breeding of the forthcoming next generation of mosquito pest (Andreadis 1986).

8.12 Advantages & Disadvantages

8.12.1 Advantages

The management of biodiversity in a variety of ecosystems could only be sustained when the insect pest hosts remained under the stress of natural enemies (protists, i.e. microsporidia) to achieve biocontrol. It was, therefore, obvious that the protozoan pathogens that acted to the detriment of such pests be conserved for the safety of human beings, and in this way, the nontarget organisms would also succeed to contribute for the benefit of a balanced ecosystem. The formal recognition to the only species, namely, *P. locustae*, the pathogen of a grasshopper species, registered by the US Protection Agency (USEPA), as a microbial insecticide, is a distinctive step towards the advantage to human beings under the biocontrol programme, after a passage of half a century of expeditious researches world over in this field.

8.12.2 Disadvantages

The typically recognized slow-acting potential of microsporidia to control insect pests quite often leads to the prolonged wait to record detectable impacts on the pest host populations. Under such an eventuality, newer methods of appropriate formulations, keeping in view the ecological fate of protozoan spores, would be required to be developed. An eye would have to be kept on the cost-effective aspects of economy for the management of protozoan organisms, particularly microsporidia to care for their potential of biocontrol in the context of IPM.

8.13 Production and Storage

8.13.1 Production

Culture media are generally available for ciliates and trypanosomatid flagellates, and there is reason to be optimistic that only minor modifications would be needed to support the growth of species, which are endoparasitic. Since the freshwater environment of the host mosquitoes would be normal habitats for the infective stages of *Lambornella* and *Tetrahymena*, the cultured ciliates could be introduced directly. The problems of preventing the ciliates being consumed by the mosquitoes, as observed by Clark and Brandl (1976), would have to be overcome. Cultured flagellates cannot be introduced in this way as they neither form cysts nor have free-swimming stages, which can survive the physical and physiological changes as do ciliates. So far no species of flagellate has been deemed pathogenic enough to warrant efforts being addressed to the problem of introduction. Other Protozoa harboured by invertebrates, suitable for development as biocontrol agents, are all obligate intracellular parasites with resistant spores or cysts for transmission. None can be cultured free of living cells, and methods for mass production are limited at present to growth in natural or experimental hosts. In vitro culture in cell lines has been achieved for a few species, but spore production is well below that which is obtained in living hosts.

In Vivo Production

The episodes of production and efficacy of the insect infested microsporidia and neogregarines were reviewed by Brooks (1980). It was hypothesized that the optimum replication rates to produce hosts and to minimize their losses should be ensured by keeping an eye on the dose applied, age of the host and the time of incubation from the standpoint of economic production. Brooks (1980) summarily described a large variety of protozoans that were produced in live hosts, but strikingly in pretty lower numbers. The protozoan infective agents, namely, *Malpighamoeba locustae*, were encountered in day-to-day faecal matter collections from the host pests. The turnover was substantial in case of *V. necatrix* from *Helicoverpa zea*. The cost analysis report by McLaughlin and Bell (1970) presented a comparable account of protozoan organisms, viz., *Mattesia grandis* (McLaughlin) and *N. gasti* (McLaughlin), from an experimental report involving pest hosts, that is, cotton boll weevils, *A. grandis*. Schwalbe et al. (1974) contemplated production of spores of *Melanoplus trogodermae* to assert biocontrol of *T. glabrum* in the live environment to store the stored grains. To envisage the production of supernumerary instars of the host pest, the use of an insect growth regulator was suggested by Brooks (1980) to trigger production of spores/larvae en masse. The efforts to produce the protozoan, *N. locustae* Canning, from the commercial standpoint in grasshoppers, *Melanoplus bivittatus*, were recognized by the US Environment Protection Agency (EPA) (Henry and Oma 1981).

In Vitro Production

The procedures have been developed to date to grow suspension cultures or monolayers of cells for the enhanced output of a variety of microsporidia as well as to ensure simultaneously the availability of the spores of these microsporidia at the required period of intervals from the culture medium. But their status of product output as on to the level of cost-effectiveness could not be ensured. Several earlier investigators (Gupta 1964; Kurtii and Brooks 1971; Bayne et al. 1975; Sohi and Wilson 1976) utilized an already infected host resource to establish primary cell cultures of these microsporidia.

The uninfected cultures were maintained to introduce sterile harvested spores. It was also not necessary to grow the microsporidia or other cells that could only exhibit closer affiliation to their host cells, whether those of invertebrate or vertebrate origin. Primary cultures of silkworms, *Bombyx mori*, were used by Ishihara and Sohi (1966) and Ishihara (1969) for growth of *N. bombycis* Nageli (Table 8.6) and cell lines were used by Undeen (1975) for *N. algerae* in pig kidney cells by Kurtii and Brooks (1977) for *N. disstriae* and by Atwell et al. (1985) for *H. zea* cells. The findings of several earlier investigators did not find consistent growth and replication rates in cultures of different kinds, though they did not report comparisons with in vivo systems to record the rates of replication beyond a few days. An improvement in the technique for infection of cells in culture was introduced by Barker et al. (1980), who centrifuged the cultures when the spores were introduced. The close contact between spores and cells, during the process of polar filament eversion, gave a higher level of infection from which the parasites could spread to other cells. Sohi and Wilson (1976) found that spore production of *N. disstriae* was depressed in culture, although the numbers of parasites remained high. This would be a serious impediment to mass production, and a similar effect has been reported for cultures of *N. algerae*, in which spore production stopped after six passages (Streett et al. 1980).

8.13.2 Storage

The resistivity of the entities that participate in transmission of certain Protozoa has a greater role to play but the most convenient and safest form of these could be to keep the cysts or spores of these in dry state. Storage presents no insurmountable problem for Protozoa with resistant stages of transmission, but the simplest method of keeping the cysts or spores in the dry state can only be used for certain categories. Dry storage is essential to prevent extrusion of the polar filaments, which occurs automatically in water. The protrusion of polar filaments can be prevented by maintaining these in dry form, which otherwise is attained automatically in the moist state in *N. whitei* Weiser, a microsporidium of flour beetles *Tribolium* spp. No apparent loss of viability was observed by Milner (1972) in the spores of *Tribolium* spp., if stored in a dry flour–yeast mixture at 4 °C to the extent of 15 months. On the other hand, the exposure to even shorter periods of desiccation was fatal to some of the microsporidia. The dried spores of *N. algerae* lost their potential to infect the larvae of *Anopheles stephensi* Liston in a study by Alger and Undeen (1970). However, an optimal survival of most of the microsporidia and a fewer variety of neogregarines were attained when these were stored in a sterile water suspension kept at a temperature above freezing point. The spores of *N. bombycis* Nageli could survive viable for as long as up to a decade, as recorded by Oshima (1964). The treatment by antibiotics was applied to maintain viability of these microsporidia as it deteriorated rapidly under the influence of these contaminants. Contrary to this, the treatment by the required concentrations of antifungal mixtures, needed to deactivate yeasts or growth of hyphae, proved damaging to a larger chunk of the microsporidia in a manner that a span of storage of *N. algarae* for 5 weeks in natamycin, fungizone and nystatin resulted in a loss of their viability (Lai 1980). Simultaneously, appreciable issues of rearing to maintain insect colonies during conduct of experiments were faced when the antibiotics, like benomyl and fumagillin, were applied to keep check on the growth and spread of microsporidia (Vavra and Maddox 1976). The noticeable reduction in viability of microsporidian spores ensued at the higher temperatures that could incapacitate these in as much as that it was reduced in most species, to weeks, hours or minutes by temperatures that have risen to 40 °C (Maddox 1977). Considerable loss in viability of microsporidian spores of a large number of protozoan species, like *N. locustae*, resulted at lowered thermal regime (3 years in distilled water at −10 °C), as well as in case these were stored dry (Henry and Oma 1981). The imminent requirement to resolve the problem of storage of such magnitude was appreciated because these protozoans have been recognized as a workable biopesticide over the years by now. But the issues like availability of these biocontrol liquids in huge quantities for applications in the fields kept the farmers baffled, because of large areas under attack by grasshoppers and locusts, and simultaneously methods to be adopted for the delivery of these biopesticides over large farm areas were cumbersome. Thus, the riddles of cost-effectiveness kept the farmers away from immediate adoption of such measures, and instead these were adopted only in cases of extreme emergencies. However, the outcome of a large number of investigations (Kramer 1970a, b; Brooks 1980) illustrated a storage regime at the freezing temperature (i.e. 4 °C) was appropriately helpful for the growth of several species. The need has been felt to enhance adequate quantities of storage of *N. locustae* at the commercial scale. The methods of lyophilization and cryopreservation in liquid nitrogen, employed to achieve long-term viability (Vavra and Maddox 1976), were not found to be cost-effective because of the high cost of liquid nitrogen.

Later, however, some apprehensions were expressed about the loss of viability of spores to a small extent immediately followed by lyophilization, but the spores that survived the onslaught retained infectivity for longer periods. The loss of viability of a negligible magnitude of *Nosema apis* Zander post-lyophilization was concluded by Bailey (1972), and, therefore, its prolonged use for storage of viable microsporidia was recommended.

8.14 Future Prospects of Protozoans Biocontrol of Insect Pests

The pests, to the extent of 98%, are known to have been controlled by biocontrol agents. Certain ecological invaders comprising newer predators and parasitoids could conform to the significant constituents of biocontrol that regulate a variety of pests. According to the modern trend, the utilization of emerging strains, biotypes, parasitic hybrids, protozoan and fungal biopesticides required utmost attention for various applications in the management of pests. The principal beneficiaries of the application of protozoan control methods to regulate populations of pests have been the aquaculturists and farmers, particularly those dedicated to modernized aquacultural techniques, like cage culture etc., organic farming and other modern agricultural methods. Newer substitutes would be added day by day by the application of modernized researches to develop formulations of newer substances. Further advancements would target commercialization and adaptability of biopesticides. These efforts would also restrict cost escalation as more and more natural means of biocontrol become acceptable to the public at large. Once more and more advancements in the development of promising candidates among natural products emanated from the farmed products, the test of their efficacy and reliability in application could be ensured by extensive checks on the target pests in a variety of cropping systems.

It is a challenging task to conduct detailed investigations on biochemical pathways in the intracellular organisms that lack mitochondria but are the potential candidates to be employed in biocontrol of insect pests.

8.15 Conclusions

Microsporidia (Protists) belong to the significant group of biocontrol agents that had the potential to keep noticeable benefits of human populations conserved as against the potential of insect pest hosts that could keep the former divested of noticeable economic gains. The unique mechanism of invasion adopted by these pathogens incorporated the use of a specialized apparatus whose functional details are yet not completely available. They are important pathogens of economically important insects and animals. It is noticeable that considerable information is available to assign (i) vertical, (ii) horizontal, as well as (iii) transovarial transmission pathways in the dynamics of the life cycle of microsporidia. But additional information on their mode of entry into the cells of insect pests, along with the mechanism of its emergence post-replicative cycle, would provide solution to the riddle of incomprehensible process of impact on the pest host. The limitations of the excessive applicability of chemicals to control or kill the insect pests have indeed not led us too far in as much as to pave safer pathways down the lane. This is, in particular, due to the well-known adverse ill-effects on the environment, nontarget species as well as public health. This does not necessarily mean that only a fast action strategy could serve meaningful purpose, particularly in the wake of a virtual slow-action reactivity exhibited by most of the known protists (microsporidia). The elements of reliable non-chemical control method could safely have the potential to long-term suppression and maintenance of low densities of insect pests, like grasshoppers and locusts, and these could well be the critical elements of control by natural enemies in terms of application of protists as tools of biocontrol. The judicious mix of ecologically safe biocides utilizing microsporidian protozoans as its constituents, to replace hazardous uneconomical chemicals for control of insect pests, is propagated.

8.16 Points to Remember

- Biocontrol elements in pathogenic interactions, other than non-chemical and other methods of control, encompassing the tiny protista (Microsporidia) and other protozoans comprise unique entomopathogenic organisms.
- The invasion of these minute pathogens primarily in the form of cyst into the cells of insect pest hosts is a commonly known event. These infective cysts have the potential to survive over unfavourable conditions as well as overwinter to survive withering conditions, during seasonal change.
- Pathogens have, of late, been recognized as suppressor of rapid breeding and feeding speed of acridid and a variety of other insect pest hosts, although their action was labelled initially as slow-paced. This could well be a successful biocontrol strategy.
- The microsporidian invaders made inroads into the intracellular environment of insect pests through cysts. Amongst *Nosema*-type group of microsporidia, the transmittance of the flagellate, *Trypanosoma* sp., did not occur by direct ingestion of true cysts but through the ingestion of faecal material. *Endoreticulatus* group

(Table 8.2) was differentiated from the other two, that is, *Nosema* type and *Vairimorpha* type, phylogenetically as well as in the status of pathogenicity.

- The horizontal transmission of microsporidians is a unique pathway in which entry of cysts through eggs, diapausing larvae, pupae or non-feeding adults is attained. The alternative vertical mode facilitated invasion of microsporidians into the contents of the eggs that terminated the process at the hatching of larvae from the contaminated egg contents.

Acknowledgements AY is thankful to The Principal, and Dr. (Mrs.) Neerja Kapoor, Head of the Department of Zoology, C.M.P.P.G. College, Allahabad, for facilities and interactions on the manuscript. NJ is grateful to Head of the Department of Zoology, Baba Bhimrao Ambedkar University, Lucknow, for facilities.

References

Alger NE, Undeen AH (1970) The control of a microsporidian *Nosema* sp., in an anopheline colony by an egg-rinsing technique. J Invertebr Pathol 15:321–327

Anderson RM, May RM (1980) Infectious diseases and population cycles of forest insects. Science 210:658–661

Anderson RM, May RM (1981) The population dynamics of microparasites within invertebrate populations. Philos Trans R Soc B 291:451–452

Anderson JR, Washburn JO, Egerter DE (1986a) Life cycle of the pathogen *Lambornella clarki* and its impact on *Aedes sierrensis*. Mosquito control research. Annual report. University of California, Davis, pp 20–21

Anderson JR, Washburn JO, Egerter DE (1986b) Mass production, storage and field release of *Lambornella clarki*, a pathogen of *Aedes sierrensis*. Mosquito Control Research. Annual Report. University of California, Davis, p. 21–22

Andreadis TG (1986) Dissemination of *Nosema pyrausta* in feral population of the European corn borer *Ostrinia nubilalis*. J Invertebr Pathol 48:335–343

Atwell BJ, Thomson CJ, Greenway H, Ward G, Waters I (1985) A study of the impaired growth of roots of *Zea mays* seedlings at low oxygen concentrations. Plant Cell Environ 8:179–188

Avery SW, Anthony DW (1983) Ultrastructural study of early development of *Nosema algerae* in *Arropheles albimanus*. J Invertebr Pathol 42:87

Bailey L (1972) The preservation of infective microsporidian spores. J Invertebr Pathol 20:252–254

Barker RJ, Smith J, Lai PF (1980) The culture of the microsporidian parasites *Nosema algerae* and *N. eurytremae* in vertebrate cells. J Protozool 27:55

Bartlett AH, Park PW (2010) Proteoglycans in host–pathogen interactions: molecular mechanisms and therapeutic implications. Expert Rev Mol Med 12:e5. https://doi.org/10.1017/S1462399409001367

Bayne CJ, Owczarzak A, Noonan WE (1975) In vitro cultivation of cells and a microsporidian parasite of *Biomphalaria glabrata* (Pulmonata, Basommatophora). Ann N Y Acad Sci 266:513–527

Becnel JJ, Johnson MA (2000) Impact of *Edhazardia aedis* (microsporidia : Culicosporidae). On a seminatural population of *Aedes aegypti*. (Diptera : Culicidae). Biol Control 18:39–48

Bidochka MJ, Khachatourians GG (1991) Review article: microbial and protozoan pathogens of grasshoppers and locusts as potential biocontrol agents. Biocontrol Sci Tech:243–259

Bohne W, Ferguson DJ, Kohler K, Gross U (2000) Developmental expression of a tandemly repeated, glycine-and serine-rich spore wall protein in the microsporidian pathogen Encephalitozoon cuniculi. Infect Immun 68:2268–2275. https://doi.org/10.1128/iai.68.4.2268-2275.2000

Bomar CR, Lockwood JA, Pomerinke MA, French JD (1993) Multiyear evaluation of the effects of *Nosema ktcustae* (Microsporida: Nosematidae) on rangeland grasshopper (Othoptera: Acrididae) population density and natural biological controls. Environ Entomol 22:489–491
Braun L, Ewen AB, Gillott C (1988) The life cycle of Malameba locustae (King and Taylor) (Amoebidae) in the migratory grasshopper *Mektnoplus sanguinipes* (F.) (Acrididae). Can Entomol 120:759–772
Brooks WM (1968) Transovarian transmission of *Nosema heliothidis* in the corn earworm *Heliothis zea*. J Invertebr Pathol 11:511–512
Brooks WM (1974) Chapter 5. Protozoan infections. In: Cantwell GE (ed) Insect diseases, vol 1. Marcel Dekker, New York, pp 237–300
Brooks WM (1979) Production and efficacy of protozoa. Biotechnol Bioeng 22:91–116
Brooks WM (1980) Production and efficacy of protozoa. Biotechnol Bioeng 22:1415–1440
Brooks WM (1988) Entomogenous protozoa. In: Ignoffo CM (ed) CRC handbook of natural pesticides. Microbial insecticides. Part A. Entomogenous protozoa and fungi, vol 5. CRC Press, pp 1–149
Brosson D, Kuhn L, Prensier G, Vivares CP, Texier C (2005) The putative chitin deacetylase of Encephalitozoon cuniculi: a surface protein implicated in microsporidian spore-wall formation. FEMS Microbiol Lett 247:81–90. https://doi.org/10.1016/j.femsle.2005.04.031
Bulla LA Jr, Cheng TC (1976) Comparative pathobiology, 1. In: Biology of the microsporidia. Plenum Press, New York
Burri L, Williams BA, Bursac D, Lithgow T, Keeling PJ (2006) Microsporidian mitosomes retain elements of the general mitochondrial targeting system. Proc Natl Acad Sci U S A 103:15916–15920. https://doi.org/10.1073/pnas.0604109103
Cali A, Becnel JJ, Takvorian PM (2017) Microsporidia. Hand Protists:1559–1618. https://doi.org/10.1007/978-3-319-28149-0_27
Canning EU (1953) A new microsporidian, Nosema locustae n. sp., from the fat body of the African migratory locust, Locusta migratorioides R & F. Parasitology 43:287–290
Canning EU (1962a) The life cycle of *Nosema locustae* Canning in *Locusta migratoria migratorioides* (Reiche & Fairmaire), and its infectivity to other hosts. J Insect Pathol 4:237–241
Canning EU (1962b) The pathogenicity of *Nosema locustae* Canning. J Insect Pathol 4:248–256
Canning EU (1982) An evaluation of protozoal characteristics in relation to biological control of pests. Parasitology 84:119–149
Canning EU, Lom J (1986) The microsporidia of the vertebrates. Academic Press Inc, New York City, NY, p 289
Canning EU Barker RJ (1981) Transmission of microsporidia between generations of winter moth Operophtera brumata. Parasitology (in press)
Cali A, Takvorian PM (1999) Developmental morphology and life cycles of the microsporidia. In: Wittner M, Weiss LM (eds) The microsporidia and microsporidiosis. AMS Press, Washington, DC, pp 85–128
Canning EU, Wigley PJ, Barker RJ (1983) The taxonomy of three species of microsporidia (Protozoa: Microspora) from an oakwood population of winter moths Operophtera brumata (L.) (Lepidoptera: Geometridae). Syst Parasitol 5:147–159
Cali A, Takvorian PM (2014) Developmental morphology and life cycles of the Microsporidia. In: Weiss LM, Becnel JJ (eds) Microsporidia-pathogens of opportunity. Wiley Blackwell Press, Hoboken NJ, pp 71–133. https://doi.org/10.1002/9781118395264.ch2
Capinera JL, Hibbard BE (1987) Bait formulations of chemical and microbial insecticides for suppression of crop-feeding grasshoppers. J Agric Entomol 4:337–344
Chen J, Geng L, Long M, Li T, Li Z, Yang D et al (2013) Identification of a novel chitin-binding spore wall protein (NbSWP12) with a BAR-2 domain from Nosema bombycis (microsporidia). Parasitology 140:1394–1402. https://doi.org/10.1017/S0031182013000875

Chen L, Li R, You Y, Zhang K, Zhang L (2017) A novel spore wall protein from Antonospora locustae (microsporidia: Nosematidae) contributes to sporulation. J Eukaryot Microbiol 64:779–791. https://doi.org/10.1111/jeu.12410

Clark TB, Brandl DG (1976) Observations on the infection of *Aedes sierrensis* by a tetrahymenine ciliate. J Invertebr Pathol 28:341–349

Corliss JO, Coats DW (1976) A new cuticular cyst-producing tetrahymenid ciliate, *Lambornella clarki* n.sp. and the current status of ciliatosis in culicine mosquitoes. Trans Am Microsc Soc 95:725–739

Dean P, Sendra K, Williams T, Watson A, Major P, Nakjang S, Kozhevnikova E, Goldberg AV, Kunji ERS, Hirt RP, Embley TM (2018) Transporter gene acquisition and innovation in the evolution of microsporidia intracellular parasites. Nat Commun 9:1709. https://doi.org/10.1038/s41467-018-03923-4

Dolgikh VV, Senderskiy IV, Pavlova OA, Naumov AM, Beznoussenko GV (2011) Immunolocalization of an alternative respiratory chain in *Antonospora* (Paranosema) locustae spores: mitosomes retain their role in microsporidial energy metabolism. Eukaryot Cell 10:588–593. https://doi.org/10.1128/EC.00283-10

Egeter DE, Anderson JR, Washburn JO (1986) Dispersal of the parasitic ciliate *Lambornella clarki*: implications for ciliates in the biological control of mosquitoes. Proc Natl Acad Sci U S A 83:7335–7339

Englund PT, Hajduk SL, Marini JC (1982) The molecular biology of trypanosomes. Annu Rev Biochem 51:695–726

Fine PEM (1981) Vertical transmission of pathogens of invertebrates. In: Bulla LA, Cheng TC (eds) Comparative pathobiology, vol 6, 178 pp

Fine PEM (1975) Vectors and vertical transmission: an epidemiologic perspective. Ann N Y Acad Sci 266:173–194

Flint ML, Dreistadt SH (eds) (1998) Natural enemies handbook: the illustrated guide to biological pest control. University of California, Division of Agriculture, National Research Publication, Oakland, p 3386

Franz JM, Huger AM (1971) Microsporidia causing the collapse of an outbreak of the green trotrix (Tortrix viridana L.) in Germany. Proc 4th Int Colloq Insect Pathol:48–53

Frixione E, Ruiz L, Santillán M, De Vargas LV, Tejero JM, Undeen AH (1992) Dynamics of polar filament discharge and sporoplasm expulsion by microsporidian spores. Cell Motil Cytoskeleton 22:38–50. https://doi.org/10.1002/cm.970220105

Gaugler RR, Brooks WM (1975) Sublethal effects of infection by *Nosema heliothidis* in the corn earworm, *Heliothis zea*. J Invertebr Pathol 26:57–63

Goertz D, Hoch G (2008) Vertical transmission and overwintering of microsporidia in the gypsy moth, *Lymantria dispar*. J Invertebr Pathol 99(1):43–48

Goldberg AV, Molik S, Tsaousis AD, Neumann K, Kuhnke G, Delbac F, Vivares CP, Hirt RP, Lill R, Embley TM (2008) Localization and functionality of microsporidian iron–sulphur cluster assembly proteins. Nature 452:624–628. https://doi.org/10.1038/nature06606

Gupta KS (1964) Cultivation of *Nosema mesnili* Paillot (microsporidia) *in vitro*. Curr Sci India 33:407–408

Han B, Weiss LM (2017) Microsporidia: obligate intracellular pathogens within the fungal kingdom. Microbiol Spectrum 5:97–113

Han B, Polonais V, Sugi T, Yakubu R, Takvorian PM, Cali A, Maier K, Long M, Levy M, Tanowitz HB, Guoging P, Delbac F, Zhou Z, Weiss LM (2017) The role of microsporidian polar tube protein 4 (PTP4) in host cell infection. PLoS Pathog 13:e1006341. https://doi.org/10.1371/journal.ppat.1006341

Han B, Ma Y, Tu V, Tomita T, Mayoral J, Williams T, Horta A, Huang H, Weiss LM (2019) Microsporidia interact with host cell mitochondria via voltage-dependent anion channels using sporoplasm surface protein 1. MBio 10:e001944–e001919. https://doi.org/10.1128/mBio.01944-19

Han B, Takyorian PM, Weiss LM (2020) Invasion of host cells by microsporidia. Front Microbiol Rev. https://doi.org/10.3389/fmicb.2020.00172

Hans-Peter Braun RB (2009) Mitochondria. Ed. Immo E. Scheffler. Q Rev Biol 84:103. https://doi.org/10.1086/598289

Harry OG (1970) Gregarina: their effect on growth in the desert locust (*Schistocerca gregaria*). Nature 225:964–966

Hayman JR, Hayes SF, Amon J, Nash TE (2001) Developmental expression of two spore wall proteins during maturation of the microsporidian *Encephalitozoon intestinalis*. Infect Immun 69:7057–7066. https://doi.org/10.1128/iai.69.11.7057-7066.2001

Hayman JR, Southern TR, Nash TE (2005) Role of sulfated glycans in adherence of the microsporidian *Encephalitozoon intestinalis* to host cells in vitro. Infect Immun 73:841–848. https://doi.org/10.1128/iai.73.2.841-848.2005

Hazard EI, Oldacre SVV (1976) Revision of microsporidia (protozoa) close to *Thelohania* with descriptions of one new family, eight new genera and thirteen new species. United States Department of Agriculture Technical Bulletin No. 1530, pp 1–104

Heckmann RA (2020) E-book: ISBN9780367811006, molecular biology of protozoa for biological control of harmful insects, Chapter 9. In: Gunasekharan M et al (eds) 1995, molecular biology of the biological control of pests and diseases of plants. Taylor & Francis, London, p 18. https://doi.org/10.1201/9780367811006

Heinz E, Williams TA, Nakjang S, Noël CJ, Swan DC, Goldberg AV, Harris SR, Weinmaier T, Markert S, Becher D, Bernhardt J, Dagan T, Hacker C, Lucocq JM, Schweder T, Rattei T, Hall N, Hirt RP, Embley TM (2012) The genome of the obligate intracellular parasite *Trachipleistophora hominis*: new insights into microsporidian genome dynamics and reductive evolution. PLoS Pathog 8:e1002979. https://doi.org/10.1371/journal.ppat.1002979

Heinz E, Hacker C, Dean P, Mifsud J, Goldberg AV, Williams TA, Nakjang S, Gregory A, Hirt RP, Lucocq JM, Kunji ERS, Embley TM (2014) Plasma membrane-located purine nucleotide transport proteins are key components for host exploitation by microsporidian intracellular parasites. PLoS Pathog 10:e1004547. https://doi.org/10.1371/journal.ppat.1004547

Henry JE (1969) Extension of the host range of *Nosema locustae* in Orthoptera. Ann Entomol Soc Am 62:452–453

Henry JE (1981) Natural and applied control of insects by protozoa. Annu Rev Entomol 26:49–73

Henry JE (1985) Effect of grasshopper species, cage density, light intensity, and method of inoculation on mass production of *Nosema locustae* (Microsporida: Nosematidae). J Econ Entomol 78:1245–1250

Henry JE (1990) Control of insects by protozoa. In: New directions in biological control: alternatives for suppressing agricultural pests and diseases. A.R. Liss, Inc, New York, pp 161–176, 837

Henry JE, Oma EA (1981) Protozoa: pest control by *Nosema locustae*, a pathogen of grasshoppers and crickets. In: Burges HD (ed) Microbial control of pests and plant diseases, 1976–1980. Academic Press, New York, pp 573–586

Hill RE, Gary WJ (1979) Effects of the microsporidium, *Nosema pyrausta*, on field populations of European corn borers in Nebraska. Environ Entomol 8:91–95

Hinks CF, Ewen AB (1986) Pathological effects of the parasite *Malameba locustae* in males of the migratory grasshopper *Melanoplus sanguinipes* and its interaction with the insecticide, cypermethrin. Entomol Erp Appl 42:39–44

Hudson B, Vinograd J (1967) Catenated circular DNA molecules in HeLa cell mitochondria. Nature 216(5116):647–652

Hylis M, Pilarska DK, Oborní KM, Vávra J, Solter LF, Weiser J, Linde A, ML MM (2006) Nosema chrysorrhoeae n. sp. (Microsporidia), isolated from browntail moth (Euproctis chrysorrhoea L.) (Lepidoptera, Lymantriidae) in Bulgaria: characterization and phylogenetic relationships. J Invertebr Pathol 91:105–114

Ishihara R (1969) The life cycle of *Nosema bombycis* as revealed in tissue culture cells of *Bombyx mori*. J Invertebr Pathol 14:316–322

Ishihara R, Sohi SS (1966) Infection of ovarian tissue culture of *Bombyx mori* by *Nosema bombycis* spores. J Invertebr Pathol 8:538–540

Jackson LL, Baker GL, Henry JE (1968) Effect of *Malamoeba locustae* infection on the egg lipids of the grasshopper *Melanoplus hivitattus*. J Insect Physiol 14:1773–1778

Jaroenlak P, Boakye DW, Vanichviriyakit R, Williams BA, Sritunyalucksana K, Itsathitphaisarn O (2018) Identification, characterization and heparin binding capacity of a spore-wall, virulence protein from the shrimp microsporidian. Enterocytozoon hepatopenaei (EHP). Parasit Vectors 11:177. https://doi.org/10.1186/s13071-018-2758-z

Johnson MA, Becnel JJ, Undeen AH (1997) A new sporulation sequence in Edhazardia aedis (Microsporidia: Culicosporidae), a parasite of the mosquito Aedes aegypti (Diptera: Culicidae). J Invertebr Pathol 70:69–75

Keithly JS, Langreth SG, Buttle KF, Mannella CA (2005) Electron tomographic and ultrastructural analysis of the *Cryptosporidium parvum* relict mitochondrion its associated membranes and organelles. J Eukaryot Microbiol 52:132–140

Knell JD, Allen GE (1978) Morphology and ultrastructure of Unikaryon minutum (Microsporidia: Protozoa), a parasite of the southern pine beetle, Dendroctonus frontalis. Acta Protozool 17:271–278

King RL, Taylor AB (1936) *Malpighamoeba locustae* sp. n. (Amoebidae), a protozoan parasitic in the malpighian tubes of grasshoppers. Trans Am Microscopists Soc 55:6–10

Koella JC, Sorensen FL, Anderson RA (1998) The malaria parasite, *Plasmodium falciparum*, increases the frequency of multiple feeding of its mosquito vector, Anopheles gambiae. Proc Biol Sci 265(1398):763–768. https://doi.org/10.1098/rspb.1998.0358

Kohlmayr B, Weiser J, Wegensteiner R, Handel U, Zizka Z (2003) Infection of Tomicus piniperda (Col., Scolytidae) with Canningia tomici sp. n. (Microsporidia, Unikaryonidae). Anz Schädlingskunde 76:65–73

Kramer JP (1959a) Some relationships between *Perezia pyraustae* Paillot (Sporozoa, Nosematidae) and *Pyrausta nubilalis* (Hiibner) (Lepidoptera, Pyralidae). J Insect Pathol 1:25–33

Kramer JP (1959b) Observations on the seasonal incidence of microsporidiosis in European corn borer populations in Illinois. Entomophaga 4:37–42

Kramer JP (1970a) Longevity of microsporidian spores with special reference to *Octosporea muscaedomesticae* flu. Acta Protozool 8:217–224

Kramer JP (1970b) Longevity of microsporidian spores with special reference to *Octosporea muscae domesticae*. Flu. Acta Protozool 8:127–135

Kurtii TJ, Brooks MA (1971) Growth of a microsporidian parasite in cultured cells of tent caterpillars *(Malacosoma)*. Curr Top Microbiol Immunol 55:204–208

Kurtii TJ, Brooks MA (1977) The rate of development of a microsporidian in moth cell culture. J Invertebr Pathol 29:126–123

Kurtti TJ, Munderloh UG, Noda H (1990) *Vairimorpha necatrix*: infectivity for the development in a lepidopteran cell line. J Invertebr Pathol 55:61

Kyei-Poku G, Gauthier D, Schwarz R, van Frankenhuyzen K (2011) Morphology, molecular characteristics and prevalence of a Cystosporogenes species (Microsporidia) isolated from Agrilus anxius (Coleoptera: Buprestidae). J Invertebr Pathol 107:1–10

Lacey LA, Undeen AH (1986) Microbial control of blackflies and mosquitoes. Annu Rev Entomol 31:265–296

Lai PF (1980) Microsporidia in *Schistosoma mansoni* and their propagation in vivo and in vitro. Ph. D. thesis, University of London, p 252

Lange CE, Azzaro FG (2008) New case of long-term persistence of *Paranosema locustae* (microsporidia) in melanopline grasshoppers (Orthoptera: Acrididae) of Argentina. J Invertebr Pathol 99:357–359

Leonard CA, Hayman JR (2017) Role of host cell integrins in the microsporidium *Encephalitozoon intestinalis* adherence and infection in vitro. Fed Eur Microbio Soc Microbiol Lett 364:fnx169. https://doi.org/10.1093/femsle/fnx169

Levine ND, Corliss JO, Cox FEG, Deroux G, Grain J, Honigberg BM, Leedale GF, Loeblich AR, Lom J, Lynn D, Merinfeld EG, Page FC, Poljansky G, Sprague V, Vavra J, Wallace FG (1980) A newly revised classification of the protozoa. J Protozool 27:37–58

Li Y, Wu Z, Pan G, He W, Zhang R, Hu J, Zhou Z (2009) Identification of a novel spore wall protein (SWP26) from microsporidia *Nosema bombycis*. Int J Parasitol 39:391–398. https://doi.org/10.1016/j.ijpara.2008.08.011

Li Z, Pan G, Li T, Huang W, Chen J, Geng L et al (2012) SWP5, a spore wall protein, interacts with polar tube proteins in the parasitic microsporidian Nosema bombycis. Eukaryot Cell 11:229–237. https://doi.org/10.1128/EC.05127-11

Lipa JJ (1976) Microsporidians parasitizing the green tortrix in Poland and their role in the collapse of the tortrix outbreak in Puszcza Niepoilonicka during 1970–1974. Acta Protozool 1(5):529–536

Lom J, DyKova I (1992) Protozoan parasites of fishes. In: Developments in aquaculture and fisheries science, vol 26. Elsevier Publishers, New York, pp 125–154

Lukes J, Guilbride DL, Votypka J, Zikova A, Benne R, Englund PT (2002) Kinetoplast DNA network: evolution of an improbable structure. Eukaryot Cell 1(4):495–502

Maddox JV (1973) The persistence of the microsporidia in the environment. Misc Publ Entomol Soc Am 9:99–104

Maddox JV (1977) Stability of entomopathogenic protozoa. Misc Publ Entomol Soc Am 10:3–18

Maddox JV (1987) Protozoan diseases. In: Fuxa James R, Tanada Y (eds) Epizootiology of insect diseases. John Wiley & Sons, New York, pp 417–452

Maddox JV, Solter L (1996) Infective microsporidian spores in liquid nitrogen. J Eukaryot Microbiol 43(3):221–225. https://doi.org/10.1111/j.1550-7408.1996.tb01395.x

Maddox JV, Brooks WM, Fuxa JR (1981) *Vairimorpha necatrix* a pathogen of agricultural pests: potential for pest control. In: Burges HD (ed) Microbial control of pests and plant diseases 1976–1980. New York, Academic Press, pp 587–594

Maddox JV, McManus ML, Solter LF (1998) Microsporidia affecting forest Lepidoptera. In: McManua ML, Liebhold AM (eds) Proceedings of population dynamics, impacts, and integrated management of forest defoliating insects, USDA Forest Service general technical report NE-247, pp 187–197

Maddox JV, Baker MD, Jeffords MR, Kuras M, Linde A, Solter LF, McManus ML, Vavra J, Vossbrinck CR (1999) *Nosema Portugal*, n.sp., isolated from gypsy moth (*Lymantria dispar* L.) collected in Portugal. J Invertebr Pathol 73:1–14

Major P, Sendra KM, Dean P, Williams TA, Watson AK, Thwaites DT, Embley TM, Hirt RP (2019) A new family of cell surface located purine transporters in microsporidia and related fungal endoparasites. elife 8:e47037. https://doi.org/10.7554/eLife.47037

Malone LA (1990) In vitro spore hatching of two microsporidia, *Nosema costelytrae* and *Vavraiaon coperae*, from New Zealand pasture insects. J Invertebr Pathol 55:441–443. https://doi.org/10.1016/0022-2011(90)90091-j

Martin PA, Johnson DL, Forsyth DJ, Hill BD (1997) Indirect effects of the pyrethroid insecticide, deltamethrin on reproductive success of chestnut-collared longspurs. Ecotoxicology 7:89–97

Mclaughlin RE (1971) Use of protozoa for microbial control of insects. In: Burges HD et al (eds) Microbial control of insects and mites. Academic Press, New York, NY, p 861

McLaughlin RE (1973) Protozoa as microbial control agents. Misc Publ Entomol Soc Am 9:95–98

McLaughlin RE, Bell MR (1970) Mass production *in vivo* of two protozoan pathogens, *Mattesia grandis* and *Glugea gasti,* of the boll weevil, *Anthonomis grandis.* J Invertebr Pathol 16:84–88

McManus ML, Solter L (2003) Microsporidian pathogens in European gypsy moth populations. In: Proceedings: ecology, survey, and management of forest insects. USDA Forest Service, Northeast Research Station Gen Tech Rep: NE-311, pp 44–51

Micieli MV, Garcia JJ, Andreadis TG (2009) Factors affecting horizontal transmission of the microsporidium Amblyospora albifasciati to its intermediate copepod host Mesocyclops annulatus. J Invertebr Pathol 101:228–233

Milner RJ (1972) Nosema whitei, a microsporidan pathogen of some species of Tribolium: III. Effect on T. castaneum. J Invertebr Pathol 19:248–255. https://doi.org/10.1016/0022-2011(72)90216-9

Munderloh VG, Kuritti TJ, Ross SE (1990) Electrophoretic characterizations of chromosomal DNA from two microsporidia. J Invertebr Pathol 55:243

Mussgnug GL (1980) Integration of *Nosema locustae* with chemical insecticides and entomopox virus for control of grasshoppers. Ph.D. thesis, Montana State University, p 133

Nordin GL, Maddox JV (1974) Microsporida of the fall webworm Hyphantria cunea. 1. Identification, distribution, and comparison of Nosema sp. with similar Nosema spp. from other Lepidoptera. J Invertebr Pathol 24(1–1):3

Oshima K (1964) Method of gathering and purifying active spores of *Nosema bombycis* and preserving them in good condition. Annotationes Zoologicae Japonenses 37:94–101

Paillot A (1927) Sur deux protozoaires nouveaux parasites des chenilles de *Pyrausta nubilalis* Hubner. CR Acad Sci Paris 185:673–675

Patti JM, Allen BL, Mcgavin MJ, Höök M (1994) MSCRAMM-mediated adherence of microorganisms to host tissues. Annu Rev Microbiol 48:585–617

Peuvel-Fanget I, Polonais V, Brosson D, Texier C, Kuhn L, Peyret P et al (2006) EnP1 and EnP2, two proteins associated with the Encephalitozoon cuniculi endospore, the chitin-rich inner layer of the microsporidian spore wall. Int J Parasitol 36:309–318. https://doi.org/10.1016/j.ijpara.2005.10.005

Pickford R, Randell RL (1969) A non-diapause strain of the migratory grasshopper, *Melanoplus sanguinipes* (Orthoptera: Acrididae). Can Entomol 101:894–896

Polonais V, Mazet M, Wawrzyniak I, Texier C, Blot N, El Alaoui H et al (2010) The human microsporidian Encephalitozoon hellem synthesizes two spore wall polymorphic proteins useful for epidemiological studies. Infect Immun 78:2221–2230. https://doi.org/10.1128/IAI.01225-09

Purrini K (1975) On the distribution of microorganisms infecting the larvae of Euproctis chrysorrhoea L. (Lep., Lymantriidae) near Kosova, Yugoslavia. Anz Schädlingskde Pflanzenschutz Umweltschutz 48:182e183

Purrini K, Weiser J (1975) Natural enemies of Euproctis chrysorrhoea in orchards in Yugoslavia. Anz Schädlingskde Pflanzenschutz Umweltschutz 48:11–12

Purrini K, Weiser J (1985) Ultrastructural study of the microsporidian Chytridiopsis typographi (Chytridiopsida: Microspora) infecting the bark beetle, Ips typographus (Scolytidae: Coleoptera), with new data on spore dimorphism. J Invertebr Pathol 45:66–74

Quednau FW (1968) Natural control of larch sawfly at Carleton Bonaventure County, Quebec Province, Canada. (Thelohania, Pristophorae). Can Dept Forest Rural Develop Bi-Mon Res Notes 24:14

Raina SK, Rai MM, Khurad AM (1987) Grasshopper and locust control using microsporidian insecticides. In: Maramorosch K (ed) Biotechnology in invertebrate pathology and cell culture. Academic Press, New York, pp 345–365

Raina SK, Ewen AB (1979) In vitro infection of the fat body cells of the grasshopper *Melanoplus sanguinipes* (Fab.) by the microsporidian Nosema locustae Canning. In: Proceedings of the Vth international conference on invertebrate tissue culture, developments and applications. Rigi Kalbad, Switzerland

Schwalbe CP, Burkholder WE, Boush GM (1974) *Mottesia trogodermae* infection rates as influenced by mode of transmission, dosage and host species. J Stored Prod Res 10:161–166

Shapas TJ, Burkholder WE, Boush GM (1977) Population suppression of *Trogoderma glabrum* by using pheromone luring for protozoan pathogen dissemination. J Econ Entomol 70:469–474

Siegel JP, Maddox JV, Ruesink WG (1988) Seasonal progress of *Nosema pyrausta* in the European corn borer, *Ostrinia nubilalis*. J Invertebr Pathol 52:130–136

Smirnoff WA (1966) Thelohaniu pristiphorae sp. n., a microsporidian parasite of the larch sawfly, Prktiphorn erichsonii (Hymenoptera: Tenthredinidae). J Invertebr Pathol 8:360–364

Smirnoff WA (1967) Diseases of the larch sawfly, Pristiphora erichsonii, in Quebec. J Invertebr Pathol 10:417–424

Sneller VP (1979) Inhibition of *Dirofilaria immitis* in gregarine-infected *Aedes aegypti:* preliminary observations. J Invertebr Pathol 34:62–70

Sohi SS, Wilson GG (1976) Persistent infection of *Malacosoma disstria* (Lepidoptera, Lasiocampidae) cell cultures with *Nosema (Glugea) disstriae* (microsporidia: Nosematidae). Can J Zool 54:336–342

Southern TR, Jolly CE, Lester ME, Hayman JR (2007) EnP1, a microsporidian spore wall protein that enables spores to adhere to and infect host cells in vitro. Eukaryot Cell 6:1354–1362. https://doi.org/10.1128/ec.00113-07

Steinhaus EA (1951) Report on diagnoses of diseased insects 1944–1950. Hilgardia 20:629–678

Streett DA, Ralph D, Hink WF (1980) Replication of *Nosema algerae* in three insect cell lines. J Protozool 27:113–117

Tachezy J (2019) Hydrogenosomes and mitosomes: mitochondria of anaerobic eukaryotes. Springer, Berlin

Takyorian P, Weiss L, Cali A (2005) The early events of *Brachiola algerae* (microsporidia) infection: spore germination, sporoplasm structure, and development within host cells. Folia Parasitol 52:118–129

Takyorian PM, Buttle KF, Mankus D, Mannella CA, Weiss LM, Cali A (2013) The multilayered interlaced network (min) in the sporoplasm of the *microsporidium anncaliia algerae* is derived from golgi. J Eukaryot Microbiol 60:166–178. https://doi.org/10.1111/jeu.12019

Tanada Y (1964) Incidence of microsporidiosis in field population of the armyworm, *Pseudaletia unipuncta* (Haworth). Proc Hawaiian Entomol Soc 3:435

Tanada Y (1976) Epizootiology and microbial control. In: Bulla LA Jr, Cheng TC (eds) Biology of the microsporidia, vol 1. Plenum, New York, pp 127–135

Tanada Y, Chang GY (1962) An epizootic resulting from a microsporidian and two virus infections in the army-worm, *Pseudaletia unipuncta* (Haworth). J Insect Pathol 4:129–131

TerHorst CP (2010) Evolution in response to direct and indirect ecological effects in pitcher plant inquiline communities. Am Nat 176:675–685

TerHorst CP (2011) Experimental evolution of protozoan traits in response to interspecific competition. J Evol Biol 24:36–46

Thomson HM (1955) Perezia fumiferanae n. sp., a new species of microsporidia from the spruce budworm Choristoneura fumiferana (Clem.). J Parasitol 41:416–423

Thomson HW (1958) The effect of a microsporidian parasite on the development, reproduction and mortality of the spruce budworm, *Choristoneura fumiferana* (Clem). Can J Zool 36:499–511

Thomson HM (1959) A microsporidian parasite of the forest tent caterpillar Malacosoma disstria Hbn. Can J Zool 37:217–221

Tovar J, Fischer A, Clark CG (1999) The mitosome, a novel organelle related to mitochondria in the mitochondrial parasite *Entamoeba histolytica*. Mol Microbiol 32:1013–1021. https://doi.org/10.1046/j.1365-2958.1999.01414.x

Tovar J, León-Avila G, Sánchez LB, Sutak R, Tachezy J, Van Der Giezen M, Hernandez M, Muller M, Lucocq JM (2003) Mitochondrial remnant organelles of *giardia* function in iron-sulphur protein maturation. Nature 426:172. https://doi.org/10.1038/nature01945

Tsaousis AD, Kunji ER, Goldberg AV, Lucocq JM, Hirt RP, Embley TM (2008) A novel route for ATP acquisition by the remnant mitochondria of *Encephalitozoon cuniculi*. Nature 141:553. https://doi.org/10.1038/nature06903

Undeen AH (1975) Growth of *Nosema algerae* in pig kidney cell cultures. J Protozool 22:107–110

Van Denburgh RS, Burbutis PB (1962) The host parasite relationship of the European corn borer, *Ostrinia nubilalis,* on the protozoan, *Perezia pyraustae,* in Delaware. J Econ Entomol 55:65–67

Van der Paauw (1990) The microsporidia and microsporidiosis. In: Wittner M, Weiss LM (eds) American Soc Microbiol, Washington, DC, USA, 1999, pp 129–171

van der Paauw H, Johnson DL, Maiga B (1990) Evaluation of a large-scale application of *Nosema locustae* bait against grasshoppers in the Malian Sahel. Report to the U.S. Agency for International Development, Service Nationale de la Protection des Vegetaux, Mali, and Ciba-Geigy AG, Basel, p 116

Van Frankenhuyzen K, Ebling P, McCron B, Ladd T, Gauthier D, Vossbrinck C (2004) Occurrence of Cystosporogenes sp. (Protozoa: Microsporidia) in a multispecies insect production facility and its elimination from a colony of the eastern spruce budworm, Choristoneura fumiferanae (Clem.) (Lepidoptera: Tortricidae). J Invertebr Pathol 87:16–28

Vavra J, Larsson JIR (2014) Structure of the microsporidia. In: Becnel LM, Weiss JJ (eds) Microsporidia pathogens of opportunity. Wiley Blackwell, Oxford, pp 1–70. https://doi.org/10.1002/9781118395264.ch1

Vavra J, Maddox JV (1976) Methods in microsporidiology. In: Bulla LA et al (eds) Comparative pathobiology, biology of the microsporidia, vol 1, pp 281–231

Wang Y, Dang X, Ma Q, Liu F, Pan G, Li T et al (2015) Characterization of a novel spore wall protein NbSWP16 with proline-rich tandem repeats from Nosema bombycis (microsporidia). Parasitology 142:534–542. https://doi.org/10.1017/S0031182014001565

Wallace FG (1979) Biology of the Kinetoplastida of arthropods. In: Lumsden WHR et al (eds) Biology of the Kinetoplastida, vol 2, pp 213–240

Waraich NF, Jain S, Colloms SD, Stark WM, Burton NP, Maxwell A (2020) A novel decatenation assay for DNA topoisomerases using a singly-linked catenated substrate. Biotechniques. https://doi.org/10.2144/btn-2020-0059

Washburn JO, Anderson JR (1986) Distinction of *Lambornella clarki* (Ciliophora: Tetrahymenidae) and other mosquito parasites in California treeholes. J Invertebr Pathol 48:296–309

Watts JG, Huddleston EW, Owens JC (1982) Rangeland entomology. Annu Rev Entomol 27:283–311

Wegensteiner R, Weiser J, Führer E (1996) Observations on the occurrence of pathogens in the bark beetle Ips typographus L (Col., Scolytidae). J Appl Entomol 120:199–204

Weidner E (1970) Ultrastructural study of microsporidian development. 1. *Nosema* sp. Sprague, 1965 in *Callinectes sapidus* Rathbun. Z Zellforsch 105:33–54

Weidner E (1972) Ultrastructural study of microsporidian invasion into cells. Z Parasitenkd 40:227–242

Weiser J (1969) Phylum microsporida sprague. In: Lee JJ, Hunter SH, Boree EC (eds) Illustrated guide to the protozoa. Allen Press, Lawrence, KS, pp 375–383

Weiser J (1970) Three new pathogens of the Douglas fir beetle, Dendroctonus pseudotsugae: Nosema dendroctoni n. sp., Ophryocystic dendroctoni n. sp., and Chytridiopsis typographi n. comb. J Invertebr Pathol 16:436–441

Weiser J (1971) Three new pathogens of the Douglas fir beetle, Dendroctonus pseudotsugae: Nosema dendroctoni n. sp., Ophryocystic dendroctoni n. sp., and Chytridiopsis typographi n. comb. J Invertebr Pathol 16:436–441

Weiser J, Veber J (1975) Die mikrosporidie Thelohania hyphantriae Weiser des Weissen barenspinner (Hyphantria cunea) und anderer mitghiden seiner luaconose. Z Angew Ent 40:55–70

Weiser J, Wegensteiner R, Zizka Z (1997) Ultrastructures of Nosema typographi Weiser 1955 (Microspora: Nosematidae) of the bark beetle Ips typographus L. (Coleoptera; Scolytidae). J Invertebr Pathol 70:156–160

Weiser J, Holusa J, Zizka Z (2006) Larssoniella duplicati n.sp. (Microsporidia, Unikaryonidae), a newly described pathogen infecting the double-spined spruce bark beetle, Ips duplicatus (Coleoptera, Scolytidae) in the Czech Republic. J Pest Sci 79:127–135

Williams BAP, Hirt RP, Licocq JM, Embley M (2002) A mitochondrial remnant in the microsporidian *Trachipleistophora hominis*. Nature 418:865–869

Williams BA, Dolgikh VV, Sokolova YY (2014) Microsporidian biochemistry and physiology. In: Weiss LM, Becnel JJ (eds) Microsporidia: pathogens of opportunity. Wiley Blackwell, Hoboken, NJ, pp 245–260. https://doi.org/10.1002/9781118395264.ch9

Wilson GG (1975) Occurrence of Thelohania sp. and Pleistophora sp. (Microsporida: Nosematidae) in Choristoneura fumiferana (Lepidoptera: Tortricidae). Can J Zool 53:1799–1802

Wilson GG, Burke JM (1971) Nosema thomsoni n. sp., a microsporidian from Choristoneura conflictana (Lepidoptera: Tortricidae). Can J Zool 49:786e788

Wilson GG, Burke JM (1978) Microsporidian parasites of Archips cerasivoranus (Fitch) in the district of Algoma, Ontario. Proc Entomol Soc Ontario 109:84e85

Wu Z, Li Y, Pan G, Tan X, Hu J, Zhou Z et al (2008) Proteomic analysis of spore wall proteins and identification of two spore wall proteins from Nosema bombycis (Microsporidia). Proteomics 8:2447–2461. https://doi.org/10.1002/pmic.200700584

Wu Z, Li Y, Pan G, Zhou Z, Xiang Z (2009) SWP25, a novel protein associated with the *Nosema bombycis* endospore. J Eukaryot Microbiol 56:113–118. https://doi.org/10.1111/j.1550-7408.2008.00375.x

Xu J, Pan G, Fang L, Li J, Tian X, Li T, Zhou Z, Xiang Z (2006) The varying microsporidian genome: existence of long terminal repeat retrotransposon in domesticated silkworm parasite Nosema bombycis. Int J Parasitol 36:1049–1056

Yang D, Dang X, Peng P, Long M, Ma C, Qin JJG et al (2014) NbHSWP11, a microsporidia Nosema bombycis protein, localizing in the spore wall and membranes, reduces spore adherence to host cell BME. J Parasitol 100:623–633. https://doi.org/10.1645/13-286.1

Zimmack HL, Brindley TA (1957) The effect of the protozoan parasite *Perezia pyraustae* Paillot on the European corn borer. J Econ Entomol 50:637–640

Yang D, Pan G, Dang X, Shi Y, Li C, Peng P et al (2015) Interaction and assembly of two novel proteins in the spore wall of the microsporidian species Nosema bombycis and their roles in adherence to and infection of host cells. Infect Immun 83:1715–1731. https://doi.org/10.1128/IAI.03155-14

Zimmack HL, Arbuthnot KD, Brindley TA (1954) Distribution of the European corn borer parasite *Perezia pyraustae* and its effect on the host. J Econ Entomol 47:641–645

Zuk M (1987) The effects of gregarine parasites on longevity, weight loss, fecundity and developmental time in the field crickets *Gryllus veletis* and *G. pennsybanucus*. Ecol Entomol 12:349–354

Entomopathogenic Nematodes 9

Ashok Kumar Chaubey and Aasha

Contents

Abstract

Entomopathogenic nematodes (EPN) are found in all inhabited continents except Antarctica (no report yet) and a range of ecologically diverse habitats, from cultivated fields to deserts. *Steinernema* and *Heterorhabditis* are the well studied

A. K. Chaubey (✉) · Aasha
Nematology Laboratory, Department of Zoology Institution: Chaudhary Charan Singh University, Meerut, Uttar Pradesh, India

Omkar (ed.), *Microbial Approaches for Insect Pest Management*,
https://doi.org/10.1007/978-981-16-3595-3_9

genera which belong to the family Steinernematidae and Heterorhabditidae, associated with symbiotic bacteria *Xenorhabdus* and *Photorhabdus* respectively. As far as entomopathogenicity is concerned genus *Oscheius* is less studied, also having ability to kill the host insect due to the mutually associated with efficacious bacterial genera of *Pseudomonas*, *Enterococcus* and *Serratia*. The bacterial complex of these nematodes makes them a prominent mediator for the bio-management of many insect pests. Many studies have proven the active and main involvement of nematode's bacterial partner by releasing secondary metabolites in causing septicemia and oenocytoids. Globally, both the genera, that is, *Steinernema* and *Heterorhabditis,* are represented by 100 and 16 species, respectively, while *Oscheius* is represented by 45 species, out of which 17 are from the Indian subcontinent. The information on EPN diversity is limited in India. EPN show high potential for plant protection and can play a major role in Integrated Pest Management (IPM) of insects.

Some species-specific EPN-based formulations, like Biovector, Sanoplant, Helix, Magnet and Entonem, are available in the developed countries, which are being used by the farmers. In India, only two formulations, that is, green commandos and soil commandos, were developed using exotic EPN species but these nematodes were not efficacious against insects probably because of their poor adaptability to Indian environmental conditions. These products were withdrawn from the market. Currently, the latest formulation developed by Multiplex Biotech Pvt. Ltd. is marketed in the name of *Soldier* (contains *Heterorhabditis indica*) and Bouncer (contains *Steinernema carpocapsae*) but these formulations are not much in use by the farmers. In most cases, there is no need for special application equipment. Most nematode species are compatible with pressurized, mist, electrostatic, fan and aerial sprayers. Hose-end sprayers pump sprayers, and watering cans are effective applicators as well. Nematodes can even be applied through irrigation systems in agricultural fields during crop growing seasons.

Keywords

Steinernema · *Heterorhabditis* · *Oscheius* · Biological control · Formulation and pathogenicity

Learning Objectives

1. To isolate entomopathogenic nematodes and screen their virulence against the insect pests.
2. To develop cost effective EPN based biopesticide.
3. To increase the shelf-life of developed biopesticide for long duration storage in Indian agriculture and farmers' perspectives.

9.1 Introduction

9.1.1 Nematodes and Entomopathogenic Nematodes (EPNs)

Nematodes are highly diverse and ubiquitous creature on the planet Earth. Till date more than 28,000 species of nematodes have been described from across the world and the global species richness has been estimated between 500,000 and one million species (Hugot et al. 2001; Gaugler and Bilgrami 2004). Nematodes represent an especially abundant and spacious phylum Nematoda, with many free-living and parasitic species. They are found in all the conceivable habitats, but are often overlooked because most of them cannot be seen by the naked eyes. Many species are highly specialized parasites of vertebrates, including humans or of insects and other invertebrates. The most studied nematode, *Caenorhabditis elegans* (Rhabditidae), is currently considered a eukaryote model organism for research in biology and biomedical fields. However, many other nematode species are emerging at present as powerful model organisms for studying diverse disciplines, including ecology, physiology, developmental biology and evolution. On the other hand, nematodes are also utilized as biocontrol agents of insect pests. Natural enemies of insects and arthropods fall into three major categories: predators, parasitoids and pathogens. The nematodes fall into the category of pathogens because of causing sickness to insect, which ultimately cause death of the host insect. Among the vast variety of parasitic nematodes, some have evolved an association with insect pathogenic bacteria. Together the bacteria and nematode are a lethal duo. These nematodes are called 'entomopathogenic nematodes' (EPNs). EPNs are lethal obligatory insect parasites inhabiting soil.

The term entomopathogenic nematodes (EPNs) originated from the Greek vocables, 'Eντομος' (entomos, 'insect') 'πάθη' (pathê, 'disease') and 'γένος' (guenos, 'producing'), means a group of nematodes having ability to cause disease and severity to kill the insects. Dillman et al. (2012) clarified 'entomopathogenicity' as '*EPN must rapidly kill their hosts with the aid of bacterial partners and must pass on the associated bacteria to future generations*'. EPNs live inside the body of their hosts. These nematodes are taxonomically grouped under two families, the Steinernematidae consisting of two genera, that is, *Steinernema* (100 valid species) and *Neosteinernema* (01 species only, i.e. *N. longicurvicauda*); and Heterorhabditidae, which contains only one genus, *Heterorhabditis* (16 valid species) globally (Bhat 2019). They are also commercially produced and used in the biological control of insects.

To become efficacious biocontrol agent, the selection of the organisms should be based on host range, host finding or foraging strategy, tolerance to environmental factors (temperature, moisture, soil type, exposure to ultraviolet light, salinity and organic content of soil, means of application, agrochemicals and others) and their effects on survival and efficacy. In this context, EPNs have great potential and positive attributes for the biocontrol (Kaya and Gaugler 1993; Shapiro-Ilan and Grewal 2008). They have a wide host range, where some of nematode species have been reported to infect dozens of insect species (Poinar 1979; Klein 1990). EPNs are also amenable to mass production under in vivo or in vitro conditions (Shapiro-Ilan and Gaugler 2002; Shapiro-Ilan et al. 2014).

Application of EPNs is safe for environment, humans and other non-targeted organisms (Akhurst and Smith 2002; Ehlers 2005) and are exempted for pesticide registration in many countries (Ehlers 2005) with a few exceptions, such as *Steinernema scarabaei* (Koppenhöfer and Fuzy 2003a, 2003b). Grewal et al. (1994) have observed two types of searching behaviours in EPNs, that is, ambusher and cruiser. In Ambusher an energy-saving strategy is utilized, where the third-stage infective juveniles (IJs) lie-in-wail (*S. glaseri* and *H. bacteriophora*) to attack mobile insects in the upper soil. In cruisers third-stage IJs actively search their hosts/target insects available at significant difference using some volatile clues and other methods to find their underground host (*S. carpocapsae*); hence, effective against less mobile insects, such as scarabid larvae. Species, such as *S. feltiae* and *S. riobrave*, showed an intermediate behaviour. Other than behaviour, there are several biotic factors, such as choice of nematode species and rate of application and abiotic factors, such as UV light and desiccation, exist in the environment, which may affect their pest control efficacy (Kaya and Gaugler 1993; Shapiro-Ilan et al. 2002, 2006, 2012).

Another genus *Oscheius* (Family: Rhabditidae) was also identified as EPN (Torres-Barragan et al. 2011). Almost all species of the genus *Oscheius* are facultative parasites (Ye et al. 2011), but a few experiments demonstrate its biocontrol characteristics. However, genus *Oscheius* is less studied with regard to its pathogenicity.

9.1.2 Morpho-Taxomerical and Molecular Identification of EPNs

Based on their morphology, *Steinernema* and *Heterorhabditis* are very similar to each other, making them undistinguishable for a non-expert eye. However, systematic feature keys are used for the identification of EPN species (Hominick et al. 1997). Poinar (1990) gave the detailed morphological comparison and emphasized clear differences between these two genera, including family that is, Steinernematidae and Heterorhabditidae. Based on a few important attributes, like (i) position of excretory pore anterior to nerve ring in case of *Steinernema* (Fig. 9.1a) and posterior in case *Heterorhabditis* (Fig. 9.2a), (ii) colour variation in infected cadavers (produce a large number of IJs/Nematodes), which in case of former appears black or no colour change, while in later brick red (Fig. 9.3a–c), (iii) cadavers infected with *Heterorhabditis* show bioluminescence, which is not seen in case of *Steinernema*, (iv) bacteria associated are *Xenorhabdus* in *Steinernema* and *Photorhabdus* in *Heterorhabditis* (Fig. 9.5a, b).

Adults of first-, second-generations and third-stage IJs of *Steinernema* and *Heterorhabditis* possess some distinctive morphological features, which are very important from the taxonomic point of view. However, it became a monotonous task to categorize the increasing number of species with these taxonomic characteristics.

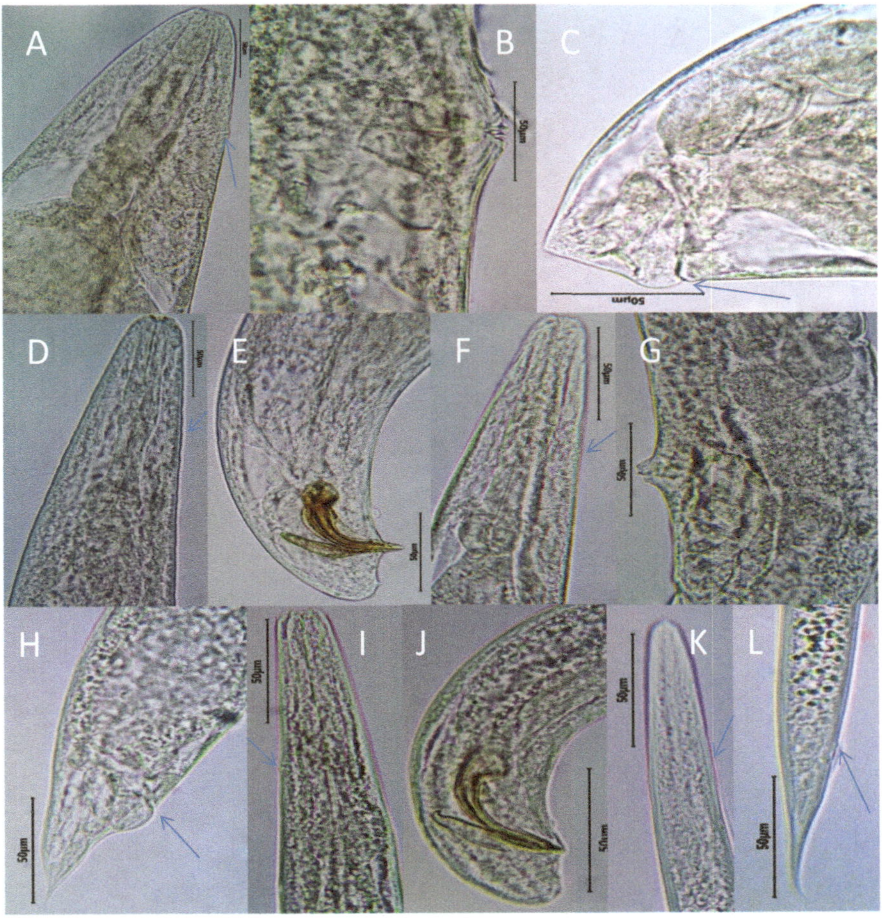

Fig. 9.1 Photomicrographs of *Steinernema*. (**a–c**) first generation Female, (**d** and **e**); first generation male, (**f–h**); second generation female, (**i** and **j**); second generation male; (**k** and **l**); third stage juvenile. (**a, d, f, i, k**); anterior region, (**b, g**); vulval region, (**c, e, h, j, l**); posterior region. Scale bars: (**a–-l**) = 50 μm (Bhat 2019)

Therefore, certain ratios and De Man Indices were created in order to delineate the species more appropriately (de Man 1880). These ratios are based on the following characteristics, viz., tail length; position of excretory pore, nerve ring and pharynx length (Figs. 9.1a, c, d, f, h, i, k, l and 9.2a, c, d, f, h, i, j). Besides these, males acquire some prominent characters, such that spicules and gubernaculums (Figs. 9.1e, j and 9.2e). Analysis and measurement of these characters play an essential role in the identification of species. Vulva is well known in the females of the entomopathogenic nematodes and its position in addition to the associated structure of the vulva gives the taxonomists a comprehensible way in recognition of the species (Figs. 9.1b, g and 9.2b, g). In case of IJs lateral field, tail shape and

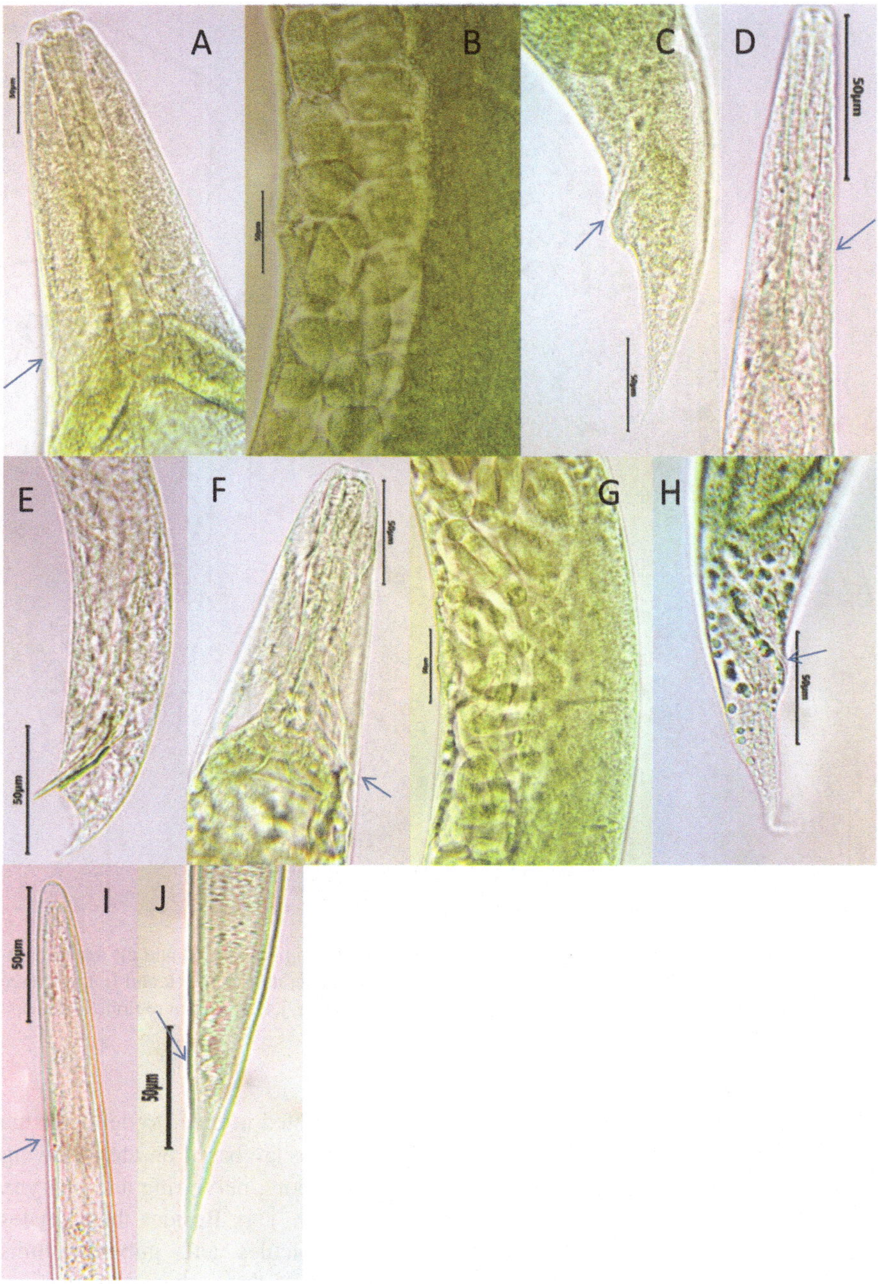

Fig. 9.2 Photomicrographs of *Heterorhabditis* species. Hermaphrodite: (**a**) Pharyngeal region; (**b**) Vulval region. (**c**) Anal region. Male: (**d**) Pharyngeal region; (**e**): Posterior region. Amphimictic female: (**f**): Pharyngeal region; (**g**): Vulval region; (**h**): Anal region. third stage juvenile: (**i**): Anterior region; (**j**): Anal region. Scale: **a–j** = 50 µm. (Bhat 2019)

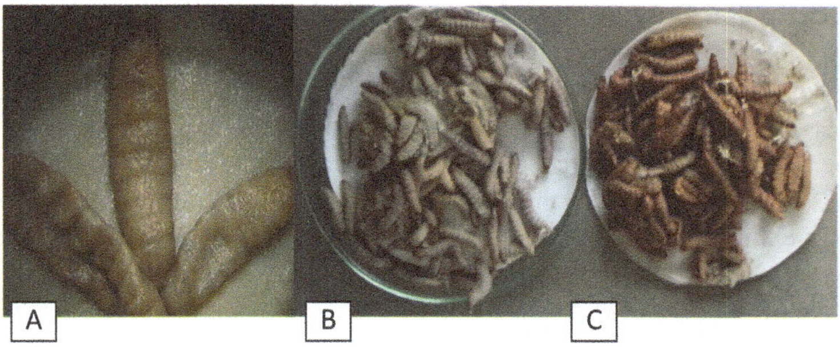

Fig. 9.3 Infected *Gallaria mellonella* larvae with entomopathogenic nematodes. (**a**) Emergence of EPN from infected larvae and larva infected with *Steinernema* spp., (**b**) larvae infected with *Heterorhabditis indica* (*Photorhabdus asymbiotica*) (**c**) larvae infected with *Heterorhabditis indica* (*P. laumondi*)

length, head contour, cephalic horns etc., are some of the important characteristics of taxonomic importance (Mráček and Bednarek 1991).

In case of *Steinernema*, based on IJ length, four 'species group' are created, that is, *glaseri group* (IJ L > 1000 μm), *feltiae group* (IJ L = 700–1000 μm), *intermedium group* (IJ L = 600–700 μm) and *corpocapsae group* (IJ L < 600 μm). There is another group '*bicornutum*' in which small morphology-based cluster of species was found and is diagnosed on the basis of the presence of horn-like structures on the anterior region. The reproductive apparatus of male, protruding from male body called as spicule, is one of the most discriminative features in the identification of steinernematids. The spicules of second-generation male are more separated from each other (Adams and Nguyen 2002). Some measurements of females, males and third-stage juveniles (J3) of EPNs allow discrimination among species (Nguyen and Smart 1990). Generally females are always longer than males, although the size ratio varies between species (Adams and Nguyen 2002). Accurate identification of the nematodes is not a trivial task. It is both labour-intensive and time-consuming, and these nematodes are morphologically conservative. Morphology is entirely dependent on the external features of the specimen; however, some genes have the tendency to not express themselves in the form of phenotype, although they possess the phylogenetic relationships of the species with the other species of a genus and with other orders are also established by utilizing the modern and advanced molecular tools (some conserved regions), which are very important from the taxonomic point of view. Moreover, the 18S and 28S rDNA are found to be conserved genes as they evolve slowly and are used to compare the distant taxa that had diverged a long time ago. Besides this, spacer sequences, viz., ITS1, ITS2 and ETS, are being used to compare the phylogeny of closely related species as they evolve at a faster rate as compared to 18S and 28S rDNA genes (Subbotin and Moens 2006). Highly variable D2 and D3 expansion segments of 28S rDNA have been used for molecular characterization among the nematodes to infer the phylogenetic relationships among the species.

9.1.3 Life Cycle of EPN and their Bacterial Symbionts

The nematode family Steinernematidae and Heterorhabditidae share several common characteristics, such as mutualistic association with enteric bacteria within their intestine and need to parasitize the insect host and also almost similar life cycle pattern (Fig. 9.4). EPNs have three preadaptations, which allow them to evolve the life styles exhibited by the genera *Steinernema* and *Heterorhabditis:* (i) they have evolved species that have a variety of associations with insects; (ii) they produce the dauer juvenile stage, which confers the capacity to enter an insect and persist in the absence of food and (iii) they are bacterial feeders and are pre-adapted to enter a mutualistic relationship with entomopathogenic bacteria (EPB) inside the insect haemocoel (Sudhaus 1993). The IJ of EPNs is compatible with other biological and chemical pesticides, fertilizers and soil amendments (Krishnayya and Grewal 2002).

EPNs are globally distributed, have the ability to live in drastic conditions and tolerance against harsh environmental conditions including anoxybiosis, thermobiosis and desiccation (Grewal 2000). The free-living stage of these nematodes is carrying bacterial symbionts in their alimentary canal (Fig. 9.5a, b) once encounter the suitable host insect, it penetrates host through mouth, anus and spiracle or by tearing the skin making the way to haemocoel (Kaya and Gaugler

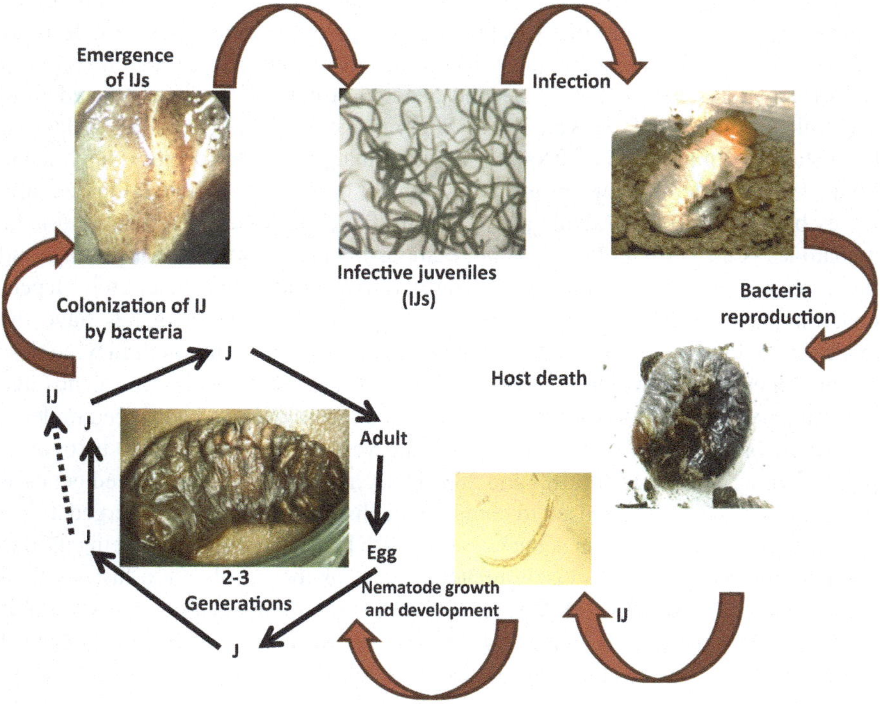

Fig. 9.4 Life cycle of entomopathogenic nematodes (Modifiied after Ffrench-Constant et al. 2003)

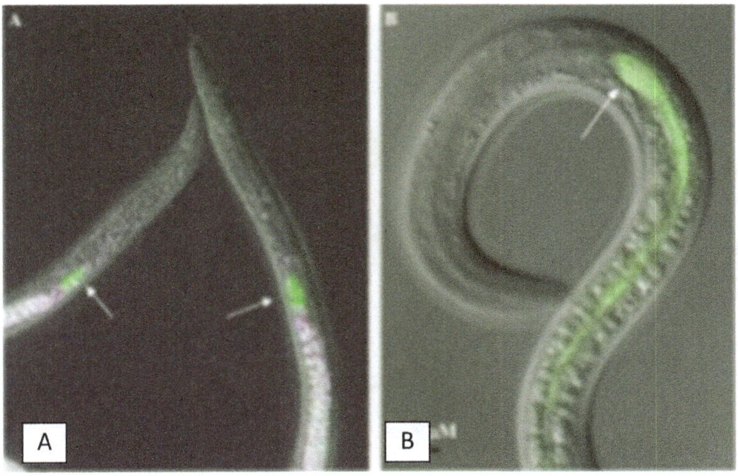

Fig. 9.5 Location of symbiotic bacteria in the intestine of entomopathogenic nematodes (**a**) *Xenorhabdus* with *Steinernema*, (**b**) *Photorhabdus* with *Heterorhabditis* (reproduced from Ciche et al. 2006)

1993). Presence of buccal cuticular tooth in heterorhabditid nematodes helps them to penetrate the cuticle of insect (Bedding and Molyneux 1982). The routes taken by EPN to enter into the body of insect hosts are reproduced from Castillo et al. (Castillo et al. 2011; Fig. 9.6). Once the IJ reach to the haemocoel, they release bacteria in the haemocoel, which start to colonize therein by releasing toxins. The toxins in turn kill insect hosts within 24–48 h. When the food conditions decline, the new generations of IJs coming from the last adult generation, abandon the dead body of insect to seek a new alive host (Poinar 1990).

Xenorhabdus and *Photorhabdus* are entomopathogenic bacteria (EPB) vectored by *Steinernema* and *Heterorhabditis* genera of EPNs, respectively. The life cycles of both *Xenorhabdus* and *Photorhabdus* are similar, revolving around the free-living infective form of their specific nematode, which acts as a vector for transferring the bacteria from host to host. Following penetration of the nematodes into the insect, bacteria are regurgitated directly into the haemocoel of insect host. If the infection is successful, nematodes resume development and start feeding. One of the three generations occurs inside the host and over the course of infection quality of the resource diminishes (Kaya and Gaugler 1993; Adams and Nguyen 2002). Subsequently, the bacteria grow unrestricted by the insect immune system (Daborn et al. 2001) releasing toxins to kill the insect host (Ffrench-Constant et al. 2003) and also serving as a food source for their nematode symbionts (Forst and Clarke 2002). As conditions decline, IJs are produced that leave the insect to seek new hosts (Ciche and Ensign 2003). The IJ is the only free-living stage and actively seeks hosts using a species specific strategy along a continuum from ambush to cruise foraging (Campbell and Gaugler 1993, 1997; Campbell and Kaya 2002; Grewal et al. 1994). Among the newly produced nematodes, the IJ re-associates with the

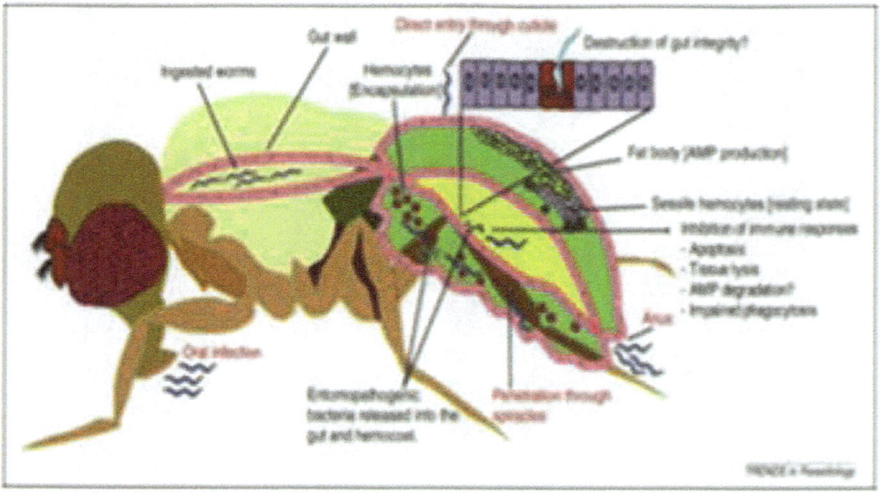

Fig. 9.6 A generalized depiction of entry routes used by entomopathogenic nematodes (reproduced from Castillo et al. 2011)

mutualistic bacteria before leaving the insect cadaver (Martens et al. 2003). The nematodes carry their bacterial symbionts monoxenically in a special vesicle of the IJ known as 'vesicle of Bird and Akhurst' (Bird and Akhurst 1983) in Steinernematidae and throughout the whole intestine of Heterorhabditidae (Endo and Nickle 1991) (Fig. 9.5a, b).

The free-living stage of the EPN is the most infective and operative stage where the free-living infective juveniles can search and move into the soil in search of insects to infect (Spence et al. 2008). As per the present scenario, the tools and techniques of molecular biology and genomics can be used to improve the efficiency and ability of EPNs by understanding their ecology, behaviour and how they respond to climate change through screening of the related genes (Segal and Glazer 2000). The life cycle pattern of entomopathogenic nematode/bacterium complex and their residing sites within insect host is depicted in Fig. 9.6.

9.1.4 Mode of Action

The majority of the eggs are retained inside the EPN maternal body after mating. The offspring then develops and feeds in the maternal body. This process is called endotokia-matricida [from the Greek word ενδο ('endo', inside) and τοκοσ ('tocos', birth) and from the Latin 'mater', mother and 'caedere', kill. This term was coined by Maupas (1919) when he described *Caenorhabditis elegans*. This phenomenon has the advantage to protect the offspring and in the case of EPN, to provide it with a high lipid food source, especially when the infected insect cadaver is about to be exhausted. If endotokia-matricida is promoted in case of scarce food supply, this phenomenon occurs for the first generation of juveniles even if the food source is still abundant (Fig. 9.6). It becomes obvious that the size of the susceptible insect will affect the development and survival of EPN. Some authors reported the inefficiency of *Steinernema* IJ to control micro-insect pests (Ebssa et al. 2004). Recently, Bastidas et al. (2014) clearly demonstrated the impaired development of EPN in micro-insects and the complete lack of infectivity of four different *Steinernema* species in insects smaller than 5 mm in length and concluded that *Steinernema* and *Heterorhabditis* nematodes cannot persist for a long time in the environment if no larger insects are available to them for completing their life cycles.

9.2 Target Insects Pests and their Control

Generally the term pest is derived from the French word '*Peste*' and the Latin word '*Pestis*', which means contagious disease. Pest may be defined as '*any organism whose population increases to cause economic injury to the crops, stored products, and nuisance or health hazards either to humans or to the livestock*'. Pests are sufficiently numerous to cause economic damage by causing injury or death to agricultural crops, destruction of stored products, which include insects, nematodes, mites, snails, slugs, fungus, bacteria, rats and birds. Insects are the most important competitors of food and fibres, as they cause millions of the economy loss to the humans. They have adverse and damaging impacts on agricultural production and market access, the natural environment and our lifestyle, and cause damage to the crops and food production, parasitizing livestock, or being a nuisance and health hazards to the human being.

9.2.1 Insect Pests in Abroad

Insect pests are the major destructive agents harbouring the fruits and vegetables with negative impact over their quality. This issue is a major threat for reduction in trade of fruits and vegetables in international market. Insect infestations reduce the economic value of the product, and product returns resulting from the discovery of live insects cause loss of consumer confidence (Johnson et al. 2013). There are more than six million pests on this planet earth, of which very few are serious pests. The important insect pests of commodities abroad are presented in Table 9.1.

9.2.2 Insect Pests in India

Indian economy is based on agriculture but most of the agri-products are damaged by pests. A large chunk of the agri-products (standing crops or stored grains) are damaged by various types of insects (Table 9.2). In India, estimated crop losses due to the pests and diseases are worth Rs. 50,000 crores annually (DACFW 2020) in which insects contribute~20%. Problem can easily be assessed, viz. *Helicoverpa armigera* alone causes loss to the tune of Rs. 1000 crore in crops, like cotton, tomato,

Table 9.1 Some of the important insect pest commodities in abroad

Pest common name	Pest scientific name	Family	Crops (targeted)
Armyworm	*Spodoptera frugiperda*	Noctuidae	Vegetables
Artichoke plume moth	*Platyptilia carduidactyla*	Pterophoridae	Artichokes
Banana moth	*Opogona sacchari*	Tineidae	Ornamentals
Banana root borer	*Cosmopolites sordidus*	Curculionidae	Bananas
Billbug	*Sphenophorus* parvulus	Curculionidae	Turf
Black cutworm	*Agrotisipsilon*	Noctuidae	Turf, vegetables
Black vine weevil	*Otiorhynchus sulcatus*	Curculionidae	Berries, ornamentals
Borer	*Synanthedon* spp. and other sesiids	Sesiidae	Fruit trees, ornamentals
Citrus root weevil	*Pachnaeus* spp.	Curculionidae	Citrus, ornamentals
Codling moth	*Cydiapomonella*	Tortricidae	Pome fruit
Corn earworm	*Helicoverpa zea*	Noctuidae	Vegetables
Corn rootworm	*Diabrotica virgifera*	Chrysomelidae	Vegetables
Cranberry girdler	*Chrysoteuchi atopiaria*	Crambidae	Cranberries
Crane fly	*Valdiviana edwardsina*	Tipulidae	Turf
Diaprepes root weevil	*Diaprepes abbreviates*	Curculionidae	Citrus, ornamentals
Fungus gnat	Arachnocampa spp.	Sciaridae	Mushrooms
Grape root borer	*Vitace apolistiformis*	Sesiidae	Grapes
Iris borer	*Macronoctua onusta*	Noctuidae	Iris
Large pine weevil	*Hylobius abietis*	Curculionidae	Forest plantings
Mole cricket	*Scapteriscus* spp.	Gryllotalpidae	Turf
Navel orangeworm	*Amyeloistra nsitella*	Pyralidae	Nut and fruit trees
Navel orange-worm	*Amyeloistr ansitella*	Pyralidae	Nut and fruit trees
Plum curculio	*Conotrachelus nenuphar*	Curculionidae	Fruit trees
Plum curculio	*Conotrachelus nenuphar*	Curculionidae	Fruit trees
Root weevil	*Otiorhynchus ovatus*	Curculionidae	Berries strawberry
Scarab grub	*Acanthonitis* spp., *Endrodius* spp.	Scarabaeidae	Turf, ornamentals
Serpentine Leafminer	*Liriomyza brassicae*	Agromyzidae	Vegetables, ornamentals
Shore fly	*Scatella* spp.	Ephydridae	Ornamentals
Small hive beetle	*Aethinatumida*	Nitidulidae	Bee hives
Sweet potato weevil	*Cylasformic arius*	Convolvulaceae	Sweet potato

pigeonpea, groundnut, sorghum, pearl, millet and other crops of economic importance. The behaviour of the insects is also changing due to climate change/cropping pattern from minor to major or from secondary to primary. Such changes are due to change of genotypes/impact of transgenics, injudicious use of pesticides, modification of cultural practices/tillage and the insects are likely to become serious and invasive (Rathee and Dalal 2018; Table 9.3a, b).

Table 9.2 Some of the commodities wise important insect pests in India

Common name	Insect name	Family	Commodity
Banana stem weevil	*Odoiporus longicollis*	Curculionidae	Banana
Black cutworm	*Agrotisip silon*	Noctuidae	Potato
Common cutworm	*A. segetum*	Noctuidae	Potato
Diamondback moth	*Plutella xylostella*	Plutellidae	Crucifers
Oriental armyworm	*Pseudaletias eparata*	Noctuidae	Rice
Potato tuber moth	*Phthorimaea operculella*	Gelechiidae	Potato
Red hairy caterpillar	*Amsacta albistriga*	Arctiidae	Groundnut
Rice green semi-looper	*Tryporyza xylostella*	Pyralidae	Rice
Shining leaf chafers	*Anomala* sp.	Scarabaeidae	Vegetables
Spotted stem borer	*Chilo zonellus*	Pyralidae	Maize
Tobacco beetle	*Lasioderma serricorne*	Curculionidae	Egg plant
Tobacco cutworm or cotton leaf-worm	*Spodoptera litura*	Noctuidae	Tobacco
Yellow stem borer	*Tryporyza incertulas*	Pyralidae	Rice
White fly	*Bemisia tabaci*		Cotton & tobacco
Serpentine leaf miner	*Liriomyzatri folii*		Cotton, tomato, cucurbits
Pink stem borer	*Sesamia inferens*	Noctuidae	Wheat, maize and sorghum
Hoppers	*Nilaparvata lugens*	Delphacidae	Rice & mango
Papaya mealybug	*Paracoccus marginatus*	Pseudococcidae	Papaya, cotton, mulberry
Citrus brown mite	*Eutetranychu sorientalis*	Tetranychidae	Apple, ber, citrus, cucurbits
Cotton bollworm	*Helicoverpa armigera*	Noctuidae	Vegetables and pulses
Tobacco cutworm or cotton leafworm	*Spodoptera litura*	Noctuidae	Vegetables, cotton, oilseeds
Sweet potato whitefly	*Bemisiata baci*	Aleyrodidae	Cotton

9.2.3 Application of Pesticides in India

To reduce the huge loss in crop yields, farmers apply tons of agrochemicals in their agriculture fields, as a tool to control the pests. It has been estimated that about Rs. 1200 crore worth of pesticides were used in India to control the bollworm of cotton only. The consumption of pesticides and biopesticides in India for Rabi and Kharif seasons was recorded to be 60,599 and 7804 MT, respectively, during 2019–2020. There is tremendous increase in the use of pesticides from 2014–2015

Table 9.3 A and B Recent imerging trends of insect pests in India

A Insect pests likely to become serious (with changes in climate/cropping patterns)		
Whitefly	*Bemisia tabaci* (Gennadius)	Cotton and tobacco
Fruit fly s	*Bactrocera* spp.	Fruits and vegetable
Mealybugs	*Paracoccus marginatus* (Williams and Granara de Willink) and *Phenacoccus solenopsis* Tinsley	Field and horticultural crops
Thrips	Several species *Scirtothrips dorsalis* Hood, *Frankliniella schultzei* Trybom, *Thrips tabaci* L., *Scirtothrips citri* (Moulton)	Groundnut, cotton and citrus
Wheat aphid	*Macrosiphum miscanthi* Takahashi	Wheat, barley and oat
Rice gall midge	*Orselia oryzae* (wood Mason)	Rice
Serpentine leaf miner	*Liriomyza trifolii* burgess	Cotton, tomato and cucurbits
Hoppers	*Nilaparvata lugens* Stal and *Nephotettix* spp.	Rice and mango
Pyrilla	*Pyrilla perpusilla* Walker	Sugarcane
Pink stem borer	*Sesamia inferens* Walker	Wheat, maize and sorghum
Newly emerging major pests (during last decade in Northern Plains)		
Common name	Scientific name	Crops
White fly	*Bemisia tabaci*	Cotton & Tobaccon &
Fruit fly	*Bactrocera* spp.	Fruits & vegetables
Mealy bugs	*Paracoccus marginatus, Phenacoccus solenopsis*	Field & horticulture crop
Thrips	*Scirtothrips dorsalis, Frankliniella schultzei, Thrips tabaci, Scirtothrips citri*	Groundnut, cotton & citrus
Wheat aphid	*Macrosiphum miscanthi*	Wheat, barley and oat
Rice gall midge	*Orselia oryzae*	Rice
Serpentine leaf miner	*Liriomyza trifolii*	Cotton, tomato and cucurbits
Hoppers	*Nilaparvata lugens*	Rice & mango
Pyrilla	*Pyrilla perpusilla*	Sugarcane
Pink stem borer	*Sesamia inferens*	Wheat, maize and sorghum

B Insect pest invasions

Common name	Scientific name	Crop(s)	Reference(s)
Tomato leaf Miner	*Tuta absoluta* (Meyrick)	Tomato	Sridhar et al. (2014)

(continued)

Table 9.3 A and B (continued)

B Insect pest invasions			
Common name	Scientific name	Crop(s)	Reference(s)
Western flower Thrips	*Frankliniella occidentalis* (Pergande)	Fruits and vegetables	Tyagi and Kumar (2015)
Coffee berry Borer	*Hypothenemus hampei* (Ferrari)	Coffee	Singh and Ballal (1991)
Coconut Eriopyhid mite	*Aceria guerreronis*	Coconut	Sathiamma et al. (1998)
Coconut leaf Beetle	*Brontispa longissima* (Gestro)	Coconut	CPCRI (2016)
Eucalyptus gall Wasp	*Leptocybe invasa* Fisher and La Salle	Eucalyptus	Jacob et al. (2007)
Papaya Mealybug	*Paracoccus marginatus* (William Granara de Willink)	Papaya, cotton and mulberry	Muniappan et al. (2008)

Source: Rathee and Dalal (2018)

to 2019–2020 (Table 9.4). Data revealed that the consumption of chemical and biopesticide increased 7.8% and 51.5%, respectively, while there was tremendous decrease in imported chemicals and biopesticides to the tune of 63.3% and 100% from 2014–2015 to 2019–2020, respectively (http://ppqs.gov.in/statistical-database). The data also indicate the increasing and decreasing trend of 38.3% and 19.8%, respectively, for chemical and bio-pesticide applications in agriculture fields during Rabi and Kharif crop growth seasons year 2014–2015 and 2019–2020 (Table 9.4) (http://ppqs.gov.in/statistical-database). The negative trend in the application of biopesticides indicates that the biopesticides available in the markets are either not efficacious/species specific or not much publicized amongst the actual beneficiaries. Therefore, there is a need to develop efficacious biopesticides and must be given publicity at farmers' fields, so that the farmers could be motivated to apply the biopesticides without any hesitation. A global report of transparency market research indicated that the market of biopesticide is bright and has been expected that the biopesticide market would be valued at US$4.17 million by 2023, which was US$1.72 million in 2014 and North America is expected to dominate global biopesticide market demand (http://www.transparencymarketresearch.com/biopestides-market.html).

Table 9.4 Consumption of pesticides in India during 2014–2015 and 2019–2020 for Rabi and Kharif crops (Unit: M.T)

Types of pesticide		Year		Quantity increased
		2014–2015	2019–2020	
Pesticide	Chemical	56,268.00	60,599.00	+4331.00
	Bio-pesticide	5152.00	7804.00	+2652.00
Indigenous	Chemical	17,859.00	24,627.00	+6820.00
	Bio-pesticide	2942.00	2359.00	−583.00
Imported	Chemical	1269.62	466.82	−802.80
	Bio-pesticide	2.00	0.00	−2.00

Source: http://ppqs.gov.in/statistical-database

9.3 Insect Pest Management

Chemical fertilizers and pesticides are presently accumulating in the environment harming the ecosystem, causing pollution and spreading some of the diseases (Gerhardson 2002). In the sixteenth century, Chinese were the first to use natural enemies, ants (*Oecophyllasma ragdina*) to control citrus insect pests (*Tesseratoma papillosa*) near Canton. Similarly, colonies of predaceous ants were collected from date groves from North to control various pests (Bellows Jr. and Fisher 1996).

9.3.1 EPN in Insect Pest Management

Use of chemicals has been one of the conventional methods to reduce losses by insect pests; but nowadays due to various unwarranted side effects, pest management is relied upon many other options along with pesticides. The integration of all these options is called Integrated Pest Management (IPM). IPM is a strategy to manage pests on the basis of a system's approach that looks at the whole orchard ecosystem. Synthetic chemical pesticides have various disadvantages, like crop and soil contamination; killing of beneficial flora and fauna; development of resistance in insects and adverse effects due to contamination in food chain and other environmental issues (Smart 1995; Bailey et al. 2009). To minimize pesticides contamination, EPNs were identified as potent biocontrol agents and are most suitable natural enemies of problematic insects because they reduce risk to humans and other related vertebrates (Athanassiou et al. 2010; Campos-Herrera et al. 2012).

The first EPN species introduced as biocontrol agent was *Steinernema glaseri* used against Japanese beetle, *Popillia japonica*, in the USA and thereafter, scientists focused in this area (Smart 1995). In the 1960s and 1970s, these creatures re-emerged as effective biocontrol agents with *S. carpocapsae* (named as *Neoaplectana carpocapsae*) as the main biocontrol agent (Pye and Burman 1977). With the advancement of fermentation technology, several species of EPNs (including *S. carpocapsae, S. scapterisci, S. feltiae, S. glaseri* and *H. megidis*) have been mass produced commercially and are sold in the market for use by growers in formulations suitable for short-term storage (Ehlers 2001; Dillon 2003). The mass production of IJs of EPNs is easy and cost-effective. The preferred method of application is inundative release (Feaster and Steinkraus 1996; Dillon et al. 2007).

Some of the well-known, most common and successfully applied nematodes as biopesticides, include *Steinernema carpocapsae, S. feltiae, S. kraussei, S. glaseri, S. riobrave, Heterorhabditis bacteriophora* and *H. megidis* and this is due to their easy mass production in liquid culture (Abate et al. 2017). Culture in live insect hosts (in vivo) requires low start-up costs, low level of technology and high nematode quality but low-cost efficiency; however, in vitro solid or liquid culture is cost-efficient method in which liquid culture requires the largest start-up capital.

At commercial level in Asia, Europe and North America at least 13 different species, namely *H. bacteriophora, H. indica, H. marelata, H. megidis, H. zealandica, S. carpocapsae, S. feltiae, S. glaseri, S. kushidai, S. kraussei, S. longicaudum, S. riobrave* and *S. scapterisci* (Lacey et al. 2015) have been commercialized. The products based on EPNs are species specific, efficacious and commercially available globally (Table 9.5).

9.3.2 Bio-Formulations Using EPNs

Mass production of EPNs, especially *Steinernema,* was first successfully started by Rudolph Glaser on synthetic medium for controlling the larvae of the Japanese beetle (*Popillia japonica*) in the USA (Glaser 1931). He was not aware of the presence of symbiotic bacterium in the gut of *Steinernema*, which was responsible for high virulence, nutrient supply for IJs and production of antibiotics to prevent secondary invaders. The rearing of EPNs was, therefore, mostly made primarily with insect hosts (Dutky et al. 1964) and most of the laboratories and cottage industries working on EPNs still employ this method of production. The presence of the bacterium was first discovered in *S. feltiae* by Bovien (Bovein 1937); however, the importance of the bacterium *X. nematophilus* for the reproduction of *S. carpocapsae* was evaluated by Poinar and Thomas (Poinar and Thomas 1966). This research of Poinar and Thomas laid the foundation of in vitro mass production of EPNs. Solid-state production was followed by growing IJs on Petri dishes using different agar media (House et al. 1965; Wouts 1981). Bedding (1981) made further advancement by growing *Steinernema* sp. on a three-D medium in flasks, using polyether-polyurethane sponge as an inert medium carrier and this method is still used in developing countries for in vitro mass production of IJs.

9.3.3 Formulation Technology in Aboard

The first effort at formulating these EPNs was commenced in 1979 and first nematode-based formulation was placement of third-stage juveniles on moist substrate, such as vermiculite, polyether-polyurethane sponge and peat or cedar

Table 9.5 Globally commercial use of entomopathogenic nematodes against some insect pest

Pest common name	Pest scientific name	Crops (targeted)	Effective nematodes[a]
Artichoke plume moth	*Platyptilia carduidactyla*	Artichokes	Sc
Armyworm	*Spodoptera frugiperda*	Vegetables	Sc, sf, Sr
Banana moth	*Opogona sacchari*	Ornamentals	Hb, Sc
Banana root borer	*Cosmopolites sordidus*	Bananas	Sc, sf, sg
Billbug	*Sphenophorus* spp.	Turf	Hb, Sc
Black cutworm	*Agrotisipsilon*	Turf, vegetables	Sc
Black vine weevil	*Otiorhynchus sulcatus*	Berries, ornamentals	Hb, Hd, Hm, Hmeg, Sc, sg
Borer	*Synanthedon* spp. and other sesiids	Fruit trees, ornamentals	Hb, Sc, sf
Cat flea	*Ctenocephalides felis*	Home yard, turf	Sc
Citrus root weevil	*Pachnaeus* spp.	Citrus, ornamentals	Sr, Hb
Codling moth	*Cydia pomonella*	Pome fruit	Sc, sf
Corn earworm	*Helicoverpa zea*	Vegetables	Sc, sf, Sr
Corn rootworm	*Diabrotica* spp.	Vegetables	Hb, Sc
Cranberry girdler	*Chrysoteuchi atopiaria*	Cranberries	Sc
Diaprepes root weevil	*Diaprepes abbreviates*	Citrus, ornamentals	Hb, Sr
Fungus gnat	Diptera: Sciaridae	Mushrooms	Sf, Hb
Grape root borer	*Vitace apolistiformis*	Grapes	Hz, Hb
Iris borer	*Macronoctua onusta*	Iris	Hb, Sc
Large pine weevil	*Hylobius abietis*	Forest plantings	Hd, Sc
Leafminer	*Liriomyza* spp. (dip: Agromyzidae)	Vegetables, ornamentals	Sc, sf
Mole cricket	*Scapteriscus* spp.	Turf	Sc, Sr, Sscap
Cat flea	*Ctenocephalides felis*	Home yard, turf	Sc
Navel orangeworm	*Amyeloistran sitella*	Nut and fruit trees	Sc
Plum curculio	*Conotrachelus nenuphar*	Fruit trees	Sr
Scarab grub[b]	Coleoptera: Scarabaeidae	Turf, ornamentals	Hb, Sc, sg, Ss, Hz
Shore fly	*Scatella* spp.	Ornamentals	Sc, sf
Root weevil	*Otiorhynchus ovatus*	Berries strawberry	Hm
Small hive beetle	*Aethina tumida*	Bee hives	Hi, Sr
Sweet potato weevil	*Cylasformic arius*	Sweet potato	Hb, Sc, sf
Navel orange worm	*Amyeloistran sitella*	Nut and fruit trees	Sc
Plum curculio	*Conotrachelus nenuphar*	Fruit trees	Sr

Source: Gozel and Gozel (2016). doi: https://doi.org/10.5772/63894

[a]Abbreviations of nematode species; Hb: *Heterorhabditis bacteriophora*, Hd: *H. downesi*, Hi: *H. indica*, Hm: *H. marelata*, Hmeg: *H. megidis*, Hz: *H. zealandica*, Sc: *Steinernema carpocapsae*, Sf: *S. feltiae*, Sg: *S. glaseri*, Sk: *S. kushidai*, Sr: *S. riobrave*, Sscap: *S. scapterisci*, Ss: *S. scarabaei*

[b]Efficacy against various pest species within this group varies among nematode species

shavings. They were used during 1979–1984 and had shelf-life of 0.5–1.1 months. In this formulation, IJs were active and had high requirement of oxygen (oxygen used/10^6 nematodes/day = 3.2 ml); thus, they were not considered good. In 1985, the motion of nematodes was made low by formulating them on absorbent activated charcoal, and it increased their shelf-life to about 1.2 months. During 1987–1990, IJs were made immobilized by placing them on moist gel materials, such as calcium alginate, xanthene or polyacrylamide, which increased their shelf-life to 2.6 months and reduced their oxygen (0.67 ml) requirements (Georgis 1990), but still nematodes need energy requirements to perform different activities, which decrease in their viability. To reduce their energy requirements and prolonging their shelf-life, the IJs were partially desiccated by placing them on or between two layers of absorbent clays and these were used during 1988. In 1993, the IJs of EPNs were further partially desiccated by encasement in 10–20 mm diameter water dispersible granules consisting of silica, clays, cellulose compounds, lignin and starches, which extended their shelf life to about 5.1 months and further reduced their oxygen requirements. With the more focus on increasing their shelf-life, nematologists recognized morphological, behavioural and biochemical differences of certain infective stages of steinernematids and heterorhabditids. The high lipid content in IJs was responsible for their prolonged survival rates and adaptation against environmental stresses (Selvan et al. 1993).

The formulation in polyurethane sponges was later developed and in these, IJs achieved a survival time of 1–3 months at 5–10 °C (Grewal 2002). Clay and powder formulations were developed by Bedding (1988) in which the IJs of EPNs were encapsulated in a hygroscopic attapulgite clay formulation with survival time of 8 weeks at 23 °C, which was called 'sandwich' type formulation. Later, other formulations, like alginate gels, aqueous suspensions and calcium alginate gels, were developed, but all these had issues of shelf-life, storage problems and application. Shaprio-Ilan et al. (2001) developed the cadaver formulation technology for the control of pest populations using larvae of *Galleria mellonella* and *Tenebrio molitor*. These cadavers were coated with different coating materials to prevent rupture of cadavers and their sticking together. Later, these infected cadavers were tape formulated via an automatic packaging machine to wrap EPNs-infected cadavers in masking tape by Shapiro-Ilan et al. (2010) and Morales-Ramos et al. (2013) (Fig. 9.7). Desiccation approach was used to facilitate the transport and handling of EPNs-infected cadavers by Spence et al. (2011). Zhu et al. (2011) established an automated system for the cadavers' application in the field conditions. The techniques for desiccation and cold storage of EPN-infected cadavers were developed by Wang et al. (2014) to promote their large-scale production and application in the fields. Gumus et al. (2015) followed a new approach of directly releasing live insect hosts in the fields that were pre-infected with EPNs.

Different high quality EPN products were prepared in several developed countries for the control of insect pests affecting agricultural crops. The different companies, like Enema (Germany), Bionema (UK), Biosys (USA) and Koppert (the Netherlands), have prepared several high quality EPN products, which have given

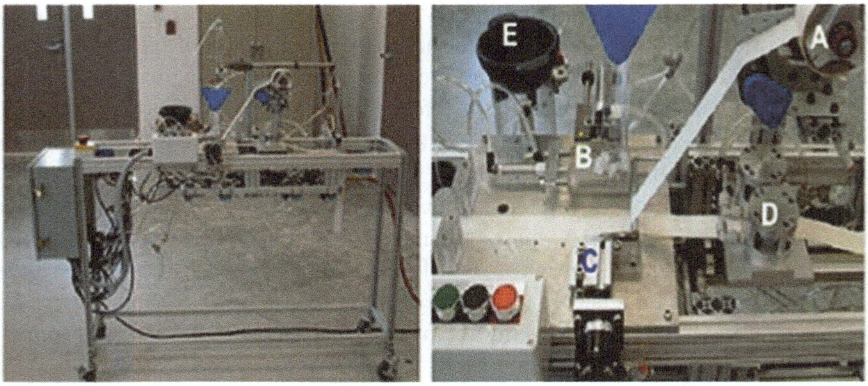

Fig. 9.7 Automatic formulation and packaging machine for enclosing nematode infected hosts in tape. (**a**) Tape Dispenser, (**b**) Cadaver positioning machine, (**c**) Cadaver wrapper, (**d**) Moving piston, (**e**) Cadaver holding bowl (Shapiro-Ilan et al. 2010)

spectacular results not only in laboratory conditions but also in fields. The EPN formulations are in commercial use in developed countries (Table 9.6).

9.3.4 Formulation Technology in India

Several national-level research laboratories, such as National Institute of Plant Health Management (Hyderabad), Indian Agricultural Research Institute (New Delhi), National Bureau of Agricultural Insect Resources (Bengaluru), Chaudhary Charan Singh University (Meerut) and Tamil Nadu Agriculture University (Coimbatore), and a few agrochemical companies, such as Multiplex Biotech Pvt. Ltd., Ajay Biotech (India) Ltd. and Pest Control (India) Pvt. Ltd., are conducting research to bring EPN-formulated products with enhanced shelf-life. Various researches were conducted to advance the formulations with respect to storage, shelf-life, application technology, pathogenicity etc. The efforts were made to develop formulations using different local species, viz. *H. indica, S. abbasi, S. bicornutum, S. carpocapsae* and *S. riobrave* in different carrier materials, like talc, alginate capsule, wheat bran pellets, sodium alginate pearl, vermiculite, spray-adjuvants or hydrogel (Hussaini et al. 2001, 2003; Gupta 2003; Vyas et al. 2001). However, till date it is under research phase only (Tables 9.7 and 9.8).

Water-based formulations of IJs showed better survival than the talc and alginate-based formulations of *Steinernema* sp. (Hussaini 2003); however, talc-based formulation of *H. indica* showed good viability. Ganguly et al. (2008) developed Nema Gel-based hydrogel formulation of *S. abbasi* (= *S. thermophilum*), which could not provide good results in field conditions; hence discontinued (Divya and Sankar 2009).

Table 9.6 Entomopathogenic nematode-based products available globally

EPN species	Product Name	Country
Steinernema carpocapsae	ORTHO biosafe	USA
S. carpocapsae	Biovector	USA
S. carpocapsae	Exhibit	USA
S. carpocapsae	XGNAT	USA
S. carpocapsae	Helix	Germany
S. carpocapsae	BodenNiitzlinge	Switzerland
S. carpocapsae	Sanoplant	USA
S. carpocapsae	Proactant	USA
S. carpocapsae	Green commandos	India[a]
S. Feltiae	Manget, Entonem, Nemasys	USA
S. Feltiae	Stealth	UK
S. Riobrave	Vector MG	USA
S. Kushidai	SDS biotech	Japan
Heterorhabditis bacteriophora	Otinem	USA
H. bacteriophora	E-nema	Germany
H. bacteriophora	Soil commandos	India[a]

[a]Products withdrawn from the market due to inconsistent results in field conditions

Table 9.7 Entomopathogenic Nematode based formulations developed and tested in India

EPN species	Production/ formulation	Reference
Heterorhabditis indica	Talc	Hussaini et al. (2003)
Steinernema abbasi	Talc	Hussaini et al. (2003)
S. abbasi = S. thermophilum	Hydrogel	Ganguly et al. (2008), Divya and Sankar (2009)
S. bicornutum	Bait as alginate capsule	Hussaini et al. (2001)
S. carpocapsae	Talc	Hussaini et al. (2003)
S. carpocapsae	Alginate capsule	Hussaini et al. (2001)
S. carpocapsae	Wheat bran pellets	Hussaini et al. (2001)
S. carpocapsae	Pearl (sod. Alginate)	Gupta (2003)
S. carpocapsae	Vermiculite	Hussaini et al. (2003)
S. Riobrave	Spray-adjuvants	Vyas et al. (2001)

Table 9.8 Entomopathogenic Nematodes and their efficacy against various insect pests in India

S. No.	EPN species	Insect	Author
1.	*H. bacteriophora*	*Amsacta albistriga*	Bhaskaran et al. (1994)
2.	*H. indica*	*Helicoverpa armigera, Leucinodes orbonalis*	Hussaini et al. (2003)
3.	*Steinernema carpocapsae*	*Spodoptera litura*	Hussaini et al. (2003), Sitaramaiah et al. (2003)
4.	*S. carpocapsae*	*A. Albistriga*	Bhaskaran et al. (1994)
5.	*S. carpocapsae*	*Holotrichia longipennis*	Hussaini et al. (2005)
6.	*S. glaseri*	*S. litura, H. consanguinea*	Vyas and Yadav (1993)
7.	*S. Riobrave*	*Agrotisip silon*	Mathasoliya et al. (2004)
8.	*Steinernema* sp.	*Papilio* sp.	Singh (1993)
9.	*Heterorhabditis* sp.	*H. armigera*	Vyas et al. (2002)

9.3.5 Globally Available Formulations and their Application

The performance of EPNs has got more success to control soil-borne insect pests when compared to foliar pests. The major reason for the lack of success of foliar application of EPNs is the intolerance of juveniles to extreme of desiccation (Lello et al. 1996), temperature (Grewal et al. 1994) and ultraviolet radiation (Gaugler et al. 1992). Schroer and Ehlers (2005) used *S. carpocapsae* on cabbage against foliar insect, *Plutella xylostella* formulated with surfactant Rimulgan and polymer xanthan to provide optimal conditions that support nematode invasion on the foliage surface. EPN efficacy was also found to be compatible with chemical insecticides, fungicides and acaricides (Ishibashi 1993) and, therefore, can often be tank mixed and applied with other pesticides as integrated pest management tool. A few pesticides, such as Imidacloprid (Koppenhöfer et al. 2000), tefluthrin (Nishimatsu and Jackson 1998), neonicotinoid (Koppenhöfer et al. 2002) and *Bacillus thuringiensis* (Koppenhöfer and Kaya 1997), were found to be synergistic with EPNs. Therefore, formulation and integration of EPNs with pesticides and surfactants should be properly evaluated before it is released into the field. Non-availability of potential nematicides or nematode-based biopesticides is expensive in field application, and the present-day trend for eco-friendly approaches for pest and disease control is the compelling reason as nematode control is considered on priority.

In the present scenario of globalization demand of pesticide-free agri-product is increasing day by day, EPN-based formulation is one of the best options in the agricultural fields. The use of nematicides is particularly prohibited in organic farming and is likely to become unavailable in future. In India, a large number of crops, such as vegetables, fruits and pulses, are affected by plant parasitic nematodes (PPNs) specially by root-knot nematode, which causes significant damage to the crops. Since EPN has already established to control insect pests, it would be highly economical if the same could also be used for the control of PPN. There are numerous reports which show the application of EPNs in the suppression of PPN (Grewal et al. 1997). The effect of *Steinernema* and *Heterorhabditis* and their associated bacteria *Xenorhabdus* spp. and *Photorhabdus* spp., respectively, have suppressed selected species of PPN including root-knot nematode in green house experiments (Ishibashi and Choi 1991). However, the existing literature has limited information as to which stage(s) of PPN is affected by EPNs. Tests conducted in the fields have demonstrated PPN population suppression for up to 8 weeks after application of EPN products (Grewal et al. 1997).

Driven by the desirability of reducing pesticide usage, the past decades extensive research has led to the discovery of many efficacious isolates/strains and significant advances in mass production and formulation technology (Askary and Ahmad 2017). Currently, *S. carpocapsae, S. feltiae, S. kraussei, S. glaseri, S. riobrave, H. bacteriophora* (CAB Reviews 2018) and *H. megidis* are the most commonly and successfully applied nematodes due to their easy production in liquid culture (Abate et al. 2017). Today, EPNs are mainly used in environments where chemical pesticides fail, that is, in the soil; in the galleries of boring insects, in cases where resistance to insecticides has developed or where hazardous pesticides are banned (Ehlers 2001).

In India, various workers have used EPNs against cutworms, ragi pink borer, stem borer, white grubs etc., in laboratory and field conditions (Singh 1977). In field trails, a few EPNs, viz. *S. corpocapsae* (strain DD-136), *H. bacteriophora* (strain Burliar) and *Heterorhabditis* species, were reported against the fourth instar larvae of *Amsacta albistrigata* on ground nut (Bhaskaran et al. 1994). Two EPN-based formulations, viz. green commandos (*S. corpocapsae*) and soil commandos (*H. bacteriophora*), were developed by Ecomax Company in 1980 using the exotic species; however, both the products were not effective against the insect pests because of their poor adaptability under environmental conditions in India or due to problems in formulations; therefore, both the products were withdrawn from the market.

9.3.6 Formulations Developed at Chaudhary Charan Singh University, Meerut

Since EPNs entered at the commercial use for biocontrol of insects; therefore, to maintain survival, it is necessary to keep a balance between reduction of metabolism and water availability. EPNs require a film of water around their body for optimum metabolism, activity and movement. Keeping all such requirements for better survival, storage and efficacy, a few formulations, viz. sponge, gel, cadaver and water dispersible granules, have been developed using the best efficacious isolates and their field trials have been done against the major insect pests, that is, *Helicoverpa armigera, Spodoptera litura* and *Pieris brassicae* (Aasha 2020).

Gel Formulation

EPN slurry was inoculated in 1.0% nutrient agar gel. Further, the gel was transferred into polythene bags. EPN slurry of each isolate prepared and inoculated into the gel separately. These sealed and labelled polythene bags were stored at 15 °C in BOD incubator (Fig. 9.8). The feasibility of the gel formulation was also checked as the previous formulations (Vidhi 2012).

Water Dispersible Granule (WDG)

In granular formulation, EPNs were partially encapsulated in water dispersible granules made up of talc, soya flour, casein and tween etc. After homogeneous mixing of the above contents in different ratios, the desired genera of EPNs were sprayed over the mixed contents where granules containing EPN were made (Vidhi 2012). These granules were kept in polythene bags and stored at 15 °C in BOD incubator (Fig. 9.9). Feasibility of the formulation was noticed in the previous

Fig. 9.8 Infective Juvenile of Entomopathogenic nemadode packed in polythene bags

Fig. 9.9 Water dispersible granular formulation packed in polythene bags

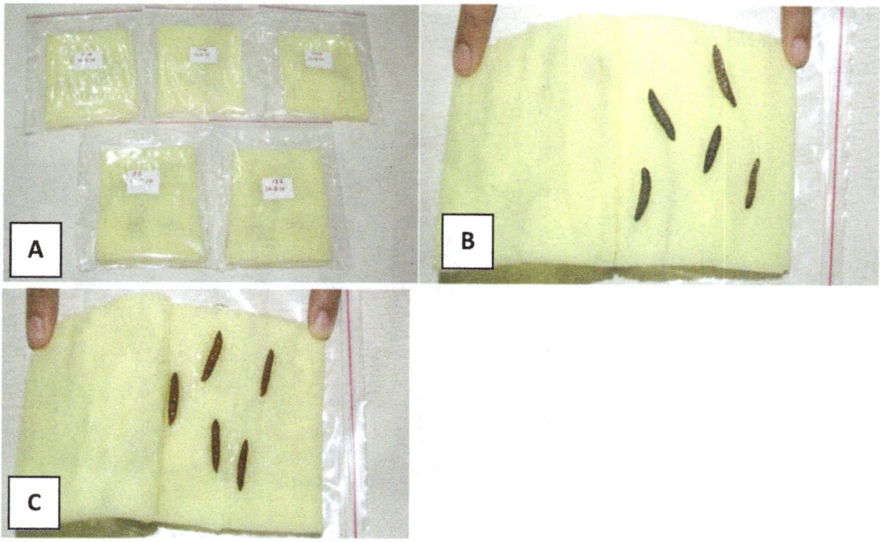

Fig. 9.10 (**a**) Sponge formulation of entomopathogenic nematodes, (**b**) *Steinernema* sp.-infected *Galleria* larvae sandwitched between polyether polyurethane sponges, (**c**) *Heterorhabditis* sp.-infected *Galleria* larvae sandwitched between sponges

formulations. The viability of the EPN isolate in WDG formulation was recorded for 6 months with full efficacy.

Sponge Formulation

The polyurethane sponge formulations were made by applying *G. mellonella* cadaver infected with *Steinernema* sp. and *Heterorhabditis* sp. to a sheet of sponge (10 × 8 cm) wet with double distilled water, placed 4–5 cadavers sponge/sheet (Fig. 9.10a–c). Further, the infected cadaver sandwiched in polyether polyurethane sponge sheet was placed in plastic bag with enough air to keep the EPN in viable state. The bags were placed in BOD incubator at 15 °C. The nematodes were removed from the sponge sheet by soaking and squeezing in water before application. Feasibility of sponge formulation for *Steinernema* sp. and *Heterorhabditis* sp. was recorded up to 90 days (Vidhi 2012). Further, this formulation was prepared by little bit modification in the formula developed by Morales-Ramos et al. (2013) to protect cadavers from breakage by formulating and packing of the infected hosts by wrapping them into masking tape (Aasha 2020).

Formulated Cadaver

For this formulation, many binding materials, that is, kaolin–starch, wheat starch gelatine, clay, wheat flour, maize flour and corn flour, were screened and tested for their best surface coating agents. The cadaver containing IJs were rolled with different materials (kaolin-starch and wheat starch) and both showed better results as compared to gelatine and kaolin starch. Few differences were observed in the

Fig. 9.11 Formulated cadavers with *Heterorhabditis indica* DD7

binding material with the surface coat of cadaver (*H. indica* DD7). Kaolin–starch combination was proved to be the most suitable and stable. But overall kaolin adhered well to the cadavers, and IJs inside the cadavers did not feel suffocated. The other materials (clay, wheat flour, maize flour and corn flour) did not stick well to cadaver's surface coat (fuller's earth). They were not producing the IJs and few showed poor productivity when transferred to the white trap. The survival rate and productivity of *H. indica* DD7 in infected cadavers was recorded for 3 months (Aasha 2020) (Fig. 9.11). Kaolin starch coating maintains the moisture and environment that may helpful in promoting nematode survival inside the cadaver. Further the formulated cadaver balls were developed and tested in field conditions and the results were very encouraging along with good soil as well as crop health (Aasha 2020).

9.3.7 EPN Harvesting Machine

The formulation machine was subjected to EPN production in liquid media. This study was different from other investigations, as we place an EPN harvesting machine. This method had an advantage that it was made by the waste material and facilitated by an automatic production machine of EPNs. An aerator of aquarium, along with filter pipe, was used for bubbling, which was attached to electricity for aeration purposes. IJs production started at 5–seventh day and end up to15–30th day into the water (Aasha 2020) (Fig. 9.12). Shapiro-Ilan et al. (2012) also suggest that the efficacy of EPNs can be enhanced by utilizing better delivery mechanisms.

Entomopathogenic nematodes have proven to be safe and effective alternatives to chemical pesticides, but in numerous field trials the efficacy is not much encouraging. They lay another avenue to expanded use in improved delivery to the target

Fig. 9.12 EPN harvesting machine developed by waste materials

pests. Due to limitations in the EPN biology, nematode delivery can only be taken so far. Few modifications are needed in context to advancement in EPN application technology come from the delivery system, optimizing entomopathogenic nematode application efficiency and increased acceptance and use by the farmers.

9.4 Conclusions

A number of crops are grown by farmers, including food crops, commercial crops, vegetables crops and plantation crops. Larval stages of various soil-borne insect pests are among the most destructive and troublesome, threatening the entire crop production, and have become a challenging subject for the farmers throughout the world. No crop is completely free from or resistant to the attack by the insect pests. Farmers use chemical pesticides to reduce insect populations, which involves various drawbacks. The nematodes are the most diverse taxonomic group having a wide range of habitats and represent most dominant metazoan phylum on the planet earth, ranging from aquatic to terrestrial environments. They are receiving a lot of interest in nematological and entomological studies because of their ability to infect and kill the insects host within 24–48 h. They are ubiquitous and exist everywhere, except Antarctica. The use of EPNs as biopesticides against the insect pests has grown rapidly in recent years. The formulations should be maintained for high efficacy, easy handling and transport, high effectiveness and ease in application in the fields. Considering these points, various EPN-based formulations i. e. cadaver balls, sponge tape were developed by using the best efficacious isolates (*H. indica* DD7, *S. abbasi* DS6 and *H. bacteriophora* DH7) and tried their applications in the laboratory as well as in field conditions. Formulation should be standardized to reduce labour in the application. But we use simple process classical methods to develop sponge tape formulation. This process needs more tactic applications to develop more standardized type of tape formulation, which allows ease of handling that would be beneficial for Indian farmers. This formulation will also be amenable to mass production of EPNs and will facilitate fruitful commercialization. However, this formulation was used only for the storage purpose not for the experimentation. Additionally, this formulation should be useful against selective pests, such as *Helicoverpa armigera* and *Pieris brassicae* at higher infestations. In summary, we have developed a novel formulation and EPN production system for better harvesting of entomopathogenic nematodes. The developed formulation creates an option for hand application by home owners or small farmers to avoid touching a dead insect. But, further research is required to determine if the formulated tape can be stored prior to application (e.g., under refrigeration or partial desiccation) without loss of IJ yield or efficacy. Furthermore, to determine their efficacy under field conditions, studies are required.

9.5 Points to Remember

- Nematodes are the most diverse taxonomic group on the planet Earth. They belong to the most dominant metazoan phylum ranging from aquatic to terrestrial environments, except Antarctica. They possess the ability to infect and kill the insect hosts within 24–48 h.
- Nematodes may be harmful (Plant Parasitic Nematode) or beneficial (Entomopathogenic Nematodes). The beneficial nematodes belong to the family Steinernematidae (*Steinernema*) and Heterorhabditidae (*Heterorhabditis*).
- *Steinernema*, IJs invade the insect larvae through natural openings, such as the mouth, anus, spiracles and wounds, whereas *Heterorhabditis* is also able to penetrate the insect body by directly scratching their cuticle. After entering into the host body, IJs release their symbiotic bacteria (EPB) and together kill the insect, whereas nematodes complete their life cycle or after depleting their food sources, search their new host.
- Both *Steinernema* and *Heterorhabditis* females lay eggs in the insect cadaver after mating with males. The life cycle of most *Steinernema* involves both sexually differentiated partners, first-generation males and females, while all *Heterorhabditis* IJs develop into self-fertilizing hermaphrodite females after insect infection.
- The third-stage juveniles of *Steinernema* and *Heterorhabditis* move freely in soil in search of the host and have been distinguished into three categories on the basis of their host finding behaviours-: (i) cruisers, (ii) ambushers and (iii) intermediates.
- All species in genus *Heterorhabditis* are cruisers, whereas *Steinernema* displays all three behaviours but *S. carpocapsae* displays ambush behaviour.

- EPN formulation is a process of the transformation of living entities into a product that can be applied for the suppression of pest populations. Few factors affect their application in field conditions, which include market value; crop and target insects; formulation type and shelf life; usage directions; technical support; cost and others.
- Entomopathogenic nematodes have proven to be safe and effective alternatives to chemical pesticides, but often fail under field conditions. They can also be used in combinations with *Bt* or synthetic pesticides.

Acknowledgements Authors are thankful to the Head, Department of Zoology, Chaudhary Charan Singh University, Meerut, for providing the laboratory facility to conduct the experimental work. Department of Science and Technology, New Delhi, India, is also thankfully acknowledged for providing financial assistance to Aasha through DST WOS-A (SR/WOS-A/LS-1083/2014).

References

Aasha (2020) Development of effective bio-formulation of entomopathogenic nematodes and the evaluation of their pathogenicity against white grub (*Holotrichia serrata*) infesting sugarcane and their field trials. Ph. D. Thesis. Chaudhary Charan Singh University, Meerut, pp 486

Abate BA, Wingfield MJ, Slippers B, Hurley BP (2017) Commercialisation of entomopathogenic nematodes: should import regulations be revised? Biocontrol Sci Tech 27(2):149–168

Adams BJ, Nguyen KB (2002) Taxonomy and systematics. In: Gaugler R (ed) Entomopathogenic nematology. CABI Publishing, New York, NY, pp 1–34

Akhurst R, Smith K (2002) Regulation and safety. In: Gaugler R (ed) Entomopathogenic nematology, vol 1. CABI, New York, pp 311–332

Askary T, Ahmad MJ (2017) Entomopathogenic nematodes: mass production. Formul Appl 261:–86

Athanassiou CG, Kavallieratos NG, Menti H, Karanastasi (2010) Mortality of four stored product pests in stored wheat when exposed to doses of three entomopathogenic nematodes. J Econ Entomol 103(3):977–984

Bailey AS, Bertaglia M, Fraser IM, Sharma A, Douarin (2009) Integrated pest management portfolios in UK arable farming: results of farmer's survey. Pest Manag Sci 65:1030–1039

Bastidas B, Portillo E, San-Blas E (2014) Size does matter: the life cycle of *Steinernema* spp. in micro-insect hosts. J Invertebr Pathol 121:46–55. https://doi.org/10.1016/j.jip.2014.06.010

Bedding RA (1981) Low cost *in vitro* mass production of *Neopectana* and *Heterorhabditis* sp. for field control of insects pests. Nematologica 27:109–114

Bedding RA (1988) Storage of entomopathogenic nematodes. WIPO Patent No. WO 88/08668

Bedding RA, Molyneux AS (1982) Penetration of insect cuticle by infective juveniles of *Heterorhabditis* spp. (Heterorhabditidae: Nematoda). Nematologica 28:354–359

Bellows TS Jr, Fisher TW (1996) Handbook of biological control. Academic Press, San Diego. CA

Bhaskaran RKM, Sivakumar CV, Venugopal MS (1994) Biocontrol potential of entomopathogenic nematode in control of red hairy caterpillar, *Amsacta albistriga* of groundnut. Indian J Agric Sci 64:655–657

Bhat AH (2019) Taxonomic diversity of entomopathogenic nematodes in Meerut Division. Ph. D. Thesis. Chaudhary Charan Singh University, Meerut, pp 1–442

Bird AF, Akhurst RJ (1983) The nature of the intestinal vesicle in nematodes of the family Steinernematidae. Int J Parasitol 13:599–606

Bovein P (1937) Some types of association between nematodes and insects. Videnskabelige Meddelelser Fra Dansk Naturhistorisk Forening, Kobenhavn 101:1–114

CAB Reviews 2018 13, No. 058, http://www.cabi.org/cabreviews

Campbell JF, Gaugler R (1993) Nictation behaviour and its ecological implications in the host search strategies of entomopathogenic nematodes (Heterorhabditidae and Steinernematidae). Behaviour 126:154–169

Campbell JF, Gaugler R (1997) Inter-specific variation in entomopathogenic nematodes foraging strategy: dichotomy or variation along a continuum. Fundamental of Applied Nematology 20:393–398

Campbell JF, Kaya HK (2002) Variation in entomopathogenic nematode (Steinernematidae and Heterorhabditidae) infective-stage jumping behaviour. Nematologica 4:471–482

Campos-Herrera R, Barbercheck M, Hoy CW, Stock SP (2012) Entomopathogenic nematodes as a model system for advancing the frontiers of ecology. J Nematol 44:162–176

Castillo JC, Reynolds SE, Eleftherianos I (2011) Insect immune responses to nematode parasites. Trends Parasitol 27(12):537–547. https://doi.org/10.1016/j.pt.2011.09.001

Ciche TA, Ensign JC (2003) For the insect pathogen *Photorhabdus luminescens*, which end of a nematode is out? Appl Environ Microbiol 69:1890–1897

Ciche TA, Darby C, Ehlers R-U, Forst S, Goodrich-Blair H (2006) Dangerous liaisons: the symbiosis of entomopathogenic nematodes and bacteria. Biol Control 38:22–46

CPCRI (2016) Annual report 2015–2016. ICAR-Central Plantation Crops Research Institute, Kasaragod, p 210

Daborn PJ, Waterfield N, Blight MA, Ffrench-Constant RH (2001) Measuring virulence factor expression by the pathogenic bacterium *Photorhabdus luminescence* culture and during insect infection. J Bacteriol 183:5834–5839

DACFW (2020) Annual report of Department of Agriculture, Cooperation and Farmer's Welfare, Ministry of Agriculture, Govt. of India. http://agricoop.nic.in/sites/default/files/ACFW%20English%20%20Annual%20Report%202019-20

de Man JG (1880) Die einheimischen, frei in der reinen Erde und imsüssen Wasserleben den Nematoden. Tijdschrift van der Nederlands chedierkundige Vereeniging 5:1–104

Dillman AR, Guillermin ML, Lee JH, Kim B, Sternberg PW, Hallem EA (2012) Olfaction shapes host-parasite interactions in parasitic nematodes. Proc Natl Acad Sci U S A 109(35):E2324–E2333. PubMed: 22851767

Dillon A (2003) Biological control of the large pine weevil, *Hylobius abietis* L., (Coleoptera: Curculionidae) using entomopathogenic nematodes. Dissertation, National University of Ireland–Maynooth, Maynooth, Ireland

Dillon AB, Downes MJ, Ward D, Griffin CT (2007) Optimizing application of entomopathogenic nematodes to manage large pine weevil, *Hylobius abietis* L. (Coleoptera: Curculionidae) populations developing in pine stumps, *Pinus sylvestris*. Biol Control 40:253–263

Divya K, Sankar M (2009) Entomopathogenic nematode in pest management. Indian J Sci Technol 2:53–60

Dutky SR, Thompson J, Cantwell GE (1964) A technique for the mass propagation of the DD-136 nematode. J Insect Pathol 6:417–422

Ebssa L, Borgemeister C, Semrau J, Poehling H (2004) Efficacy of entomopathogenic nematodes against western flower thrips *Frankliniella occidentalis* at different pupation depths. Nematology 6:495–505

Ehlers R-U (2001) Mass production of entomopathogenic nematodes for plant protection. Appl Microbiol Biotechnol 56:623–633

Ehlers R-U (2005) Forum on safety and regulation. In: Grewal PS, Ehlers R-U, Shapiro-Ilan DI (eds) Nematodes as biocontrol agents. CABI, Wallingford, pp 107–114

Endo BY, Nickle WR (1991) Ultrastructure of the intestinal epithelium, lumen and associated bacteria in *Heterorhabditis bacteriophora*. J Helminthol 58:202–212

Feaster MA, Steinkraus DCS (1996) Inundative biological control of *Helicoverpa zea* (Lepidoptera: Noctuidae) with the entomopathogenic nematode *Steinernema riobravis* (Rhabditida: Steinernematidae). Biol Control 7:38–43

Ffrench-Constant R, Waterfield N, Daborn P, Joyce S, Bennett H, Au C, Dowling A, Boundy S, Reynolds S, Clarke D (2003) *Photorhabdus*: towards a functional genomic analysis of a symbiont and pathogen. FEMS Microbiol Rev 26:433–456

Forst S, Clarke D (2002) Bacteria-nematode symbiosis. In: Gaugler R (ed) Entomopathogenic nematology. CAB International, London, pp 57–77

Ganguly S, Anupama, Kumar A, Parmar BS (2008) Nemagel—a formulation of the Entomopathogenic nematode *Steinernema thermophilum* mitigating the shelf-life constraint of the tropics. Nematol Mediterranean 36:125–130

Gaugler R, Bilgrami A (2004) Chapter 4: feeding behaviour. In: Gaugler R, Bilgrami A (eds) Nematode behaviour. CABI Publication, Wallingford, pp 2–3. https://doi.org/10.1079/9780851998183.0000

Gaugler R, Campbell JF, Selvan S, Lewis EE (1992) Large-scale inoculative releases of the entomopathogenic nematode *Steinernema glaseri*: assessment 50 years later. Biol Control 2:181–187

Georgis R (1990) Formulation and application technology. In: Gaugler R, Kaya HK (eds) Entomopathogenic nematodes in biological control. CRC Press, London, Boca Raton, pp 173–191

Gerhardson B (2002) Biological substitutes for pesticides. Trends Biotechnol 20(8):338–343. https://doi.org/10.1016/s0167-7799(02)02021-8

Glaser RW (1931) The cultivation of a nematode parasite of an insect. Science 73:614–931

Gozel U, Gozel C (2016) Entomopathogenic nematodes in pest management. In: Integrated Pest management (IPM): environmentally sound Pest management, Chapter 3, p 56. https://doi.org/10.5772/63894

Grewal PS (2000) Enhanced ambient storage stability of an entomopathogenic nematode through anhydrobiosis. Pest Manag Sci 56:401–406

Grewal PS (2002) Formulation and application technology. In: Gaugler R (ed) Entomopathogenic nematology. CABI Publishing, Wallingford, UK, pp 265–288

Grewal PS, Selvan S, Gaugler R (1994) Thermal adaptation of entomopathogenic nematodes, niche breadth for infection, establishment, and reproduction. J Thermal Biol 19:245–253

Grewal PS, Martin WR, Miller RW, Lewis EE (1997) Suppression of plant- parasitic nematode populations in turfgrass by application of entomopathogenic nematodes. Biocontrol Sci Tech 7:393–399

Gumus A, Karagoz M, Shapiro-Ilan D, Hazir S (2015) A novel approach to biocontrol: release of live insect hosts pre-infected with entomopathogenic nematodes. J Invertebrate Pathol 130:56–60

Gupta P (2003) Entomopathogenic nematodes-work done at Allahabad agriculture institute, Allahabad. In: Hussaini SS, Rabindra RJ, Nagesh M (eds) Current status of research on entomopathogenic nematodes in India. Project Directorate of Biological Control, Bangalore, pp 161–166

Hominick WM, Reid AP, Bohan DA, Briscoe BR (1997) Entomopathogenic nematodes : biodiversity, geographical distribution and the convention on biological diversity. Biocontrol Sci Tech 6:317–332

House HL, Welch HE, Cleugh TR (1965) A food medium of prepared dog biscuit for the mass production of the nematode DD-136 (Nematoda: Steinernematidae). Nature 206:847

Hugot J, Baujard P, Morand S (2001) Biodiversity in helminths and nematodes as a field study: an overview. Nematology 3:199–208

Hussaini SS (2003) Progress of research work on entomopathogenic nematodes in India. In: Hussaini SS, Rabindra RJ, Nagesh M (eds) Current status of research on entomopathogenic nematodes in India. Project Directorate of Biological Control, Bangalore, India, pp 27–68

Hussaini SS, Ansari MA, Ahmad W, Subbotin SA (2001) Identification of some Indian populations of *Steinernema* species (Nematoda) by RFLP analysis of ITS region of rDNA. Int J Nematol 11:73–76

Hussaini SS, Kavitha J, Satya, Hussain MA (2003) Survival and pathogenicity of indigenous entomopathogenic nematodes in different UV protectants. Indian J Plant Protect 31:12–18
Hussaini SS, Nagesh M, Dar M, Rajeshwari R (2005) Field evaluation of entomopathogenic nematodes against white grubs (Coleoptera: Scarabaeidae) on turf grass in Srinagar. J Plant Protect Sci 13(1):190–193
Ishibashi N (1993) Integrated control of insects pest by *Steinernema carpocapsae*. In: Bedding R, Akhurst R, Kaya HK (eds) Nematodes and biological control of insects. CSIRO, East Melbourne, Australia, pp 105–113
Ishibashi N, Choi DR (1991) Biological control of soil pests by mixed application of entomopathogenic and fungivorous nematodes. J Nematol 23:175–181
Jacob JP, Devraj R, Natrajan R (2007) Outbreak of the invasive gall inducing wasp *Leptocybe invasa* on eucalypts in India. Invasives 8:4
Johnson SN, Mitchell C, McNicol JW, Thompson J, Karley AJ (2013) Downstairs drivers - root herbivores shape communities of above-ground herbivores and natural enemies via changes in plant nutrients. J Anim Ecol 82:1021–1030
Kaya H, Gaugler R (1993) Entomopathogenic nematodes. Annu Rev Entomol 38:181–206
Klein MG (1990) Efficacy against soil-inhabiting insect pests. In: Gaugler R, Kaya HK (eds) Entomopathogenic nematodes in biological control. CRC, Boca Raton, FL, pp 195–214
Koppenhöfer AM, Fuzy EM (2003a) Ecological characterization of *Steinernema scarabaei*, a scarab-adapted entomopathogenic nematode from New Jersey. J Invertebrate Pathol 83:139–148
Koppenhöfer AM, Fuzy EM (2003b) *Steinernema scarabaei* for the control of white grubs. Biol Control 28:47–59
Koppenhöfer AM, Kaya HK (1997) Additive and synergistic interaction between entomopathogenic nematodes and *Bacillus thuringiensis* for scarab grub control. Biol Control 8:131–137
Koppenhöfer AM, Grewal PS, Kaya HK (2000) Synergism of imidacloprid and entomopathogenic nematodes against white grub: the mechanism. Entomol Exp Appl 94:283–294
Koppenhofer AN, Cowles RS, Cowles EA, Fuzy EM, Baumgartner L (2002) Comparison of neonicotinoid insecticidesas synergists for entomopathogenic nematodes. Biol Control 24:90–97
Krishnayya PV, Grewal PS (2002) Effect of neem and selected fungicides on viability and virulence of the entomopathogenic nematode *Steinernema feltiae*. Biocontrol Sci Tech 12:259–266
Lacey LA, Grzywacz D, Shapiro-Ilan DI, Frutos R, Brownbridge M, Goette MS (2015) Insect pathogens as biological control agents: back to the future. J Invertebrate Pathol 132:1–41
Lello ER, Patel MN, Mathews GA, Wright DJ (1996) Application technology for entomopathogenic nematodes against foliar pests. Crop Protect 15:567–574
Martens EC, Heungens K, Goodrich-Blair H (2003) Early colonization events in the mutualistic association between *Steinernema carpocapsae* nematodes and *Xenorhabdus nematophila* bacteria. J Bacteriol 185:3147–3154
Mathasoliya JB, Maghodia AB, Vyas RV (2004) Efficacy of *Steinernema riobrave* against *Agrotis ipsilon Hufnagel* (Lepidoptera: Noctuidae) on potato. Indian J Nematol 34(2):177–189
Maupas E (1919) Essaisd'hybridation chez des nématodes. Bull Biol Fr Belg 52:466–498
Morales-Ramos JA, Tedders WL, Dean CB, ShapiroIlan DI, Rojas MG (2013) Apparatus for packaging arthropods infected with entomopathogenic nematodes. U.S. Patent No. USA. vol. 8, pp. 505–236
Mráček Z, Bednarek A (1991) The morphology of lateral fields of infective juveniles of entomogenous nematodes of the family Steinernematidae (Rhabditida). Nematologica 37:63–71
Muniappan R, Shepard BM, Watson GW, Carner GR (2008) First report of the papaya mealybug, *Paracoccusm arginatus* (Hemiptera: Pseudococcideae), Indonesia and India. J Agric Urban Entomol 25:37–40
Nguyen KB, Smart GC (1990) *Steinernema scapterisci* n. sp. (Steinernematidae: Nematoda). J Nematol 22:187–199

Nishimatsu T, Jackson JJ (1998) Interaction of insecticides, entomopathogenic nematodes, and larvae of the western corn rootworm (Coleoptera: Chrysomelidae). J Econ Entomol 91 (2):410–418

Poinar GO (1990) Biology and taxonomy of steinernematidae and heterorhabditidae. In: Gaugler R, Kaya HK (eds) Entomopathogenic nematodes in biological control. CRC Press, Boca Raton, FL, pp 23–62

Poinar GO, Thomas GM (1966) Significance of *Achromobacter nematophilus* (Achromobactericeae: Eubacteriales) in the development of the nematode, DD-136 (*Neoaplectana* sp. Steinernematidae). Parasitologica 56:385–390

Poinar GO (1979) Nematodes for biological control of insects. CRC Press, Boca Raton, p 277

Pye AE, Burman M (1977) Pathogenicity of the nematode *Neoaplectana corpocapsae* (Rhabditida: Steinernematidae) and certain microorganisms towards the large pine weevil, *Hylobius abietis* (Coleoptera, Curculionidae). Annoles Entomolgici Fennici 43:115–119

Rathee M, Dalal P (2018) Emerging insect pests in Indian agriculture. Indian J Entomol 80 (2):267–281. https://doi.org/10.5958/0974-8172.2018.00043.3

Sathiamma B, Nair CPR, Koshy PK (1998) Outbreak of nut a infesting eriophyid mite (*Eryophyes guerreronis* (K)) in coconut plantation in India. Indian Coconut J 29:1–3

Schroer S, Ehlers R-U (2005) Foliar application of the entomopathogenic nematode *Steinernema carpocapsae* for biological control of diamondback moth larvae (*Plutella xylostella*). Biol Control 33:81–86

Segal D, Glazer I (2000) Genetics for improving biological control agents: the case of entomopathogenic nematodes. Crop Prot 19:685–689

Selvan S, Gaugler R, Lewis EE (1993) Biochemical energy reserves of entomopathogenic nematodes. J Parasitol 79:167–172

Shapiro-Ilan DI, Gaugler R (2002) Production technology for entomopathogenic nematodes and their bacterial symbionts. J Indust Microbiol Biotechnol 28:137–146

Shapiro-Ilan DI, Grewal PS (2008) Entomopathogenic nematodes and insect management. In: Capinera JL (ed) Encyclopedia of entomology, vol 2. Springer, Dordrecht, pp 1336–1340

Shapiro-Ilan DI, Lewis EE, Behle RW, McGuire MR (2001) Formulation of entomopathogenic nematode-infected cadavers. J Invertebrate Pathol 78:17–23. https://doi.org/10.1006/jipa.2001.5030

Shapiro-Ilan D, Gouge D, Koppenhöfer A (2002) Factors affecting commercial success: case studies in cotton, turf, and citrus. In: Gaugler R (ed) Entomopathogenic nematology. CABI Publications, 355, p 333

Shapiro-Ilan DI, Gouge DH, Piggott SJ, Patterson FJ (2006) Application technology and environmental considerations use of entomopathogenic nematodes in biological control. Biol Control 38:124–133

Shapiro-Ilan DI, Cottrell TE, Mizell RF, Horton DL, Behleand RW, Dunlap CA (2010) Efficacy of *Steinernema carpocapsae* for control of the lesser peach tree borer, *Synanthedon pictipes*: improved aboveground suppression with a novel gel application. Biol Control 54:23–28

Shapiro-Ilan DI, Han R, Dolinksi C (2012) Entomopathogenic nematode production and application technology. J Nematol 44(2):206–217

Shapiro-Ilan DI, Brown I, Lewis EE (2014) Freezing and desiccation tolerance in entomopathogenic nematodes: diversity and correlation of traits. J Nematol 46(1):27–34

Singh SP (1977) Isolation of an entomophilic nematode from potato cutworms. Curr Sci 46:454–455

Singh SP (1993) Effectiveness of an indigenous entomophilic nematode against citrus butterfly. J Insect Sci 6:107–108

Singh SP, Ballal CR (1991) Status of coffee berry borer (*Hypothenemus hampei* Ferrari) (Coleoptera: Scolilytidae) Technical document No. 36, All India Cordinated Research Programme on biological control of crop pest and weeds, Banglore, p. 6

Sitaramaiah S, Gunneswara Rao S, Hussaini SS, Venkateswarh P, Nageshwara Rhao S (2003) Use of entomopathogenic nematodes *Steinernema carpocapsae* against *Spodoptera litura* fab. In tobacco nursery. In: Tandon PL (ed) Botanical control of lepidopteran pests. Project Directorate of Biological Control, Bangalore, pp 211–214

Smart GC (1995) Entomopathogenic nematodes for the biological control of insects. J Nematol 27:529

Spence KO, Lewis EE, Perry RN (2008) Host-finding and invasion by entomopathogenic and plant-parasitic nematodes: evaluating the ability of laboratory bioassays to predict. J Nematol 40 (2):93–98

Spence KO, Stevens GN, Arimoto H, Ruiz-Vega J, Kaya HK, Lewis EE (2011) Effect of insect cadaver desiccation and soil water potential during rehydration on entomopathogenic nematode (Rhabditida: Steinernematidae and Heterorhabditidae) production and virulence. J Invertebrate Pathol 106:268–273

Sridhar V, Chakravarthy AK, Asokan R, Vinesh LS, Rebijith KB, Vennila S (2014) New record of the invasive South American tomato leaf miner, Tuta absoluta (Meyrick) (Lepidoptera: Gelechiidae) in India. Pest Manag Hortic Ecosyst 20:148–154

Subbotin SA, Moens M (2006) Molecular taxonomy and phylogeny. In: Perry R, Moens M (eds) Plant nematology. CABI, Wallingford

Sudhaus W (1993) Die mittel symbiontis cher Bakterien entomopathogenen Nematoden Gattungen *Heterorhabditis* and *Steinernema sindkeine* Schwester taxa. Verhandlungen der deutschen Zoologischen Gesellschaf 86:146

Torres-Barragan A, Suazo A, Buhler WG, Cardoza YJ (2011) Studies on the entomopathogenicity and bacterial associates of the nematode *Oscheius carolinensis*. Biol Control 59:123–129

Tyagi K, Kumar P (2015) First report on Western flower thrips *Frankliniella occidentalis* (Pergande) (Thripidae: Thysanoptera) from India. A potential havoc to Indian agriculture. Halters 6:1–3

Vidhi T (2012) Application of entomopathogenic nematodes in biological control of insect pests. Ph. D. Thesis, Caudhary Charan Singh University, pp 475

Vyas RV, Yadav DN (1993) *Steinemema glaseri* (Steiner, 1923) Travassos, 1927, a biological control agent of root grubs *Holotrichia consanguinea* Blanchard. Pakistan J Nematol 11:11–17

Vyas RV, Patel NB, Patel PD, Patel DJ (2001) Field efficacy of entomopathogenic nematode (*Steinernema riobrave*) against cotton bollworms. National Congress on Centenary of Nematology in India. Appraisal and Future Plans. IARI, New Delhi, pp 115–207

Vyas RV, Patel NB, Parul P, Patel DJ (2002) Efficacy of entomopathogenic nematode against *Helicoverpa armigera* on pigeonpea. Int Chickpea Pigeonpea Newslett 9:43–44

Wang H, Luan JB, Dong H, Qian HT, Cong B (2014) Natural occurrence of entomopathogenic nematodes in Liaoning (Northeast China). J Asia-Pacific Entomol 17:399–406

Wouts WM (1981) Mass production of the entomogenous nematode *Heterorhabditis heliothidis* (Nematoda: Heterorhabditidae) on artificial media. J Nematol 13:467–469. www.eagri.org.> eagri50>ENTO232>lec11Lec11-

Ye WM, Torres-Barragan A, Cardoza YJ (2011) *Oscheius carolinensis* n. sp. (Nematoda: Rhabditidae), a potential entomopathogenic nematode from vermicompost. Nematology 12:121–135

Zhu H, Grewal PS, Reding ME (2011) Development of a desiccated cadaver delivery system to apply entomopathogenic nematodes for control of soil pests. Appl Eng Agric 27:317–324

Ethics and Safety Concerns

10

T. P. Rajendran

Contents

T. P. Rajendran (✉)
ICAR, New Delhi, India

ICAR-National Institute for Biotic Stress Management, Raipur, Chhattisgarh, India

Omkar (ed.), *Microbial Approaches for Insect Pest Management*,
https://doi.org/10.1007/978-981-16-3595-3_10

Abstract

Biopesticides have been an IPM tool for several decades for crop health management. Both phytochemicals and microbial pesticides are two biopesticide groups that have been regulated in India. This knowledge-intensive technique in pest management needs frequent updating of scientific information by the manufacturers of Microbial Pesticides (MP) formulations in tandem with government and private extension systems to popularize these products in integrated pest management in crops. Quality review management of MP products is essential to sustain the shelf life and field bioefficacy of the products. The biowaste management of MP production units should follow GLP and GMP standard operating procedures to prevent undesirable leakage of harmful microorganisms following the relevant National Guidelines and International Conventions. Legal compliance of label expansion of MP formulations across various crops as per good agriculture practice to manage target pests would provide farmers higher economic benefits. Risk assessment based on the perceived hazards in handling microbial biocontrol agents as MP has to be based on the global conventions and norms with regard to biological substances.

Biotechnology tools and techniques to deploy genetically modified crops as well as gene editing technology (Crisper-Casp9) for attaining pest resistance and higher commodity output with better quality parameters are promising. Ethical and practical considerations for commercialization of GM crops from *cis*-, *trans*- or sub-genic products need careful analysis for science-based assessment or 'decision tree-based' evaluation of the potential hazards. Risk assessment protocols for GM and GE crops are significant to alleviate the perceived hazards from them to humans and environment in accordance with relevant laws and rules in India. The socio-economic evaluation studies on the benefits over the possible environmental risk shall make the consumers aware of the GM and GE agriculture commodity to enable them to make informed choice for consumption of those commodities.

Keywords

Biopesticides · Microbial pesticide formulations · Quality review management · Biowaste management · Biotechnology tools · Gene editing technologies · Decision-tree based risk assessment · Socio-economic evaluation

Learning Objectives

1. Biopesticides have been an IPM tool for several decades for crop health management. Both phytochemicals and microbial pesticides are two biopesticide groups that have been regulated in India under Insecticide Act,

(continued)

1968. It is a knowledge-intensive technique in pest management and needs frequent updating of information by the manufacturers of Microbial Pesticides (MP) formulation in tandem with government and private extension system to popularize integrated pest management in crops.

2. The research and development for isolation, identification, evaluation and finalizing the bio-efficacy-based dosage of the candidate microbial pesticide needs adequate data generation under various agroclimatic conditions for managing target crop pests either individually or in integration with other pest management tools. The MP formulations may have label claim for the respective crops against given pests on which evaluation data are submitted towards registration for commercial production under the Section (3) of the Insecticide Act, 1968. However, the Rules under this act may be modified to extend label use against the same pests in other crops too.
3. The appropriate formulation technology has to be used for developing the MP formulations by adopting the Good Lab practices and Good Manufacturing Practices (GMP) for their manufacture within the global Code of Conduct for this purpose for making the MP manufacture industry viable. Risk assessment based on the perceived hazards in handling microbial biocontrol agents as MP has to be based on the global conventions and norms with regard to biological substances.
4. Genetic modification has been one of the latest technologies deployed for crop pest management by incorporating alien genes that express insecticidal proteins, such as delta endotoxin, of soil bacterium, *Bacillus thuringiensis* Crystal (Cry)gene family expressing delta endotoxins. Other biotechnology products using alien genes, such as Tm12 gene to impart resistance to whitefly in cotton crop, are in progress. Recent spurt in research on the gene editing (GE) using clustered regularly interspaced palindromic repeat(CRISPR)-CRISPR-associated proteins (Cas) (Crisper-Casp9) techniques for crop resistance against pests is in progress. Risk assessment protocols for GM and GE crops are significant to alleviate the perceived hazards from them to humans and environment in accordance with relevant laws and rules in India.
5. The socio-economic evaluation studies on the benefits over the possible environmental risk shall make the consumers aware of the GM and GE agriculture commodity to enable them to make informed choice for consumption of those commodities.

10.1 Introduction

Following various global experiences, the designing of integrated pest management (IPM)[1] in India made IPM the national norm in plant protection of crops since 1992. Options to integrate biological control agents in integrated pest management (IPM) were explored once the research in this direction provided valuable knowledge and techniques to mass produce these naturally occurring organisms for use in crops in alternation with chemical pesticides, especially in perennial crops as well as in long duration annual crops (Radcliffe et al. 2009). Out of the various biocontrol agents for crop pest management, microbial natural enemies were found to be potent weapon to suppress crop pests that cause extensive crop losses. It is a knowledge-intensive technique in pest management and needs frequent updating of information by the manufacturers of MP formulation in tandem with government and private extension system to popularize integrated pest management in crops. Farming of crops for food, fibre, fodder and feed has been part of human civilization. The crops that are chosen to be cultivated in definitive seasons have certain packages of practices to be followed in order to achieve maximum harvest of their genetic potential. The components of agro-ecologies bring about biotic stresses to crops at different phenological stages. Pests, including insects, mites, nematodes, vertebrate animals, and plant diseases weeds, cause various metabolic stresses in crops. Farmers are guided to follow certain management practices to contain and suppress damage to their crops due to herbivory from a host of organisms.

Alternate options of crop pest (insects, mites, plant pathogens, nematodes, weeds etcetera) management have been researched upon for over the last five decades to replace/supplement chemical insecticides through IPM for optimal crop production. For effective pest suppression in farmlands, intensive research on phytochemicals, microbial pest control agents and the use of advances in molecular techniques for pest suppression have contributed substantially in India as much as in the rest of the world. Reduction in chemical pesticides was aspired for while utilizing alternate pest management strategies.

This chapter delves around the microbial biocontrol agents that are presently manufactured as specific formulations for application in crops. The regulatory and ethical processes for utilizing microbial pesticides (MP) are significant aspects while planning for research and development including commercial production. Their use in agriculture is regulated through various laws and rules that guide the risk-free production and use in farms. Genetically modified (GM) crops are products from various biotechnological research and have offered better chance for growing pest-free crops. The result of introduction of GM crops to create insect-free crops, such as

[1]http://ppqs.gov.in/divisions/integrated-pest-management/ipm-glance.

cotton, soybean, maize, has been impressive in the first few years. However, both these sectors, that is, MP formulations and GM crops, have been regulated in all countries due to the perceived hazards to humans and environment. Gene editing to impart resistance to pests of crops using clustered regularly interspaced palindromic repeat (CRISPR)-CRISPR-associated proteins (Cas) (Crisper-Casp9) techniques has been intensely pursued currently. Regulatory environment for these biotechnology products also would succeed such inventions for their environmental release.

10.2 Biotic Stresses in Crops and IPM Strategies

Biotic stresses in crops have made the intensive agriculture in various cropping systems highly dependent on the use pesticides. Out of the agrochemicals that are needed to sustain crop production, synthetic chemical pesticides had an about 39% share (Subhash et al. 2017) during 2016–17 in India. There has been immense introspection about the continuing use of chemical pesticides for crop production in India as well as in many other nations. The national production of biopesticides was about 3000 MT in 2005–06 and has grown to 7890 MT, as seen in the state-wise consumption data between 2014 and 2019 (Table 10.1).

Biocontrol agents are an immense discovery for nature-friendly mitigation of biotic stresses in crops. The pest management using these naturally occurring organisms in crop fields has been a tremendous step to reduce the overdependence on chemical synthetic pesticides. The conservation and/or augmentation of natural enemies of pests could be achieved by reducing the impedance of chemical pesticides that have a number of nontarget adverse impacts on natural enemies and many other nontarget organisms. National policy on agriculture 2000[2] and 2007[3] have pronounced the need for implementing integrated pest management with emphasis on alternate pest management approaches that would conserve natural enemies in crop fields and suppress all pests including disease causing organisms. Crop protection has attained very decisive and pragmatic integrated pest management approach and has resulted in the economically beneficial harvesting of crop commodities from avoidable crop losses due to pestilence. Pests including plant pathogens are being managed presently using chemical synthetic pesticides along with biological control agents in crops by integrating MP formulations.

[2]National Policy on Agriculture (2000) Department of agriculture and cooperation. Ministry of Agriculture, Government of India. http://agricoop.nic.in/sites/default/files/npff2007%20%281%29.pdf. Accessed 22 Oct 2019.

[3]National Policy for Farmers (2007) Department of agriculture and cooperation. Ministry of Agriculture, Government of India. http://agricoop.nic.in/sites/default/files/npff2007%20%281%29.pdf. Accessed 22 Oct 2019.

Table 10.1 Consumption of biopesticides in various states during 2014–2019 (as on 13.04.2020) (MT) http://ppqs.gov.in/statistical-database

S. No.	States/UTs	2014–15	2015–16	2016–17	2017–18	2018–19	2019–20 (Prov.)
1	Andhra Pradesh	53	25	9	5	10	0
2	Bihar	252	286	314	320	350	560
3	Chhattisgarh	284	370	380	405	505	550
4	Goa	12	14	3	5	6	6
5	Gujarat	279	273	305	354	306	307
6	Haryana	330	340	380	390	410	400
7	Himachal Pradesh	15	16	2	1	2	*NR*
8	Jharkhand	3	7	11	38	41	91
9	Karnataka	530	505	473	544	544	530
10	Kerala	631	606	662	717	862	717
11	Madhya Pradesh	309	395	1063	326	322	336
12	Maharashtra	486	1173	1454	1271	1164	1082
13	Orissa	267	271	310	310	310	220
14	Punjab	136	138	134	259	246	242
15	Rajasthan	157	9	9	10	15	209
16	Tamil Nadu	286	286	294	630	500	813
17	Telangana	82	94	85	77	84	102
18	Uttar Pradesh	43	46	46	46	47	48
19	Uttarakhand	22	30	31	50	52	116
20	West Bengal	680	950	838	951	997	1017
Sub Total		**4855**	**5834**	**6802**	**6710**	**6772**	**7345**
North-Eastern							
21	Arunachal Pradesh	*NR*[a]	*NR*	*NR*	*NR*	17	18
22	Assam	130	150	188	217	234	243
23	Manipur	0.75	0.85	1	1	*NR*	*NR*
24	Meghalaya	16	23	24	75	*NR*	*NR*
25	Mizoram	*NR*	*NR*	*NR*	*NR*	*NR*	*NR*
26	Nagaland	12	12	12	14	18	19
27	Sikkim	*NR*	*NR*	*NR*	*NR*	*NR*	*NR*
28	Tripura	122	95	146	142	138	167
Sub Total		**281**	**280**	**372**	**449**	**406**	**447**
Union territories							
29	Andaman & Nicobar	0.70	*NR*	*NR*	*NR*	*NR*	*NR*
30	Chandigarh	*NR*	*NR*	*NR*	*NR*	*NR*	*NR*
31	Dadra & Nagar Haveli	*NR*	*NR*	*NR*	*NR*	*NR*	*NR*
32	Daman & Diu	*NR*	*NR*	*NR*	*NR*	*NR*	*NR*
33	Delhi	*NR*	*NR*	1.30	*NR*	13	*NR*
34	Jammu & Kashmir	0.05	0.50	1	1	2	2

(continued)

Table 10.1 (continued)

S. No.	States/UTs	2014–15	2015–16	2016–17	2017–18	2018–19	2019–20 (Prov.)
35	Ladakh	*NR*	*NR*	*NR*	*NR*	*NR*	*NR*
36	Lakshadweep	*NR*	*NR*	*NR*	*NR*	*NR*	*NR*
37	Pondicherry	16	33	14	14	11	10
Sub Total		**16**	**34**	**16**	**16**	**25**	**12**
Grand Total		**5152**	**6148**	**7190**	**7174**	**7203**	**7804**

Source: States/UTs Zonal Conferences on Inputs (Plant Protection) for Rabi & Kharif Seasons
[a]*NR* not reported

10.2.1 Biological Pesticides for Biotic Stress Management in Crops

Botanical-origin pesticides, such as Azadirachtin and other alkaloids from neem seeds, *Pongamia*spp. (*Karanj*), pyrethrum and rotenone, have been regulated under the Schedule of Insecticide Act, 1968 and Rules, 1971. The augmentation and conservation of natural enemies of crop pests are the basic approach to manage pests by utilizing their natural enemies, such as parasitoids, predators and microbial pathogens. Immense advances in knowledge on these organisms have led to robust package of practices for integrated pest management in crops (Chandler et al. 2011; Ranga Rao et al. 2007; Sinha and Biswas 2008).

Along with these augmentations of biological control, agents against insects and mite pests are parasitoids/predators/microbial pathogens of various pests as well as antagonists of plant pathogens. All these organisms are picked up from farms and from agroecological situations and identified for their specific use against crop pest spectrum for achieving crop protection. The social and environmental costs for farmers have been rationalized due to the discreet and judicious pest management plan as a national policy in the deployment of such tools in crop IPM in the country. The environmental sustainability of agriculture farms has been improved through such smart solutions (Arora et al. 2018).

10.2.2 Microbial Pesticides for Pest Management

Amongst the entomopathogenic fungi, such as *Metarhizium anisopliae*, *Metarhizium* (*Nomurea*) *rileyi*, *Beuvaria bassiana* and bacteria, such as *Bacillus subtilis*, *B. thuringiensis*, and viruses, such as nuclear polyhedrosis viruses, cytoplasmic polyhedrosis viruses, protozoan diseases and entomopathogenic nematodes have been evaluated successfully and integrated appropriately in IPM of crop insect pests occurring in soil and aerial plant parts. In respect of the antagonistic organisms that are deployed in the augmentative biocontrol of plant pathogens, fungi, such as *Trichoderma* spp., and bacteria, such as *Pseudomonas fluorescens*, have been utilized as biological pesticides to manage various fungal and bacterial diseases in crops season after season.

Microbial natural enemies of crop pests became fascinating component in the biological control of crop pests (Swati and Adholeya 2008). These could be augmented easily using various microbiological production techniques including the use of fermenters. This knowledge-intensive technique in pest management needs frequent technical knowledge updation by the manufacturers of MP formulation regarding the use and in tandem with government and private extension system for farmers to comprehend and accordingly utilize these products in their farms. The access of desirable MP formulations by farmers for use in their farms is required from either market shelves or from their own production facility. The farmers can access MP formulation technology from research institutions and go for 'own' production of the relevant microbial species under the technical supervision of the research institutions. Such production is exclusively for use in farms and cannot be for doing business. The Insecticide Act, 1968 and Rules thereon, 1971 do not prohibit farmers producing their own MP formulations for use in their farms.

There has been considerable interest amongst scientists to isolate microbial control agents from agroecosystems, identify them and multiply their pure cultures for use against insect pests in crops and other relevant systems (Gupta 2006; Rabindra 2001). Many research institutions in the country have commercialized their discoveries of candidate microbial control agents (fungi, bacteria, viruses, protozoa, nematodes and the like) along with the technology for manufacture of these microbial pesticides (MP). Pest management in crops under various cropping systems, such as paddy, wheat, maize, pulses, oilseeds, cotton, jute, spices, condiments, vegetables and orchard crops, is achieved by utilizing amongst other tools, the microbial control agents (Koul et al. 2003; Mishra et al. 2020; Rabindra 2005; Kumar et al. 2019). The MP formulations that are deployed for biological management of biotic stresses is viewed as the safety system to reduce or prevent all perceived risks due to their large-scale use of chemical pesticides (FAO 1988; Chandler et al. 2008).

Business models that offer entrepreneurships for the production of microbial biopesticides in rural India have been designed and developed (Amin 2013). There are many examples that lead village youths into technoprenuership opportunity in mass-producing the microbial control agents (Kumar et al. 2019). The grain-based (sorghum, rice etc.) dry fermentation mass-production system has been part of the technology package offered along with the candidate MP species and strain of the National Agriculture Research System (NARS) and Council of Scientific and Industrial Research (CSIR) institutions. The commercialization of these MP production technologies from these institutions needs the in-built follow-up regarding the quality insurance of the standard operating procedures laid out by the institution for their mass production.

Many public institutions under the National Agriculture Research System (NARS) and under Council of Scientific and Industrial Research (CSIR) by National Chemical Laboratory, Pune – National Collection of Industrial Microorganisms (NCIM) have discovered many candidate microbial bioagents for crop pest management. Further research proceeded to find out formulation technology using these strains for their commercial manufacture for use in agriculture farms to protect crops from various biotic stresses. However, there are no patents that have been registered in India or any other country for the manufacture process of MPs.[4] The science, technology and innovation (STI) of MP formulations in terms of research/innovation and commercial manufacture has not attained the equivalent expertise and capacities as in the case of microbial pharma processes. The critical mass that is essential in the country for this purpose is yet wanting to attain perfect manufacturing posture. This is one of the reasons for the poor spread of this technique, as an essential coordinate of IPM in crops. The convincing stand of the MP formulations for effective suppression of pests even under organic farms is shaky due to the variation in bioefficacy in the same crop season. With comparative bioefficacy of insecticides that farmers generally deploy to get 'quick-kill' effect, the acceptance of microbial pesticides become limited. Herbivory management using MP formulations may not be on firm footing in the absence of assured quality products. The national requirement of MP formulations is met with the Central Insecticide Boar-Registration Committee (CIB-RC) registered MP formulations (Table 10.2). Rabindra and Grzywacz (2010) illustrated the regulatory process in India, as on 2009 for registering three fungal entomopathogens; three fungal nematicides; three bacterial entomopathogens; two fungal antagonists and one bacterial antagonist against plant pathogens. Kumar et al. (2019) provided comprehensive data on the 306 registered microbial pesticides of 16 MP organisms and their 196 commercial formulations. It appears that there is significant variation in the data base of Directorate of Plant Protection, Government of India, in regard to the details of registered MP formulations and their manufactured quantity of the formulations. New-age innovations in formulation technology including the use of nanomaterials have intensified the vistas on improving efficiency of pest control in crops (Chhipa and Joshi 2016; Koul 2019). The regulatory machinery will then be challenged with novel registration guidance documents for such MP formulations using nanotechnological processes and substances.

India adopted the organic farming policy in 2005[5]. The organic means and methods of agriculture became a 'reinvented' wheel in the wake of increasing consumer awareness about the health advantages assumably with the consumption of organically grown commodities. The biological pesticides became strong candidates in managing pests in organic farms and Technical Bulletins on organic farming, such as that of ICAR-Central Institute of Cotton Research promoted their use (Rajendran et al. 2000). Organic cultivation in India is in an area of 3.67 million hectare[6]and the organic certification area (registered under National Programme for Organic Production) is about 2.3 m ha cultivable area. The state of Madhya Pradesh has covered largest area under organic certification followed by Rajasthan, Maharashtra, Gujarat, Karnataka, Odisha, Sikkim and Uttar Pradesh. The assessment

[4]http://www.ipindia.nic.in/advanced-search.htm accessed on 27082020.

[5]https://ncof.dacnet.nic.in/Policy_and_EFC/Organic_Farming_Policy_2005.pdf.

[6]http://apeda.gov.in/apedawebsite/organic/Organic_Products.htm#:~:text=As%20on%2031st%20March%202020,Hectare.

Table 10.2 Data on number of registrants of microbial entomopathogen biopesticides under section 9(3) in CAB & RC database & Kumar et al. (2019)

S. No.	*Product name*	No. of registrations	Number and type of commercial formulations[a] (AS,SC,WP)
1.	*Beauveria bassiana*	87	46 AS,SC,WP[b]
2.	*Beauveria brogniartii*	01	01 WP
3.	*Hirsutellathompsonii*	01	AS, WP
4.	*Isaria(=Paecerlomyceslilacinus) fumosorosea*	03	03 AS, WP
5.	*Pochoniachlamydosporia (=Verticillium chlamydosporium)*	04	02 WP
6.	*Purpurecilliumlilacinum (=Paecilomyceslilacinus)*	35	20 AS, WP
7.	*Metarhiziumanisopliae*	33	26 AS,SC,WP
8.	*Lecanicillium (Verticillium) lecanii*	62	42 AS, WP
9.	*Lecanicillium (Verticillium) lecanii +Hirsutellathomsonii*	01	01 AS
10.	*Bacillus thuringiensis kurstakii*	35	25 AS,WP,
11.	*Bacillus thuringiensis israelensis*	12	12 AS, WP, DG
12.	*Bacillus thuringiensis galleriae*	01	01 FC
13.	*Lysinibacillussphaericus*	03	01 WP
14.	*Bacillus firmus*	01	01 WP
15.	*Helicoverpa*NPV	22	11 AS
16.	*Spodoptera litura* NPV	05	5 AS
	Total	**306**	**196**

http://ppqs.gov.in/statistical-database as on 22092020

[a]Not necessarily a complete list of all products

[b]*AS* aqueous suspension, *DG* dispersible granules, *FC* flowable concentrate, *SC* suspension concentrate, *WP* ettable powder

of annual requirement of MP formulations for organic farms in the country is worthwhile to project the annual manufacturing requirement. The requirement for MP formulations in at least 10% of the 2.3 m ha of organic farmed area in the country can be around at the rate of 5 kg per ha of any one MP organism sprayable/wettable powder (the most common formulation in use) shall be 11.5 million kg, far lower than the total production quantity of biopesticides in Table 10.1.

10.3 Regulatory Process of Biopesticides in India

The regulatory framework is for registering any MP formulation product for commercial production by micro, small and medium enterprises (MSMEs) and registered companies for manufacturing in factories. The international guidance document (FAO 2012) for regulatory management of biopesticides is the one that can be used as harmonized steps. Kabulick et al. (2010) provide the global glimpse of the regulatory situation of microbial pesticides. Regulatory requirement for the commercial manufacture and marketing of MP formulations in the country was identified in the late 1980s. Mensink and Scheepmaker (2007) concluded that plant protection products with active micro-organisms are allegedly less hazardous to the environment and wildlife than synthetic chemical pesticides. In order to alleviate environmental safety concerns of possible contaminant microbials in the MP formulation, their potential toxicity and pathogenicity tests may be relevant. They also point out a 'decision tree model' as followed in the European Union through the scientific scrutiny steps of characterization, identification and efficacy and also emission, exposure, environmental effects and the environmental risk assessment. It is advisable to take up such technical scrutiny by regulators on a case-by-case basis using scientific judgement for assessing the microbial ecology, limited experience with regulatory test protocols and taxonomic status in relation to the indigenousness of active micro-organisms from the data package of the applicant. The decision tree offers regulatory guidance on the environmental safety evaluation of microbial plant protection products.

The NARS and CISR institutions that were involved in the pioneering research on identification of suitable microbial agent strains empowered with local adaptation were chartered to develop guidance document for the Central Insecticides Board to suitably incorporate in the Insecticides Rules, 1971 and Guidelines[7] for registration of the candidate formulations under the Insecticide Act, 1968. The NARS institutions that developed MP formulations commercialized them to private individuals and companies for large-scale production and marketing. These entrepreneurs including big companies have sought the registration[8] of their specific microbial strains of those fungal and bacteria species in the pesticide formulation (s) for specific crop labels in Form I after following the relevant Guidelines for microbial biopesticides as provided by the Registration Committee of the Central Insecticide Board in regard to the data requirements on bioefficacy, toxicology, packaging in addition to depositing the formulated microbe strain in any of the designated and notified national microbe depositories. The MP formulations may have label claim for the insect pests in the respective crops where evaluation data are submitted towards registration for commercial production under the Section (3) of the Insecticide Act, 1968. However, the Rules under this act may be modified to extend label use against the same pests in other crops too, based on scientific study.

[7]https://pesticides-registrationindia.nic.in. Accessed on 10 July 2020.

[8]https://pesticides-registrationindia.nic.in. Accessed on 10 July 2020.

Plate 10.1 Innovative initiative by farmer groups for on-farm mass production of microbial pesticides
With permission to reproduce
Courtesy: Foundation for Agriculture Resource Management and Environmental Remediation (FARMER), Ghaziabad, Uttar Pradesh

The Registration certificate of Microbial pesticides may carry the manufacture process that is fit for the production of respective MP formulations.

In India, there is no legal restriction to mass-produce MP formulations in agriculture farms for their own use. Farmers and Farm Producer Groups and Farm Producer Organisations can produce any MP formulation on a no-profit-no-loss basis under the guidance of the technology discovering research institution. Innovative fermenter techniques are deployed by such groups (Plate 10.1) for mass production of MP formulations for application on crops for pest management.

10.3.1 Ethical and Regulatory Concerns in the MP Formulation Sector

There are various ethical and safety issues of major concerns in the use of microbial products and molecular tools and techniques in insect pest management. Therefore, when the proper safety precautions are taken, colonies of microorganisms can be safely isolated from homes, yards, gardens etc. The majority of microorganisms are pathogenic,[9] but bacterial cultures or Petri plates that contain any type of bacterial colonies should be treated with general safety precautions (Anonymous 2007; EC 2005; Hauschild 2012; James 2008). The Good Laboratory Practices (GLP) and

[9]https://www.sciencebuddies.org/.../references/microorganisms-safety. Accessed on 12 Aug 2020.

GMP (as described in the following section) dossiers of the licensed manufacturing firm should have all the biosafety protocols recorded and those are to be meticulously followed. India being signatory to Biological Weapons Convention[10] (that came into force from 26 March 1975) the states to ensure that the abiding principles and protocols of the Convention need to be put in place and practiced. The states may have to provide the necessary undertaking to the Ministry of Home Affairs (MHA), Government of India periodically, in accordance with MHA guidelines on this.

In the interest of assuring farmers of the expected bioefficacy of the MP formulation, there is need to establish and assure quality in terms of international norms as guided by United Nations Forum for Sustainability Standards (UN-FSS) and as recommended by Quality Council of India (QCI). The ethics in the business of microbial pesticides need the following considerations.

Ethical considerations matter in the use of MP formulations that are marketed for pest management shall be:

(a) Absence of the consistent concentration (in terms of colony forming units), as prescribed microbial species/strain content of the target MP formulation,
(b) Ensuring the absence of known/unknown hazardous, dubious and dangerous microbes including the non-culturable ones,
(c) Lack of consistent bioefficacy against the target pest(s) in the crops with label claim for the MP formulation,
(d) Absence of quality regulatory management (QRM) as laid out by Quality Council of India for the manufacture, transport, storage and use of MP formulations,
(e) Use of formulants of dubious quality used in the manufacture of the MP formulations resulting in the harming of target crops and agroecology,
(f) Release of untreated objectionable effluents and laboratory/factory wastes into environment.

In order to obviate the most of the above ethical issues in the marketing and use of MP formulations, suitable Guidelines and Code of Conduct have been placed in public view for compliance and for confirming the manufacture and marketing of absolutely high quality MP formulations for pest management use in farms. As described elsewhere GMP and GLP are essential protocols for compliance by licensed manufacturers of MP formulations.

[10]https://en.wikipedia.org/wiki/Biological_Weapons_Conventioncame into force w.e.f. 26 March 1975. Accessed on 12 August 2020.

"Each State Party to this Convention undertakes never in any circumstances to develop, produce, stockpile or otherwise acquire or retain: (1) Microbial or other biological agents, or toxins whatever their origin or method of production, of types and in quantities that have no justification for prophylactic, protective or other peaceful purposes; (2) Weapons, equipment or means of delivery designed to use such agents or toxins for hostile purposes or in armed conflict."

The MP formulations for farm pest management are manufactured in unorganized sector as well as in organized factories of big industrial factories. In both situations, there is strong need for the quality review management (QRM)[10] protocol in place. This ensures the systematic assessment and control of risks during manufacture of microbial agents. In this context all manufacturing facilities shall have a state-of-the-art microbiology laboratory manned by technically experienced and talented scientific personnel. The in-house QRM documentation for batchwise production of the microbial pesticides products would be further audited by the National Accreditation Board for Testing and Calibration Laboratories/International Standardisation Organisation (NABL/ISO) system. As the microbial pesticides have become common IPM component in crop pest management, there has been regulatory issues in the manufacturing of formulations. The major concern is about the MP formulations with dubious quality (NAAS 2013) bereft of Good Manufacturing Practices (GMP).

10.3.2 Management of Quality Compliance

The formulation research for MP terminates in NARS institutions with the development of techniques for culturing the specific strains of microbial species for known bioefficacy against the target insect/mite pests in crops and using ingredient recipe for dust formulation, wettable formulation or the liquid formulation. The quality management protocols have to be set into operation for the MP formulations (Anonymous 2020; Van Lantern 2003). The step ahead for scaled up production plan and packaging did not receive much research attention. The result was that the technology to manufacture got restricted to the sharing of MP strain and the MP formulation. Their packaging, storage and transport were not given due attention and due diligence on the regulations for these aspects were not heeded to. It is also significant to observe that very few private companies came up with their own R&D set-up to develop their own microbial pesticides strains and formulations thereon; these technology packages were neither acquired from any foreign collaborators nor developed with the help of global leaders in this field.

The quality review management (QRM) of MP formulations needs greater attention in regulatory management. The regulatory entities of the states can apply the QRM principles that are put in place for drug manufacture for assuring quality production of MP formulations. Satpathy (2018) brought out the existing regulatory norms in the country and the absence of desirable quality in the MP formulations that are marketed, as prescribed by the Central Insecticide Board in its guidelines for microbial biopesticides. The use of substandard MP formulations in IPM of crops cannot only face ineffective pest suppression, but also bring in undesirable exposure to humans with health challenges from contaminant microbials.

Ultimately the most commonly formulated MP products, as dust formulation or Wettable Powder formulation is filled in polybags of suitable size that are held in the carton boxes before being stacked and transported to the designated markets. In case of liquid formulation, packaging principles to adopt polypropylene (PP) bottles need to be followed. Ideally glass bottles should be preferred. The storage stability of the spores in the shelf-life studies of the MP formulation in PP bottles in comparison to glass bottles needs to be studied.

10.3.3 Model Format for Quality Management Protocol for Microbial Pesticides Manufacture Factory

This procedure is applicable at the microbial pesticide manufacture factory. The objective of quality review management is to conduct routine reviews of the whole quality system in the production line according to a planned schedule. This includes the review of both operational and quality system review.

Operational Review based on the manufacture plan for every month as well as the capacity utilization plans based on raw material supply and production process for each batch of MP formulation.

Quality System Review shall be done by team headed by responsible personnel of the company. The quality review shall include data analysis of batchwise production, raw material mobilization and their quality reports, instances of arising problems and their resolutions.

Forms and records shall be maintained and audited periodically: Calender for Operation Review and Quality Review Meetings (Monthly).

10.3.4 Good Manufacturing Practices (GMP)

Good manufacturing practices for the production of microbial agents have been enunciated by various international agencies (WHO[11], FAO-OECD[12]). However, there is no agency that can oversee the QRM of the products that are marketed as microbial biopesticides for crop pest suppression. The desirability of empowered institutions, such as Quality Council of India (QCI), to undertake QRM of MP formulations in the country is to be introduced. Adulterants and contaminants are important parameters for the QRM of MP formulations.

As the existing manufacture of microbial pesticides and their formulations, especially using dry-fermentation techniques are yet to follow any national Guidance Document for certification of both processes and the product. There is concern about potentially harmful and hazardous microbial contamination that is to be resolved. Certified GMP batch production and their marketing would enable quality managed products for effective and efficient crop pest management. The cost of using MP formulations with poor quality is equivalent to the cost of crop loss due to pestilence.

---

[11]https://www.who.int/biologicals/areas/vaccines/Annex_2_WHO_Good_manufacturing_practices_for_biological_products.pdf?ua=1. Accessed on 10 July 2020.

[12]OECD/FAO (2016). OECD-FAO guidance for responsible agricultural supply chains. Paris, Organisation of Economic Co-operation and Development (https://mneguidelines.oecd.org/OECD-FAO-Guidance.pdf).

Hence, the farmers suffer double loss since they lose the crop even in spite of investing in MP formulations. The biohazard due to microbial contaminants to farm families and consumers of farm commodities arising out of the sub-standard MP formulations and hazardous contaminant microbes needs strong regulatory audit for verification of self-certified MP products. Self-certification shall be made mandatory for all MP production units to guarantee the absence of harmful microbes and other additives in their products. The WHO guidelines (1996) provide significance of reducing bioburden and biohazard by following GMP in the manufacturing process. The MP formulation industry in the country shall establish inter-factory audit system in which multifactor analysis of perceived risk factors can be assessed annually. India is signatory to Biological Weapons Convention (BWC) and has the obligation to report to the Convention about the peaceful purposes to handle all microbes for health, agriculture and any other national purposes and needs.

10.3.5 Good Laboratory Practices (GLP)

India has institutionalized GLP through Quality Council of India (QCI). However, the MP manufacturing sector needs to align and establish GLP norms for microbial agents that are used for the manufacture of MP formulations. Such back-end laboratory would be the pivotal set-up for the maintenance and management of microbial pure cultures without microbial contamination as well as invasion by laboratory mites (Onions 1990). It is significant to note that only about 1% of the reported microorganisms are culturable. The rest of the uncultivable microorganisms can still be the contaminants in the fungal and bacterial cultures. Molecular tools provide certain degree of tests for the obvious contaminations. However, still cautious, systematic QRM procedures need to be set up in the laboratories to obviate any possible contaminations of MP formulations. Careful consideration on the GLP practices can reduce such possibilities. Anticipatory research on the potential contaminant microbes under different fermenter operation conditions, media composition and any other extraneous factors is needed.

10.4 Hazard Perception in the Use of Microbial Biocontrol Agents

The perceived threats in the form of hazards, such as allergenicity (Ward et al. 2011; Darbro and Thomas 2009), emanating from the production and subsequent use of microbial biopesticide formulations in terms of the primary microbial species as well as potential contaminant microbials (including non-culturable) that could end up in the product during manufacture, packaging, transport and storage need clear understanding by the manufacturers. The conscious effort to avoid such introduction of hazardous microbials in the formulation product is one of the quality management protocols to be stringently followed.

Biowaste management under Basel Convention (1992) regulates transboundary movement of hazardous wastes including biological wastes (Annexure-I). Further, the GMP protocols of the manufacturing site for MP pesticides and the GLP protocols of its laboratory also stipulate the waste management and waste-processing of the spent-cultures/media and labwares. The MP formulation industry can be the self-regulatory mechanism in the country for both overseeing the compliance obligation under the National and State Pollution Control laws and Biological Weapon Convention (BWC) through self-governance protocols aligned to national and international codes of conduct.

As in the case of the United Nation Environment Program (UNEP) London Guidelines for the Exchange of Information on Chemicals in International Trade, there is the exchange of information under BWC. The national code of conduct on microbial research and manufacture shall have the database on the firms and entities who are involved in handling various live microbes for any of the microbial pesticides development and manufacture for marketing. The bioburden of microbial flora (either hazardous or otherwise) present (WHO Guidelines 1996) in raw materials for formulating MP products, in the production media, intermediate finished goods etc., as the case may be, is to evaluated by licensing authority of states in tandem with the State Pollution Control Board in all states and Union Territories.

Another major concern is the effluents that are released from these production units (OECD/FAO 2016[13]). The pollution control boards shall ensure oversight of the quality of effluents that are permitted to be released into the environment. The GMP for manufacture of MP formulations also covers the management of factory effluents. In the case of factories handling microbes, the spent-culture broth and other chemicals used in fermenters and laboratory culture systems have to be treated before being flown out into public drains. Licenses that are required in various states from respective state pollution control board to run the factory for the manufacture of microbial agents relate to effluent management; air/water/soil pollution control for chemicals and microbials. The regulatory framework includes Water (Prevention and Control of Pollution) Act,[14] 1974 and Air (Prevention and Control of Pollution) Act, 1981.

The precise and succinctly defined hazard perception is essential for each microbial pesticide formulation product, in order to attain risk assessment of this class of biological pesticides. The required hazard definition for MP formulations on the manufacture, transport, use and disposal of microbial pesticides has to be firmed up. However, the present regulations of these products do not visualize and update the regulatory requirements in this context.

---

[13]OECD/FAO, (2016). OECD-FAO guidance for responsible agricultural supply chains. Paris, Organisation of Economic Co-operation and Development.. https://mneguidelines.oecd.org/OECD-FAO-Guidance.pdf.

[14]http://moef.gov.in/about-the-ministry/organisations-institutions/boards/central-pollution-control-board/ & https://cpcb.nic.in.

An emerging issue in the context of MP formulation is the deliberate contamination with synthetic chemical pesticides in the Indian pesticides market with poor quality assurance on the colony forming unit count of microbial agent. The 'quick kill' effect of such formulations may attract the customership of farmers; however, such of these spurious and contaminated MP formulations deceives and misleads farmers on the bioefficacy and toxicology of such products and may leave undesirable chemical pesticide residues in crop commodities. Moreover, these formulations become ill-defined in hazard perception and risk mitigation assurance.

10.5 Packaging and Container Compatibility

The Code of Conduct (FAO 2015)[15]defines packaging as 'the container together with the protective wrapping used to carry pesticide products via wholesale or retail distribution to users', whereas re-packaging refers to 'the transfer of a pesticide from any authorized commercial package into any other, usually smaller, container for subsequent sale.' National laws, rules and guidelines have been aligned to this FAO document including the Code of Conduct. Accordingly, packaging of pesticides for MP formulations should conform to the safety for transport, storage, handling, and use without allowing the degradation of pesticides. Further, the packaging should not create danger to human health and that of environment. The packaging of pesticides should not resemble common packaging of consumable goods. The label for use and risk reduction measures has safety mechanism that would avoid inadvertent handling by children. Reuse of pesticide packaging containers should be banned and punishable under the relevant national law. Packaging and repackaging of pesticides can be undertaken only at the licensed premises under supervision of competent personnel. It is advised to store in cool and dry ambient conditions. However, MP formulations need stringent temperature management of the rooms, where they are stored with the shelf life prescribed on the label to be sustained for achieving very good shelf life that permits effective pest suppression due to the presence of active and live propagules under storage.

10.5.1 Packaging, Storage and Transport

It is realized that there is no globally recognized regulatory model that would obviate possible hazards in the manufacturing, packaging, transport of MP formulations for use in crop pest management (Arora et al. 2016). Continuing efforts by global agencies, such as the Codex Alimentarius[16], International Organization for

[15]http://www.fao.org/3/a-i5008e.pdf .12 August 2020.

[16]http://www.fao.org/fao-who-codexalimentarius/news-and-events/news-details/en/c/1189277/ . Accessed on 13 August 2020 **Codex looks to harmonise regulation of biopesticides (6 April, 2019).**

Biological Control (IOBC), European and Mediterranean Plant Protection Organization (EPPO) and Organization for Economic and Co-operative Development (OECD), to develop standards global standards and models for packaging and transport of biological substances and prevent any arising handling hazards in transport and storage.

10.6 Utilization of Biotechnology Procedures for Pest Management

From the late 1990s, the scientific developments in GM technology trickled in for their wide variety of applications. Progress in agriculture has been immensely benefited because of the advances in various component science and technology areas (Parekh 2004). There have been several advances in the scientific pursuit of biotechnology in agriculture and other sectors for human benevolence. The associated understanding of ethical, safety and intellectual property issues of every discovery is under constant debate in recent decades (Nambisan 2017).

The biotech products ultimately undergo appropriate risk assessment within the existing knowledge sphere, resulting in the labelling of the product for offering informed choice for consumers. However, the global debate on the worthiness and goodness of fit of the biotechnology products in agriculture has entered devious arguments related to matters other than science too (Kinderlerer and Adcock 2003). This crystal protein is found to be present in the specific strains of *Bacillus thuringiensis*, a soil inhabiting bacterium and can kill various insect pests that affect crops, such as caterpillars, maggots, grubs and so on. In the quest for tangible pest management, the idea of toxifying crop plants with the alien gene-expressed insecticidal proteins, such as delta endotoxin, was explored and commercialized globally in many crops.

Although many other biotechnological interventions were lined up for improving the quality and quantity of farm commodity output, the most favoured technology was Bt gene technology in crops to thwart insect pestilence. As a model crop, cotton became the global example. Other crops, such as soybean, maize and rice, have also been genetically modified with the target genes that express Bt delta endotoxin at given phenological stage of the crops.

India took to Bt cotton regulatory approval under the Environment (Protection) Act,[17] 1986 and Rules[18] of the Ministry of Environment, Forest and Climate Change, Government of India, to manage the environmental release of the alien Bt delta toxin Cry gene(s) in cotton since 2002. Many research institutions under CSIR, NARS and others undertake development research to get biotech crops with various features and traits. All these have the regulatory protocols under the Review Committee on Genetic Manipulations (RCGM) and Genetic Engineering Appraisal Committee

[17] http://moef.gov.in/rules-and-regulations/environment-protection/ 17 August 2020.

[18] http://moef.gov.in/wp-content/uploads/2018/03/THE_ENVIRONMENT.pdf 17 August 2020.

(GEAC) guidance documents-based evaluations before being approved for release to environment for cultivation.

There has been strong interest to alter genetic virulence of microbial pesticide organisms to improve their virulence and tools, such as Crisper-Cas 9 techniques for gene editing, and improving the existing strains for virulence is contemplated. Such gene edited organisms (GEOs) have so far not been commercialized. There is presently government ban on the environmental release of GM microbial biocontrol agents for pest management, while the environmental release of GEOs is under policy discussion by the Department of Biotechnology, Ministry of Science and Technology (NAAS 2020).

10.6.1 Biotechnology Advancements in Crop Pest Management

Over the last 30 years, the ability to modify specific genes in microorganisms has revolutionized numerous fields of biosciences, including medicine, agriculture, and basic research into life processes. Molecular tools and techniques under the modern branch of biotechnology led to the utilization of genetic transformations (both *cis* and *trans*) for integrating alien pest resistance genes and other such useful genetic trait expressing genes into certain crop species to thwart insects and pathogens. In the wake of increased consumer consciousness on the potential risks and hazards to terrestrial biomes, agroecologies and to human health due to the introduction of genetically modified (GM) crops as well as their impact on foodwebs including nontarget organisms, there has been stringent enforcement of various regulatory protocols to reduce such perceived risks while using such crops for food, fibre, feed production.

The scientific research in this area is carried on under public and private funding in NARS/CSIR institutions. The Bt brinjal with resistance to fruit and shoot borer (*Leucinodes orbonalis*) was the last instance where the regulatory moratorium was applied for the release for cultivation in the country. In the scientific research front in this realm, the latest publication is on the identification of Tma12 protein (Yadav et al. 2019) that is reported to toxify whitefly in GM cotton plants (Shukla et al. 2016). However, the common thread of scientific discussion is about the cost:benefit ratio of crop biotech products with insecticide-expressing traits that are expected to be overpowered by target insect pest species due to biological adaptation prowess. Ultimately such GM biotechnology products cannot sustain the strong adaptations of both oligophagous and polyphagous pests in crops. This has been the experience in Indian cotton crop fields.

In the present millennium, intensive application of biotechnological products globally became significant for pest management in crops. The prominent amongst these was the crop genetic modification by incorporating the alien gene expressing the delta endotoxin from the prominent MP bacteria, viz. *Bacillus thuringiensis*, manage predominantly caterpillar pests in crops, such as cotton and maize. Gene editing technology has become new tool for crop pest protection. The regulatory and ethical components for managing such developments have been taken up in various countries in order to manage the perceived hazards and risks for humans and nature once the biotechnological products were commercialized and released into environment.

10.6.2 Risk Assessment Protocols of Genetically Modified (GM) Biocontrol Agents

Novel technological discoveries and their applications in agriculture have influenced the modern crop production globally. The issues on the absorption and acceptance of the technologically driven agricultural commodities are to be brought under the category of ethics that consider the risk evaluation and hazard mitigation. CAST (2005) advocated the institutionalization of ethics in agriculture in order to evaluate and independently bring out transparent benefits and attendant hazard level for the environment and consumers.

Risk assessment as done in the regulatory system of the European Union is to bring out the potential perceived risk due to GM technology. Comparative risk assessment has been undertaken by Steinhauser (2001) between chemicals and GM products. The US Food and Drug Administration relies on the hazard identification of GM products and finalizes the risk perception based on the hazards. An essential element in the ethical evaluation of biotechnology is the analysis of the possible harms and weighing these risks against the probable benefits.

Risk evaluation protocols for genetic modifications have become a major field that has implication in the microorganisms that are immensely used in the development of more efficient products. However, the global opinion has not been unanimous on the commercial release of such GM microorganisms for agriculture purposes. In India too, there has been no acceptance of GM microbial pesticides due to evident concerns regarding the unknown effects to humans and environment. The stringent guidelines in research laboratories with Biosafety Level (BSL)4 level facility make the costs and elaborate infrastructure very high.

Genetically modified microbial biocontrol agents have been undertaken to sharpen the targeted bioefficacy as well as to improve the non-competitive performance in various agroecologies. The environmental impact and risk assessment thereon (Anonymous 2000; Migheli 2001a, b) are mandatorily undertaken in order to permit the release of such GM bioagents for use in crops. Biosafety and ethical concerns are important considerations are to be imposed in the regulatory framework (Zadoks 1998; Stemke 2004). However, this capability raises concerns about the potential hazards posed by the technology. In response to these concerns, specific protocols (Stemke 2004) have been developed to safely monitor the use of genetically modified microorganisms (GMMs). In case of approval for environmental release of GM crop plants with traits for biotic stresses, such as pest and diseases, the regulatory body under the Ministry of Environment, Forest and Climate Change, Government of India is vested with the Environmental Protection Act (EPA), 1986, has placed moratorium presently. Recent scientific advancement in gene editing technology – clustered regularly interspaced palindromic repeat (CRISPR)-

CRISPR-associated proteins (Cas)(Crisper-Cas9) techniques have stimulated research interest to develop biotic stress-tolerant crop plants (NAAS 2020).

10.6.3 Ethical and Regulatory Concerns in the Biotechnology of Crops for Pest Management

CAST (2005) advocates institutionalized agricultural ethics in which both farms and food systems need to resolve ethical conflicts and steer socio-economic advantage of the new biotechnological tools and inventions that claim better crop productivity, improved farm commodity quality, better management of biotic and abiotic stresses. Environmental ethics, socio-economic benefits and regulatory policies have strengthened as an important component in the debates on the advancement in modern biological sciences (Anonymous 2015; Southgate 2002; Kinderlerer and Adcock 2003; Gupta and Chandak 2005; Shukla et al. 2018). The anxiety for seeking answers to unknown concerns has increased over the last few decades arising out of the explosion in the information flow. While derisive about such anxieties, validation of science-based analytical processes of any new information that floats around is desirable.

Clustered regularly interspaced palindromic repeat (CRISPR)-CRISPR-associated proteins (Cas) (CRISPER - CASPER-9) system is a good tool for modifying crop genome in order to generate gene-edited crop plants to impart resistance to pests. Globally, there has been extensive debate on the ethical and regulatory policy requirements to apply this nascent and potent tool for imparting biotic stress resistance in crop plants. The National Biotechnology Development Strategy Document (http://dbtindia.gov.in/about-us/strategy-nbds) does not contain policy statement on this novel technology for agriculture. The National Academy of Agriculture Sciences (NAAS) has published Policy Brief no.7 after a consultation on the Draft Guidelines on gene edited organisms (GEOs) from the Department of Biotechnology, Government of India (NAAS 2020). The Indian research scenario in pest resistance using Crisper-Cas-9 technique is stated to be at its infancy. NAAS recommends that in the case of GEO too, in line with the EPA (1986) along with the existing Seed Act (1966), the new plant varieties developed through genome editing need to go through the regulatory processes where required, for risk analysis of biosafety and environmental safety, so that the technology applications are in compliance with the Protocol. A policy perspective on GEO techniques is in the making by the Department of Biotechnology.

The currency of perceptions has to be modified after acute effort to bring new evidences for and against any perceived threat borne out of the introduction of new technology into farming. Regulatory process of countries strives hard to undertake such intellectual steps to arrive at well-debated clarity of thoughts. In general, open debates involving every logically thinking argument would alleviate most of the apprehensions within the existing scientific and socio-economic realm. There is a tendency to approximate certain potential hazards without looking for evidence-based conclusions. Such instances leave the analysts with inconclusiveness about the technology. Major plant protection concern is the non-uniform expression of the transgenic gene expressing Bt toxin gene(s). The farmer in such cases tends to lose his investment on seeds 'with promise to suppress biotic stress' due to inconsistent protection of target pests and consequent severe crop losses.

The basic evaluation of genetically engineered products from biotechnology[19]utilizing transgenic, cisgenic- or subgenic tools to derive crops with alien genes to express insecticidal proteins is the science-based analysis of the possible harms and their most likelihood of occurrence, for weighing the risks across the anticipated benefits. Each sovereign nation introducing such GM crop needs to transparently examine the home-generated data on all aspects of safety leading to hazard definition of the GM event and the gene product in the crop plant. Biosafety assessment protocols need to be laid out for each instance of introducing GM crop bearing the genes expressing insecticidal entities in the plant, their metabolites and degradation products that may have impact to all components of agroecology and other environmental entities as defined by the regulatory system of the country.

The following ethical matters are perceived during the commercialization of GM cotton delta endotoxin gene (the only crop) permitted for environmental release in India.

(i) Whether the endotoxin expressing gene(s) technology is any more relevant for pest management in crops.
(ii) Need for transparent regulatory mechanism to oversee the claims from GM technology of crop including crop yield (Quain and Ziberman 2003) after environment release of GM crop cultivar into the environment as well as license to produce and market their seeds to farms.
(iii) Well-defined roles and responsibilities of the government agencies that deliver to farmers of all states the information on performance of the GM crop variety as well as perceived hazards due to them from time to time.
(iv) Overseeing the quality including genetic purity of seeds of the relevant traits that are marketed for cultivation.
(v) In case of cotton crop whether lint yield and fibre quality are commensurate with the label claim of the marketed seeds and acceptable for the best market price appreciation, as the crop is to produce cotton lint as industrial raw material.
(vi) Non-availability of non-GM cotton variety seeds due to non-production and marketing of these as an alternative for farmers for opting those for cultivation. The government has not made adequate provision to provide seeds of cotton varieties that are developed in public institutions. The raw material consumer industry may also support the seed availability through Corporate Social Responsibility (CSR) programmes.

[19]https://en.wikipedia.org/wiki/Genetically_modified_crops. Accessed on 10 September 2020.

(vii) All private seed companies have GM technology with monopolistic trends through hybrid seeds, as in the case of cotton crop, with which every year farmers have to depend on those seeds from them. Open pollinated cotton crop seeds have vanished from market completely. Farmers are compelled to cultivate only GM hybrid cotton seeds.
(viii) The assurance that insect pest management would be easier with low use of pesticides has not been proven as a faithful technological advantage in India.
(ix) Transparency of information on the given genetic modification and the gene product(s) expressed in host crops.
(x) The potential hazard perception to environment including agroecology.
(xi) Potential hazard for the target pest species to develop resistance to the given toxin that is expressed in crop plants.

10.7 Conclusions

The microbial pesticide formulations have been registered under the Insecticide Act, 1968 in order to regulate their manufacture, use in farms with respect to crops and target pests as well as to sustain their biological quality. There has been increasing concerns about the standards and practices in their manufacture, packing, storage, transport and handling in order to reach them to farms for application in crops. Recent spurt in low quality of these formulations as well as their contamination with chemical synthetic pesticides has alerted the regulatory system, consumers and farmers alike. Quality review management of microbial pesticide formulations needs special attention for both manufacturers under GLP/GMP regime. The effluent management of the manufacturing units also needs intense environmental audit to safeguard from perceived hazards.

Genetically modified crops are the best biotechnology derived products that target increase in both yield and quality of farm commodities. The risk assessment of these commodities in the context of hazard perception and their mitigation has grown into specialized regulatory paradigm.

New biotechnology tools and techniques, such as Crisper-Cas9 in crops for pest resistance and higher commodity output with better quality parameters along with options to reduce risks to human and environment, are promising. Ethical and practical considerations for commercialization of genetic modification of crops from *cis-trans-* or sub-genic products need careful analysis for science-based assessment of the potential hazards. Combination of microbial pesticide formulations and biotech farm crop varieties can be integrated in the overall crop health management architecture. The question of availability at farm gate of these IPM components has deeper introspective policy requirement. The agriculture farms cannot become exploitation grounds and make crop loss to bear year after year for farmers due to inefficient performance of MP formulations as well as GM varieties severe.

10.8 Points to Remember

1. The microbial pesticide formulations are useful tools for invertebrate pest management of crops in all agroecologies. Their widespread use in pest management and benefits accrued in terms of clean commodities alongside clean farm agroecology and safeguarding consumers' health have futuristic implications in regulatory principles and practice.
2. Quality review management of MP products is essential to sustain the shelf life and field bioefficacy of the products. The biowaste management of MP production units should follow GLP and GMP standard operating procedures to prevent undesirable leakage of harmful microorganisms following the relevant National Guidelines and International Conventions. The critical gaps in ethics and regulatory needs of MP formulations manufactured within GLP/GMP norms can be addressed through quality review management (QRM).
3. Biotechnology tools and techniques to deploy genetically modified crops as well as gene editing technology (Crisper-Casp9) for attaining pest resistance and higher commodity output with better quality parameters are promising. Ethical and practical considerations for commercialization of GM crops from *cis-*, *trans-* or sub-genic products need careful analysis for science-based assessment or 'decision tree-based' evaluation of the potential hazards.
4. Combination of microbial pesticide formulations can be integrated in the pest management architecture after appropriate regulatory approval.

References

Amin N (2013) Teaching of biopesticide development as a technoprenuership opportunity in plant protection. J Biol Agric Health 3:2224–3208

Anonymous (2000) Cartagena protocol on biosafety treaty. http://www.biodiversity.org/ratification.asp. Accessed on 08 August 2020

Anonymous (2007) Biosafety in microbiology and biomedical laboratories, 5th edn. Centre for Disease Control and Prevention, US Office of Health and Safety

Anonymous (2015) Pocket K No. 18: Ethics and Agricultural Biotechnology. International Service for the Acquisition of Agri-biotech Applications (ISAAA). https://www.isaaa.org/resources/publications/pocketk/18/default.asp. Accessed on 31 August 2020.

Anonymous (2020) Standard operating procedure for quality management review. https://www.pharmapathway.com/. Accessed on 1 September 2020.

Arora NK, Verma M, Prakash J, Mishra J (2016) Regulation of biopesticides: global concerns and policies. In: Arora N, Mehnaz S, Balestrini R (eds) Bioformulations: for sustainable agriculture. Springer, New Delhi. https://doi.org/10.1007/978-81-322-2779-3_16. Accessed on 12 August 2020

Arora NK, Fatima T, Mishra I, Verma M, Mishra J, Mishra V (2018) Environmental sustainability: challenges and viable solutions. Environ Sustain 1:309–318. https://doi.org/10.1007/s42398-018-00038-w. Accessed on 10 August 2020

Chandler D, Davidson G, Grant WP, Greaves J, Tatchell GM (2008) Microbial biopesticides for integrated crop management: an assessment of environmental and regulatory sustainability. Trends Food Sci Technol 19:275–283

Chandler D, Bailey AS, Tatchell GM, Davidson G, Greaves GWP (2011) The development, regulation and use of biopesticides for integrated pest management. Philos Trans R Soc Lond B Biol Sci 1573:1987–1998

Chhipa H, Joshi P (2016) Nanofertilisers, nanopesticides and nanosensors in agriculture. In: Ranjan S, Dasgupta N, Lichtfouse E (eds) Nanoscience in food and agriculture. Springer International Publishing, Cambridge, pp 247–282

Council for Agricultural Science and Technology (CAST) (2005). Agricultural ethics. Issue Paper No. 29. Ames, Iowa, USA.

Darbro JM, Thomas MB (2009) Spore persistence and likelihood of aero-allergenicity of entomopathogenic fungi used for mosquito control. Am J Trop Med Hyg 80:992–997

EC (2005) European Commission. Council directive 2005/25/EC of 14 March 2005 amending annex VI to directive 91/414/EEC as regards plant protection products containing micro-organisms. Off J Eur Union L 90:1–34

FAO (1988) Food and Agriculture Organization. Guidelines on the registration of biological pest control agents. Food and Agriculture Organization of the United Nations, Rome

FAO (2012) Food and Agriculture Organization. Guidance for harmonizing pesticide regulatory management in Southeast Asia RAP publication 2012/13. Food and Agriculture Organization of the United Nations. Regional Office for Asia and the Pacific, Bangkok, p 480

FAO (2015). International Code of Conduct on Pesticide Management Guidelines on Pesticide Legislation. 61 p.

Gupta PK (2006) Status of biopesticides-Indian scene. Toxicol Int 13:65–73

Gupta AK, Chandak V (2005) Agriculture Biotechnology in India: ethics, business and politics. In: Mehta S (ed) Bt cotton – a painful episode. Need for thoughtful policy, pp 34–50. 200 p.

Hauschild R (2012) Safety and regulation of microbial pest control agents and microbial plant growth promoters: Introduction and overview. In: Sundh I, Wilcks A, Goettel MS (eds) Beneficial microorganism in agriculture, food and the environment: safety assessment and regulation. CAB International, Wallingford, pp 67–71

James DE (2008) Nine safe practices for the microbiology laboratory. Carolina Biological Supply, Burlington, NC

Kabulick JT, Svircev AM, Goettel MS, Woo SG (eds) (2010) The use and regulation of microbial pesticides in representative jurisdictions worldwide. IOBC Global. 99 pp. Available online through www.IOBC-Global.org

Kinderlerer J, Adcock M (2003) Agricultural biotechnology, policies, ethics, and policy. Working Paper No. 3 prepared for the first meeting of the FANRPAN/IFPRI Regional Policy Dialogue on Biotechnology, Agriculture, and Food Security in Southern Africa. Johannesburg, South Africa.

Koul O (ed) (2019) Nano-biopesticides today and future perspectives. Academic Press, Cambridge, p 495

Koul O, Dhaliwal GS, Marwaha SS, Arora JK (2003) Future perspectives. In: Koul O, Dhaliwal GS, Marwaha SS, Arora JK (eds) Biopesticides and pest management, vol 1. Campus Books International, New Delhi, pp 386–388

Kumar KK, Sridhar J, Murali Bhaskaran RK, Senthil Nathan S, Kaushal P, Dara SK, Steven A (2019) Microbial biopesticides for insect pest management in India: current trends and future prospects. J Invertebrate Pathol 165:74–81

Mensink BJWG, Scheepmaker JWA (2007) How to evaluate the environmental safety of microbial plant protection products: a proposal. Biocontrol Sci Technol 17:3–20

Migheli Q (2001a) Genetically modified biocontrol agents: environmental impact and risk analysis. J Plant Pathol 83:47–56

Migheli Q (2001b) The deliberate release of genetically modified biocontrol agents Ill: Towards an alternative risk assessment paradigm. Agro-Food-Ind Hi-Technol 12:23–25

Mishra J, Dutta V, Arora NK (2020) Biopesticides in India: Technology and sustainability linkages. 3 Biotech 10:210. https://doi.org/10.1007/s13205-020-02192-7. Accessed on 10 June 2020

NAAS (2013). Biopesticides- quality assurance. Policy Paper No. 62. National Academy of Agricultural Sciences, pp 1–20.

NAAS (2020). Genome edited plants: accelerating the pace and precision of plant breeding. Policy Brief No. 7. National Academy of Agricultural Sciences, New Delhi, pp 1–16.

Nambisan P (2017) Risk analysis. In: An introduction to ethical, safety and intellectual property rights issues in biotechnology. Elsevier, pp 233–252. 392 p., ISBN 978-0-12-809231-6 392. https://www.sciencedirect.com/book/9780128092316. Accessed on 09 August 2020

OECD/FAO (2016) OECD-FAO guidance for responsible agricultural supply chains. Organisation of Economic Co-operation and Development, Paris. (https://mneguidelines.oecd.org/OECD-FAO-Guidance.pdf)

Onions AHS (1990) Prevention of mites in cultures. World J Microbiol Biotechnol 6(2):219–221. Technical Information Sheet No. 16

Parekh SR (2004) The GMO handbook. In: Parekh SR (ed) Genetically modified animals, microbes, and plants in biotechnology. Humana Press, New Jersey. ISBN 1 588 29 307 6 pp 386. https://link.springer.com/content/pdf/10.1007%2F978-1-59259-801-4_4.pdf

Quain M, Ziberman D (2003) Yield effects of genetically modified crops in developing countries. Science 299:900–902

Rabindra RJ (2001) Emerging trends in microbial control of crop pests. In: Rabindra RJ, Kennedy JS, Sathiah N, Rajasekaran B (eds) Microbial control of crop pests. Tamil Nadu Agriculture University, Coimbatore, pp 110–127

Rabindra RJ (2005) Current status of production and use of microbial pesticides in India and the way forward. In: Rabindra RJ, Hussaini SS, Ramanujam B (eds) Microbial biopesticide formulations and application, Technical document No.55, Project directorate of biological control, pp 1–12

Rabindra RJ, Grzywacz D(2010). India. In: Kabaluk, J. Todd, Antonet M. Svircev, Mark S. Goettel, Stephanie G. Woo (eds.), The use and regulation of microbial iPesticides in representative jurisdictions worldwide. IOBC Global. pp. 11–17. 99 pp. Available online through www.IOBC-Global.org

Radcliffe EB, Hutchison WD, Cancelado RE (eds) (2009) Integrated pest management: concepts, tactics, strategies and case studies. Cambridge University Press, New York

Rajendran TP, Venugoplan MV, Tarhalkar PP(2000) Organic cotton production technology. Technical Bulletin No.1, Central Institute for Cotton Research, Nagpur.

Ranga Rao GV, Rupela OP, Rao RV, Reddy YVR (2007) Role of biopesticides in crop protection: present status and future prospects. Ind J Plant Prot 35:1–9

Satpathy SN (2018) Regulatory norms and quality control of bio-pesticides in India. Int J Curr Microbiol App Sci 7:3118–3122

Shukla AK, Upadhyay SK, Mishra M, Saurabh S, Singh R, Singh H, Thakur N, Rai P, Pandey P, Hans AL, Srivastava S, Rajapure V, Yadav SK, Singh MK, Singh J, Chandrashekar K, Verma PC, Singh AP, Nair KN, Bhadauria S, Wahajuddin M, Singh S, Sharma S, Omkar RSU, Ranade SA, Tuli R, Singh PK (2016) Expression of an insecticidal fern protein in cotton protects against whitefly. Nat Biotechnol 34(**10**):1046–1051

Shukla M, Al-Busaidi KT, Trivedi M, Tiwari RK (2018) Status of research, regulations and challenges for genetically modified crops in India. GM Crops Food 9(4):173–188

Sinha B, Biswas L (2008) Potential of biopesticides in Indian Agriculture vis-a-vis rural development. India Science and Technology: 2008. NISTADS, New Delhi

Southgate C (2002) Introduction to environmental ethics. In: Bryant J, La Velle LB, Searle J (eds) Bioethics for scientists. John Wiley and Sons, New York, pp 39–55

Steinhauser KG (2001) Environmental risks of chemicals and genetically modified organisms: a comparison. Part I: Classification and characterization of risks posed by chemicals and GMOs. Environ Sci Pollut Res Int 8:120–126

Stemke DJ (2004) Genetically modified microorganisms: biosafety and ethical issues. In: Parekh SR (ed) Handbook: genetically modified animals, microbes, and plants in biotechnology ©. The GMO Humana Press Inc., Totowa, NJ, pp 85–130. pp. 386

Subhash SP, Prem Chand, Suresh Pal (2017). Pesticide use in Indian agriculture: trends, market structure and policy issues. Policy Brief 43, National Centre for Agricultural Economics and Policy Research.

Swati S, Adholeya A (2008) Biological control: alternative paradigms for commercialization. In: Maheshwari DK, Dubey RC (eds) Potential microorganisms for sustainable agriculture: a techno-commercial perspective. IK Publishing House Pvt Ltd., New Delhi

Van Lantern JC (2003) Need for quality control for mass produced biological control. In: Van Lantern JC (ed) Quality control and production of biological control agents theory and testing procedures. CABI International, UK

Ward MWD, Chung YJ, Copeland LB, Doerfler DL (2011) Allergic responses induced by a fungal biopesticide *Metarhiziumanisopliae* and house dust mite are compared in a mouse model. J Toxicol 2011:1–13

WHO (1996).Technical Report Series No.822: biological products: general recommendations. Guidelines on evaluation of similar Biotherapeutic Products (SBPs), ECBS, 19–23 October 2009. Annex 2 WHO good manufacturing practices for biological products.

Yadav SK, Archana SR, Singh PK, Vasudev PG (2019) Insecticidal fern protein Tma12 is possibly a lytic polysaccharide monooxygenase. Planta 249(6):1987–1996

Zadoks JC (1998) Risk analysis of beneficial microorganisms—wild type and genetically modified. In: *Proceedings on microbial plant protection products—workshop on the scientific basis for risk assessment*. Swedish National Chemicals Inspectorate, Stockholm, Sweden, pp 9–38

The manufacturer's authorised representative in the EU is Springer Nature Customer Service Centre GmbH, Europaplatz 3, 69115 Heidelberg, Germany. If you have any concerns regarding our products, please contact ProductSafety@springernature.com

Printed and bound by CPI Group (UK) Ltd, Croydon, CR0 4YY
30/09/2026
02225289-0012